Design of Active Key Interfaces in Fuel Cells

Toshiyuki Mori · San Ping Jiang · Zongping Shao ·
Michihisa Koyama · Zhipeng Li · Takayoshi Tanji ·
Roger Wepf · Fei Ye
Editors

Design of Active Key Interfaces in Fuel Cells

Springer

Editors
Toshiyuki Mori
Center for Green Research for Energy and Environmental Materials
National Institute for Materials Science
Tsukuba, Ibaraki, Japan

Zongping Shao
Curtin University
Perth, WA, Australia

Zhipeng Li
Zero Carbon Science and New Energy Technology Center
Northwestern Polytechnical University
Xi'an, Shaanxi, China

Roger Wepf
The Centre for Microscopy and Microanalysis
University of Queensland
Brisbane, QLD, Australia

San Ping Jiang
National Energy Key Laboratory for New Hydrogen-Ammonia Energy Technologies
Foshan Xianhu Laboratory
Foshan, China

Michihisa Koyama
Shinshu University
Nagano, Japan

Takayoshi Tanji
Nagoya University
Nagoya, Japan

Fei Ye
Southern University of Science and Technology
Shenzhen, China

ISBN 978-981-97-5705-3 ISBN 978-981-97-5706-0 (eBook)
https://doi.org/10.1007/978-981-97-5706-0

This Springer imprint is published by the registered company Springer Nature Singapore Pte Ltd.
The registered company address is: 152 Beach Road, #21-01/04 Gateway East, Singapore 189721, Singapore

Preface

In general, 'green innovation' is defined as all types of innovations that contribute to the creation of key products, services or processes to reduce the harm, impact and deterioration of the environment at the same time that optimizes the use of natural resources [1]. Based on the definition of key word 'Green Innovation', the use of fuel cells must be the best way to see it. Over the last 60 years, enormous number of materials for fuel cells have been developed for vehicular application, residential scale application and so on. However, the fuel cell business was started in small scale as compared to the business for other energy devices whose efficiency are much lower than that of fuel cells. It is because there are still many obstacles which should be overcome to use the fuel cells in our society. Nowadays, we expect that the combination work of artificial intelligence (AI) and big database will accelerate the development speed of fuel cells as well as development of other energy devices. If the conventional logic is correct, the social implementation of high-quality fuel cells with reasonable prices for users will be proposed soon. Unfortunately, however, it is unlikely to accomplish the social implementation of 'state-of-the-art fuel cells' with the quality of exceeding the limit of conventional fuel cells immediately. Following present situation, how can we take the good opportunity to make good balance between resources, production cost, high-quality performance and long operation durability for economical, ecological and environmental implementation of the 'state-of-the-art fuel cells'? In order to overcome this challenge, it is time for basic research sector, development sector and funding support sector to be lined up on useful and strategic R&D plan. (Refer to Fig. 1.)

To line up on same plan as shown in Fig. 1, we need to provide fundamental research sector, development sector and funding sector with interactive place and interactive tool to start our challenge. It is well known that interactive place can be prepared through the opportunity of open innovation. Now is the time, when all types of fuel cell scientists and engineers should come to our open innovation lounge after reading this introductory book. Then, some teams will be able to move to high level stages which are above the lounge space on the grand floor. To discuss interactive collaboration plans in the open innovation lounge at ground floor, we need useful interactive tool. What is a useful interactive tool in our challenge? That is the main

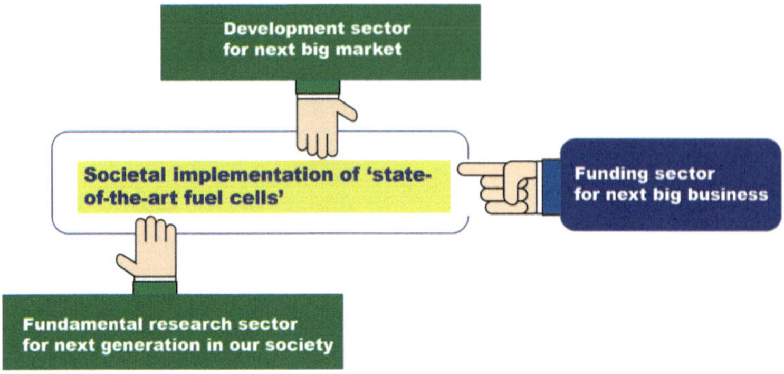

Fig. 1 Schematic diagram of important motivation for preparation of the present introductory book

topic of this introductory book. We believe that interactive tool has to be an interactive combination among fabrication, microanalysis and modeling.

As you may know, there is one major weakness in the conventional combination of AI and big databases for fuel cell application. That is the empty set (i.e., the lack of truly useful data groups) within the big database. So far, the combination teamwork of AI and big database isn't so active and productive in our fuel cell field. However, if we can put some new data based on new insights as initial input data, the combination of AI and big database will wake up and start the unique deep learning works for the basic research sector, development of sector and funding support sector without problems. Then, we will see real Green Innovation using fuel cell devices and systems.

To wake up the combination of AI and big database in our fuel cell field, the authors introduce the representative experimental data in three parts such as *'Fabrication Part'* (Chapters "Design of Interfaces of Electrolyte, Anode/Electrolyte in SOFCs and Interfaces in PEMFCs", "Design of Interface Between Cathode and Electrolyte in SOFCs" and "Design of HT-PEMFC Devices"), *'Modelling Part'* (Chapters "Atomistic Simulation of Subtle Microstructures in CeO_2-Based Electrolyte for Design of the State-of-the-Art Fuel Cells", "Computational Approach for Designing of Electrode Interfaces in Fuel Cell" and "First Principles Simulation of Microstructures for Metals/Alloy Materials in Fuel Cell") and *'Microanalysis Part'* (Chapters "*In Situ* TEM Analysis" and "TEM Characterisation of Electrode–Electrolyte Interfaces in Solid-State Fuel Cells and Batteries") of this book. For the readers of this book, I invited professional material scientists from prestigious universities, national research institutes, national institute of technology which is for education of young engineers and industry as authors. In each chapter, unique research works are explained in detail for beginners in our multidisciplinary work field. First, please read your favorite chapter and try to read all chapters.

If you are interested in our challenges, please join our virtual open innovation lounge and let's exchange the unique idea on how we can use interactive combination among fabrication, microanalysis and modeling to get our dream together. I hope that this book will provide a good opportunity for readers to take the first step towards new challenges in the near future.

Tsukuba, Japan Toshiyuki Mori, PhD in Eng.

Reference

1. Leal-Millán A, Leal-Rodríguez AL, Albort-Morant G (2017) Green innovation. In: Carayannis E (eds) Encyclopedia of creativity, invention, innovation and entrepreneurship. Springer, New York, NY. https://doi.org/10.1007/978-1-4614-6616-1_200021-1

Contents

Microanalysis

Fabrication

Design of Interfaces of Electrolyte, Anode/Electrolyte in SOFCs and Interfaces in PEMFCs

Toshiyuki Mori, John Drennan, Ding Rong Ou, Fei Ye, Zhipeng Li, Graeme Auchterlonie, Shunya Yamamoto, Hiroshi Okubo, Shigeharu Ito, and Jin Zou

Abstract Fuel cells are efficient energy conversion devices that convert the chemical energy of fuels such as hydrogen, methanol, and natural gas directly into electricity with high efficiency. Among fuel cells, solid oxide fuel cells (SOFCs) and polymer electrolyte membrane fuel cells (PEMFCs) have been actively developed as household generators. In particular, PEMFCs are being increasingly implemented in the society as vehicle engine components. To develop the 'state-of-the-art fuel cell' devices with the quality exceeding the limit of conventional SOFCs and PEMFCs, the design of defect interfaces in the fuel cells devices is key challenge. To take the opportunity to see new insight and innovative results in the challenge for develop the 'state-of-the-art fuel cell' devices and systems, multidisciplinary team work is

---

T. Mori (✉) · H. Okubo
Center for Green Research for Energy and Environmental Materials, National Institute for Materials Science, 1–1 Namiki, Tsukuba 305-0044, Ibaraki, Japan
e-mail: mori.toshiyuki@emeritus.nims.go.jp; mori.toshiyuki@nims.go.jp

J. Drennan · G. Auchterlonie · J. Zou
Centre for Microscopy and Microanalysis, The University of Queensland, St Lucia, Brisbane, QLD 4072, Australia

D. R. Ou
Hydrogen Energy Research Center, Guangzhou Power Supply Bureau, Guangdong Power Grid Co. Ltd., Guangzhou, Guangdong Province 510335, China

F. Ye
Department of Materials Science and Engineering, Southern University of Science and Technology, No. 1088, Xueyuan Road, Shenzhen 518055, Guangdong, China

Z. Li
Zero Carbon Science and New Energy Technology Center, Northwestern Polytechnical University, 127 West Youyi Road, Beilin District, Xi'an, Shaanxi 710072, China

S. Yamamoto
Takasaki Institute for Advanced Quantum Science, National Institutes for Quantum Science and Technology, 1233, Watanuki, Takasaki 370-1292, Gunma, Japan

S. Ito
Department of Creative Engineering, Chemical and Biological Course, National Institute of Technology, Tsuruoka College, 104 Sawada, Inoka, Tsuruoka 997-8511, Yamagata, Japan

T. Mori et al. (eds.), *Design of Active Key Interfaces in Fuel Cells*,
https://doi.org/10.1007/978-981-97-5706-0_1

quite important. In the present chapter, the representative data about the interactive research works of fabrication, microanalysis and modeling for the design of defect interfaces in/on rare-earth-doped ceria, stabilized zirconia, and brownmillerite-based oxides as SOFC components (i.e., electrolyte and anode) are introduced. Also, the authors introduced the representative data for improvement of Pt cathode performance and CO tolerance on Pt surface in PEMFCs in our multidisciplinary team works. For broad range readers, the general theory is introduced at first. The present chapter has the general theory part, followed by critical literature review part. In the critical review part, the authors highlight new insight and originality in our interactive research works. After literature review, we are going to take experimental part which includes advices for implementation of experiments. Then we re-introduces the representative results of our multidisciplinary team works in the results and discussion part. At the end of this chapter, the authors show the future prospect for development of 'state-of-the-art fuel cells' based on the results of our multidisciplinary challenges.

Keywords Interactive combination · Fabrication · Microanalysis · Modeling · Micro-domain · Ceria · Zirconia · Brownmillerite · SOFC · PEMFC · Co-dope effect · Pt-ceria interface

1 Introduction

Fuel cells are the electrochemical devices that generate the direct current (DC) electricity by converting the chemical energy of fuel and oxidant to electrical energy. Nowadays, these energy conversion systems (i.e., fuel cell systems) have great attention from the perspective of its high power generation efficiency and extremely low emission of pollutants. The most important advantage of fuel cells is the high theoretical efficiency of an energy conversion. It is much higher than that of thermal engines such as internal-combustion engines and thermal power generators.

In general, a fuel cell reaction is explained by inverse reaction of water electrolysis reaction. Since fuel cells consist of anode layer, electrolyte, and cathode layer, there are two types for the explanation of fuel cell reaction. When the mobile ion in electrolyte is proton (H^+), the fuel cell reaction based on inverse reaction of water electrolysis reaction is expressed by Eqs. (1) to (3);

$$\text{At the anode}: \quad 2H_2 \rightarrow 4H^+ + 4e^- \tag{1}$$

$$\text{At the cathode}: \quad O_2 + 4H^+ + 4e^- \rightarrow 2H_2O \tag{2}$$

$$\text{Overall}: \quad 2H_2 + O_2 \rightarrow 2H_2O \tag{3}$$

Namely, four electrons are transferred from the hydrogen to the oxygen (i.e., from the anode to the cathode through the electrolyte) per two water molecules.

On the other hand, when the mobile ion in electrolyte is oxide ion, the fuel cell reaction is expressed by following Eqs. (4), (5), and (3);

$$\text{At the cathode}: \quad O_2 + 4e^- \rightarrow 2O^{2-} \tag{4}$$

$$\text{At the anode}: \quad 2H_2 + 2O^{2-} \rightarrow H_2O + 4e^- \tag{5}$$

From Eq. (4) and Eq. (5), the equation of final overall reaction is the same as Eq. (3), as shown on the right. Overall : $2H_2 + O_2 \rightarrow 2\,H_2O$.

In this case, four electrons are transferred from the oxygen to the hydrogen (i.e., from the cathode to the anode through the electrolyte) per two water molecules.

In both cases, basic fuel cell reactions are exactly the same. Only mobile ion in electrolyte is different. However, note that the fuel cell reaction can be promoted when electron, mobile ion and fuel (i.e., hydrogen or various kinds of burnable gasses) or oxygen in air meet at the active interface in fuel cell devices. The active interface is named as three phase boundary (TPB). In the fuel cells, we must carefully make high TPB density in the fuel cells for development of high-quality fuel cells. This is why a design of active interface plays a key role in all R & D fuel cell works.

<<Theoretical background>>

*__Fuel cell reaction:__

To understand the high energy conversion efficiency of fuel cells, the thermodynamic analysis of this water electrolysis reaction is required. Since the overall reaction is the reaction of hydrogen combustion, the fuel cell reaction is the following exothermic reaction.

$$H_2 + 1/2O_2 \rightarrow H_2O\ (g) + 242/\text{kJmol}^{-1} \tag{6}$$

or

$$H_2 + 1/2O_2 \rightarrow H_2O\ (l) + 286/\text{kJmol}^{-1} \tag{7}$$

When the product (i.e., water) of fuel cell reaction is in the vapor form, some of energy is involved in producing water molecules in the gas phase. Then, the standard enthalpy (ΔH°) estimated by thermodynamics is 242 kJ mol^{-1}, refer to Eq. (6). This estimated value of standard enthalpy is called the hydrogen lower heating value (LHV). In contrast, when the fuel cell product is liquid water, the estimated ΔH° is 286 $kJmol^{-1}$ and the higher heat energy is formed as compared with Eq. (6), refer to Eq. (7). This value is called the hydrogen higher heating value (HHV).

The energy diagram of these reactions (i.e., Eqs. (6) or (7)) is presented in Fig. 1a.

To clarify the relationship between the thermodynamic analysis data of water electrolysis reaction and fuel cell performance, a typical fuel cell performance curve (i.e., I–V curve) is demonstrated in Fig. 1b.

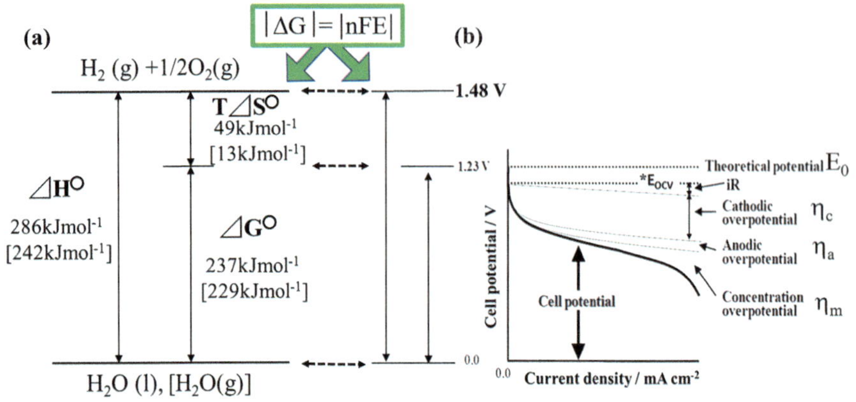

Fig. 1 Energy diagram of fuel cell reaction (**a**) and typical I–V curve (i.e., polarization curve) observed for fuel cell device (**b**)

Equations (6) and (7) correspond to Gibbs free energy equation which is expressed by Eq. (8) [1].

$$\Delta G^{\circ} = \Delta H^{\circ} - T\Delta S^{\circ} \tag{8}$$

where ΔG° is standard Gibbs free energy, ΔH° is standard enthalpy, T is absolute temperature, and ΔS° is standard entropy. The ΔG° value in the fuel cell reaction is related to the electro-motive force (i.e., emf) expressed by Eq. (9).

$$-\Delta G = nFE \tag{9}$$

where F is the Faraday constant (96,485 C mol^{-1}) and n is the number of exchanged electrons in the overall reaction ($n = 4$). The emf, which is denoted E and can be measured by the cell potential (V), corresponds to terminal potential of fuel cells. Note that Gibbs free energy and E belong to the potential energy in the broad sense. We can simply convert the thermodynamics data Δ G into the electro-motive force E using Eq. (9).

In addition, note that $T\Delta S^{\circ}$ corresponds to heat energy in the fuel cell reaction, namely, the reusable heat energy. This will be not so much if we design the high-quality fuel cell device and system.

As shown in Fig. 1, the fuel cell device can directly convert the chemical energy of fuel such as H_2 into electricity. The efficiency of energy conversion in the fuel cell reaction ($\boldsymbol{\varepsilon}_{\mathbf{fc}}$) is expressed by Eq. (10).

$$\text{Efficiency of energy conversion } (\varepsilon_{fc}) = \Delta G/\Delta H \tag{10}$$

However, when we must consider the efficiency of energy conversion expected from fuel cells (not fuel cell reaction), the fuel utilization (used fuel/applied fuel) and voltage efficiency (E_{cell}/E_{OCV}) have to be taken into account, refer to Eq. (11). Note that the fuel utilization factor strongly depends on the structure of cell stack. In the case of a single cell, the fuel utilization factor is at a negligible level.

$$\varepsilon = \varepsilon_{the}\varepsilon_{v}\alpha \tag{11}$$

where ε_{the} is the theoretical efficiency, ε_v is the voltage efficiency (E_{cell}/E_{OCV}), and α is the fuel utilization factor.

To understand the estimation process of efficiency of energy conversion of the fuel cell reaction (or fuel cell single cell), typical example of estimation of ε_{fc} is presented as follows.

When we focus on the fixed cell potential such as 0.8 V,

$$\begin{aligned}\text{Estimated energy conversion } \varepsilon_{fc} \text{ of our single cell} &= \Delta G/\Delta H \\ &= 0.8/1.48 \\ &= 0.54\end{aligned} \tag{12}$$

Consequently, we often estimate the current density at 0.8 V observed for the single cell. In this case, we can discuss the performance of our single cell at the 54% level in energy conversion efficiency.

In Fig. 1, we can see one more import point. The E value observed for single cell decreased with an increase of current density. As we can see in this figure, the observed cell potential E (E_{cell}) included some excess overpotentials.

E_{cell} is usually expressed by Eq. (13).

$$E_{cell} = E_0 - \eta_a - \eta_c - \eta_m - iR \tag{13}$$

where iR is the ohmic loss and η_a and η_c are anodic and cathodic overpotentials, respectively.

Based on Eq. (13), it is found that the performance observed for fuel cell devices is the cell potential excluded from some losses such as iR and overpotentials (i.e., cathodic, anodic, and concentration overpotentials). To observe high performance from our fuel cell device, we must learn the relationship between observed losses and interface defect structure in the next step.

*__Ohmic loss:__

Since the mobile ion such as oxide ion or proton in the electrolyte needs energy in its hopping process, iR (i.e., ohmic loss) is observed as the internal resistance of the fuel cell. This loss is explained using typical diffusion concept of mobile ion in electrolytes. Usually, the diffusion process in the crystal is explained by the relationship between potential energy and distances between cations (or anions) in the crystal, refer to Fig. 2. A detailed general explanation about the diffusion of mobile ion is given elsewhere [2].

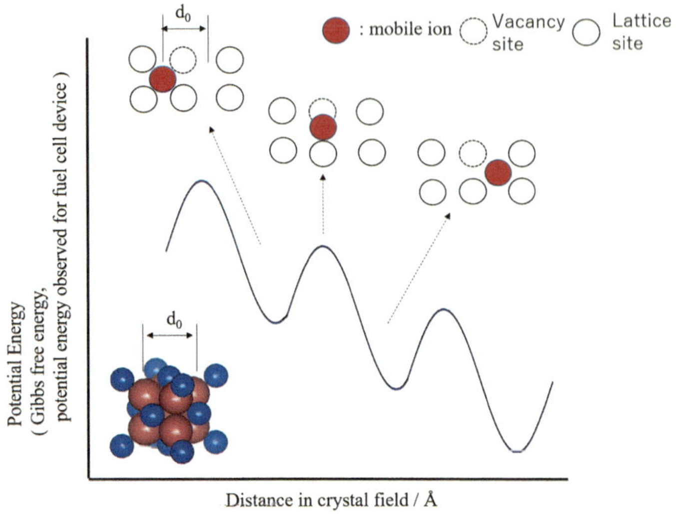

Fig. 2 Schematic diagram of diffusion process by mobile ion in crystal such as oxide

As demonstrated in Fig. 2, long distance diffusion from the cathode side (or anode side) to anode side (or cathode side) through the electrolyte uses the potential energy in the fuel cell device. Note that potential energy is a function of distance in some fields. If the number of mobile ion is at a high level (i.e., in high current density region) or thickness of the electrolyte is thick, iR will be high. In addition, iR becomes high level, if the mobile ion pathway in the crystal is not well formed in the electrolyte material. Besides carrier density or thickness of the electrolyte, the observed iR is strongly influenced by the defect interface structure in the electrolyte and/or defect structure at the electrolyte/electrode interface in the fuel cell devices [3]. Consequently, the observed activation energy which is estimated by temperature dependence of conductivity has shown variation as seen in previously published papers.

To highlight why we must focus on the interface microstructure in the fuel cells so carefully, we must learn the diffusion process of mobile ion in the materials in more detail (not only by the schematic diagram in Fig. 2). According to general theory of ion diffusion, diffusion constant (**D**) is expressed by the following equation [4a,b].

$$\begin{aligned}\mathbf{D} &= D_0 \exp(-\Delta G/kT) \\ &= \gamma(1 - N_c)Zd_0{}^2\nu_0 \exp(\Delta S/K)\exp(-E/kT) \qquad (14)\end{aligned}$$

where d_0 is jump distance in hopping process, T is the absolute temperature, k is Boltzmann's constant, ν_0 indicates the attempt frequency, E is activation energy, the $(1-N_c)Z$ factor means the number of neighboring unoccupied sites in the crystal, γ defines geometric and correlation factor. Note that correlation factor is 0.65 in the case of cubic lattice (ex. stabilized zirconia and doped ceria). Since the mobile ion cannot diffuse easily over the potential barrier in the crystal, the correlation factor is below unity.

Based on the Nernst-Einstein relation (i.e., $\mu = qD/kT$), an ionic conductivity (σ) which is the most important physical property to determine the iR in a solid electrolytes is described by Eq. (15).

$$\sigma = N_c Cq\mu \tag{15}$$

where N_c is the fraction of occupied sites, C means the density of mobile ion sites in the lattice.

In the case of defect free materials, N_c is 1 (i.e., all occupied sites: fully occupied) or 0 (i.e., all interstitial sites: vacant). In contrast, an ionic material without ordering under ideal condition, N_c is considered to be 0.5.

Substituting Eqs. (14) in (15), we obtain Eq. (16).

$$\begin{aligned}\sigma &= N_c Cq(qD/kT)\\ &= \gamma\left(Cq^2/kT\right)N_c(1 - N_c)Zd_0{}^2\nu_0 \exp(\Delta S/k)\exp(-E/kT)\\ &= \sigma_0/T\exp(-E/kT)\end{aligned} \tag{16}$$

where σ_0 is named as pre-exponential factor. Note that the term ΔS is in the pre-exponential factor.

If we can conspecuously increase the number of disordered sites for the diffusion of mobile ion in the lattice, the term ΔS will be increased and σ_0 obtains a higher level in our measurement. The typical example of this significant increase can be observed in the brownmillerite-based materials ($Ba_2In_2O_5$), bismuth sesquioxide (δ-Bi_2O_3), silver iodide (AgI), and so on. In the case of proton conducting materials such as Nafion® membrane electrolyte, oxide proton conductors, the mobile proton ion is introduced by vapor in the measurement atmosphere. Consequently, large increments of σ_0 can be observed in the measurement.

In contrast, the conventional electrolytes such as Y_2O_3 stabilized ZrO_2 or $LaGaO_3$-based electrolyte do not show large increments of ΔS. Instead of the ΔS effect, we can see changes of ΔH in those electrolytes. The activation energy is usually expressed using the term Δ H, refer to Eq. (17) [5, 6].

$$\begin{aligned}E &= (E_m + (n/2)E_a)\\ &= (\Delta H_m + (n/2)\Delta H_a)\end{aligned} \tag{17}$$

where E_m and ΔH_m are migration energy and migration enthalpy, respectively. E_a and ΔH_a are association energy and association enthalpy, respectively. Also, n is equal to 1 or 2 for di- or trivalent dopants, respectively.

Usually, the term ΔH_m (migration enthalpy) depends on the free space for diffusion of mobile ions in the crystal. The critical radius is fixed by the crystal structure, **refer to** Fig. 3a. On the other hand, the term ΔH_m (association enthalpy) is changed

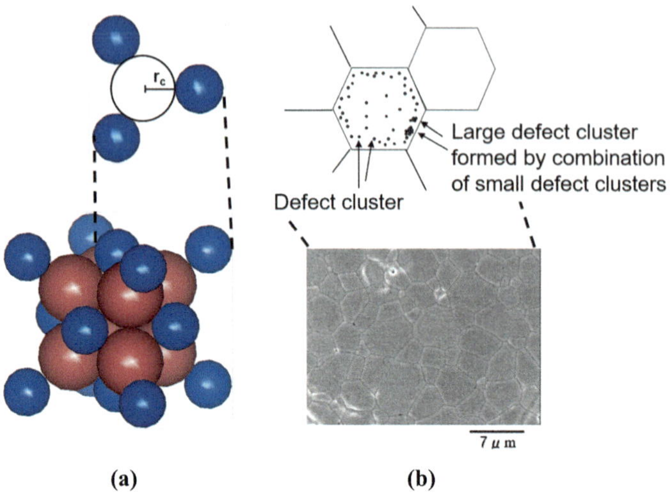

Fig. 3 Schematic diagram of free space, r_C is critical radius in free space: (**a**) and defect clusters in the grain of oxide electrolyte: (**b**)

by the interface defect structure in the electrolyte, We often observe formation of large defect clusters in the grain and grain boundary area in the oxide electrolyte (Fig. 3b). Needless to say, we can observe similar defect cluster formation at the interface into the anode layer which consists of the active metal (Ni) and the oxide electrolyte. That is why the observed activation energy in the previously published papers shows a large variation, while the composition of oxide electrolyte is similar. This clearly indicates that the defective interface microstructure in the fuel cell materials and devices should be taken into account for the design of high quality electrolyte or low iR in the fuel cell devices.

***Overpotentials (Activation Overpotential and Concentration Overpotential):**

Besides the iR term, Eq. (13) includes important losses such as cathodic overpotential, anodic overpotential, and concentration overpotential, refer to Fig. 1. Usually, there are two representations in the definition of overpotential. One is activation overpotential, and the second is the concentration overpotential. In addition, we must remember that all electrode reactions are promoted through the electrical double layer on the electrode materials. In all types of fuel cells, the basic principle is the same. All electrodes basically obey Faraday's law. Consequently, the theoretical background of the electrode reaction is concisely re-introduced in this chapter.

As well as the transportation of mobile ions in the electrolyte, the electrode reaction occurs when enough potential energy is applied at the electrode surface.

To understand the kinetics of the electrode reaction, we must remember why the electrode reaction is promoted under the applied potential energy, even though the measurement temperature is room temperature at first.

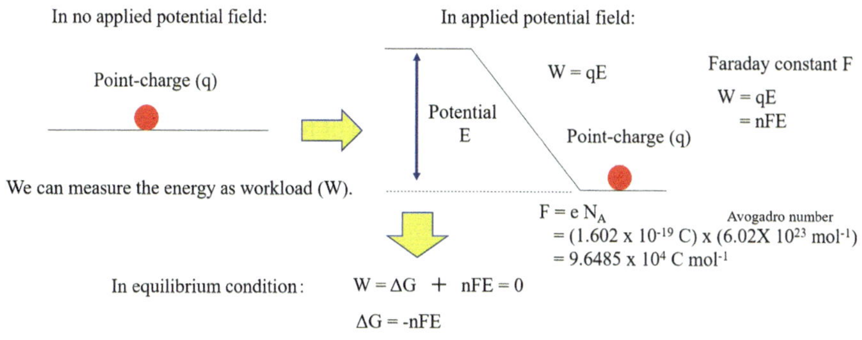

Fig. 4 Schematic diagram for explanation of two potential energies (i.e., ΔG and E)

To understand this point, we must touch on the key point of Eq. (9) again, refer to Fig. 4.

When no applied potential is on the measurement field, the point charge is just on the equipotential surface. After the cell potential is applied on the electrode surface, a potential gradient is formed. Note that we measure the change of potential energy by the observed workload (W) from the surface. This phenomenon is expressed by Eq. (18).

$$\begin{aligned} W &= qE \\ &= nFE \end{aligned} \tag{18}$$

In the equilibrium condition, we expect Eq. (19).

$$W = \Delta G + nFE \tag{19}$$

Then, Eq. (9) (i.e., $-\Delta G = nFE$) is obtained.

When the potential energy in the electrical field is in an equilibrium state, the potential energy can be described by Eq. (20).

$$\begin{aligned} \Delta G &= -\,nFE \\ E &= -\,\Delta G/(nF) \\ E^{\bigcirc} &= -\,\Delta G^{\bigcirc}/(nF) \end{aligned} \tag{20}$$

where superscript '○' means standard reference condition.

Now, we consider the following general electrode reaction.

$$aA + bB \leftrightarrow cC + dD$$

Also, we apply the law of mass action to Eq. (20).

Then, Eq. (21) is obtained as Nernst's equation.

$$E = -\Delta G^{\bigcirc}/nF - RT\ln\left(\left(X_C^c X_D^d / X_A^a X_B^b\right)\right)/nF$$
$$E = E^{\bigcirc} - RT\ln\left(\left(X_C^c X_D^d / X_A^a X_B^b\right)\right)/nF \tag{21}$$

where R is gas constant, X presents concentration of reactants or products.

When no applied potential is on the surface of electrode, the equilibrium potential which is subject to the Nernst equation appears on the electrode surface. Note that the observed equilibrium potential becomes low because of the rest potential of the measurement system. On the other hand, the electrons move on the lowest unoccupied molecular orbital (LUMO) of the adsorbed molecule, from active metal surface when the Fermi level of active metal is increased by the sweep of the applied potential. This charge transfer process (i.e., reduction reaction on the electrode) occurs on the cathode. The reduction current is observed on the cathode as a negative value. In contrast, the charge transfer from the highest occupied molecular orbital (HOMO) of adsorbed molecule to active metal surface occurs when the Fermi level of the active metal is decreased by the sweep of the potential. This charge transfer process (i.e., oxidation reaction on electrode) occurs on the anode. The oxidation current is observed as a positive value. The charge transfer on the anode or cathode stably proceeds under applied potential. Note that we must carefully confirm that observed electrochemistry is in equilibrium state at first. Then, we have a chance to proceed with the electrode reaction to the oxidation side or reduction side by using potential energy, refer to Fig. 5.

To think about kinetics, we must remember the difference between conventional chemical reaction and an electrochemical reaction, refer to Fig. 6. In both cases, the kinetic constant can be described by the activation energy (E_a or E). The equations of kinetic constant for the oxidation reaction or the reduction reaction clearly indicate that oxidation and reduction on the respective electrodes can proceed when the applied potential is over the activation energy in the equilibrium condition.

Usually, this electrochemistry is explained by the Butler–Volmer equation, refer to Fig. 7. Note that this equation basically consists of kinetics for oxidation and

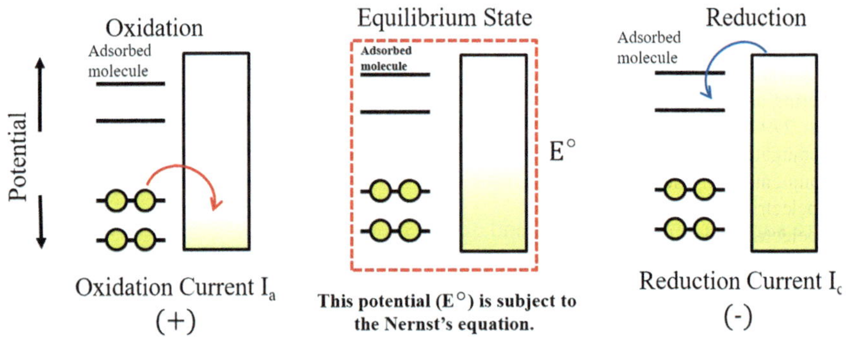

Fig. 5 Schematic diagram for explanation of basic principle of electrode reaction. Yellow color level indicates Fermi level of active metal

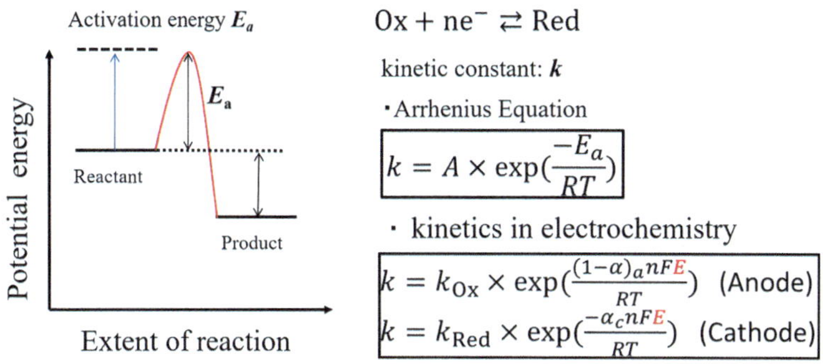

Fig. 6 Schematic diagram of comparison of kinetics between conventional chemical reaction and electrochemical reaction

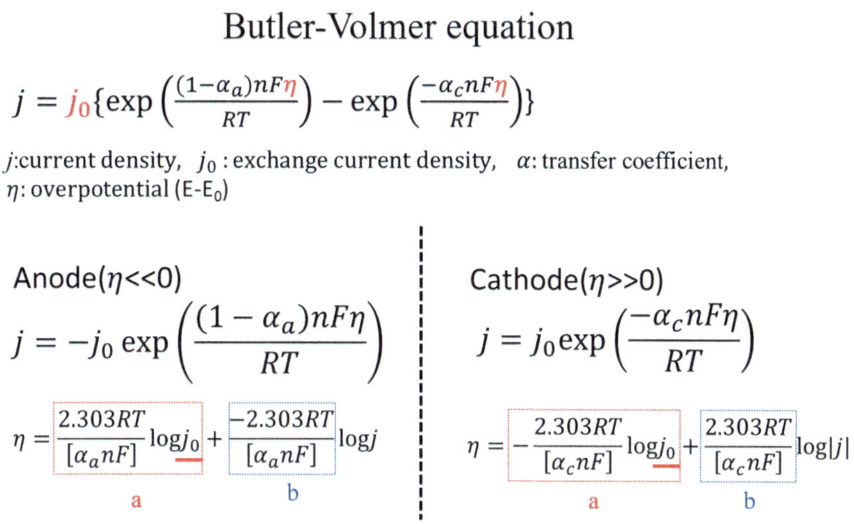

Fig. 7 Relationship between Butler–Volmer equation and Tafel equation

reduction reactions on the electrode. The cathodic and anodic reaction will proceed when the applied potential is over the theoretical potential and overpotential. That is to say, overpotential (η) is described by the following Eq. (22)

$$\text{Overpotential}(\eta) = E - E_0 \tag{22}$$

Note that, this kinetics can be discussed when the diffusion effect is at negligible level. If the diffusion phenomenon which is drastically changed by the applied potential, we cannot discuss this kinetics in our work.

In Fig. 7, we can see one more important equation, refer to Eq. (23)

$$\eta = a + b \log j \tag{23}$$

Equation (23) is known as the Tafel equation. Accordingly, the region which can be covered by conventional kinetics is called the **Tafel region or activation overpotential region**.

From the viewpoint of analysis, specifically concerning the kninetics in our experiment, we must use this Tafel plot. For instance, the level of oxygen reduction reaction (ORR) activity observed for the cathode corresponds to the intersection point of the horizontal axis of Tafel line. The level of exchange current density (i.e., kinetics constant of ORR) corresponds to the intersection point of the vertical axis of the Tafel line. This means that the intersection points between both axes (i.e., vertical and horizontal axes) and the Tafel line derived from the electrochemistry on high quality cathodes is seen to shift to the higher side on both axes as compared with conventional cathodes in our experiment. This analysis point is important for design of high quality electrode materials.

Note that, the Tafel region is not so wide in our measurement of the electrochemistry on the electrode which influences the performance of fuel cell devices, refer to Fig. 8. Also, we must remember that the electrode reaction proceeds through the electrical double layer. This electrical double layer on the electrode is relatively wide on Pt/ C systems in membrane electrode assembly (MEA) of PEMFC devices. Under large applied potential field, the diffusion process is conspicuously promoted, refer to Fig. 8. We can then see large current densities from the electrochemistry on electrode and from single cells (or MEA) of fuel cells. This region is named the **concentration overpotential region** which is affected by three-dimentional porus structure in electrode materials or electrode layer in single cells. In order to design high qulaity fuel cell devices, we must examine two regions in our experiment. One is the activation of overpotenatial region. The other is the concentration overpotential region. Note that the electrode reaction analysis in the concentration overpotential region (i.e., electrochemistry analysis) is not enough for a design of fuel cell devices. To understand this basic key point in our research, we must learn what is the three phase boundary (or two phase boundary) in the electrode layer of fuel cell devices, refer to Fig. 9.

The fuel cell reaction (i.e., Eqs. (1) and (2) or (4) and (5)) occurs at the TPB in the electrode layers of fuel cell devices. In the case of PEMFC device, Eq. (1), in the anode layer and Eq. (2) in the cathode layer of MEA proceeds at the TPB area which is indicated by using green colored region in Fig. 9. In the anode layer, proton (i.e., H_3O^+) comes to the TPB through ionomer solution. Electrons come to the same site through the conductive carbon channel in the anode layer. In addition, hydrogen gas diffuses to the TPB sites through the open space in the anode layer. At this point, protons, electrons, and H_2 gas meet at TPB area. Finally, we can see the fuel cell reaction in its entirety, which is described by Eq. (1). On the other hand, oxygen gas diffuses to TPB site and the proton (i.e., H_3O^+) comes to TPB through the ionomer solution in a similar fashion to the anode side. After electrons reach the TPB site through the conductive carbon, a cathodic reaction (ORR reaction) in the cathode layer proceeds in MEA. Note that, the TPB area is not so wide in the high temperature PEMFC devices. The reason for this is that the operation temperature of

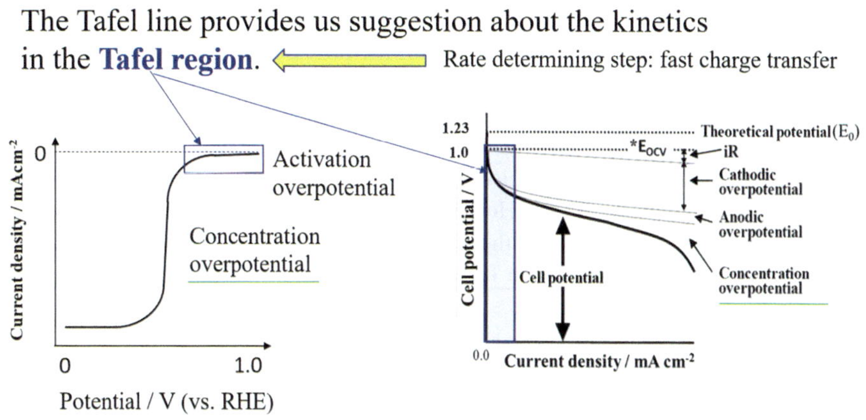

Fig. 8 Tafel region in electrochemistry analysis and fuel cell performance test

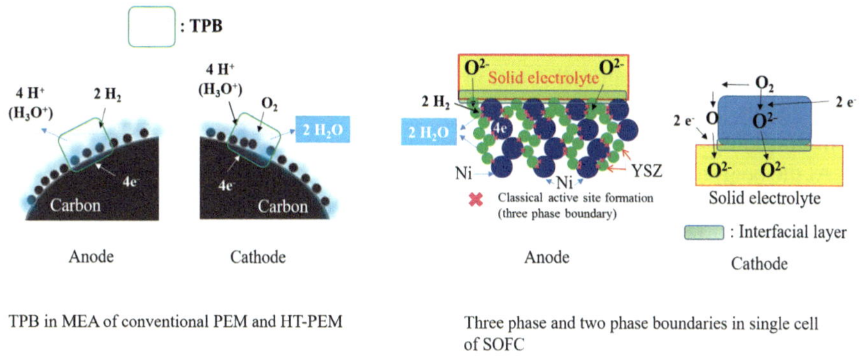

Fig. 9 Schematic diagram of TPB (i.e., three phase boudary or two phase boudary) in the elecrode of fuel cell devices

conventional PEMFC is below 100 °C. In this challenge, the interfaces in electrode layer should be optimally desinged well. In addition, TPB density in the anode layer or cathode layer of MEA has to be higher than a conventional electrode layer. We must try to maximize the Pt activity or minimize the Pt content in electrode layer of MEA. In this challenge, the active interface area (i.e., high-quality TPB area) has to be optimally designed. In both cases, the interface design is the key challenge and focus of our research works.

In the case of the SOFC device, the components in the single cells are ceramics materials. During fabrication process for the cathode and anode layers on the solid electrolyte, high temperature heat treatment (i.e., 1200 to 1300 °C heat treatment for fabrication of anode layer on the solid electroyte, 1100 °C heat treatment for fabrication of cathode layer on the solid electrolyte) is required in air. In this process, the thin interfacial layer between the electrode and electrolyte is formed, refer to Fig. 9.

On the anode side of the SOFC single cell, oxide ions (O^{2-}) diffuse into the anode layer through the solid electrolyte. However, the diffusion of O^{2-} is affected in the anode/electrolyte interface at first. In addition, O^{2-} diffuses into the anode layer through the channels of the solid electrolyte particles and charge tarsnfer proceeds through the channel of Ni particles into the anode layer. Then, hydrogen (H_2) is converted from fuel gas, $O^{2-}$and electrons meet at the interface (i.e., TPB) between solid electrolyte particle and Ni. However, the narrow interfacial region is formed between two kinds of particles at the TPB area. To maximize the fuel cell reaction (i.e.,Eq. (5)) at TPB in the anode layer, careful observation and design based on microanalysis results are required.

On the other hand, oxygen (O_2) in air adsorbs on the cathode material such as ((LaSr)MnO_3: LSM) and is dissociated on LSM. After disassociation of O_2, one of disassociated O ion diffuses on the surface of cathode and reaches the interface between cathode and solid electrolyte (i.e., TPB). Another one reacts with electron and O^{2-} is formed at the two phase boundary on LSM. Then, mobile ions (O^{2-}) which are formed at three phase boundary and two phase boudary diffuse into the electrolyte and move on to the anode side. However, the narrow interfacial region is formed between LSM and solid electrolyte at the TPB area as well as the anode case. To promote the fuel cell reaction (i.e.,Eq. (4) at the TPB in the cathode layer, microanalysis data of the interfacial layer should be taken into account for a design of high-quality cathode layers in the single cells.

At the beginning of this chapter, we highlighted the importance of interface design in fuel cell devices through the explanation of basic theory of fuel cell reactions. Note that interfacial layer is thin, such as below 100 nm and the characterization of the interface defect structure is not so easy. The reason for this is that modification at this level is not simple and requires a multidisciplinary approach. However, the readers must recognize that there are significant opportunities to find out something new in the interface of our fuel cell devices.

Now, we move on to the literature review part in this chapter. To highlight the important point of this chapter, we focus on the relationship between fabrication, microanalysis and modeling and try to review critically the previously published works for the readers.

Over the last 60 years, a large number of materials have been developed for electrolytes and electrodes for development of various kinds of fuel cells such as solid oxide fuel cells (SOFCs), polymer electrolyte membrane fuel cells (PEFCs), molten carbonate fuel cells (MCFCs), alkaline fuel cells (AFCs), and phosphoric acid fuel cells (PAFCs). At the same time, the compatibility of proposed materials for fuel cell devices and systems has been examined on a 'trial and error' basis.

The present conventional methods do not allow for significant innovation in developing the next generation of fuel cells. As a consequence, a new approach is required to re-design key interface structures in order to develop 'state-of-the-art fuel cells'. The authors will review the present status of systems with a view of developing a modern novel strategy.

1.1 Key Interfaces in SOFC Materials

First of all, the authors highlight the relationship between interface defect structure and electrolytic and electrocatalytic properties based on microanalysis data. As a consequence, this is a critical review about the influence of defect interface modification on properties of fuel cell materials.

The SOFC single cells include the following key interfaces, refer to Fig. 10.

The key interfaces which can be observed in SOFC devices are the second phase inclusion: (a), micro-domains or nano-domain with ordered structure of oxygen vacancies: (b), liquid phase accumulation induced by impurity: (c), segregation effect: (d), precipitation in the matrix: (e), or second phase at grain boundary: (f). Usually, the performance of the intermediate temperature (IT)-SOFC device is influenced by the change of interface structures at the cathode/ electrolyte, in the electrolyte, and at the electrolyte/ anode and in the cermet anode layer. Therefore, the authors begin by summarizing key interfaces as shown in Fig. 10 [7].

The starting products for SOFC devices consist of nano-sized and well-dispersed component powders. Generally, these are oxide components and are rounded in shape. Even though great care is taken to produce homogeneous powders, it is difficult to obtain full compositional homogeneity. During the sintering process, to fabricate dense sintered bodies, nano-level dopant segregation which is formed in the grain boundaries of electrolyte sintered bodies seriously influenced by the formation of the micro-domain or nano-domains in the electrolytes [8–12].

1.1.1 Stabilized Zirconia

In the first stage of this review, the authors concentrate on previously published works on stabilized zirconia systems. The reason for this is that Y_2O_3 stabilized ZrO_2, and CaO stabilized ZrO_2 are representative electrolyte material and component of cermet anode of SOFC single cells.

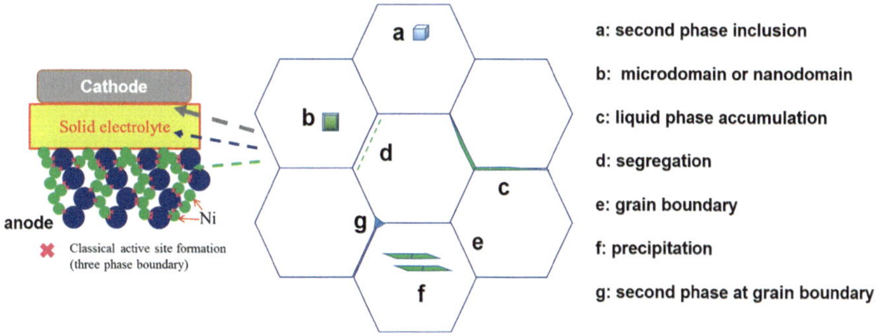

Fig. 10 Schematic illustration of the key interface regions in SOFC materials

Cubic ZrO_2 consists of fluorite structure (i.e., CaF_2 structure) that can be stabilized with dopant Y_2O_3 or CaO at room temperature. In the cubic ZrO_2, each cation is surrounded by eight equivalent nearest-neighbor lattice oxygen (O) atoms each of which in turn is surrounded by a tetrahedron of four equivalent cations, refer to Fig. 11. In this fluorite type crystal, we must remember that the large interstitial sites exist and anion Frenkel pairs are the dominant intrinsic defects in the fluorite type compounds [13–15].

Anion-deficient fluorites have been examined for their transport and thermodynamic properties. The reason for this is that these anion-deficient compounds exhibit high anion mobility. In the early stage of defect structure analysis for this anion-deficient fluorites, it was difficult to understand the details. The reason for this is that the defect structure observed by X-ray, microscopy was much more complicated as compared with typical simple defect structure models of nonstoichiometric fluorite oxides presented. The combination of X-ray, selected area electron diffraction analysis, and microscopy analysis suggested that very short-range ordering of cations is key to characterizing the defect structural feature of anion-deficient fluorites [16]. Unfortunately, the microanalysis techniques in the early stage of these studies did not reach the adequate level of sophistication to describe fully the observed short-range ordering in the fluorite and we had to wait until more advanced microanalysis techniques had been established.

On the other hand, the transport properties observed for all fluorite type materials were several orders of magnitude higher than those of other cation-based systems. Especially, high oxide ion diffusion coefficients were observed in the anion-deficient systems. In the case of systems that show low oxide ion diffusion coefficients was supported by calculation of oxygen vacancy activation energies [13]. In the present literature review section, we must focus on the transport studies based on the simple defect model of fluorite as well as previously published works. Then, important points outlined in the future work will be clear. That is the important aim of the present critical review section.

For a doped fluorite type oxide such as CaO stabilized ZrO_2, the association effects between dopant cations and oxygen vacancies have been discussed in many previously published papers. In the representative model, the analysis of Arrhenius plot (i.e., the relationship between electrical conductivity and reciprocal temperature) for stabilized ZrO_2 was analyzed and three separate regions were demonstrated, refer to Fig. 12 [16]. In this analysis, no dopant precipitation or phase changes was assumed.

Region A: This region A occurs at high measurement temperatures and is a result of the intrinsic lattice defects (Frenkel type defect or Schottky type defect) in the fluorite type crystal. Namely, this region A is characterized by oxygen vacancies which are generated in the electrolyte.

Region B: The electrical conductivity in this region is controlled by the population of charge carrying defects which are determined by an aliovalent dopant or impurities. This means that the oxygen vacancies are generated by an introduction of an aliovalent dopant in the electrolyte and the conducting property is influenced by the amount of generated oxygen vacancies.

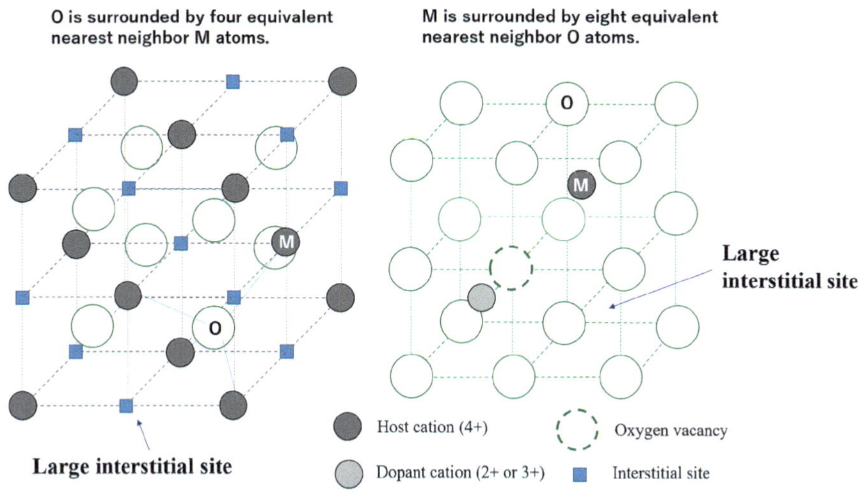

Fig. 11 Crystal structures of fluorite showing the position of host cation, dopant cation, oxygen, and oxygen vacancy [13–15]

Region C: In this region, the electrical conductivity is influenced by dopant – vacancy interactions. Also, the number of free oxygen vacancies without paring or clustering is reduced at this low temperature.

Since the crystals of fluorite type oxides usually involves the large amount of extrinsic vacancies, it is unlikely to observe the region A clearly in the experiment [16].

Needless to say, the relationship between electrical conducting property and lattice defects is observed in other type oxide structures. Consequently, the defect interfaces in the microstructure of fuel cell materials should be taken into account for a design of high-quality fuel cells.

In the previously published works, the curvature between the Region B and Region C was observed for Y_2O_3 stabilized ZrO_2 solid electrolyte sintered bodies. However, it was not always observed for all kinds of stabilized ZrO_2 solid electrolytes. For example, the Arrhenius plot derived from the conductivity measurement data observed for CaO stabilized ZrO_2 and Yb_2O_3 stabilized ZrO_2 solid electrolytes exhibited a straight line up to 1000 °C [17]. The relationship between vacancy association enthalpy calculation and diffusion coefficient observations suggested that improved understanding of the concentration dependence of the defect interactions is required. In addition, defect model calculation revealed that the defect association situation is more complex than previously described [18].

Conventional defect association in ZrO_2 fluorite lattice was quite simple, refer to Eqs. (24)–(26).

$$\mathrm{Ca_{Zr}}'' + \mathrm{Vo}^{\cdot\cdot} \Rightarrow [\mathrm{Ca_{Zr}}''\,\mathrm{Vo}^{\cdot\cdot}]^{x} \tag{24}$$

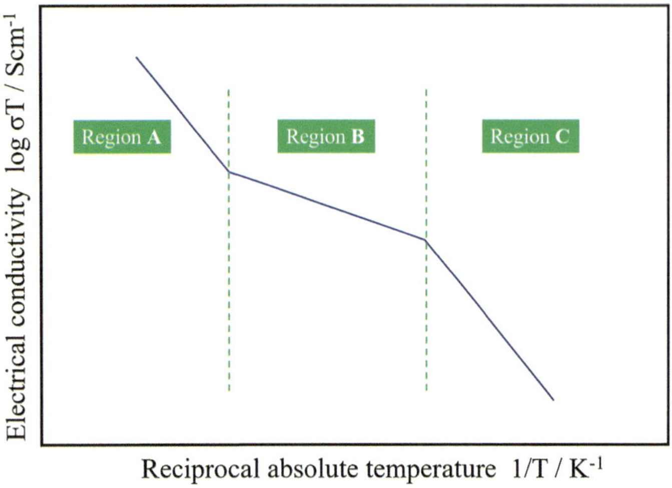

Fig. 12 Schematic representation of electrical conductivity as a function of reciprocal absolute temperature for the oxide electrolyte including fluorite type oxide electrolyte [16]

$$Y_{Zr}{}' + Vo^{\cdot\cdot} \Rightarrow [Y_{Zr}{}'Vo^{\cdot\cdot}]^{\cdot} \tag{25}$$

$$Yb_{Zr}{}' + Vo^{\cdot\cdot} \Rightarrow [Yb_{Zr}{}'Vo^{\cdot\cdot}]^{\cdot} \tag{26}$$

However, the real SOFC single cells which consist of solid electrolyte and electrodes (i.e., cathode and cermet anode) are not described by such a simple model. The reason for this is that the microstructure of the interfaces of SOFC materials strongly depends on fabrication route and conditions. In the case of solid electrolytes, we must carefully observe the non-equilibrium defect structure which appears as a result of the slow cooling during the sintering process. This type of non-equilibrium defect should be carefully discussed based on microanalysis using equilibrium phase diagrams. In the case of Sc_2O_3 stabilized ZrO_2 system (7.8 mol% Sc_2O_3 content), the conductivity deterioration observed for the Sc_2O_3 stabilized ZrO_2 sintered bodies held at 1000 °C was inexorably tied to the formation and subsequent decomposition of a high temperature tetragonal phase, designated as the t'-phase. This phase is formed by diffusion less transformation from the cubic phase on cooling from the sintering temperature [7, 19]. Y_2O_3 - ZrO_2 system (3 mol% Y_2O_3 content) shows similar conductivity deterioration by annealing between 600 and 1500 °C. TEM microanalysis showed a complicated substructure in grains of the sintered body. The microanalysis indicated large yttrium concentration gradients and tetragonal phase variants with respect to the solute and regions of monoclinic phase [20]. The previously published microanalysis works clearly suggest that interactive combination work of fabrication condition optimization, microanalysis, and modeling is important for the design of high-quality ZrO_2-based solid electrolytes for SOFC application.

In addition, the resistivity of grain boundary control is also an important issue for design of high-quality solid electrolytes, refer to Fig. 10. In the early stages of R & D of stabilized ZrO_2 solid electrolyte, trace amount of siliceous film remained in the grain boundary region of Y_2O_3 stabilized ZrO_2 sintered bodies. Usually, commercial available Y_2O_3 stabilized ZrO_2 powders are prepared by using hydrolysis reaction under gentle hydrothermal condition around 100 °C [21]. The starting materials used are $ZrOCl_2{\cdot}8H_2O$ and YCl_3 powders. The homogeneous Y_2O_3 stabilized ZrO_2 powders are prepared based on Eq. (27). In this process, the reaction product is in acidic solution whose pH is less than unity.

$$ZrOCl_2 + YCl_3 + (n+1)H_2O \rightarrow Y_2O_3 - ZrO_2 \cdot nH_2O + 2H^+ + 2Cl^- \quad (27)$$

It means that trace amount of impurity such as non-crystalline siliceous material would be leached into the product from glass vessel during hydrothermal process. Another possibility is that $ZrOCl_2{\cdot}8H_2O$ which was prepared by using the starting material, the mineral $ZrSiO_4$ would include trace amount of SiO_2. Non-crystalline siliceous material as trace amount impurity increased grain boundary resistivity in the Y_2O_3 stabilized ZrO_2 sintered body in the early days of this research activity. Even though the content of siliceous material is in the ppm level, the grain boundary resistivity conspicuously was increased and it was strongly influenced by the grain size dependence of electrical conductivity of Y_2O_3 stabilized ZrO_2 solid electrolytes [22–26].

To decrease the grain boundary resistivity in Y_2O_3 stabilized ZrO_2 or CaO stabilized ZrO_2 sintered bodies, small amount Al_2O_3 addition [27–34] or pre-heat treatment [35–38] before sintering were examined. Eventually, the grain boundary resistivity was conspicuously decreased by their scavenging effects. However, with present day materials, the impurity level of siliceous material in Y_2O_3 stabilized ZrO_2 powder is now under detectable level. Nowadays, it is unlikely to use the scavenging effect of other phases or heat treatments to improve electrolyte performance of Y_2O_3 stabilized ZrO_2 systems. However, if the similar impurity is in the Sc_2O_3 stabilized ZrO_2 solid electrolytes, the scavenging effect will be useful approach to minimize the grain boundary resistivity in a similar way as observed in the Y_2O_3 stabilized ZrO_2 system.

As mentioned above, the most important point for the design of good conducting pathways in stabilized ZrO_2 systems is the interactive combination between fabrication route design and modeling based on advanced microanalysis of interfaces in the products. Note that the non-equilibrium defect structure variation depends on the distribution of residual thermal stress in the sintered bodies which strongly depends on slow cooling condition in the fabrication process. In order to design the stabilized ZrO_2 electrolyte with top level quality, the scientists and developer in R & D field must analyze the interface defect structure in their products. Then, they must use the modeling technique based on microanalysis data for optimization of the fabrication condition. It is most important point in the future R & D of fuel cell materials.

1.1.2 Doped CeO_2-Based Materials

M_2O_3 doped CeO_2 (M: Gd, Sm, Y, Yb, La, Ca, and so on) systems have received considerable attention for development of intermediate temperature operation of SOFC (i.e., IT-SOFC) devices. It was reported in 2000, Gd_2O_3 doped CeO_2 can be used as a solid electrolyte for IT-SOFC whose operation temperature is below 500 °C [39]. A representative electrical conductivity and activation energy diagram observed for M_2O_3 doped CeO_2 systems was demonstrated in Fig. 13.

In this Fig. 13, there are two types of the composition dependence of the observed activation energy. One is monotonous change observed for Y_2O_3-CeO_2 system and Yb_2O_3-CeO_2 system. Another type has local minimum value of activation energy observed for Gd_2O_3-CeO_2 system and Sm_2O_3-CeO_2 system. Conventional defect association model based on simple fluorite crystal structure cannot satisfactorily

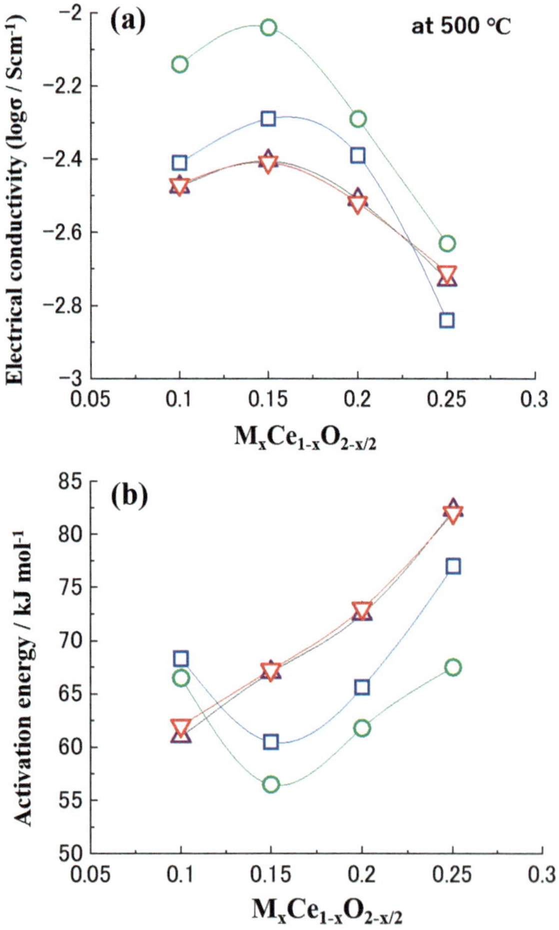

Fig. 13 Composition dependence of electrical conductivity (**a**) and activation energy (**b**) observed for Gd_2O_3-CeO_2 system (○) [40], Sm_2O_3-CeO_2 system (□) [41], Y_2O_3-CeO_2 system (△) [42], and Yb_2O_3-CeO_2 system (▽) [43]. Activation energy was derived from the measured electrical conductivity data in the temperature region from 350 °C to 600 °C using DC three-probe method

explain this experimental result well. Of course, the grain size dependence of the electrical conductivity and the activation energy in M_2O_3-doped CeO_2 systems was pointed in a similar fashion as is the case in stabilized ZrO_2 systems [15]. In the case of Fig. 13, both of electrical conductivity and activation energy were observed for the sintered samples with relatively large average grain size (i.e., 1 to 3 μm) to ensure a reasonable comparison. If the reader compared the observed values (i.e., electrical conductivity and activation energy) with various kinds of composition of M_2O_3 doped CeO_2 (M: Gd, Sm, Y, Yb, La, Ca, and so on) system, different results will be observed across all systems in each separate experiment. The reason for this is as a result of the observed data being strongly dependents on the microstructure at the nano-scale. This microstructure variation occurs as a consequence of the modification of non-equilibrium defect structures which occur during the various cooling regimes of the fabricated sample. Again, we can see same important key point which is mentioned in 'Sect. 1.1.1 stabilized zirconia section'. The suggested important key point is interactive combination between optimization of fabrication condition and modeling based on microanalysis for interfaces in the fabricated samples.

In addition, M_2O_3 doped CeO_2 system has another important factor for design of high-quality fuel cell materials and devices. It is well known that cerium dioxide (CeO_2) which involves multivalent cations (i.e., Ce^{4+} and Ce^{3+}) deviates strongly from stoichiometry at elevated temperature and reduced partial pressure of oxygen (Po_2). Consequently, CeO_2 shows n-type semiconducting property. In contrast, addition of di- and trivalent cation impurities to CeO_2 serves to enhance its oxide ionic conductivity orders of magnitude as confirmed by electrical conductivity versus Po_2 [44, 45] and concentration cell measurements [46]. Doped CeO_2 becomes a predominantly ionic conductor at high P_{O2} atmosphere and at temperatures below 700 °C.

As illustrated in Fig. 11, the fluorite lattice has large interstitial sites in the crystal. Multivalent cations (i.e., Ce^{4+} and Ce^{3+}) would provide interstitial sites with numerous opportunities for defect structure variations during slow cooling step of fabrication. Consequently, many co-doped effects with a view to improving electrical conductivity and lowering of activation energy were observed for doped CeO_2 systems [47–51]. In addition, the phase boundary composition between fluorite CeO_2 and Y_2O_3 which consists of C type rare earth structure, $[La_aSr_bCe_{0.25}Ye_{(1\text{-}a\text{-}b)}]_2O_{3+\delta}$ (a = 0.05, 0.1, and 0.15, b = 0, 0.006, and 0.0125) shows improved electrical conductivity in high P_{O2} at elevated temperature [52]. Note that different phase boundary compositions between fluorite cubic ZrO_2 and pyrochlore (i.e., $(M^{3+}{}_{1\text{-}x}M'^{3+}{}_x)_2Zr_2O_7$ (M: Gd, Sm, or Nd; M': Y, Yb, or Gd) systems also reveals high electrical conductivity in high P_{O2} at elevated temperature [53]. In addition, another phase boundary composition between Y_2O_3-ZrO_2 system and Sc_2O_3-ZrO_2 system (i.e., Y_2O_3-Sc_2O_3 co-doped ZrO_2) [54] shows higher electrical conductivity under similar measurement condition as well as $(M^{3+}{}_{1\text{-}x}M'^{3+}{}_x)_2Zr_2O_7$ (M: Gd, Sm, or Nd; M': Y, Yb, or Gd) systems. The improved electrical conductivities observed for co-doped (or multi-doped) fluorite systems are most enlightening. This observation strongly suggests that formation of non-equilibrium defect interface in the solid electrolytes is key for a design of high quality fuel cell materials. Almost all previously reported data tell us that the interactive combination of optimization of fabrication condition, microanalysis, and modeling provides us with good opportunity to develop 'state-of-the-art fuel cell' materials and devices in the future.

1.1.3 $LaGaO_3$-Based Materials and Brownmillerite-Based Materials

The perovskite type oxides which are expressed by ABO_3 structure consist of a trivalent inner transition ion in the A-site coupled with a trivalent transition metal ion in the B-site. The ABO_3 perovskite structure has a three-dimensional net of corner sharing $[BO_6]$ octahedra with A-site cation in the 12-fold cavities in between the polyhedra, refer to Fig. 14. The ABO_3 perovskite structure is highly flexible to tailoring by cation doping on both the A- and B-sites. Consequently, numerous co-doping systems have been reported for various kinds of applications.

In the case of solid oxide fuel cells, lanthanum strontium manganite ($La_{0.8}Sr_{0.2}MnO_3$ (LSM)) perovskite-based material has been used as the most important cathode material in a SOFC device [55–57]. It is because the LSM has an excellent track record of microstructural stability and long-term performance stability.

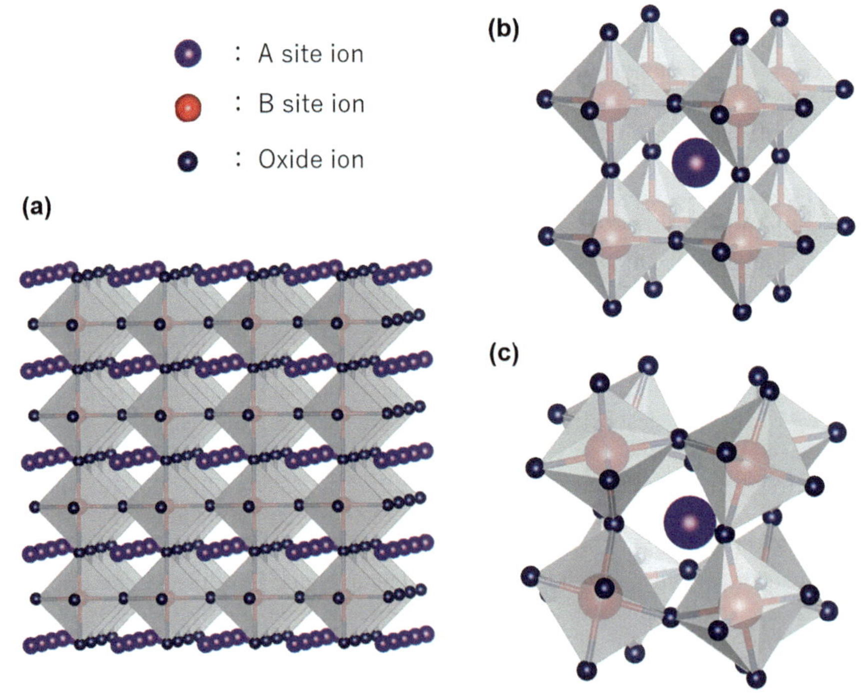

Fig. 14 Crystal structures of ABO_3 perovskite cubic structure (**a**), magnified illustration of ABO_3 perovskite cubic structure (**b**), and magnified illustration of ABO_3 perovskite orthorhombic structure (**c**)

The design concept for both LSM based cathode interface and LSM-based cathode/solid electrolyte interface is described by Professor San Ping Jiang and Dr. Shuai He (i.e., Part 1 chapter "Design of Interface Between Cathode and Electrolyte in SOFCs") in detail, refer to chapter "Design of Interface Between Cathode and Electrolyte in SOFCs". As solid electrolyte, $LaGaO_3$-based material has been known as one of promising electrolyte materials. $LaGaO_3$ is a distorted perovskite with tilting of the GaO_6 octahedra (Fig. 14c).

The crystal structure of $LaGaO_3$ at room temperature belongs to an orthorhombic symmetry. Above 145 °C, $LaGaO_3$ transforms from orthorhombic phase into a rhombohedral phase (space group: $R\bar{3}c$) [58]. Since doped $LaGaO_3$ perovskite type oxide was discovered as an oxide ionic conductor [59, 60], $La_{0.9}Sr_{0.1}Ga_{0.8}Mg_{0.2}O_{2.85}$ (LSGM1020) whose electrical conductivity is comparable to scandia-doped zirconia system has been of particular interest as a candidate electrolyte in the development of solid electrolytes in intermediate temperature operation solid oxide fuel cells (IT-SOFCs). The crystal structure of $LaGaO_3$ and LSGM1020 is summarized in Table 1 and Fig. 15. Neutron diffraction analysis [61] and TEM microanalysis (high resolution observation, selected area electron diffraction analysis, and precession electron diffraction analysis) [62–65] refined the defect interfaces in $LaGaO_3$ and LSGM1020 at room temperature and at elevated temperature (>145 °C) with high accuracy. TEM microanalysis clearly indicates the formation of nano-sized domain with ordered structure of oxygen vacancy in the sintered samples [63]. These defect interfaces would be formed in slow cooling process as non-equilibrium defects in a similar fashion to the fluorite-based oxides (i.e., stabilized ZrO_2 system and doped CeO_2 system).

In the previously published work, the larger In^{3+} species was progressively substituted onto the Ga^{3+} site of LSGM1020 [66]. The relationship between electrical conductivity and microstructure modification was examined by using XRD and TEM. In^{3+} specimens successfully substituted onto the Ga^{3+} site of LSGM1020 sintered bodies. The lattice constant was clearly increased by doping of In^{3+} cation into Ga^{3+} site of LSGM1020. In this case, the trace amount level of two isolated secondary phases such as $LaInO_3$ and $LaSrGaO_4$ was observed up to $x = 0.2$ in the composition of $La_{0.9}Sr_{0.1}Ga_{0.8-x}In_xMg_{0.2}O_{2.85}$ (x: 0.05, 0.1, 0.2, and 0.4) system. It is unlikely that the observed electrical conductivity was significantly affected by co-existence of trace amount of impurities in the sintered samples. Usually, the estimated free-volume would be increased and migration of oxide ion from one hopping site to another hopping site should be improved if we simply used the crystal structure of the LSGM1020 system. However, the observed electrical conductivity decreased

Table 1 Crystal structure of LaGaOs and LSGM1020

	Room temperature	Above 145 °C
$LaGaO_3$	Orthorhombic (Pbnm)[61–63]	Rhombohedral ($R\bar{3}c$)[61,63]
LSGM1020 ($La_{0.9}Sr_{0.1}Ga_{0.8}Mg_{0.2}O_{2.85}$)	Monoclinic (I2/a)[61,63,64]	Rhombohedral ($R\bar{3}c$)[61,63,65]

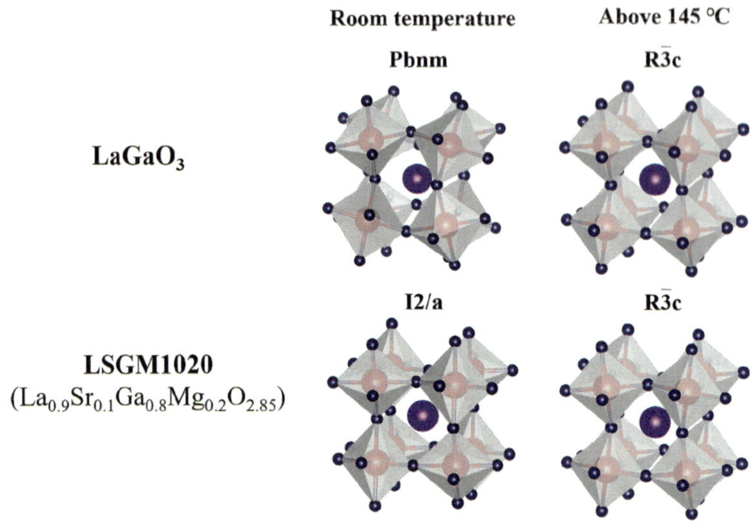

Fig. 15 Crystal structures of $LaGaO_3$ (space group: Pbnm (room temperature), R3c (>145 °C)) and LSGM1020 (space group: I2/a (room temperature), R3c (>145 °C)) (refer to Table 1)

with an increase of In^{3+} cation content. In addition, the activation energy derived from electrical conductivity observed for $La_{0.9}Sr_{0.1}Ga_{0.8-x}In_xMg_{0.2}O_{2.85}$ (x: 0.05, 0.1, and 0.2) increased with increasing In^{3+} cation content, refer to Fig. 16.

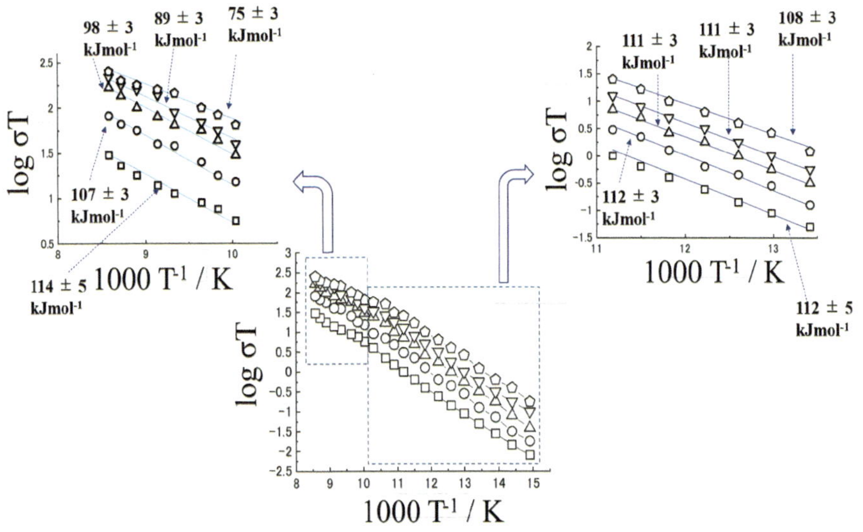

Fig. 16 Temperature dependence of electrical conductivity observed for $La_{0.9}Sr_{0.1}Ga_{0.8-x}In_xMg_{0.2}O_{2.85}$ (x: 0: (⬠), 0.05: (▽), 0.1: (△), 0.2: (○), and 0. 4: (□)) [66, 67]

While the conventional XRD analysis did not show clearly the hint to conclude a lowering of electrical conductivity observed for the $La_{0.9}Sr_{0.1}Ga_{0.8-x}In_xMg_{0.2}O_{2.85}$ (x: 0.05, 0.1, and 0.2) samples, the TEM microanalysis showed the increasing levels of diffuse scatter and superstructure formation in the selected area electron diffraction pattern observed for the $La_{0.9}Sr_{0.1}Ga_{0.8-x}In_xMg_{0.2}O_{2.85}$ (x: 0.05, 0.1, and 0.2) samples. This indicates that short-range ordering of oxygen vacancies was observed in specific crystallographic planes [67]. As a consequence, the volume of free hopping sites would be decreased by doping of In^{3+} cation into Ga^{3+} site of the In doped LSGM system.

In addition, LSGM1020 is $La_{0.9}Sr_{0.1}Ga_{0.8}Mg_{0.2}O_{2.85}$. However, the smallest unit cell of this composition is $La_{18}Sr_2Ga_{16}Mg_4O_{57}$. The unit cell of LSGM1020 is quite large. It corresponds to an array of 7 × 12 × 10 of the $La_{0.9}Sr_{0.1}Ga_{0.8}Mg_{0.2}O_{2.85}$ unit cells. This structure is 5.3 nm × 6.6 nm × 5.5 nm. LSGM1020 samples involve a fine micro-domain which is visible at the edges of these thinned samples, refer to Fig. 17. HRTEM image indicates that size of these micro-domains is around 5 nm in width and these micro-domains are in the individual twins. This is comparable in size to the large unit cell of LSGM1020. Each of these micro-domains with oxygen vacancy ordering is not exactly aligned with its neighboring micro-domain, with slight misfit between each. Their length was the full length of the grain, the order of microns. However, these micro-domains are not visible in lower magnifications observation, as shown schematically in Fig. 17 [63].

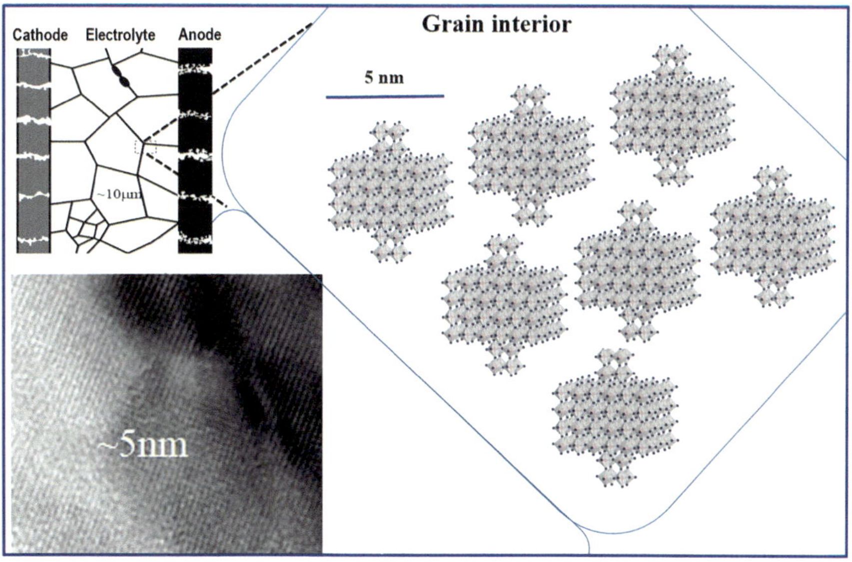

Fig. 17 Schematic diagram of microstructure with micro-domain with ordered structure of oxygen vacancy observed for LSGM1020 single cell [63]. (By courtesy of Dr. Graeme Auchterlonie, The University of Queensland, Australia)

As mentioned in the fluorite-based materials section, the distribution and content of micro-domains or nano-domains (i.e., tiny region of short-range ordering of oxygen vacancies) would strongly depend on slow cooling step condition in the fabrication process. This also strongly suggests that an interactive combination between fabrication route design and modeling based on microanalysis is key for design of high-quality fuel cell materials.

Brownmillerite type oxides which are expressed by the $A_2B_2O_5$ structure is considered to be an ordered ABO_3 structure. The oxygen vacancy sites in the unit cell of $A_2B_2O_5$ structure are ordered along the [101] direction [68]. The representative brownmillerite type oxide $Ca_2(FeAl)O_5$ consists of orthorhombic symmetry (space group: Ibm2). One-sixth of all the oxygen ions are missing. The general composition is also expressed as $ABO_{2.5}$. The unit cell of $A_2B_2O_5$ structure consists of octahedral and tetrahedral sites in the unit cell, refer to Fig. 18 [68].

In the SOFC field, the electrical conductivity of $Ba_2In_2O_5$ which is one of the brownmillerite type oxides were examined by Goodenough et al. for the first time. The relationship between phase transformation from orthorhombic (a = 16.79 Å, b = 6.08 Å, c = 5.89 Å) to disordered perovskite (ABO_3)-type cubic phase was investigated [68, 69]. A drastic improvement of the electrical conductivity was observed for $B_2In_2O_5$ oxide at 920 °C which is order/disordered transition temperature (T_t) between orthorhombic phase with ordering of oxygen vacancy and cubic phase with disordering of oxygen vacancies [68–71]. However, the large volume change (approximately 3.4 vol%) of the lattice appears between the ordered state and disordered state of $Ba_2In_2O_5$ [72]. This big volume change of the unit-cell of $Ba_2In_2O_5$ hinders the application of $Ba_2In_2O_5$ for solid electrolytes.

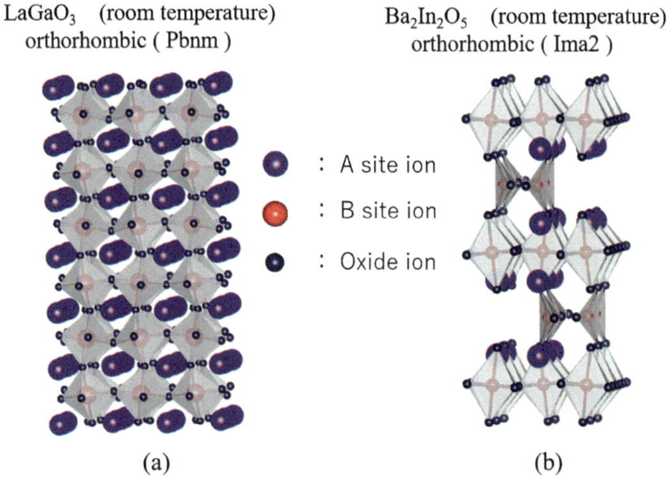

Fig. 18 Crystal structures of $LaGaO_3$ (space group: Pbnm): **a** and $Ba_2In_2O_5$ (space group: Ima2): **b** at room temperature

To use the electrical property observed for $Ba_2In_2O_5$ with disordered oxygen vacancies as solid electrolytes, many attempts were explored to try to decrease T_t of the order–disorder transition of oxygen vacancies in $Ba_2In_2O_5$. In the previously reported works regarding co-doped $Ba_2In_2O_5$ based compounds, the relationship between the electrical conductivity and T_t was examined using $Ba_2(In, M)_2O_5$ (M: Ga [73–76], Al [74, 77], Sc [78, 79], Y [78, 79], Yb [78], Dy [78], Sm [78]) system, $Ba_2(In,(Sc,Lu))_2O_5$ system [80], $Ba_2(In,M)_2O_{5+\delta}$ (M: Zr [69, 81], Ce [82, 83], Ti [84, 85], Sn [86], Si [87], V[88], Mo [88], W [89]) system, $(Ba,M)_2(In,Pb)_2O_5$ system [90], $(Ba,M)_2InM_2O_{5+\delta}$ (M: La [91], (Sr, La) [92]) system, $(Ba,M)_2(In,N)_2O_5$ [(M, N): (La, Ga) [93], (La, Ce) [94], (La, Zr) [94], (La, Mg) [94]] system, fluorine doped $Ba_2In_2O_{5+\delta}$ system [95] and oxyanions (i.e., PO_4^{3-}, SO_4^{2-}) doped $Ba_2In_2O_{5+\delta}$ system [96]. Those works suggest that the electrical conductivities observed for $Ba_2In_2O_5$-based materials were certainly improved by fabrication of co-doped (or multi-element doped) $Ba_2In_2O_5$-based materials. T_t of the order–disorder transition of oxygen vacancies in aforementioned co-doped (or multi-element doped) $Ba_2In_2O_5$-based materials shifted towards the low temperature side. However, the electrical conductivity values observed for co-doped (or multi-element doped) $Ba_2In_2O_5$-based materials were lower than that of disordered state of $Ba_2In_2O_5$ with cubic phase above 920 °C.

This indicates that the microstructure of the defect interface in the fabricated co-doped (or multi-element doped) $Ba_2In_2O_5$-based materials has to be taken into account for the design of high-quality $Ba_2In_2O_5$-based oxides. In the results section of the present chapter, the authors reintroduce one representative example for multi-disciplinary work (i.e., interactive combination of fabrication, microanalysis, and modeling). In order to see new breakthrough in the fabrication of fuel cell materials and devices field, we believe that the interactive combination between fabrication and modeling based on microanalysis is absolutely necessary.

1.1.4 Key Interfaces of Anode/Electrolyte and in Cermet Type Anode Layer

To develop the 'state-of-the-art' intermediate temperature SOFC (IT-SOFC) devices and systems exceeding the limit of conventional IT-SOFC, the design of good balance between the performance of SOFC and its stability is required. In this endeavor, the interface design of anode/electrolyte and cathode/electrolyte plays a key role for development of 'state-of-the-art' IT-SOFC devices and systems. In terms of recent progress and future work for design of cathode/electrolyte, the approach is explained in detail by San Ping Jiang and Shuai He, refer to Part 1 chapter "Design of Interface Between Cathode and Electrolyte in SOFCs". The authors focused on the interface design of anode/electrolyte in the present this chapter.

The anode of SOFC must exhibit the following properties to minimize excess overpotential at the three phase boundary (TPB) area.

a. Design of high activity on electro-catalyst towards the fuel
b. Interface design for an increase in TPB density
c. Open porous structure for gas diffusion and flow in the anode layer
d. Spatial contiguity between electrolyte particle and electrocatalyst for maximization of TPB area
e. Percolation of electronically conducting phase in the anode layer
f. Thermal expansion compatibility with the electrolyte.

In the present section, the authors focused on terms a. and b. to highlight key points for design of interfaces in the anode layer. In previously published works, Ni shows the best electrochemical activity for H_2 oxidation reaction as compared to other metals such as Co, Fe, Mn, Pt, Rh, Ru, Pd, and Au [97, 98]. Ru also has better electrocatalytic activity for H_2 oxidation and methane steam reforming as compared to other precious metals such as Pt, Rh, and Pd, however, the example of the use of Ru is limited as a result of its high cost. In addition, it is subject to evaporation as RuO_x and the stability of performance is reduced at high temperatures. In contrast, Ni is an abundant element with low production costs. Ni offers high catalytic activity for the fuel (i.e., hydrogen, hydrocarbon, methanol, and so on) oxidation, good chemical stability, high electronic conductivity, and small thermal expansion mismatch with an arrangement of solid electrolytes. In addition, the electrochemical oxidation reaction of hydrocarbon on Ni-based anodes can be improved by surface and structural modification [99–104].

In general, natural gas which is the main component of town gas is odor-free. In its pure form, town gas would be a suitable fuel for SOFCs. However, for the safety reasons, tert-butyl mercaptan (TBM, 2-Methyl-2-propanethiol, $C_4H_{10}S$) or dimethyl sulfide (DMS, C_2H_6S) are added to provide a detectable odor. These compounds contain sulfur which is a problem for Ni-based electrodes. Ni-based anodes show poor stability against sulfur even when the concentrations are at the ppm level [105, 106]. Consequently, the new cermet anodes whose components are the 'mixed ionic and electronic conducting (MIEC)' particles such as Y-doped $SrTiO_3$, co-catalysts such as CeO_2 and solid electrolyte particles have been actively examined [106]. Ru (0.4 vol%) and nano-sized CeO_2 as co-catalysts were added to the cermet anode with MIEC particle such as Y-doped $SrTiO_3$ using an infiltration method. As a consequence, the chemical stability against H_2S in H_2 fuel and device performance of SOFCs were conspicuously improved. In addition, it was reported that infiltration of $La_{0.3}Sr_{0.7}TiO_3$ together with CeO_2 and Pd (0.5 wt%) [107], impregnation of $La_xSr_xCr_{1-y}Mn_yO_{3-d}$ (LSCM) into YSZ framework anode [108], and infiltration of CeO_x-Cu into YSZ framework anode [108] improved both of anode performance and chemical stability against small amount of impurity H_2S in the fuel.

Moreover, 20 wt% Cu and 1.5 wt% Pd impregnation and a small amount of Ni or Ni/CeO_2 impregnation on LSCM-based anodes exhibited the improved performance in both H_2 and CH_4. No carbon deposition (i.e., 'coking') was observed on the impregnated anodes after CH_4 oxidation on anode in SOFC devices [108]. The previously reported infiltration/impregnation techniques using MIEC nanoparticle, co-catalysts such as CeO_2 and precious metal (less than 1 wt%) or Cu additives were useful for improvement of the H_2S tolerance, carbon deposition (i.e., 'coking') and a lowering of excess overpotential on the cermet anodes. In addition, a distributed charge-transfer model analysis of Cu- and Ce-based IT-SOFCs was able to simulate the polarization curves and impedance spectra observed for Cu–Pd mixture of CeO_2 and ZrO_2 anode /Sm doped CeO_2 (SDC) solid electrolyte/$La_{0.6}Sr_{0.4}Co_{0.2}Fe_{0.8}O_{3-\delta}$ (LSCF)-Gd doped CeO_2 (GDC) cathode single cell [109].

The conventional modeling on the basis of kinetics and thermodynamics can simulate polarization experiments and impedance experiments successfully. This approach would be helpful for optimization of complicated compositions of composite cermet anodes.

Recently, the endeavor to develop the infrastructure for hydrogen-based society is focused on R & D technology in the fuel cell field. Usage of syngas such as CO/H_2 or CH_3OH/H_2O is receiving attention as potential fuels for SOFC. In these cases, the sulfur poisoning can be a negligible issue. The design of anode/electrolyte interface is simplified. Instead of modeling for cermet anodes with complicated compositions, the interactive combination of fabrication of Ni-based cermet anode and the modeling on the basis of microanalysis data should be performed in order to propose the design paradigm for fabrication of 'state-of-the-art' IT-SOFC devices and systems with quality exceeding the limit of conventional IT-SOFC devices and systems.

Latest advancement in the nano-structure architecture, contaminant tolerance, and interface optimization in the Ni-based cermet anodes is introduced in detail in the review paper of Professor San Ping Jiang [110]. In this chapter, the authors touch on the essential features of design for the readers. In general, the anode cermet precursors are mechanically mixed by commercial NiO and YSZ (or doped ceria such as Sm_2O_3 doped CeO_2 (SDC), Gd_2O_3 doped CeO_2 (GDC), and so on) powder in ball mill or mortar in wet or dry condition. The finer powders are beneficial for the fabrication of cermet anode with higher performance. In this case, the fine YSZ particles covered the NiO particle which prevents the grain growth of NiO during the high temperature calcination of over 1200 °C. To fabricate well-dispersed fine powders, the readers must learn basic fabrication process and typical feature of well dispersed and round shaped powders for fabrication of cermet anode. In the case of YSZ fine powders, a typically standard commercially available YSZ is produced by TOSOH Company. This powder is fabricated by using hydrolysis of $ZrOCl_2{\cdot}8H_2O$ aqueous solution [21, 111]. When $ZrOCl_2{\cdot}8H_2O$ is dissolved into an aqueous solution, tetramer complexes (i.e., $[Zr(OH)_2{\cdot}4H_2O]_4{}^{8+}$) are created in the first step of the formation process of hydrous zirconia particles. Then, $[Zr(OH)_{2+x}{\cdot}(4\text{-}x)H_2O]_4{}^{(8-4x)+}$ forms in the aqueous solution through the deprotonation process, refer to Eq. (28).

$$\left[Zr(OH)_2 \cdot 4H_2O\right]_4^{8+} \xrightarrow{-H^+} [Zr(OH)_{2+x} \cdot (4-x)H_2O]_4^{(8-4x)+} \tag{28}$$

The precipitation which consists of primary and secondary particles are fabricated through polymerization, super-saturation, and nucleation process in the hydrolysis process. The precursor of YSZ solid solution is fabricated with similar process for YCl_3 in aqueous solution. This suggests that the YSZ precursor consists of hydrous zirconia and hydrous yttria. In this YSZ precursor formation process, the nano-scale heterogeneity of hydrous yttria in hydrous zirconia would be maintained by polymerization process of hydrates such as $[Zr(OH)_{2+x}\bullet\ (4\text{-}x)H_2O]_4^{(8-4x)+}$. Then mixture of hydrous zirconia and hydrous yttria is fabricated as precipitants. These precursor particles mainly consist of well dispersed and round shaped. On the other hand, it is hard to observe the crystal facets from round shaped particles. The reason being, non-crystalline or partial crystalline thin film covers the surface of the particles. After drying and rinsing of precipitants, the dried powers are calcined at around 900 °C for fabrication of Y_2O_3 stabilized ZrO_2 solid solution particles. Even using this well-studied fabrication process of Y_2O_3 stabilized ZrO_2, it is unlikely that perfect solid solution of Y_2O_3 stabilized ZrO_2 powders without heterogeneity of Y_2O_3 in nano-scale would be produced, refer to Fig. 19a.

Similar nano-scale heterogeneity of Sm_2O_3, Gd_2O_3, Y_2O_3 in CeO_2 matrix can be observed in the fabrication process of well dispersed and round shaped SDC, GDC and Y_2O_3 doped CeO_2 (YDC) particles which were reported in previously published works. The fabrication process of Re_2O_3 (Re: Sm [112–115], Gd [112, 115–117], Y [112, 118, 119], La [112], Yb [43], Tb [120, 121, 122], Nd [112], Dy [112, 123], Ho [112, 124], Er [112], and Pr [125]) doped CeO_2 solid solutions using carbonate co-precipitation method has been systematically examined by Li and Wang.

Based on X-ray diffraction analysis obtained from dried precursors which are products of co-precipitation method, the representative precursor compositions which strongly depend on the difference of precipitant were reported as follows.

(Case A): Precipitant: ammonium carbonate ($(NH_4)_2CO_3$).
Formation temperature of precipitation: 60 °C

$$\left((NH_4)_x CeRe_{0.25}(CO_3)_{1.875+x/2}\cdot y\,H_2O,\ x\le 1\ \text{and}\ y\le 2\right)$$

(Re = rare earth cation such as Sm, Gd, Y, La, Nd, Dy, Ho, Er, and Pr)

(Case B): Precipitant: ammonium hydrogen carbonate (NH_4HCO_3).
Formation temperature of precipitation: 60 °C

$$2[(Ce,Sm)(OH)(H_2O)_{n-1}]^{2+} + 2CO_3^{2-} \rightarrow (Ce,Sm)_2(OH)_2(CO_3)_2\cdot H_2O + (2n-3)H_2O \quad (29)$$

(Case C): Precipitant: urea ($CO(NH_2)_2$).
Formation temperature of precipitation: 60 °C

mixture of crystalline $(Ce, Sm)_2(CO_3)_2(OH)_2\cdot H_2O$ and $(Ce, Sm)_2O(CO_3)_2\cdot H_2O$

In general, thermal decomposition temperature of hydroxide precipitation (i.e., Case B) is lower than carbonate precipitation (i.e., Case A and Case C). In those series of co-precipitation methods at elevated temperature, all kinds of precursors mainly consist of well dispersed and round shaped particles when the precipitation temperature is optimized. To fabricate the dense solid electrolytes or good anode layer (refer to necessary properties of anode a ~ f), well-dispersed and round shaped solid solution particles were fabricated in previously published papers. However, the crystal facets were not clearly observed for almost all well-dispersed particles so far. The reason for this is that the non-crystalline thin film covers the surface of well-dispersed precursor particles. It suggests that nano-scale heterogeneity is found that both in the co-precipitation method as well as hydrolysis process, refer to Fig. 19b.

Figure 19a and b clearly suggests that we must carefully characterize the interface of cermet anode layers and optimize the defect interface composition and structure for development of high-quality anode materials for SOFC device and system application.

Both NiO/YSZ cermet composition [126] and NiO/SDC cermet composition [127] particles grow significantly after sintering at temperature above 1200 °C in air. After being reduced from 800 to 1000 °C in hydrogen, the overall porosity of the anode is increased by 25% as compared with before reduction in hydrogen. This overall porosity change is attributable to the volume reduction of NiO to Ni. In general, the porosity increases with an increase in the Ni volume fraction in cermet anode. The overall porosity of cermet anode whose composition is 40 vol% Ni/60 vol% YSZ is typically in the range of 40%.

Previously published works suggest that stability of the microstructure of Ni-based cermet anodes strongly depends on the Ni content and spatial contiguity of Ni and the solid electrolyte particles in the cermet anode layer [128, 129]. The Ni/YSZ ratio is expected to have a significant impact on the observed anode overpotential (i.e., anode polarization resistance (R_E)) by affecting the TPB density and other structural features. Usually, a Ni volume fraction of 40–45% (corresponding to ~60 wt% NiO) gives us minimum R_E [130]. For anode-supported thin film single cells, 50:50 to 60:40 wt% NiO:YSZ yielded the low R_E and contributed to observation of relatively large power densities [131–134]. The relationship between the spatial contiguity of Ni and solid electrolyte particles and the electrochemical properties was observed for varied Ni/YSZ functional-layer compositions in anode-supported SOFCs [135–137]. The three-dimensional reconstruction of a Ni-YSZ composite anode was demonstrated based on SEM observation data [135]. In this work [135], approximately 63% of TPB were interconnected and 19% were unconnected small sections (i.e., poor spatial contiguity between Ni and YSZ particles) in the anode layer. This observation suggests that TPBs on such networks should be considered to have unknown electrochemical factor for a design of high quality anode material. To analyze the transport property observed for porous cermet anode layer, both a two-dimensional numerical model as well as a one-dimensional analytical model were developed [138]. This theoretical approach is helpful for conclusion of observed electrochemical data. This theoretical analysis suggests that much benefit is to be obtained by fabrication of a microstructure as fine as possible. Also, the conclusion

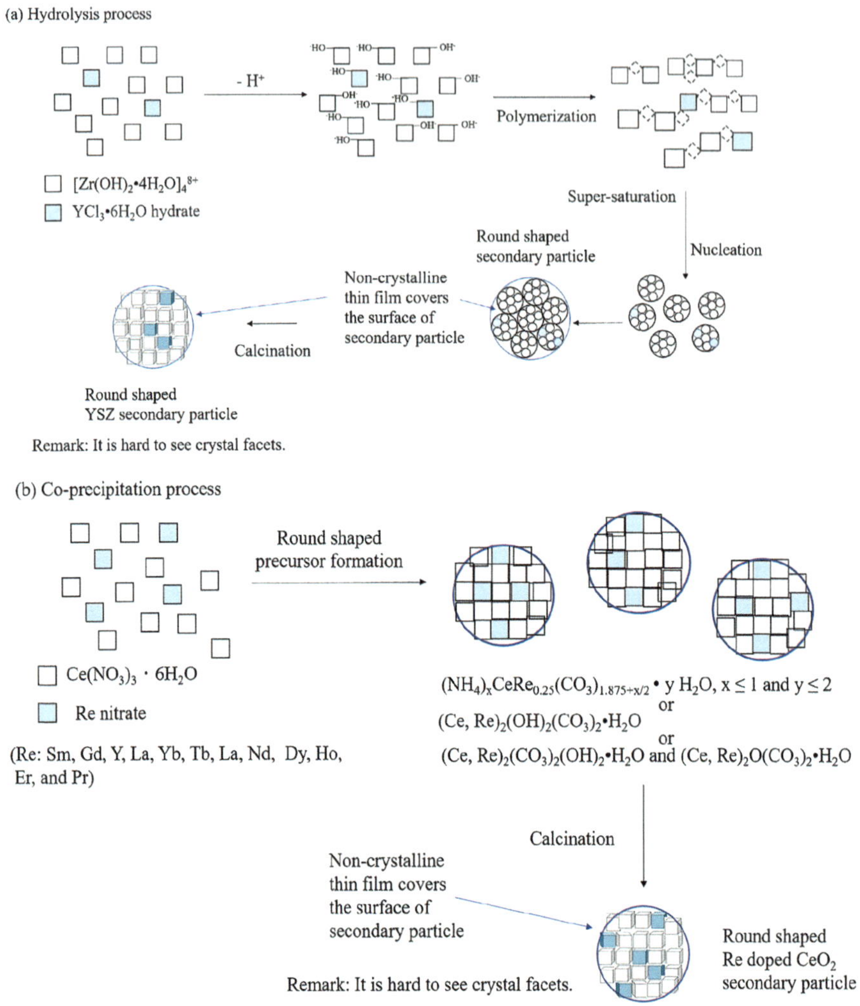

Fig. 19 **a** Schematic diagram of fabrication process of round shaped YSZ particles which involve nano-heterogeneity using hydrolysis process, refer to [21, 111]. **b** Schematic diagram of fabrication process of round shaped Re doped CeO_2 particles which involve nano-heterogeneity using hydrolysis process, Re = Sm, Gd, Y, La, Yb, Tb, Nd, Dy, Ho, Er, and Pr

section in this theoretical analysis work tells us that appropriate interface design in anode layer is important to obtain low excess overpotential. Namely, there is now something new in this suggestion through the 1D and 2D theoretical analyses.

Recently, microstructure design using nano-scale materials has paid attention to maintaining the balance between performance and durability of SOFCs [139–141]. It is well-know that the sintering and agglomeration of Ni phase is a key factor to determine the stability of performance observed for Ni-based cermet anodes. As the authors introduced in this section, an infiltration method is often used for design of nano-scale microstructure in cermet anode of SOFC. For example, various Ni nitrate solution have been infiltrated into porous YSZ tube as anode support layer. As a consequence, good distribution of the Ni particles in the YSZ tube and a low polarization resistance were observed in that work [142]. In addition, very small polarization resistance (i.e., 0.1 Ω cm^2 per electrode at 800 °C) was observed for Ni infiltrated YSZ scaffolds which consist of finer particle and small pore sizes (~200 nm each) in the anode layer [143]. On the other hand, similar approaches were examined for Ni-SDC and Ni-GDC cermet anodes. The reason for this is that the overpotential observed for Ni-SDC and Ni-GDC in the direct electrochemical oxidation of hydrocarbon fuels was lower than the Ni-YSZ cermet anode [144–147]. The following example is typical of this approach. GDC nano-sized particles which are impregnated into Ni anodes significantly improved the electrocatalytic activity on the Ni anode. In addition, impregnated GDC nanoparticles improved the carbon tolerance (i.e., coking tolerance) at 800 °C in the fuel composition of 97% CH_4/ 3% H_2O [148]. Also it was reported that CeO_2-modified Ni/YSZ fully suppressed carbon deposition in methane with 10% H_2O fuel gas [149, 150].

As the authors touched on in this section, the surfaces of fine particles of YSZ and Re doped CeO_2 (RE: Sm, Gd, Y, La, Yb, Tb, Nd, Dy, Ho, Er, Pr, and so on) are covered by non-crystalline thin film which mainly consists of precursor compositon of electrolyte particle. Nano-scale heterogeneity remains in the interface of the cermet anode layer which is calcined at temperature above 1200 °C. In addition, non-equilibrium defect interface structure would be formed at the interface of cermet anode layer in slow cooling process, refer to Sect. 1.1.2 for doped ceria-based materials and Sect. 1.1.3 $LaGaO_3$-based materials and brownmillerite-based materials. To design the fine microstructure in the cermet anode layer, the authors believe that modeling based on TEM microanalysis is key challenge in R & D of fuel cell field.

Based on the previously reported works, the authors would like to highlight that interactive combination of fabrication and modeling based on microanalysis is key for development of 'state-of-the-art' IT-SOFC devices and systems. The authors reintroduce some important data in the Sect. 3 Representative Results section to support this point of view.

1.2 Key Interfaces in PEMFC Materials

Polymer electrolyte membrane fuel cells (PEMFCs) have been developed for use in vehicular applications, residential scale cogeneration systems, and combined energy conversion/storage systems [151–153]. PEMFC device also consists of an electrolyte membrane (i.e., proton conducting membrane), anode, cathode, and gas diffusion layer (GL). The electrodes (i.e., anode and cathode) must consist of porous structure because the reactant gases (i.e., O_2 and H_2) are fed into the system and must reach the interface between the electrodes and the proton conducting membrane. In addition, both the anode and cathode layers consist of the wet-proofed carbon papers (i.e., gas diffusion layer: GDL) with catalyst ink. The catalyst ink can be prepared using conventional Pt/ C and ionomer (i.e., 5 wt% Nafion® solution). To fabricate membrane electrode assembly (MEA), commercially available proton conducting membrane (i.e., Nafion® membrane) is sandwiched by two electrodes and pressed at elevated temperature. In the MEA, three phase boundary (TPB) consists of the proton conducting phase (i.e., proton conducting membrane and ionomer), electron conducting phase (electrocatalyst and conductive carbon), and gas phase (including porous structure). Consequently, the performance observed for MEA strongly depends on the proton conducting membrane/ electrodes (i.e., anode and cathode) interface structures, interface structures in the electrode layers (i.e., anode layer and cathode layer), and TPB density in MEA.

Inside of MEA, the acid–base interface design much attention has been paid to the development of high quality MEA which is operated under 90 °C. In this endeavor, various kinds of ionomeric membrane systems have been developed and demonstrated to have high performance as compared with conventional MEA which consists of Nafion® membrane and Nafion® ionomer [154, 155]. Moreover, high temperature PEMFCs (HT-PEMFCs) whose conductivity can be stable from 100 °C to 120 °C have been actively developed for miniaturization as part of the radiator system in the vehicle. The reason for this is that reducing the radiator footprint will be useful for improvement of weight-saving and gas mileage of the vehicle. In this challenge, the design of interfaces in the ionomeric membrane systems is key [156–160]. Also, the interface model in MEA has been developed for high and low humidity operation [161]. The simulated I-V curve (i.e., electro-polarization curve) using non-equilibrium interface model [161] agrees with the I-V curve observed for MEA. The recent advances of HT-PEMFC are introduced by Professor Z.Shao et al., refer to Part 1 chapter "Design of HT-PEM Devices".

While the design of interface in ionomeric membrane systems for HT-PEMFC application is still active in the academic field, the maximization of activity on Pt/ C cathode and anode in PEMFC device which is operated under 90 °C is strongly required in vehicular application. In the Pt/C electrocatalysts, the interface of Pt and conductive carbon has been examined using subnano-sized Pt particles deposition on the noble conductive carbons such as nitrogen doped graphene [162]. Pt particles whose average particle size is under 0.5 nm were dispersed on graphene nano-sheet (GNS). Subnano-sized Pt particles on GNS revealed strong CO oxidation property.

On the other hand, the longtime stability of electrocatalytic performance observed for the nano-sized Pt particles whose average particles sizes are 2, 3, and 4 nm supported on carbon black was examined using standard potential step cycles between 0.6 V and 1.0 V (vs. RHE) in N_2-purged 0.1 M $HClO_4$ solution at 65 °C [163]. The reason for this testing regime is that the standard potential step cycles test simulates load cycles for fuel cell vehicles. In this durability test, the kinetically controlled mass activity observed on all Pt/C electrocatalysts decreased with a decrease of electrochemical active surface area (ECSA) of Pt up to 30,000 potential cycles. This clearly indicates that nano-sized and subnano-sized Pt particles are easily agglomerated under the applied potential. In conjunction with this, the activity of electrocatalyst and performance of MEA is reduced. Consequently, the design of interface between Pt and oxide promotor has attracted much attention to keep the preferable balance between performance and its stability observed for Pt electrocatalyst and MEA device.

In this section, the authors briefly review the Pt with oxide promotor/conductive electrocatalysts as anode and cathode. We would like to highlight the importance of interface design in anode layer and cathode layer for the fuel cell application.

1.2.1 Interface on Anode

In general, the observed overpotential on the anode is much smaller than that on the cathode in MEA performance tests using humidified pure H_2 gas and humidified oxygen (or air). However, it is well-known that trace amount of carbon monoxide (CO: more than 3 ppm) will be strongly adsorbed on Pt surface and inhibit the activity of Pt electrocatalyst (i.e., CO poisoning). To increase CO tolerance on Pt, Pt-based alloy electrocatalysts [164–188] such as Pt alloyed with Ru [164–185], Co [172–174], Mo [175–177], Sn [176–179], Bi [178], Pb [178], Sb[178], W [176], Nb [177], Ta [177], Ni [180, 181–183], Mn [186], Zn [188] have been developed for the improvement of CO tolerance of Pt anodes. Of those candidates, PtRu alloy shows the best improvement in performance of the Pt-based alloys examined. This is a result of the lower potential required for formation of Ru-OH pair as compared with the Pt-OH formation while activated OH on Ru can additionally oxidize CO to CO_2 on Pt. This phenomenon is known as bifunctional mechanism. However, the PtRu alloy is still not sufficiently CO tolerant for effective application and PtRu alloy electrocatalyst still involves a high platinum content which increases environmental burden on our planet.

To provide the bifunctional mechanism with the surface of Pt anode, oxide supports were added to Pt/carbon. Also, the effect of oxide supports such as RuO_x [182, 185], SnO_x [182], $InSnO_x$ [182], MoO_x [187], WO_3 [189], NiO [190], TiO_2 [191], and $NbRu_xO_y$ [192], $RuSnO_x$ [193], on the electro-alcohol oxidation has been examined for an improvement of CO tolerance on Pt/C anode. In the previously published works for the Pt-oxide support series, Pt-cerium oxide (CeO_x)/carbon (C) [193–211] is a unique system because of electrochemical redox behavior of CeO_2 (i.e., redox reaction between Ce^{3+} and Ce^{4+} under applied potential at room temperature) in the anodic fuel cell reaction. Since the electrochemical redox behavior on Pt-CeO_x supports would contribute to decreasing the strong adsorption of CO on Pt through the interface of Pt and CeO_x [211], the design of defect structure of Pt and CeO_x interface is key in the challenge of improvement of CO tolerance on Pt in fuel cell devices. For this challenge, the fabrication of key defect structure using the guidance of the microanalysis based on atomistic simulation is useful for improvement of fuel cell device performance, not only improvement of on-set potential of electro-oxidation reaction of alcohol on anode. The authors reintroduce representative data in the results section of this chapter. To improve the CO tolerance on Pt anode, the authors would like to point out that the readers should use combination of fabrication and microanalysis based on atomistic simulation. Then, we believe that the readers can see technical innovation in their challenging works without exception.

1.2.2 Interface on Cathode

To improve the gas mileage of fuel cell vehicles, a design of highly active Pt/C cathode is required. Recently, the non-platinum cathodes such as carbon-based materials (i.e., N-doped carbon nanotube array [212], N-doped graphene [213–218], N-doped mesoporous carbon [219], graphene-based carbon nitride nano-sheets [220], S-doped graphene [221], graphene oxide [222], N-doped carbon nanocable [223]) have been examined as cathode in PEMFC applications. However, the current density observed for MEA with carbon-based materials was observed to be lower than the more conventional systems and it seems to be difficult to use these cathode types instead of Pt. The most suitable material design is combination of Pt-oxide support/conductive carbon-based material from the viewpoint of providing reliable cathode performance and a stable performance of the PEMFC device. In this challenge, the design of key interface between Pt and oxide support is effective to reduce the excess overpotential loss on Pt cathode surfaces.

The oxygen reduction reaction (ORR) activity on Pt-oxide supports such as Pt-WO_3/conductive carbon (C) [224], Pt-TiO_2/C [224], Pt-Ta_2O_5/C [225], Pt-NbO_2/C [226], and Pt-CeO_x/C [227–235] has been actively examined. In the previously reported Pt-oxide supports type cathode series, Pt-CeO_x/C type cathode is a unique system as well as the anode. The reason for this is that the electrochemical redox behavior of CeO_x contributes to the reduction of excess overpotential on Pt for ORR was observed by using in-situ electrochemical X-ray absorption fine structure (XAFS) analysis [234]. The in-situ XAFS analysis indicates that the electrochemical redox reaction of Ce ($Ce^{3+} \leftrightarrow Ce^{4+}$) on Pt-$CeO_x$/C cathode inhibited Pt-OH formation on the Pt surface in high potential region and contributed to the improvement of ORR activity on the Pt surface in the cathodic condition. To obtain the preferable balance between high performance of MEA and the long-term stability, analysis of electrochemistry on Pt-CeO_x/C cathode is not sufficient. The Pt-CeO_x/C cathode which is fabricated by conventional fabrication method cannot be used as active cathode in MEA because of low three-phase boundary density in the cathode layer of MEA. Defect interface on Pt-CeO_x/C cathode has to be designed by the combination work of fabrication, microanalysis, and modeling. The authors reintroduce representative results in the results section of this chapter. The authors expect that our combination work will provide the readers with great inspiration for development of 'state-of-the-art' fuel cell devices.

2 Experimental

2.1 *Fabrication of CeO_2 Nanoparticles, CeO_2 Nanowire, and CeO_x Thin Film*

The ammonium carbonate co-precipitation method at elevated temperature was used for fabrication of nano-sized round shape CeO_2 particles. The typical fabrication conditions are re-introduced in this section as follows.

The **un-doped CeO_2 (i.e., pure CeO_2) powder** was fabricated by a hot ammonium carbonate precipitation method. The starting materials used were commercially available $Ce(NO_3)_3 \cdot 6H_2O$ and $(NH_4)_2CO_3$ powders. The $Ce(NO_3)_3 \cdot 6H_2O$ powder was first dissolved in distilled water. To fabricate the fine precipitation, a $(NH_4)_2CO_3$ aqueous solution was heated at 58 °C in a thermostatic chamber. The control of this temperature is first key point for fabrication of round shaped pure CeO_2. If the temperature of $(NH_4)_2CO_3$ aqueous solution is higher than the above setting temperature, precipitation will consist of rod like shaped particles. If the setting temperature of the $(NH_4)_2CO_3$ aqueous solution is below 55 °C, we will see an agglomerated precipitation after the drying process.

The $Ce(NO_3)_3$ aqueous solution was dropped into the $(NH_4)_2CO_3$ aqueous solution over a 1 h period, and this mixture was continuously stirred using magnetic stirrer at 58.0 °C for 24 h. Note that this part of the process is the second key point for fabrication of fine round shaped pure CeO_2. If the $(NH_4)_2CO_3$ aqueous solution dropped into $Ce(NO_3)_3$ aqueous solution for fabrication of CeO_2 particles (i.e., the opposite of the above process), the readers will see huge agglomerated particles in your products. The size and shape of precipitate will be changed by the repeat of dissolution and re-precipitation process in the mixture at 58.0 °C for 24 h.

After filtration and rinsing, the precipitate was dried at room temperature in a N_2 gas flow. The third key point is this drying process. Do not add heat energy to precipitate in drying process. The agglomeration will be enhanced by heat energy. The reason for this is that water molecules will be trapped in the precipitate after drying at elevated temperature, and hydrogen bonds will be created among the secondary particles. This will lead to strong agglomeration in the dried powders. If you can remember the aforementioned three key important points, the control of the size and shape of pure CeO_2 is easy to produce. The dried powders were calcined at 400 °C for 2 h in an O_2 gas flow. This calcination condition is just a recommendation from the authors. The aim is to remove burnable impurities from the powders, and improve the fabrication of nano-sized pure CeO_2 particles.

The **RE_2O_3 doped CeO_2 (RE: Y, Sm, Gd, Yb, La, Ho, Tb and multiple dopants by combination of two kinds of 'RE' elements) powder** was fabricated by a hot ammonium carbonate precipitation method using different conditions to those described above. The starting salts are Ce and RE nitrate hexahydrates ($RE(NO_3)_3 \cdot 6H_2O$, RE: Y, Sm, Gd, Yb, La, Ho, or Tb) with a purity of at least 99.99%. The salt solution was mixed for preparation of $RE_xCe_{1-x}O_{2-x/2}$ powders (x = 0.10, 0.15, 0.20, and 0.25) and was heated at 50 to 60 °C at first. Then, this hot mixed solution was dropped slowly into the ammonium carbonate solution and stirred mildly at the reaction temperature of 60 to 70 °C. In this process, the important point is the control of the reaction temperature for fabrication of the precursor of $RE_xCe_{1-x}O_{2-x/2}$ (RE: Y, Sm, Gd, Yb, La, Ho, or Tb). Simultaneously, the condition of the fabrication of pure CeO_2 powders must be controlled. The preferable reaction temperature should be checked in advance. The reason for this is that is the preferable reaction temperature is different for the fabrication of different $RE_xCe_{1-x}O_{2-x/2}$ (RE: Y, Sm, Gd, Yb, La, Ho, or Tb) systems. The mixture was then aged for 1 h at the reaction temperature to fabricate nano-sized round shape precursors. Note that the control of shape and size of precursors is the second key point in this process. If the precursor particles did not consist of nano-sized round shaped particles, it will be hard to fabricate the well-crystalline nano-sized round shaped $RE_xCe_{1-x}O_{2-x/2}$ (RE: Y, Sm, Gd, Yb, La, Ho, or Tb) particles after the calcination process. The precursors were then collected by suction filtering, washed with distilled water twice and ethanol just one time. In this process, rinsing process using ethanol is also important to minimize any remaining water molecules among the precursor particles. Finally, the precursor particles are dried at room temperature with flowing nitrogen gas. The dried precursors were subsequently calcined at 600 to 700 °C for 2 h in an oxygen environment to yield the well-crystalline nano-sized round shaped $RE_xCe_{1-x}O_{2-x/2}$ (RE: Y, Sm, Gd, Yb, La, Ho, or Tb) powders.

The CeO_2 Nanowires (NW) were fabricated using the soft-template method. The starting materials used were commercially available cerium chloride $CeCl_3 \cdot H_2O$ (99.9% pure), cetyl tetra ethyl ammonium bromide $C_{19}H_{42}BrN$ (98% purity), and urea $(H_2N)_2CO$ (99.99% purity) powders. $C_{19}H_{42}BrN$ and $(H_2N)_2CO$ were used as the surfactant components for micelle formation and as a base for control of pH in the micelle formation reaction, respectively. Those powders were dissolved into the dehydrated ethanol and mixed using magnetic stirrer at room temperature for 1 h in dry nitrogen atmosphere. In this process, the first key point is the use of dehydrated ethanol. To fabricate CeO_x NW, commercially available super dehydrated ethanol has to be used in this process. If any remaining water is in the ethanol, micelle which consists of cerium precursor is easily oxidized and soft template cannot be used for fabrication of CeO_x NW. The mixed solution was filtered by using ultrafiltration. Half amount of filtrate was transferred into 25 ml Teflon autoclave. This transferred volume is the second important factor. If the transferred volume is over the half level, the pressure in the vessel cannot reach the required level. Then, the formation of CeO_x NW does not occur. The formation reaction of CeO_x NW takes place within the 25 ml solution in the Teflon autoclave maintained at 80 °C for 4 days in an electric oven. The control of this temperature is third important point. From the viewpoint of safety, the reaction temperature should be less than 85 °C. In addition, if the reaction temperature goes below 60 °C, the growth speed of CeO_x NW is very slow. It is hard to expect fabrication of CeO_x NW in this process. After filtration and rinsing of all components, the mixture was dried at room temperature in a N_2 gas flow for 48 h.

The CeO_x Thin Film was fabricated using pulsed laser deposition (PLD) method. The laser ablation was carried out on dense CeO_2 target (purity of CeO_2 powder, 99.9%; disk size, 50 mm in diameter and 5 mm thickness). The produced CeO_x

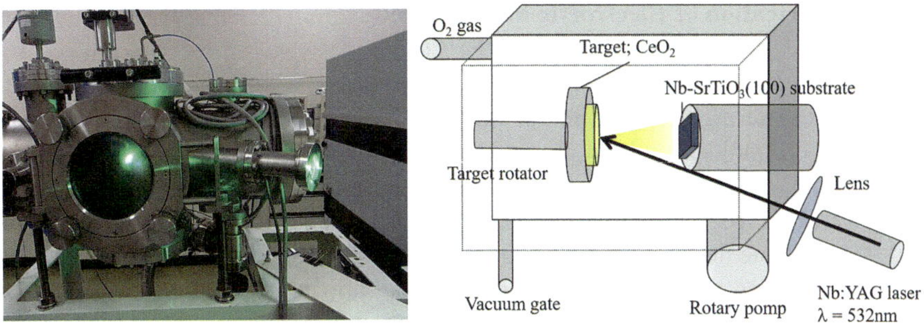

Fig. 20 Photograph and schematic diagram of entire PLD equipment at QST

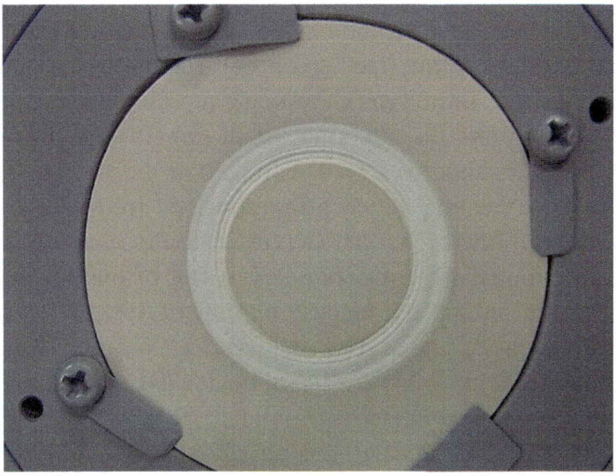

Fig. 21 Photograph of the target of dense un-doped CeO_2 sintered body in PLD equipment

film was deposited on the conductive single crystal substrate by using this PLD method. The commercially available epi-polished $SrTiO_3$ single crystal doped with Nb (Nb content, 0.5 wt %; specific resistance, 3 to 7 × 10^{-3} Ω·cm) in which crystal orientation is (100) was used as the conductive single crystal substrate. The size of conductive Nb-doped $SrTiO_3$ single crystal substrate was 10 × 10 × 0.5 mm. PLD was performed by the second harmonic Q-switched Nd/YAG laser with the wavelength of 532 nm, 8 ns in pulse width, and 5 Hz in frequency, refer to Fig. 20.

During the deposition, the laser was focused onto the continuous rotating CeO_2 target (rotation speed: 5 rpm). The substrate was separated from the target by approximately 5 cm, refer to Fig. 21. In the present PLD method, the substrate temperature was set at 570 °C under residual O_2 gas pressure of 5.3 Pa. The deposition time was 30 min with the laser energy of 30 mJ $pulse^{-1}$. The estimated thickness of CeO_x film on 0.5 wt % Nb-doped $SrTiO_3$ (100) single crystal substrate was approximately 50 nm. After the PLD process, the CeO_x film was annealed at 1000 °C for 1 h to improve its crystallinity.

2.1.1 Fabrication of Electrolyte and Electrode

To fabricate the electrolyte, the authors used conventional sintering and pulsed electric current sintering methods. In the **conventional sintering method**, the calcined powders were molded under a pressure of 500 kg cm^{-2} and subjected to a rubber press 2 t cm^{-2} (1 t = 10^3 kg) to obtain a green body. The conventional sintering was performed from 1000 to 1450 °C for 4 to 6 h. The heating and cooling rate is 200 °C h^{-1}. Note that the slow cooling rate such as 200 °C h^{-1} causes the formation of heterogeneous microstructure during sintering process. The micro-domain with ordered structure of oxygen vacancy is formed during the slow cooling process. The reason for this is that the non-equilibrium defects are formed in the samples. The observation of these non-equilibrium defect structures at interfaces or bulk regions will provide us with a good opportunity to observe breakthroughs in the fuel cell field. Consequently, the readers should observe their own materials after sintering and design the interface and bulk structure based on the data of microanalysis and modeling. The authors show some representative examples in the results section.

To depress the abnormal grain growth, **the pulsed electric current sintering (PECS)** method was used. The bulk of the sintered sample could be heated evenly by Joule heating between the powders. In this case, a two-step sintering process was performed. The first step uses the PECS that is followed in the second step by conventional sintering. For PECS, 2 g of the powders were placed into a 15 mm graphite die and an electric current of 1000 A was applied under a pressure of 600 kg cm^{-2}. The heating rate was 500 °C min^{-1}. The sintering temperature ranged from 1000° to 1200°C. To minimize the penetration of carbon from the graphite die into the specimens, no holding time at the aforementioned sintering temperature was used. The current was simply shut off at temperature and the pressure was immediately released after the sintering. Following the PECS process, the specimens were sintered at the temperature ranging from 1000 to 1200 °C for up to 48 h in air. Note that fast heating rate process can depress the abnormal grain growth during sintering process. However, it is hard to fabricate the large size sintered sample which has high density using this method. In this chapter, just one example which was fabricated by this combined method is re-introduced to demonstrate the effect of microstructure modification on improvement of bulk property such as electric property in the solid electrolyte. Since the fabrication of large size sample is required for industrial application, the authors show other options in this chapter.

To fabricate the anode for solid oxide fuel cells (SOFCs), the authors used home-made $RE_xCe_{1-x}O_{2-x/2}$ (RE: Y, Sm, Gd, La or Dy) powders, commercially available YSZ powder (grade name; TZ-8Y, TOSOH, Japan) and commercially available nickel oxide (NiO) powder (99.9%, Wako Laboratory Chemicals, Japan). Also, we used the commercially available $La_{0.85}Sr_{0.15}MnO_3$ slurry (LSM20-1, Fuel Cell Materials, USA) for fabrication of the cathode layer, refer to Sect. 2.1.2 'Fabrication of SOFC and PEFC devices' in this chapter.

For fabrication of functional interfaces on the cathode in polymer electrolyte membrane fuel cells (PEMFCs), the authors used Pt loaded CeO_x nanoparticles or Pt loaded CeO_x nanowire. **Pt-CeO_x nanoparticle/C cathodes** were prepared in a stepwise process. The pure CeO_2 powder was synthesized by a hot ammonium carbonate precipitation method. The preparation detail of pure CeO_2 was introduced in 'Sect. 2.1 Fabrication of CeO_2 nanoparticles, CeO_2 nanowire, and CeO_x thin film'. To impregnate the fine Pt particles on the prepared CeO_2 nanoparticles, $H_2PtCl_6 \cdot 6H_2O$ powder was dissolved into ethanol (C_2H_5OH) at the glovebox where the moisture level was kept below 1 to 10 ppm. The previously prepared pure CeO_2 nano-sized round shaped powders were dispersed in the H_2PtCl_6 ethanol solution and dried in the glovebox. The dried mixture was calcined at 400 °C for 1 h in a 10% H_2/He mixed gas flow. The prepared Pt-CeO_x and carbon black (C; Vulcan XC−72R, Cabot Co.) were dispersed in C_2H_5OH and dried in a N_2 gas flow.

Pt-CeO_x Thin Film Model Cathode was prepared using combination of pulsed laser deposition (PLD) and impregnation methods. The experimental setup used for PLD is described in 'Sect. 2.1 Fabrication of CeO_2 nanoparticles, CeO_2 nanowire, and CeO_x thin film'. The laser ablation was carried out on the target of dense CeO_2 sintered body. The CeO_x thin film was deposited on the conductive single crystal substrate by using PLD method. The commercially available epi-polished $SrTiO_3$ single crystal doped with Nb (Nb content: 0.5 wt%; specific resistance: from 3 to $7{*}10^{-3}$ Ω·cm) in which crystal orientation is (100) was used as the conductive single crystal substrate. After the PLD process, the CeO_x film was annealed at 1000 °C for 1 h to improve its crystallinity. To impregnate fine Pt particles on both Nb-doped $SrTiO_3$ single crystal substrate and the prepared CeO_2 thin film, which was deposited on Nb-doped $SrTiO_3$ single crystal substrate, $H_2PtCl_6 \cdot 6H_2O$ powder was dissolved in C_2H_5OH under the glovebox conditions where the moisture level was kept from 1 to 10 ppm. To fabricate the electrode layer (area; 19.6 mm^2), which contains 2 μg of Pt on the Nb- $SrTiO_3$ single crystal substrate, a predetermined amount of H_2PtCl_6 ethanol solution was dispersed on both Nb-doped $SrTiO_3$ single crystal substrate and CeO_x thin film deposited on the substrate. After drying in the glovebox, the aforementioned H_2PtCl_6 loaded CeO_x thin film on the substrate and the H_2PtCl_6 loaded on the substrate were calcined at 400 °C for 1 h in a 10% H_2/He gas flow.

Pt-CeO_x Nanowire (NW)/C Cathodes were fabricated in a stepwise process as well. The CeO_x NW was synthesized by an alco-thermal process, refer to 'Sect. 2.1 Fabrication of CeO_2 nanoparticles, CeO_2 nanowire, and CeO_x thin film'. To fabricate the CeO_x NW with nano-sized Pt particles, $K_2PtCl_4 \cdot H_2O$ powder was dissolved into distilled water and stirred for 1 h. The aqueous K_2PtCl_4 solution was added to a suspension of the previously prepared CeO_x NW which was dispersed in distilled water. The obtained mixture was allowed to stand at room temperature for 12 h. The K_2PtCl_4 in the mixture was reduced by addition of aqueous sodium borohydride ($NaBH_4$) solution and that mixture was mixed at room temperature for 1 h. Then, the mixture was rinsed with water and C_2H_5OH. This mixture was dried at room temperature in a N_2 atmosphere. In the last stage, the prepared Pt-CeO_x NW and carbon black were dispersed in C_2H_5OH and dried in a N_2 gas flow.

To enhance the interaction between Pt and CeO_x NW at the interface on the Pt-CeO_x NW/C cathode, the authors used proton beam irradiation or electron beam irradiation method.

Fig. 22 Photograph of entire proton beam irradiation system at RIKEN

An ion irradiation system in liquid was constructed using a taped glass capillary placed at the end of a vacuum beamline, refer to Fig. 22. The glass capillary was connected to a liquid container. CeO_x NW, $K_2PtCl_4{\bullet}xH_2O$, and C_2H_5OH (C_2H_5OH / Pt molar ratio = 5) were dispersed into distilled water in the liquid container. This mixture was irradiated by a proton beam for 60 min (estimated Dose: 5 kGy). Protons were accelerated with an electrostatic tandem accelerator at the RIKEN Institute. Note that the weight ratio of Pt and C in Pt-CeO_x NW which is fabricated using proton beam irradiation method is 0.02. It is just one-tenth as compared to that of conventional Pt/C.

Note that it is hard to fabricate large amount of samples whose surface is modified using proton beam irradiation method. As demonstrated in Fig. 22, the volume of reaction vessel is small size. To fabricate the large amount of sample whose surface is modified using quantum beam, the authors used the electron beam irradiation method as well. The electron beam was irradiated using 2 MV, 60 kW electron accelerator of the Cockcroft-Walton type, refer to Figs. 23 and 24. Pt loaded CeO_x NW samples in an aqueous solution were irradiated with a 2 MeV electron beam with 10 kGy min^{-1}. To minimize the electron beam damage of the sample surface, a cooling stage was used. CeO_x NW and $K_2PtCl_4{\bullet}xH_2O$ dissolved in distilled water and the mixture was allowed to stand at room temperature for 12 h.

And two kinds of samples, whose surfaces were modified using proton beam or electron beam, were mixed with carbon black in C_2H_5OH. In addition, the mixed suspension was filtrated and dried in a N_2 gas flow for 2 days. Eventually, proton beam irradiated Pt-CeO_x NW/ C (weight ratio between Pt and C: = 0.02) and electron beam irradiated Pt-CeO_x NW/ C (weight ratio between Pt and C: = 0.2) were fabricated.

2.1.2 Fabrication of SOFC and PEMFC Devices

SOFC single cells were fabricated by screen printing anode and cathode materials on both sides of a variety of solid electrolyte sintered bodies such as RE_2O_3 doped CeO_2 (RE: Gd, Sm, La, or Dy) and 8 mol% yttria stabilized zirconia sintered bodies

Fig. 23 Photograph of generation source of electron beam at QST

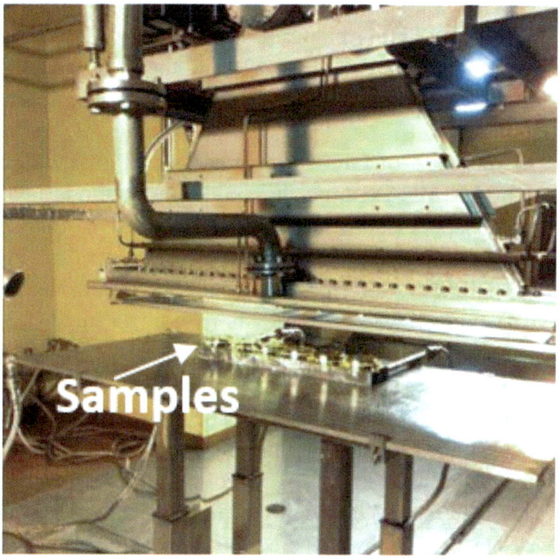

Fig. 24 Photograph of electron beam irradiation table with cooling stage. This sample stage is located underneath the generation source of the electron beam

(diameter, approximately 13 mm; thickness, 0.5 mm; relative density, ~95% of the theoretical density). A slurry containing the mixture of nickel oxide (99.9%, Wako Laboratory Chemicals, Japan) and solid electrolytes powder, such as RE_2O_3 doped CeO_2 (RE: Gd, Sm, La, or Dy) (home-made samples), and YSZ powder (grade name; TZ-8Y, TOSOH, Japan) with the mass ratio of NiO: solid electrolytes powder = 4:1 was used as the cermet anode material.

To examine the promotion effect of brownmillerite-based promotor in the anode, the prepared promotor powder (amount of promoter: from 0.1 to 0.4 wt % in whole anode), the cermet anode powder, and small amount of ethanol (i.e., several drops of ethanol) were mixed by using agate mortar and pestle for 30 min. The anode slurry was printed on the solid electrolyte disk in a five-step repeated process. The coated material was calcined at 1200 and 1300 °C for 1 h. The thickness of anode layer after the calcination was approximately 40 μm.

To examine the promotion effect of trace amount of PtO_x, platinum family metal oxides (i.e., RhO_x, PdO_x, or RuO_x), and FeO_x promotor in the anode, the authors deposited trace amount of active PtO_x, platinum family metal oxides (i.e., RhO_x, PdO_x, or RuO_x), and FeO_x on the calcined anode layer (calcination condition: 1300 °C for 1 h, thickness: approximately 40 μm) by using magnetron sputtering which is a suitable method for deposition of the thin oxide promotor films with precise control of the thicknesses (i.e., trace amount of the precious metal is well controlled), refer to Figs. 22 and 23. The deposition was carried out using metal (99.99% purity) targets (i.e., Pt, Rh, Pd, Ru, or Fe) in oxygen atmospheres at room temperature. With the constant pressure in magnetron chamber of 0.17 Pa and RF power of 15 W, the deposition rate was approximately 120 nm h^{-1}, refer to Figs. 25 and 26.

For the cathode, the $La_{0.85}Sr_{0.15}MnO_3$ slurry (LSM20-1, Fuel Cell Materials, USA) was used. In a process similar to that described previously, the cathode material was applied in five-step process and then calcined at 1100 °C for 1 h. The thickness of cathode layer was approximately 50 μm. The area of electrodes was 6 mm in diameter.

In the case of fabrication process of PEMFC device (i.e., membrane electrode assembly (MEA)), both the anode and cathode layers consist of the wet-proofed carbon papers (i.e., gas diffusion layer: GDL) embedded by using catalyst ink. To fabricate the catalyst ink, the cathode powder (i.e., Pt-CeO_x NW/ C or conventional Pt/C (HiSPEC3000, Johnson Matthey Corporation Japan)) and Nafion® solution (5 wt% solution) were mixed by homogenizer mixer from 1 to 2 min. Then, additional mixing was performed using shaker at 2500 rpm for 30 min. The anode powder (i.e., conventional Pt/ C) and the Nafion® solution were also mixed in the same way as the preparation of cathode ink. The formed catalyst inks were directly spread on the GDL using a brush. The amount of Pt in Pt/ C cathode ink is 0.15 $mg_{Pt}cm^{-2}$ on GDL. Also, the amount of Pt in Pt/ C anode ink is 0.15$mg_{Pt}cm^{-2}$ on the GDL. To make MEA, a commercially available Nafion® membrane (NR212, DuPont Japan) was sandwiched by two electrodes and pressed at 150 °C for 20 min. The pressing pressure was 2.5 MPa.

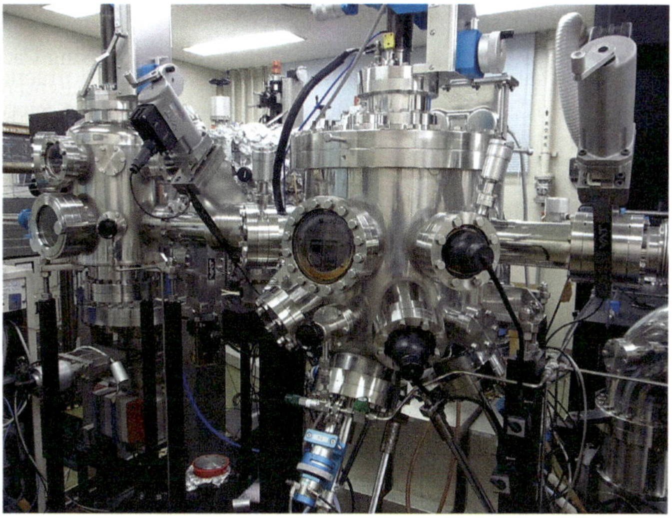

Fig. 25 Photograph of entire magnetron sputtering equipment at QST. Left hand side chamber is pre-chamber which is connected with main reaction chamber

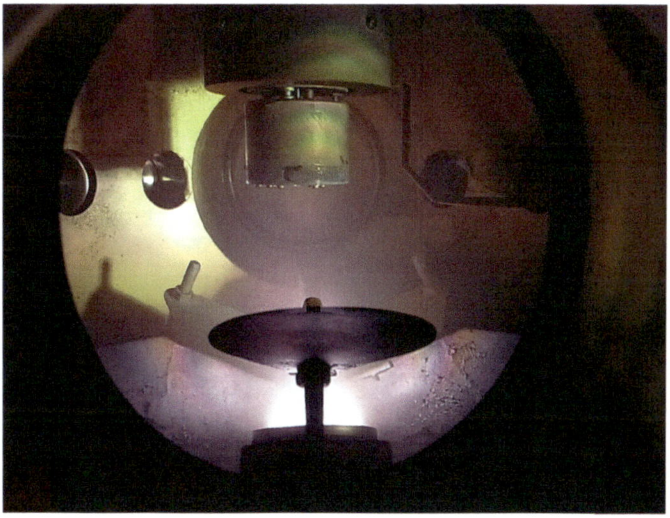

Fig. 26 Photograph of inside-main chamber of magnetron sputtering equipment. This photograph was taken during sputtering process

2.1.3 Sample Characterization

The crystal phases of samples were identified using power X-ray diffraction (XRD) analysis. In the case of XRD analysis, the authors made use of the Rietveld analysis method. Detailed measurement conditions of Rietveld analysis of XRD are as follows.

Rietveld analysis program: RIETAN-FP, X-ray source: CuKα (wavelength λ: 1.5418 Å), X-ray tube voltage: 40 mV, X-ray tube current: 40 mA, Optical system: concentration method, Monochromator: Flat, Step size: 0.02 deg. or 0.05 deg. and Scanning field (2θ): 10 to 140 deg.

The detailed microstructural features, crystal phases, and compositions at the sub-micrometer scale and at the nano-scale were analyzed by transmission electron microscopy (TEM) and selected area electron diffraction (SAED) pattern, refer to Figs. 27 and 28. Those two representative TEM examples which were used in our works.

The aperture size used for SAED was about 0.2 μm on TEM images. To examine the valence state of Ce and the local ordering of oxygen vacancies in the electrolyte film, electron energy loss spectroscopy (EELS) was performed in an analytical TEM operated at 200 or 300 kV equipped with imaging filter. The regions selected for EELS study were made to deviate from the strong Bragg reflections and were chosen to have a thickness of 0.1 to 0.3 inelastic mean-free paths to minimize the influence of multiple scattering. The power-law method was applied for a background subtraction of the EELS spectrum. The cross-sectional TEM specimens were prepared by mechanical polishing and dimpling and were finished by low angle ion-beam thinning. To analyze the cross section in details, the cross-sectional TEM specimens were prepared and analyzed using focused ion beam (FIB).

Fig. 27 Photographs of TEM (electron accelerating voltage: 200 kV (HF 5000, Hitachi High Technologies Company, Japan) and TEM operation room. These photographs are taken at Center for Microscopy and Microanalysis, Faculty of Science, the University of Queensland, Australia. The operator of TEM: Dr. Graeme Auchterlonie who is one of co-authors of the present chapter

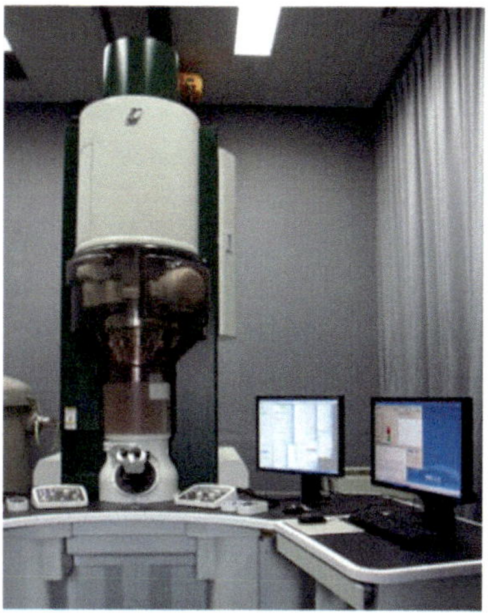

Fig. 28 Photograph of TEM (200 kV, JEM-ARM 200F, JEOL Ltd., Japan) at NIMS

For comparison of microanalysis data, local structure in Y_2O_3 doped CeO_2 sample was characterized using synchrotron-based EXAFS (Extended X-ray Absorption Fine Structure) spectroscopy as well. Quantitative curve fitting was performed for radial structure function typically from 1.0 to 3.3 Å (in R-space) using the theoretical phase shifts and amplitude functions derived from the FEFF code. The lattice parameters and the atomic coordinates used for the FEFF calculation were referred from the crystal structure in the ICSD database. All fits utilized k^3-weighed data over k range from ~2 to ~14 $Å^{-1}$ for room temperature EXAFS and ~2 to ~12 $Å^{-1}$ for high temperature EXAFS [236].

The surface chemical states of prepared samples were characterized using soft X-ray photoelectron spectroscopy (SXPS). SXPS spectra were observed by using a standard laboratory Al Kα X-ray source ($h\nu = 1486.6$ eV). To characterize the average chemical states of deep surface regions of the sample, we also used hard X-ray photoelectron spectroscopy (HXPS). HXPS experiments were performed at the undulator beamline BL15XU of the synchrotron facility, SPring-8. The photon energy was set to 5953 eV (inelastic mean free path of photoelectrons from Ce 3d of CeO_2 was estimated to be 7.1 nm). The photoelectrons were detected and analyzed for their kinetic energy using a high-resolution electron analysis. Total energy resolution was set to 240 meV. Binding energies of Ba 3d, In 4f, Zr 4f, Zn 3d, Ni 2p, Ce 3d, Pt 4f, and O 1 s core level peaks were corrected using the binding energy of C 1 s core level peak as 284.5 eV.

Also, the particle size distribution observed for the cermet anode was examined by using linear intercept technique. The number of intercepts between the test lines and grain boundaries was counted using 400 grains in SEM pictures taken from the cross section of the reduced anode layer of SOFC single cells.

The relationship between sputtering time and thickness of the deposited PtO_x layer on anode of SOFC single cells was determined by measuring the step height difference between sputtered and non-sputtered areas on the 8YSZ substrate, using

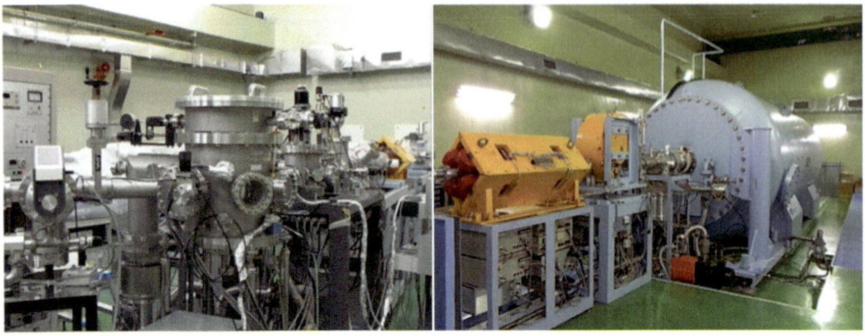

Fig. 29 Photograph of entire RBS system at QST

a surface profiler (DEKTAK3ST, Veeco Company, USA). The element content in PtO_x and platinum family metal layers was estimated by Rutherford Backscattering Spectrometry (RBS), refer to Fig. 29.

Since the RBS analysis contains an uncertainty of approximately $\mp 3\%$, the aforementioned analysis uncertainties were noted in the results [237]. In addition, to ensure accuracy of characterization across all methods, the compositional analysis of Pt, Rh, Ru, Pd, Ni, Y, and Zr in representative samples was performed by inductively coupled plasma mass spectrometry (ICP-MS, ELAN DRC II, Perkin Elmer Company, Japan). For ICP-MS analysis in our work, the anode samples containing trace amounts of Pt and Pt group metals were collected from the single SOFC cell. These samples were dissolved in hot aqua regia (hydrochloric acid: nitric acid = 3:1). The resulting solution was filtered and analyzed using ICP-MS. In this chemical analysis process, four calibration curves for quantitative analysis of samples with different concentrations were prepared in advance using standard solutions containing a certain concentration of each element. Based on the aforementioned calibration curves, the amount of trace elements in the electrode was calculated as volume concentration. To confirm the presence of elements in the residue from the filtration process, potassium sulfide (K_2S) and the residue were mixed and melted at elevated temperature. The molten solution was cooled and filtered, and the elemental content in the filtered solution was analyzed by ICP-MS. As a result, no target elements were found in the residue. Based on the above two ICP-MS analysis data (filtrate and residue), the final content was determined. Since ICP-MS analysis involves a minute level of uncertainty, the uncertainty is noted in the result parts.

2.1.4 Bulk or Surface Atomistic Simulation

To support the microanalysis data for design of functional interfaces in fuel cell devices, the authors performed bulk or surface atomistic simulation using General Utility Lattice Program (GULP) code. In terms of experimental procedure of bulk

Fig. 30 Photographs of workstations for bulk and surface atomistic simulation at NIMS

atomistic simulation, the readers can refer to Part 2 chapter "Atomistic Simulation of Subtle Microstructures in CeO_2-Based Electrolyte for Design of the State-of-the-Art Fuel Cells" (Authors: Fei Ye et al.) of this book. To minimize overlap in each chapter, the authors explain the surface atomist simulation process of our work in the present chapter.

To highlight the defect structural feature on the top surface of electrodes or interfaces on electrode/ electrolyte specimens, the authors used the surface atomistic simulation method using several workstations, refer to Fig. 30.

The surface cluster energy is calculated by a two-region approach coded in the GULP. In this approach, the lattice for energy minimization is partitioned into two regions: a spherical inner region I at the center where the defects are introduced and an outer region II, which extends to infinity. To ensure a smooth transition between region I and region II, an interfacial region IIa is introduced. Note that, the defect model has to be placed in the center region for simulation. In addition, the readers must assume large super cell region as compared to bulk atomist simulation to minimize affection of boundary from region II.

The surface model of fluorite cubic ZrO_2 (111) is stacked by repeating units of O−Zr layers. In this case, the size of the supercell containing the cubic ZrO_2 surface is **7 × 7**. Region I consists of approximately 10 repeating units with thicknesses about 22.9 Å. Region II consists of approximately nine repeating units with thicknesses of 21.7 Å. Also, the surface model of fluorite related cubic CeO_2 (111) surface is constructed by stack repeating units of O-Ce–O layers. The size of the super cell containing the (111) surface is **7 × 7**. Region I consists of approximately six repeating units with thicknesses of about 19.6 Å. Region II consists of approximately five repeating units with thicknesses of 13.3 Å. The atomistic interactions for the energy minimization are described within the framework of the Born model of an ionic crystal, which includes a long-range Coulomb interaction and a short-range interaction including the electronic attractive van der Waals force and electronic repulsion between electron clouds. The short-range term is described by the Buckingham potential as well as bulk atomistic simulation. The short-range potential parameters and the shell model parameters which were used in the present surface atomistic simulation are shown in the results section of this chapter.

The interactions resulting from the polarizations of Ce^{4+}, Zr^{4+}, Pt^{2+}, Ba^{2+}, In^{3+}, Zn^{2+}, and O^{2-}, are described by the shell model. The short-range interactions are set to zero beyond a cutoff of 20 Å. In this model, the polarized ions are described by a massive core with charge X|e| connected by a massless shell with charge Y|e|, resulting in an overall ion charge of (X + Y)|e|. The core and shell are connected by an isotropic harmonic spring with the force constant k of O^{2-}. The shell parameters are presented in the results section of this chapter. To investigate the preference and stability of the surface defect clusters, the binding energy (ΔE_b) is defined using Eq. (30).

$$\Delta E_b = \left(\sum E_{isolated} - E_{cluster}\right)/\text{number of defects in cluster} \quad (30)$$

where $\Sigma E_{isolated}$ is the sum of the formation energy for isolated defects and $E_{cluster}$ is the formation energy of a lattice defect cluster. The calculated energies of isolated point defects are shown in the result section. Equation (30) indicates that a positive ΔE_b predicts a preference of the formation of surface defect clusters over its individual components. Also, when the calculated ΔE_b, $\Sigma E_{isolated}$, and $E_{cluster}$ are positive values, the simulation for the proposed model is reasonable.

For detailed simulation for the relationship between defect structure and performance of fuel cell materials, the authors performed surface first-principles simulation using ABINIT code. The electron–ion interaction was described by the projector augmented wave (PAW) method and the PAW atomic data were taken from the JTH PAW atomic dataset table. In this large size simulation, we used NIMS supercomputer, refer to Fig. 31.

Fig. 31 Photograph of supercomputer for surface first-principles calculation at NIMS

2.1.5 Measurement of Electrolyte, Electrode, and Fuel Cell Performance

The electric conductivity in solid electrolyte was measured using three-probe and four probe measurement methods. The ionic transport number of oxygen in solid electrolytes was estimated using concentration cell method or Webb-Wagner direct current (DC) polarization.

Webb-Wagner direct current (DC) polarization measurements were performed for estimation of the ratio of oxide ionic conductivity and n-type electronic conductivity in the samples. For this electrochemical measurement, a dense sintered sample with a density greater than 93% of the theoretical value was used; the Webb-Wagner DC polarization measurement requires one electrode to be an ion-blocking electrode and the other a reversible (non-blocking) electrode. A platinum plate (0.1 mm thick), the ion-blocking electrode, was placed on a pellet 13 mm in diameter and 0.5 mm thick. On the other side of the sample pellet, a thin Pt paste was applied and calcined at 900 °C for 1 h at a temperature increase rate of 10 °C min^{-1}. This process was used to fabricate porous non-blocking electrodes on the dense pellets.

The oxygen partial pressure (P_{O2}) dependence of the DC conductivity measurements was performed at 700 °C. The P_{O2} was controlled at 10^{-17} to 10^{-20} atm by humidified hydrogen at room temperature, and this P_{O2} was confirmed by online 8 mol% Y_2O_3 stabilized zirconia oxygen sensor. The DC conductivity was measured by four-point probe method. The steady state current was measured in the potentiostatic mode. The potential was increased from 0.05 to 1.2 V in 0.1 V steps. Each step was maintained at fixed time, and the steady state current data at each P_{O2} were collected.

Note that the readers should use the potential sweep method to see the steady state condition in the samples. If the readers waited for steady state condition in each P_{O2} level at fixed temperature, it will take a long time to reach the steady state. The reason for this is that the difference of P_{O2} inside and outside of the sample cannot be balanced quickly. To reach the steady state condition smoothly, above potential sweep method is better choice in our experiment.

The electrochemistry of Pt-CeO_x nanoparticle (NP)/C and Pt-CeO_x nanowires (NW)/C was examined using a standard three-electrode glass cell, refer to Fig. 32.

The anodic properties of the Pt-CeO_x NP/C and Pt-CeO_x NW/C were investigated by cyclic voltammetry (CV) in a mixed aqueous solution of 0.5 M CH_3OH and 0.5 M H_2SO_4 in the potential ranging from 0 to 1.5 V (vs. Ag/AgCl). Electrochemical measurements were performed at 28 °C. Samples for the electrochemical measurements were prepared from Pt-CeO_x NP/C and Pt-CeO_x NW/C samples (2 mg ml^{-1}) suspended in C_2H_5OH solution. The suspension was spread on an Au electrode (area: 0.197 cm^2) using a micropipette containing 5 ml of suspended solution. The measured potentials were converted to the reversible hydrogen electrode (RHE) scale. Prior to the electrochemical measurements, electrochemical pre-treatment for conditioning of the anode surfaces was performed by 30 (vs. Pt-CeO_x NP/C) or 120 (vs. Pt-CeO_x NW/C) cycle sweeps in the potential range 0 to 1.5 V (vs. Ag/AgCl). The sweep rate during the conditioning process was 50 mV s^{-1}. To estimate the onset potential of

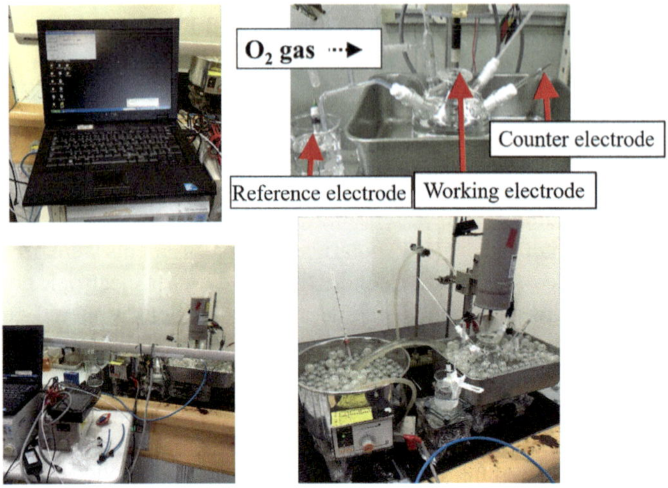

Fig. 32 Photographs of three-electrode glass cell at NIMS

methanol oxidation reaction, a slow sweep rate (i.e., 1 mV s^{-1}) was selected because fast sweeps affect the baseline drift of the CV curves. The electrochemically active surface area (ECSA) was estimated using the amount of charge associated with the electro-desorption of hydrogen from the Pt surface at 28 °C.

To estimate the ECSA of Pt on the sample surface, the sweep rate of 20 mV s^{-1} was used. The current observed in the CV measurements was normalized by the geometric surface area calculated using Levich's equation (sweep rate: 10 mV s^{-1}). To compare the anodic properties observed with commercially available materials, the electrocatalytic activity of methanol electro-oxidation observed with Pt/C (Johnson Matthey Co., HiSPEC 3000) was also examined.

The cathodic properties on the Pt-CeO_x NP/C and Pt-CeO_x NW/C were also investigated by voltammetry in a 0.5 M H_2SO_4 aqueous solution over a potential range from 0.8 to 0 V (vs. Ag/AgCl). The water was purified using a RO (reverse osmosis)-EDI (electro-deionization) system, and the ultrapure reagent grade H_2SO_4 was used for preparation of electrolyte solution. The electrochemical measurements were performed in a standard three-electrode glass cell with a rotating disk electrode after conditioning process at 28 °C. A Pt foil and Ag/AgCl were used as counter and reference electrodes, respectively. The samples for the electrochemical measurement were prepared from a suspension of the cathode materials (2 mg ml^{-1}) in a C_2H_5OH solution. The suspension was applied to the surface of an Au electrode (area: 0.20 cm^2) using a micropipette as well as for the anode sample. The amount of cathode material on the Au electrode was 5 μl. The amount of Pt loaded in the conventional Pt/C, Pt-CeO_x NP/C, and Pt-CeO_x NW/C samples was 2 μg. To evaluate the ORR activity, the hydrodynamic voltammograms were observed in O_2-saturated 0.5 M H_2SO_4 aqueous solution. The final hydrodynamic voltammograms were obtained by subtracting the hydrodynamic voltammogram in N_2-saturated 0.5 M H_2SO_4 aqueous solution from the aforementioned observed hydrodynamic voltammogram in O_2-saturated 0.5 M H_2SO_4 aqueous solution at the same rotation speed (2000 rpm). In the electrochemical measurements, the potential was swept from the positive to the negative side. The sweep rate was 10 mV s^{-1}. For a fair comparison, the same amount of each electro-catalyst by mass was loaded on an Au rotation disk electrode, and the geometric surface area calculated using Levich's equation (sweep rate: 10 mV s $^{-1}$) and this result was used to calculate the current density. The measured potentials were converted to the reversible hydrogen electrode (RHE) scale and this was done for the anode case as well.

Note that the sample thickness on the working electrode in both experiments for measurement of anode and cathode activities should be very thin and uniform. If the thickness of sample on the working electrode is very thick, the effect of concentration overpotential becomes an issue in the samples. Then it will be hard to discuss the kinetics on the samples. In order to analyze the kinetics on the samples, the effect of concentration overpotential in the sample has to be at a negligible level.

Electrode performance of the fabricated SOFC single cells was investigated in a SOFC evaluation system with gas-sealed cement, refer to Fig. 33. To minimize the interfacial contact resistance, a platinum mesh was attached to both cathode and anode surfaces. In addition, two platinum wires (500 μm in diameter), one for current supply and the other for potential detection, were connected to the platinum mesh to complete a circuit. Prior to the measurements, each SOFC cell was exposed to a reducing atmosphere at 800 °C for 3.5 h with a dilute hydrogen mixture (H_2 5 ml min^{-1} + N_2 45 ml min^{-1}) on the anode side and a gas mixture similar to air (O_2 10 ml min^{-1} + N_2 40 ml min^{-1}) on the cathode side. During the electrode performance measurements, the humidified hydrogen (80 ml min^{-1}) and oxygen (80 ml min^{-1}) flowed into the anode and cathode sides, respectively. The IR-free polarization characteristics observed in the single cells were evaluated by a current interrupt method at 700 °C. The potentiostat/galvanostat as a fast solid-state switch was used for this measurement. Measurement data were collected using a digital storage oscilloscope. To conclude the correct values of anode overpotential, the cathode overpotential was measured using a reference electrode Pt (diameter, 0.2 mm). The IR included polarization characteristics (i.e., I-V curves and I-P curves) recorded from the single cell were also examined using the potentiostat/galvanostat and the electrochemical measurement system. To measure the cell voltages between OCV and zero, the current was varied by adjusting the resistance of the external load. To obtain higher current density data, an external voltage was applied to the cell using function generator. And all I-V curves were obtained under the steady state condition.

The electrode performance of the prepared samples was investigated in the membrane electrode assembly (MEA). Both the anode and cathode layers consist of the wet-proofed carbon papers (i.e., gas diffusion layer: GDL) embedded with electro-catalyst ink. To prepare the electro-catalyst ink, the cathode powder (i.e., 20 wt% Pt-CeO_x NW/C or conventional Pt/C (HiSPEC3000, Johnson Matthey Corporation Japan)) and Nafion® solution were mixed using homogenizer mixer for 1 to 2 min.

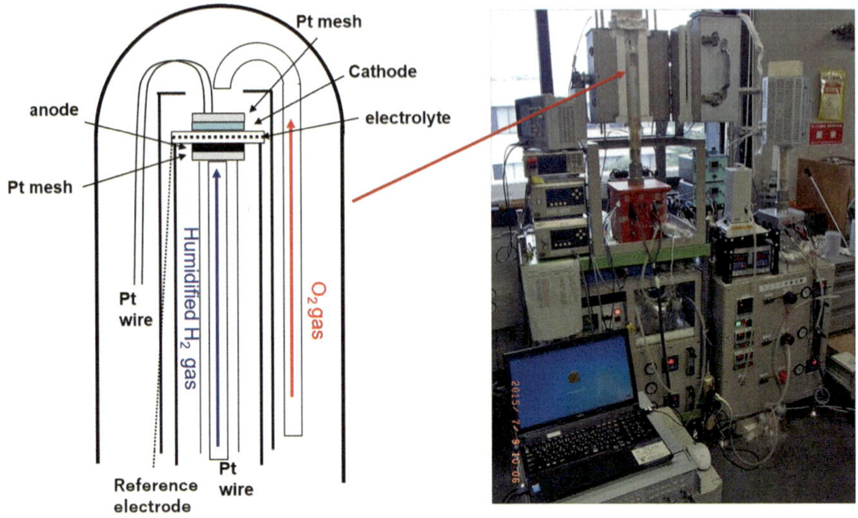

Fig. 33 Photograph and schematic diagram of test equipment of SOFC single cell at NIMS

Additional mixing was then performed using a shaker at 2500 rpm for 30 min. The anode powder (i.e., conventional Pt/ C) and the Nafion solution were also mixed in a similar manner to the preparation of cathode ink. The formed electro-catalyst inks were applied directly on the GDL using a brush. The amount of Pt in Pt/C cathode ink is 0.15 $mg_{Pt}cm^{-2}$ on the GDL. The amount of Pt in Pt/C anode ink is 0.15 $mg_{Pt}cm^{-2}$ on the GDL as well. To fabricate the MEA, a commercially available Nafion® membrane (NR212, DuPont Japan) was sandwiched between two electrodes and pressed at 150 °C for 20 min. The pressing pressure was 2.5 MPa.

The performances of the cathode and anode layers of the MEA were examined using the MEA test holder and performance test apparatus shown in Fig. 34. During the electrode performance measurements, a hydrogen gas humidified at 70 °C (80 ml min^{-1}) and oxygen gas humidified at 70 °C (80 ml min^{-1}) flowed into the anode and cathode sides, respectively. The IR-free polarization characteristics (i.e., cell potential (IR-free) vs. cell current density curves) and IR included polarization characteristics (i.e., I (current density)-V (cell potential) curves and I (current density)-P (power density) curves) observed for the single cells were evaluated by a current interrupt method at 70 °C. The potentiostat/galvanostat as a fast solid-state switch was used for this measurement. Data were collected using a digital storage oscilloscope.

To measure the cell potentials between OCV and zero, the current was varied by adjusting the resistance of the external load. To obtain data at higher current densities, an external voltage was applied to the cell using a function generator. In addition, all I-V curves were obtained under the steady state condition with reference to the NEDO fuel cell measurement protocol [238].

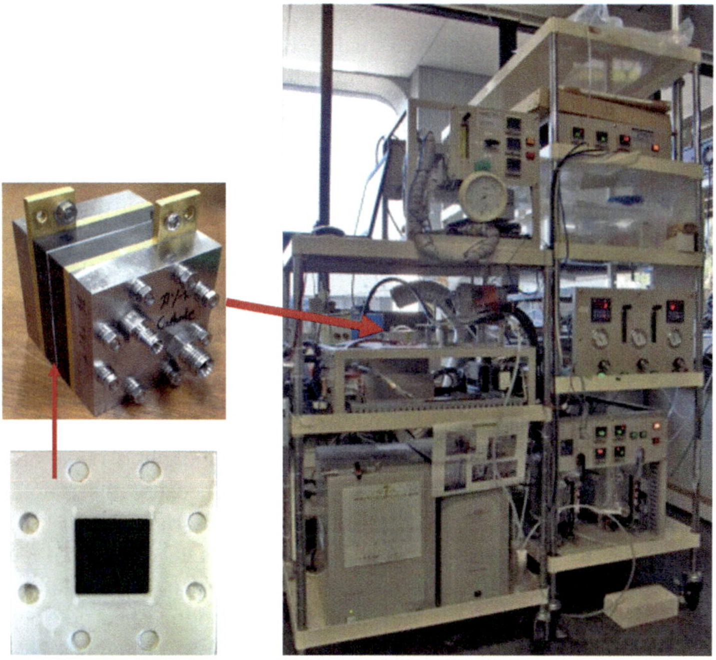

Fig. 34 Photographs of the MEA, PEMFC test holder, and test system of MEA performance at NIMS

3 Results and Discussion

3.1 Interactive Combination Work for Improvement of Electrical Conductivity in Solid Electrolytes

3.1.1 Fluorite Related Doped CeO_2-Based System

Approximately 20 years ago, we firstly examined the electrolytic property using co-doped CeO_2 electrolyte such as $(Sm_{0.5}Ca_{0.5})_{0.225}Ce_{0.775}O_{1.84}$ [8(a, b)] and $(La_{0.75}Sr_{0.2}Ba_{0.05})_{0.175}Ce_{0.825}O_{1.891}$ [8(a, b)] for development of high-quality intermediate temperature solid oxide fuel cell (IT-SOFC) devices. In this case, we assumed that the high-speed oxide ion diffusion pathway is designed based on the crystallographic knowledge. In this light, the authors attempted to understand all the crystallographic information using only X-ray diffraction data. In this classical approach, ionic radius ratio between dopant and host cation, the amount of calculated oxygen vacancies based on charge neutrality condition, and other crystallographic information were key factors for design of the present solid electrolyte [47–50, 52]. Some progress has been made in improving electrolytic properties and IT-SOFC performance, but we were not able to make significant breakthroughs in designing 'state-of-the-art IT-SOFC' devices with greatly improved quality, exceeding the limit of conventional IT-SOFC devices. To seek out new breakthroughs, we started the interactive cooperation between fabrication, microanalysis, and modeling techniques using co-doped CeO_2 solid electrolytes.

In the initial stages of our microanalysis, we carefully observed subtle changes of defect structure which can be detected using selected area electron diffraction pattern (SAEDP) using TEM. In our observation, very weak diffuse scattering was clearly observed in the background of SAEDP which is observed for all samples. In the solely doped CeO_2 solid electrolyte such as La_2O_3 doped CeO_2 system [8(b)] or Sm_2O_3 doped CeO_2 [8(a, b)], the intensity of diffuse scattering was high. In contrast, its intensity observed for co-doped CeO_2 systems was much lower. In general, it is well-know that the diffuse scattering appeared in the background of SAEDP is attributable to the formation of the short-range ordering of lattice defect in the crystal [239]. However, nobody noticed the influence of the formation of short-range ordering of lattice defects on the performance of fuel cell materials, because it is a subtle change of the microstructure seen at the interface of fuel cell materials.

Our first SAEDP data observed for solely doped CeO_2 and co-doped CeO_2 sintered bodies is presented in reference No. [8(b)]. Also, high resolution TEM images observed for same observation area of solely doped and co-doped CeO_2 sintered bodies were shown in the same reference (refer to reference No. [8(b)]). That SAEDP analysis indicated that the intensity of diffuse scattering in the background of SAEDP recorded from co-doped CeO_2 sintered body was much lower than solely doped CeO_2 sintered body. In addition, high-resolution TEM images suggest that the micro-domain with ordered structure of oxygen vacancies is formed in the samples and the volume of micro-domain area in co-doped CeO_2 sintered bodies is smaller than that of solely doped CeO_2 samples. Those microanalysis data suggest that the dopant cations are likely to locate in the local area and create the micro-domain with ordered structure of oxygen vacancies in the samples. Also, the local short-range ordering of oxygen vacancies was suppressed in co-doped ceria.

After we observed this microanalysis data, we concluded that similar phenomena occurred in other co-doped solid electrolytes such as **Lanthanum strontium gallium magnesium oxide** (LSGM), co-doped perovskite related material, and stabilized ZrO_2 systems. Consequently, we concluded that the following three import questions based on our microanalysis data should be clarified in order to see the new breakthrough point for design of 'state-of-the-art IT-SOFC' devices.

Question (i):
Why the micro-domains are formed in the fuel cell samples?

Question (ii):
How can we suppress the formation of micro-domains in the samples?

Question (iii):
What model can we develop for design of superior quality fuel cell materials based on microanalysis data?

To conclude why the micro-domains are formed in the fuel cell samples, we must start careful microanalysis using solely doped CeO_2 such as Gd doped CeO_2 which is one of the representative samples in the fuel cell field. As the authors touched on in Fig. 19b, nano-scale heterogeneity is remnant in the easily sinterable Gd doped CeO_2 particles. In addition, non-equiribrium defect structure would be easily formed in the slow cooling process of fabrication of fuel cell materials. To make sure of these points, the grain boundry area in dense Gd doped CeO_2 sintered body was observed using high resolution TEM analysis. The authors reintroduce some of that representative microanalysis data.

The in-focus grain was tilted along (110) zone to make sure the grain boundary plane was approximately parallel to the incident electron beam. The diffuse scattering in selected area electron diffraction pattern was selected for dark field imaging, and the corresponding dark field image is presented in Fig. 35b. Figure 35b suggests that the grain boundary with enhanced brightness is enriched in constituents different from those in neighboring grain interiors. Figure 35c is a magnified image at the grain boundary denoted by a dashed square in Fig. 35b. Note that separated tiny micro-domain with brighter contrast, denoted by arrows in Fig. 35c, appear at the highlighted grain boundary. This microanalysis data clearly indicate that the dopant segregation is observed at the grain boundary regions. The origin of the dopant segregation at this nano-scale would be nano-heterogeneity of easily sinterable Gd doped CeO_2 powders.

Moreover, our high-resolution TEM image shows that there is a slight expansion of the microstructure at the grain boundary, refer to Fig. 36.

From the inserted diffractogram of this grain boundary, two sets of diffraction spots can be distinguished, each of which belongs to the two adjacent grains. These spots were filtered and assigned by black in the diffractogram, refer to Fig. 36a. The related Fourier-filtered image is demonstrated in Fig. 36b. The slightly expanded region was clearly observed in the grain boundary. Moreover, white arrows in Fig. 36b indicate that small defect clusters can also be observed with brighter contrast than

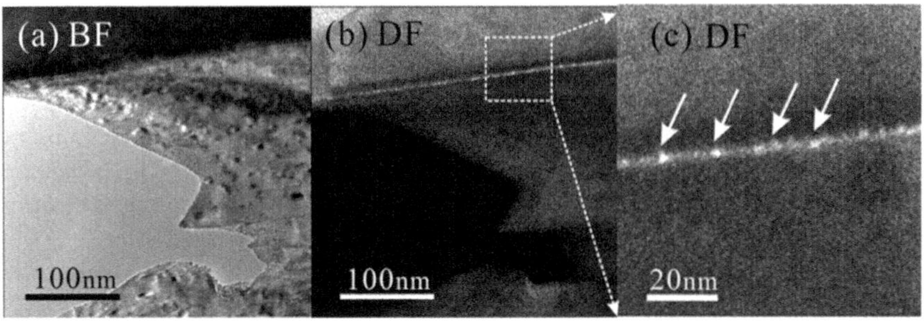

Fig. 35 Bright field (BF) TEM image of one typical intergranular region observed in Gd doped CeO_2 sintered body (**a**); **b** is the corresponding dark field (DF) TEM image revealing that the grain boundary with bright contrast is ascribed to the diffuse scattering in selected area diffraction pattern; **c** is the zoomed in region denoted by dashed square in (**b**), clearly demonstrating the dopant segregation at the grain boundary [adopted from Ref. 240]

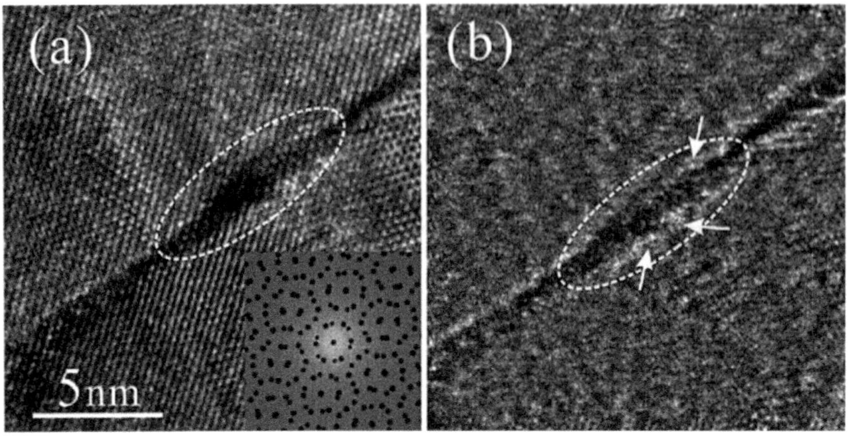

Fig. 36 High resolution TEM image observed for grain boundary in Gd doped CeO_2 sintered body (**a**), inset is the related diffractogram and diffraction spots arising from two adjacent grains have been blocked by black dots. The corresponding reconstructed image is presented in (**b**) [adopted from Ref. 241]

neighboring grain interior. Our microanalysis tells us that segregation is observed at grain boundaries in dense Gd doped CeO_2 sintered bodies.

Figure 37a shows a clear snapshot of the moment that tiny micro-domains are created from the grain boundary region. The black dashed line area in Fig. 37a can be clearly distinguished along the different lattice image of the two adjacent grains. Also, Fig. 37b presents the corresponding diffractogram of such a grain boundary region. Diffraction spots from two grains are assigned by black dots and white solid circles, respectively. Besides diffraction spots arising from fluorite related ceria lattices, the extra diffraction reflections can also be observed, refer to Fig. 37b. The filtered diffractogram (Fig. 37c) clearly shows that the different microstructural feature is a result of the neighboring grain interior, resulting in the appearance of extra diffraction features rather than typical fluorite reflections. The authors observed the similar formation process of micro-domains with ordered structure of oxygen vacancy in di- [242–246] and trivalent [247–257] doped CeO_2 sintered samples. The detailed microanalysis results of defect structure of micro-domain region are also present in Part 2 chapter "Atomistic Simulation of Subtle Microstructures in CeO_2-Based Electrolyte for Design of the State-of-the-Art Fuel Cells" (Fei Ye et al.). The authors strongly recommend that the readers refer to the another interactive combination work between modeling and microanalysis by Professor Fei Ye et al.

In the second step of our interactive combination challenge, we must consider how we can suppress the large micro-domain formation in order to maximize the electrolytic properties in IT-SOFC devices. As we already described in this chapter, the size of micro-domain which is formed by the segregation of dopant can be suppressed

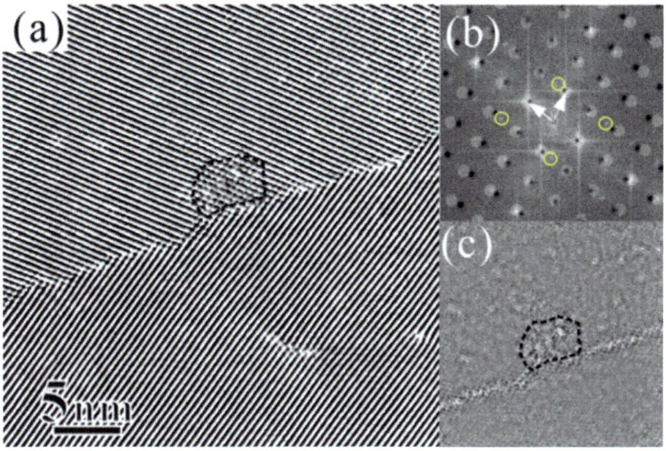

Fig. 37 High resolution TEM image of a tiny micro-domain which was formed at a grain boundary in Gd doped CeO_2 sintered body (**a**), the corresponding diffractogram is demonstrated in (**b**). Diffraction spots belonging to these two adjacent grains were blocked by black dots and solid circles in (**b**), the corresponding reconstructed image is represented in (**c**) [adopted from Ref. 241]

using co-doping effect. In addition, we examined the influence of fabrication conditions (i.e., heating rate in sintering process, and particle morphology) on the electrolyte property. Note that we will examine our model to gain an understanding of the micro-domain size control mechanism in the co-doped CeO_2 sintered bodies later in the text.

According to the conventional sintering theory, the densification process is strongly dependent on the homogeneity of grain boundary condition (i.e., amount of lattice defect, chemical composition, and so on) [258–262]. Heterogeneous grain growth during the sintering process would depress the densification of the sintered body. In addition, the slow cooling process will create the non-equilibrium defects in the bulk. As demonstrated in Fig. 38 [263], the high temperature sintering condition using conventional sintering method promotes both heterogeneous grain growth and growth of micro-domain in the sintered samples.

To minimize the size of the micro-domains and improve the electrolyte property in doped CeO_2 sintered sample, we performed pulsed electric current sintering (PECS). This sintering method is also known as spark plasma sintering (SPS). Note that it is just a brand name of the equipment of PECS. No plasma is generated during sintering process at all. In this process, however, we can use high speed heating conditions under applied pressure. In addition, we can select rapid cooling conditions as well as high speed heating conditions, refer to Fig. 39.

As we can see in Fig. 39, micro-domain size in doped CeO_2 dense sintered body can be minimized using high speed heating and high-speed cooling condition which can be created for the fabrication of dense sintered bodies by only PECS. Note that this sintering method will be useful for fabrication of very small-sized pellets. If the size of samples required is of a large scale, the reader will see quite different

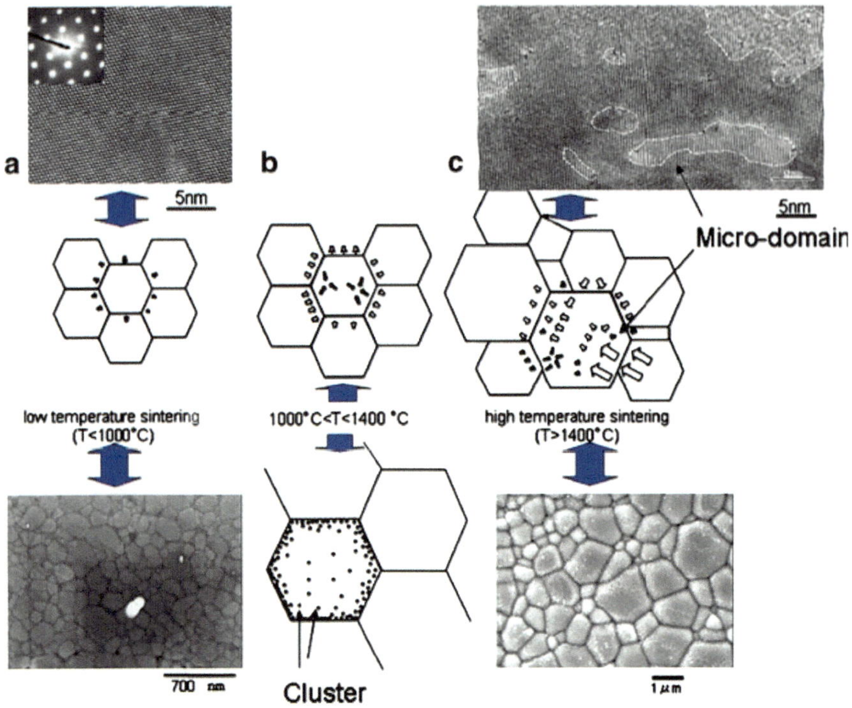

Fig. 38 Schematic diagram of relationship between sintering temperature of conventional sintering method and microstructural feature in micro and nano-scales [adopted from Ref. 263]

structures within the samples. It is not so easy to fabricate dense and large specimen using the joule heating effect among the particles in the compact body using PECS.

Another factor that will influence micro-domain size is the original particle morphology. The reason for this is that round shaped particles are covered by thin amorphous layer, refer to Fig. 19a and b. To provide evidence for the influence of particle morphology difference on micro-structural features which will influence the electrolyte property, we fabricated rod like shaped doped CeO_2 particles and round shaped doped CeO_2 particles. We then examined the effect of micro-domain size on the electrical conducting properties [10]. Figure 40 presents electrolyte conductivity vs. selected area electron diffraction patterns recoded from Sm doped CeO_2 sintered bodies which were made from round shaped particles and rod like shaped particles.

The intensity of the diffuse scattering in the background of selected area electron diffraction pattern recorded from sintered samples which were prepared from round shaped particles was definitely lower than that from samples which were prepared from rod like shaped particles. This result indicates that the micro-domain size in the sintered samples which were prepared from round shaped particles were smaller than that in the samples which were prepared from rod like shaped particles. In our

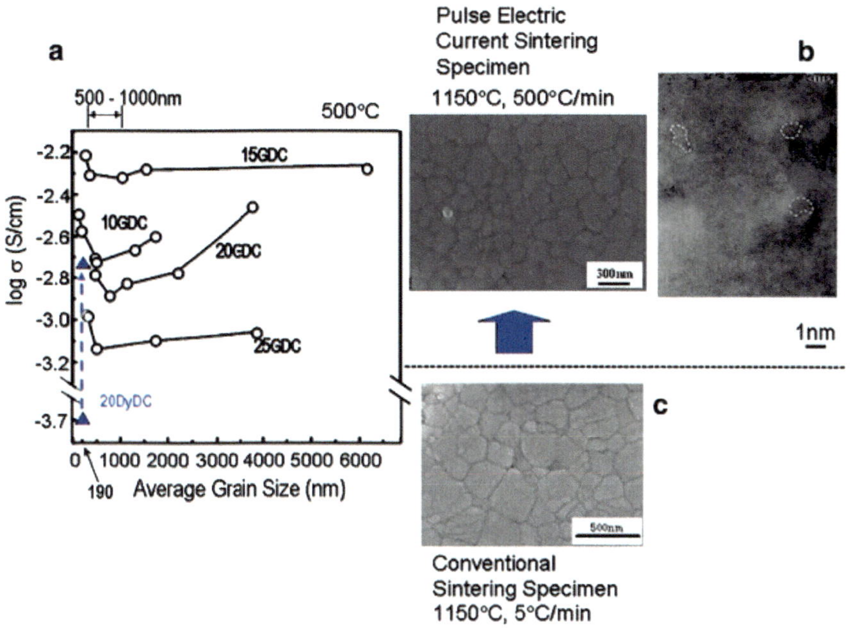

Fig. 39 Schematic diagram of the relationship between electrolytic properties and microstructural features observed for Gd doped CeO_2 sintered by conventional sintering method and Dy doped CeO_2 sintered by PECS. DyDC: Dy doped CeO_2, GDC: Gd doped CeO_2. Measurement temperature of electric conductivity: 500 °C.Cooling rate up to 500°C from sintering temperature is almost same to heating rate [adopted from Ref. [263]]

preliminary TEM microanalysis, we made sure that our conclusions were reasonable. While BET surface area observed for rod like shaped particles was lower than that of round shaped particles, the surface of rod like shaped particles is covered by an amorphous in a similar fashion to the to the round shaped particles. In addition, the heterogeneity around the grain boundary region of sintered samples which were prepared from rod like shaped particles were in high level as compared to that prepared from round shaped particles. As a result, the large sized micro-domain in the sintered samples which were prepared from rod like shaped particles resulted in the electric conductivity to be lower and in the activation energy to be higher as compared to the samples which were prepared from round shaped particles.

As described above, we can minimize the micro-domain size and increase the electrical conductivity in doped CeO_2 electrolytes through the optimization of the fabrication conditions. In conclusion, it is important to use round shaped particles in order to produce homogeneity around the grain boundaries, and in order to fabricate high-quality solid electrolyte with high electric conductivity and low activation energy in the intermediate temperature region. However, co-doping or multiple doping effect was a more effective means of fabricating high-quality solid electrolytes. To conclude why co-doping or multiple doping is so useful for minimization of micro-domain

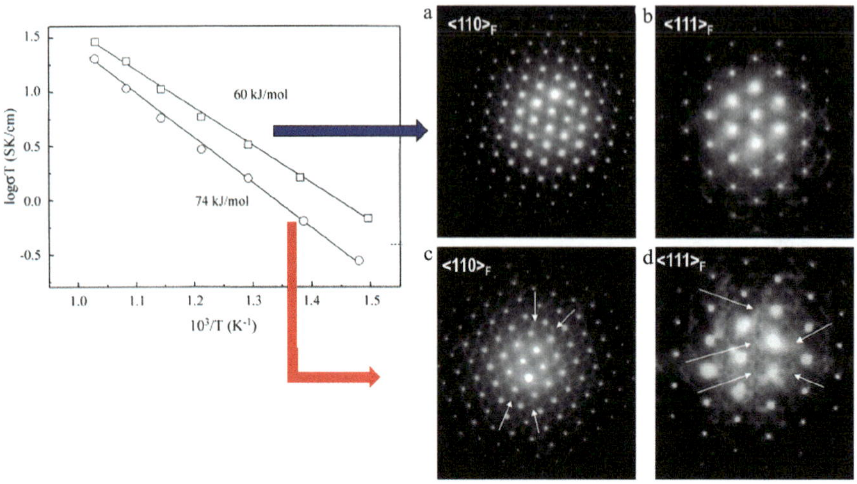

Fig. 40 Relationship between electrolyte conductivities and selected area electron diffraction patterns recorded from $Sm_{0.2}Ce_{0.8}O_{1.9}$ sintered bodies which were made from round shaped particles (**a**, **b**) and rod like shaped particles (**c**, **d**). The measurement temperature range of electrical conductivity: 400 to 650 °C. The white arrows indicate the diffuse scattering [adopted from Ref. 10]

size and enhancement of electrical properties, the authors performed bulk atomistic simulation based on microanalysis data, refer to Table 2 and Fig. 41.

In our modeling, we assumed the defect clusters were formed in the bulk. The reason for this is that short-range ordering effects can be considered using this model and the diffuse scattering in the background of selected area electron diffraction pattern confirms the formation of this type of local ordered defect structure. In addition, it is expected that the extra reflection in the selected area electron diffraction patter recorded from our samples will appear when the defect cluster size is on the large scale.

To estimate the growth potential of the defect clusters in the micro-domain, the energy change on increasing the size of the defect clusters, e.g., increasing the number of oxygen vacancies from n to n + 1, was calculated. In this calculation, a higher ΔE_{n+1} value indicates that the growth of clusters could be restrained. Namely, a large minus ΔE_{n+1} value means that the growth of clusters is thermodynamically preferred. The micro-domain growth would be suppressed when minus ΔE_{n+1} value is closed to zero or calculated ΔE_{n+1} value > 0. Based on our defect cluster model, our ΔE_{n+1} values calculated from Sm solely doped CeO_2, Sm-Gd co-doped CeO_2 and Sm-Y doped CeO_2 systems are as follows [264].

$$\Delta E_{n+1} \text{ calculated for } 6M_{Ce}{}'\,3Vo^{\cdot\cdot} \text{ in Sm solely doped } CeO_2 \text{system}: \quad -0.45\,\text{eV}$$

$$\Delta E_{n+1} \text{ calculated for } 6(M_1,M_2)_{Ce}{}'\,3Vo^{\cdot\cdot} \text{ in Sm-Gd co-doped } CeO_2 \text{ system}: \quad -0.38\,\text{eV}$$

$$\Delta E_{n+1} \text{ calculated for } 6(M_1,M_2)_{Ce}{}'\,3Vo^{\cdot\cdot} \text{ in Sm-Y co-doped } CeO_2 \text{ system}: -0.27\,\text{eV}$$

Table 2 List of structural features of the thermodynamically preferred defect clusters. $V_O^{\bullet\bullet}$-M_{Ce}' shows the separation between oxygen vacancies and their associated dopant cations and $V_O^{\bullet\bullet}$–$V_O^{\bullet\bullet}$ shows the separation between oxygen vacancies and neighboring oxygen vacancies. The Miller index <hkl> indicates the set of all equivalent lattice vectors [adopted from Ref. 264]

Dopant	Structure	$2M_{Ce}'\ 1V_O^{\bullet\bullet}$	$4M_{Ce}'\ 2V_O^{\bullet\bullet}$	$6M_{Ce}'\ 3V_O^{\bullet\bullet}$	$8M_{Ce}'\ 4V_O^{\bullet\bullet}$
Y	$V_O^{\bullet\bullet}$- M_{Ce}'	(111) / 4	(111) / 4	(111) / 4	(111) / 4
	$V_O^{\bullet\bullet}$- $V_O^{\bullet\bullet}$	—	(110) / 2	(110) / 2	(110) / 2
	$V_O^{\bullet\bullet}$ chain	—	—	Curved	Curved
Sm	$V_O^{\bullet\bullet}$- M_{Ce}'	(113) / 4	(113) / 4	(113) / 4	(113) / 4
	$V_O^{\bullet\bullet}$- $V_O^{\bullet\bullet}$	—	(111) / 2	(110) / 2	(110) / 2
	$V_O^{\bullet\bullet}$ chain	—	—	Curved	Curved
Gd	$V_O^{\bullet\bullet}$- M_{Ce}'	(113) / 4	(113) / 4	(113) / 4	(113) / 4
	$V_O^{\bullet\bullet}$- $V_O^{\bullet\bullet}$	—	(111) / 2	(110) / 2	(110) / 2
	$V_O^{\bullet\bullet}$ chain	—	—	Curved	Curved
Y - Sm	$V_O^{\bullet\bullet}$- M_{Ce}'	(113) / 4	(111) / 4	(113) / 4	(113) / 4
	$V_O^{\bullet\bullet}$- $V_O^{\bullet\bullet}$	—	(110) / 2	(110) / 2	(110) / 2
	$V_O^{\bullet\bullet}$ chain	—	—	Curved	Curved
Y - Gd	$V_O^{\bullet\bullet}$- M_{Ce}'	(113) / 4	(111) / 4	(111) / 4	(113) / 4
	$V_O^{\bullet\bullet}$- $V_O^{\bullet\bullet}$	—	(110) / 2	(110) / 2	(110) / 2
	$V_O^{\bullet\bullet}$ chain	—	—	Curved	Curved
Sm - Gd	$V_O^{\bullet\bullet}$- M_{Ce}'	(113) / 4	(113) / 4	(113) / 4	(113) / 4
	$V_O^{\bullet\bullet}$- $V_O^{\bullet\bullet}$	—	(111) / 2	(110) / 2	(110) / 2
	$V_O^{\bullet\bullet}$ chain	—	—	Curved	Curved

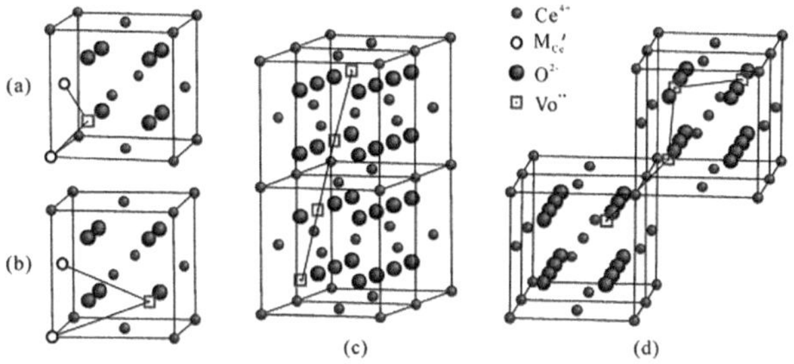

Fig. 41 The defect cluster model of $2M_{Ce}'1V_O^{\cdot\cdot}$ in which oxygen vacancies and dopant cations are separated by **a** <111>/4 and **b** <113>/4. And the defect cluster model of oxygen vacancies in the $8M_{Ce}'4\ V_O^{\cdot\cdot}$ arranged as **c** a straight chain and **d** a curved chain, respectively [adopted from Ref. 264]

Our calculated ΔE_{n+1} for solely doped CeO_2 and co-doped CeO_2 sintered sample clearly indicates that dopant cations in solely doped CeO_2 system are likely to locate in the outer region of the defect cluster. Then, the micro-domain region will be easily developed in the bulk of sintered samples. In contrast, the co-doping system could restrain the growth of large clusters and reduce the lattice strain on the atomic level. Then, the local ordering of oxygen vacancies in co-doped system is suppressed and the size of micro-domain can be of smaller scale. (refer to calculated ΔE_{n+1} values for solely doped CeO_2 and co-doped CeO_2 systems).

In the selected area electron diffraction patterns recorded from co-doped CeO_2 sintered samples, the diffuse scatterings and extra spots can also be observed. However, the intensity of extra reflection and diffuse scattering pattern recorded from co-doped sintered samples (i.e., Sm-Gd co-doped CeO_2 and Sm-Y co-doped CeO_2 system) became very weak as compared to solely doped sintered sample (i.e., Sm solely doped CeO_2 system), **refer to** Fig. 42. These microanalysis data indicate that the short-range ordering of oxygen vacancies can be drastically restrained by the co-doping effect, and therefore the micro-domains prefer to remain in a small size and make more disordered state of oxygen vacancy in the co-doped CeO_2 system.

As described above, our bulk atomistic simulation supported microanalysis data very well. Also, the electrolytic property observed for our sintered samples in intermediate temperature region agreed with microanalysis and bulk atomistic simulations. Therefore, it is concluded that the improvement in microstructural homogeneity contributed to a higher ionic conductivity observed for co-doped or multiple doped CeO_2 sintered bodies in the intermediate temperature region.

3.1.2 Brownmillerite Type-Based System

Since co-doped fluorite related CeO_2 based oxides suppressed the growth of micro-domain with ordered structure of oxygen vacancies and promoted the local disordering on the microscopic scale, the authors examined the influence of co-doping effect on electrical conductivity in brownmillerite type $Ba_2In_2O_5$ based oxide alongside the fluorite systems with a view to developing an overall view of oxide ion conducting systems. It is well known that high ionic conductivity in $Ba_2In_2O_5$ appears above the transition temperature (from orthorhombic phase to cubic phase, T_t: 930 °C) of an order–disorder transition of oxygen vacancies. However, the large volume change (approximately 3.4 vol %) of $Ba_2In_2O_5$ lattice occurs before and after this order–disorder transition of oxygen vacancies. This large volume change of the $Ba_2In_2O_5$ lattice may become an obstacle for the application of $Ba_2In_2O_5$ to solid-state devices.

To expand the applicability of $Ba_2In_2O_5$-based oxides as a solid electrolyte in solid oxide fuel cells, a number of studies have tried to decrease T_t which is temperature of the order–disorder transition of oxygen vacancies in $Ba_2In_2O_5$ lattice. In the previously published papers, the relationship between electrical conductivity and T_t was examined using $Ba_2In_2O_5$ [68], $Ba_2((Zr_{0.5}, Zn_{0.5})_xIn_{1-x})_2O_5$ (x: 0.1, 0.2 and 0.3) [265, 266], $Ba_2(In_{0.8},Ga_{0.2})_2O_5$ [75], $Ba_2(In_{0.9},Ga_{0.1})_2O_5$ [76], $Ba_2(In_{0.9},Al_{0.1})_2O_5$

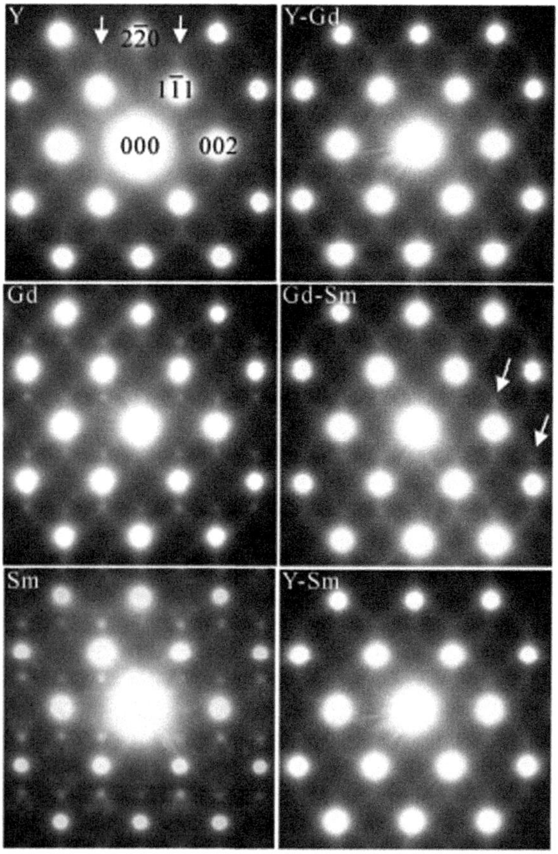

Fig. 42 Selected area electron diffraction patterns recorded from solely doped and co-doped CeO_2 sintered bodies [adopted from Ref. 264]

[267], $Ba_2(In_{0.9},Sc_{0.1})_2O_5$ [78], $Ba_2(In_{0.9},Y_{0.1})_2O_5$ [78], $Ba_2(In_{0.9},Yb_{0.1})_2O_5$ [78], $Ba_2(In_{0.9},Sm_{0.1})_2O_5$ [78], $Ba_2(In_{0.85},Y_{0.15})_2O_5$ [268], $Ba_2(In_{0.9},Ce_{0.1})_2O_{5+\delta}$ [82], $Ba_2(In_{1.95},Sn_{0.05})_2O_{5+\delta}$ [269], $Ba_2(In_{1.9},Mo_{0.1})_2O_{5+\delta}$ [88], $Ba_2(In_{1.9},V_{0.1})_2O_{5+\delta}$ [88], $(Ba_{0.475},La_{0.025})_2(In_{0.475},Ga_{0.025})_2O_{5+\delta}$ [93], and $Ba_2In_2O_{4.99}F_{0.02}$ [270] and oxides. In those series of co-doped $Ba_2In_2O_5$-based oxides, $Ba_2((Zr_{0.5}, Zn_{0.5})_{0.3}In_{0.7})_2O_5$ [265, 266] presented the best performance out of all $Ba_2In_2O_5$ based oxide series. The volume change of $Ba_2((Zr_{0.5}, Zn_{0.5})_{0.3}In_{0.7})_2O_5$ from room temperature to 1000 °C (0.89 vol %) was much smaller than $Ba_2In_2O_5$ without dopant (3.4 vol %). Also, the electrical conductivity observed for $Ba_2((Zr_{0.5}, Zn_{0.5})_{0.3}In_{0.7})_2O_5$ was two times higher than that of the disordered state of none doped $Ba_2In_2O_5$, refer to Fig. 43.

Since $Ba_2((Zr_{0.5}, Zn_{0.5})_{0.3}In_{0.7})_2O_5$ can be a good example indicating the usefulness of the interactive combination of fabrication, microanalysis, and modeling for development of fuel cell materials, the authors reintroduce the representative data observed for $Ba_2((Zr_{0.5}, Zn_{0.5})_{0.3}In_{0.7})_2O_5$ in this chapter.

The crystal phase analysis using X-ray diffraction suggested that the ordered structure of oxygen vacancy was changed by co-doping effect. As shown in Fig. 44a,

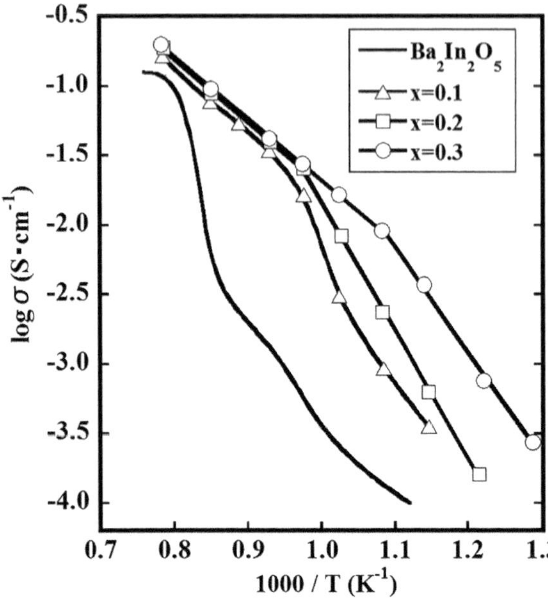

Fig. 43 Temperature dependence of electrical conductivity observed for $Ba_2((Zr_{0.5}, Zn_{0.5})_xIn_{1-x})_2O_5$ (x: 0, 0.1, 0.2 and 0.3) [adopted from Ref. 266]

the peak intensity recorded from super lattice peak (0 2 0) plane of the brownmillerite type structure around 30 degree clearly decreased with an increase in amount of dopant addition. In addition, other super lattice peaks such as (611), (512) (721), and so on follow the change of peak intensity of (0 2 0). And whole peak shape recorded from $Ba_2((Zr_{0.5}, Zn_{0.5})_{0.3}In_{0.7})_2O_5$ with orthorhombic phase became similar to cubic phase $Ba_2In_2O_5$ which is observed over T_t of the order–disorder transition of oxygen vacancies in its lattice.

However, the lattice constants observed for $Ba_2In_2O_5$-based oxides, which were refined using Rietveld analysis method, showed continuous change which is hard to explain in terms of the effect of co-doping on the crystal structure modification, refer to Fig. 44b.

To conclude why the electrical conductivity observed for $Ba_2((Zr_{0.5}, Zn_{0.5})_{0.3}In_{0.7})_2O_5$ was improved by co-doping effect, the defect interface structure of $Ba_2((Zr_{0.5}, Zn_{0.5})_{0.3}In_{0.7})_2O_5$ was observed using TEM. Figure 45a shows <101> zone images observed for $Ba_2((Zr_{0.5}, Zn_{0.5})_{0.3}In_{0.7})_2O_5$ with orthorhombic structure.

To clarify the defect structural features in the observed high resolution images, the fast Fourier transform (FFT) patterns were obtained from corresponding high resolution images, refer to Fig. 45b. The FFT patterns which are divided into four areas (i.e., from (i) to (iv)) taken from $Ba_2((Zr_{0.5}, Zn_{0.5})_{0.3}In_{0.7})_2O_5$ which reveals the heterogeneous defect interface structural features as shown in Fig. 45b. In the area (iv) of Fig. 45b, the intensities of some reflections in the area (iv) were under detectable level as compared to area (i) to (iii). In addition, the inverse fast Fourier transform (IFFT) lattice contrast image taken from the area (iv) has different lattice fringe contrast as compared with area (i) which is representative example of IFFT of

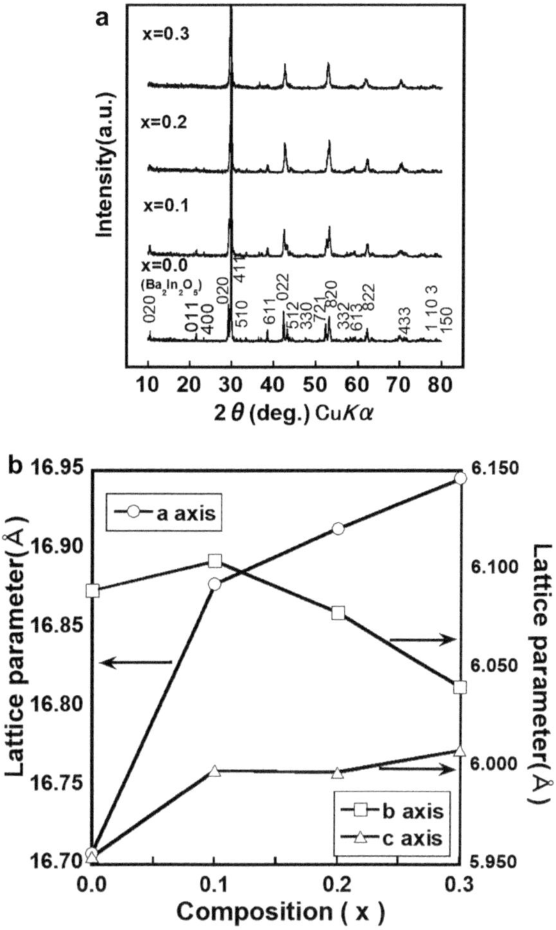

Fig. 44 **a** XRD patterns of $Ba_2In_{2-x}(ZnZr)_xO_5$ (x = 0.0–0.3) [adopted from Ref. 266]. **b** Lattice parameters (in a axis, b axis, and c axis directions) obtained from co-doped $Ba_2In_{2-x}(ZnZr)_xO_5$ as a function of x (x = 0 to 0.3) [adopted from Ref. 266]

areas (i) to (iii), refer to Fig. 45c. This result clearly indicates that the local disordering of oxygen vacancies in the grain is induced by co-doping effect.

To support the microanalysis works, the authors performed bulk atomistic simulation. In this case, the following defect structure formation reaction was assumed.

$$ZnO + ZrO_2 + 3In_{In}{}^{x} + 4O_O{}^{x} + 7V_i{}^{x} \xrightarrow{Ba_2In_2O_5} 3V_{In}{}'''In_i^{\cdot\cdot\cdot}Zn_i^{\cdot\cdot}Zr_i^{\cdot\cdot\cdot\cdot}4O_i{}''4V_O^{\cdot\cdot} + In_2O_3 \quad (31)$$

where the Kröger–Vink notation is used for simple explanation of both defect formation and solid solution formation reactions. Equation (31) predicts that seven interstitial sites are formed by Zn^{2+} and Zr^{4+} cations doping into brownmillerite type $Ba_2In_2O_5$ lattice. Note that oxygen vacancy sites are simultaneously formed by co-doping of Zn^{2+} and Zr^{4+} cations. Also, In^{3+} interstitial defect, Zn^{2+} interstitial defect,

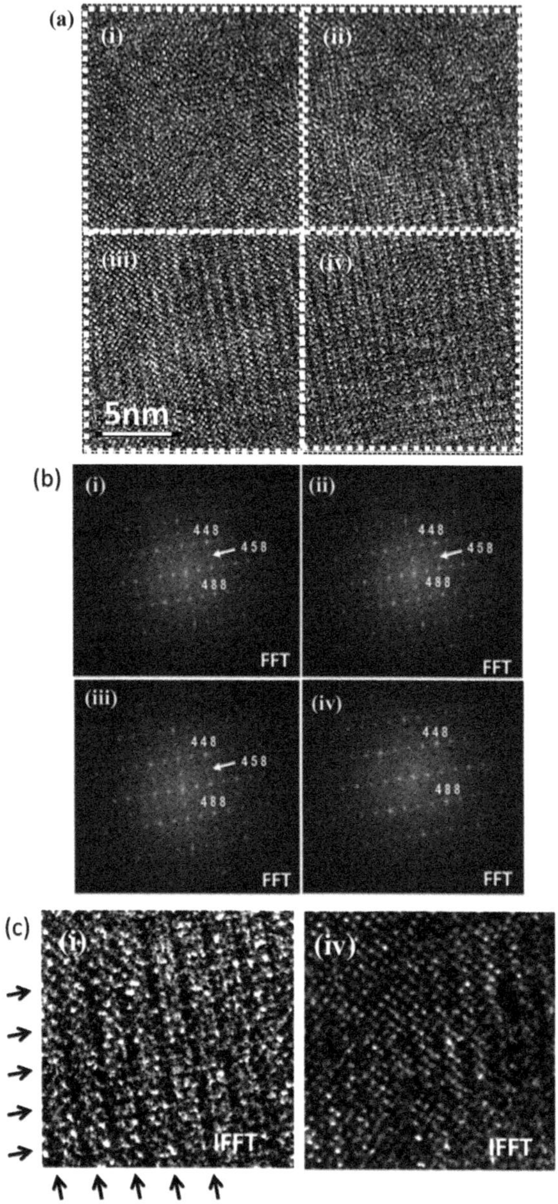

Fig. 45 High resolution TEM <101> zone images (**a**), FFT patterns from high solution images (**b**), and IFFT lattice contrast images from FFT patterns (**c**) [adopted from Ref. 266]

and Zr^{4+} interstitial defect sites are formed as well as defect sites of oxygen vacancies. In our bulk atomistic simulation works, we calculated the energies of isolated point defects in $Ba_2In_2O_5$. Our calculated isolated energies agreed with previously reported values of $Ba_2In_2O_5$ as well. The highest binding energy (ΔE_b) of defect clusters which consists of seven interstitial sites based on Eq. (31) was calculated using the lowest defect cluster formation energy.

The calculated highest binding energy (ΔE_b) was 1.2 eV. We concluded that it is high enough to encourage the formation of stable defect clusters in $Ba_2In_2O_5$-based lattice. The configuration of stable defect cluster was presented in Fig. 46. In our defect cluster model, the two dopants (i.e., Zr^{4+} and Zn^{2+}) were located at interstitial sites which are close to four coordination of oxygen around In sub-lattice site. The oxygen vacancy sites were formed at four coordination of oxygen. Also, the six coordination of oxygen sites were apparently formed along a axis in the orthorhombic $Ba_2In_2O_5$ to maintain the local charge neutrality in the $Ba_2In_2O_5$-based lattice, refer to Fig. 46. Eventually, the both sites of local disordering and local ordering of oxygen vacancies were formed at positions adjacent to each other in the lattice. Thus, the additional hopping sites and continuous pathway for oxide ion diffusion would be formed into $Ba_2((Zr_{0.5}, Zn_{0.5})_{0.3}In_{0.7})_2O_5$ by formation of large defect clusters and electrical conductivity observed for $Ba_2((Zr_{0.5}, Zn_{0.5})_{0.3}In_{0.7})_2O_5$ reached higher level as compared to none doped $Ba_2In_2O_5$.

According to the theory proposed by Rice et al. [271], the interaction of interstitial cation defects with the strain field promotes the order–disorder transition of cations in the ionic materials. The order–disorder transition of cations in super-ionic conductor is appeared when the following condition is satisfied in the ionic materials:

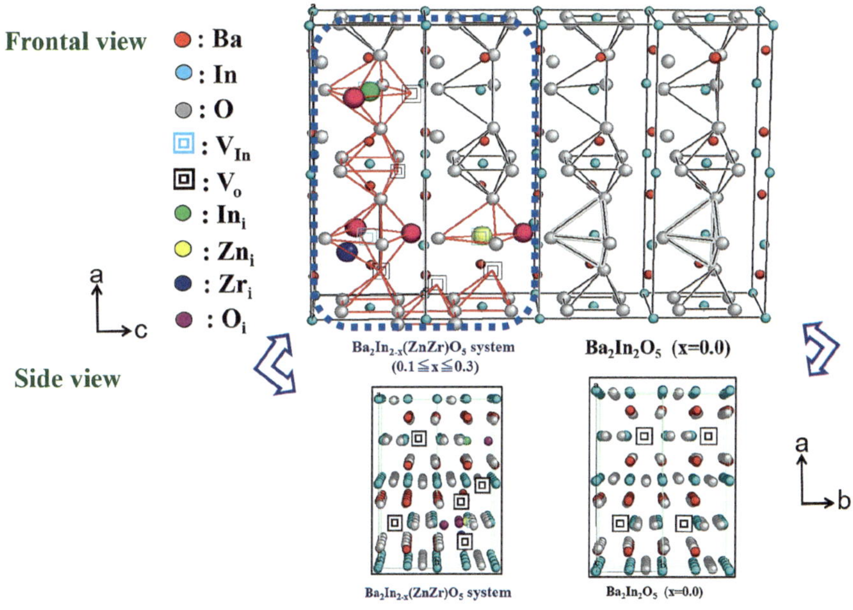

Fig. 46 Illustration of defect interface model for $Ba_2((Zr_{0.5}, Zn_{0.5})_xIn_{1-x})_2O_5$ (x: 0, 0.1, 0.2 and 0.3) [adopted from Ref. 266]

$$\lambda/\varepsilon_0 > 0.98 \tag{32}$$

where λ is interaction parameter, ε_0 is required energy for migration of cation to interstitial sites. This theory indicates that large number of interstitial defects should be existed in the ordered state of ionic materials below the first-order transition temperature. Pasternak et al. experimentally made sure the correctness of the aforementioned theory by a lowering of order–disorder transition temperature of silver ion in Ag_2Se super-ionic conductor [272]. Our bulk atomistic simulation which is supported by microanalysis suggests that T_t of the order–disorder transition of oxygen vacancies in $Ba_2In_2O_5$ lattice can be lowered by introducing large number of interstitial sites in the lattice in an analogous way as seen in the Ag_2Se superionic conductor.

As shown in this Sect. '3.1.2 Brownmillerite type-based system', the authors highly recommend the interactive combination of fabrication, microanalysis, and modeling to the readers if the reader wants to discover new insights for the design of active defect interfaces in solid electolytes.

3.2 Interactive Combination Work for Lowering of Overpotentials on Anode in SOFC Device

The most important advantage of SOFCs is the capability of direct utilization of various kinds of fuels such as the aqueous methanol (methanol/ water molar ratio = unity). Aqueous methanol can be fabricated using both of the hydrogen which is produced using primary energy and the greenhouse gas (i.e., CO_2 gas) which is produced using non-edible biomass and/or stored CO_2 gas recovered from exhausted gas. To establish the sustainable society through the formation of a hydrogen economy, the design of active interface on the anode is important as well as design of interface on the cathode. In terms of the recent progress of interface design on anode side, refer to the latest review paper by Professor San Ping Jiang et al. [273]. In this chapter, the authors described key challenge from the viewpoint of interactive combination work of fabrication, microanalysis and modeling. We believe that our key concept will provide the broad readers with new inspiration for technical innovation in the energy and environmental field.

3.2.1 Defect Interface Between Fluorite Related Doped CeO_2 and Ni in Anode Layer

The important defect structure of the anode is in the interface between oxide electrolyte and Ni. So far, a number of research projects have paid much attention to the suppression of segregation phenomenon at the interface between perovskite type cathode and oxide electrolyte with a view to a lowering the operation temperature of conventional SOFCs. For establishment of a hydrogen economy using the 'state-of-the-art IT-SOFC' devices and systems, however, we should examine more closely improvement in the kinetics of the anode reaction. The reason for this is that carbon build up is one of the problems observed in fuel cells operating at intermediate temperatures [3]. In this case, we must remember the defect structure changes in oxide electrolyte. To design the 'state-of-the-art IT-SOFC' stack, the design of high-quality active sites at the interface of electrolyte/ anode, and interface of electrolyte particles and Ni particles are required.

At first, the authors reintroduce microanalysis results of the interface of electrolyte and anode layer for readers. Our research team was the first in the field of fuel cell materials research to clearly observe changes in the defect structure at the interface between the Sm doped high-density CeO_2 film and the anode layer as a function of distance from the interface and the degree of oxygen vacancy ordering, and published its microanalysis data in an international journal (refer to reference [127]). Our STEM-EELS analysis clearly indicates that local ordering of oxygen vacancies has occurred around the interface area of dense Sm doped CeO_2 thin film and the anode layer. The authors already reported that a large value of the peak intensity ratio between peak B and peak C (I_B/I_C) indicates a higher degree of the local ordering in doped CeO_2 samples [274, 275]. The value of I_B/I_C of O K-edge spectra increased from interface of thin film/anode layer to the bulk region of Sm doped CeO_2 thin film. This microanalysis clearly suggests that the formation of oxygen vacancy ordering becomes easier in the interfacial region because the segregation of Ni has favorably occurred there. In addition, the reduction in Ce can greatly increase the concentration of oxygen vacancies in this region. So far, many scientists have focused on the role of Sm doped CeO_2 as a buffer layer to minimize the interaction between anode layer and oxide electrolyte film. However, all material scientists should recognize the importance of careful microanalysis of the defect interface using this type of sample. The level of local ordering of oxygen vacancies becomes significant in the interfacial region between the oxide electrolyte film and the anode layer. Note that similar segregation phenomena in the interfacial region between Gd doped CO_2 and the anode layer is also observed. Our bulk atomistic simulation predicts that defect clusters $(Ni_{Ce}''V_O^{\cdot\cdot})^x$ were formed at the interfacial region by segregation of Ni through the grain boundary (i.e., interfusion of Ni from anode layer to the bulk of solid electrolyte thin film) [276]. In addition, the authors concluded that a super-lattice with highly ordered oxygen vacancies was formed in the interface between anode layer and oxide solid electrolytes such as Sm doped CeO_2 and Gd doped CeO_2.

To conclude how we can promote oxide ion diffusion and charge transfer through this interfacial region (i.e., local ordering region) for development of 'state-of-the-art IT-SOFC', the authors carefully observed the defect structures in the oxide solid electrolyte/anode interface and in the interfaces inside of anode layer using STEM-EELS and STEM-EDS analyses [277–280].

Our STEM-EDS line scan analysis indicates that not only Ni can diffuse into GDC grains as previously reported but also Ce and Gd can diffuse into metallic Ni particles with equal diffusion lengths as that of Ni diffusion as shown in Fig. 47. This equal length diffusion between both oxide electrolyte and anode layer seems to be independent of ionic radii of components of electrolyte and anode. Namely, all the diffused cations had similar diffusion lengths in the interfacial region. In addition, the formation of a superstructure with highly ordered oxygen vacancies was suggested by appearance of extra reflection in the selected area electron diffraction pattern recorded from one single grain at the interfacial region [277].

Figure 48a to c presents EELS spectra of the O K-edge which were recorded from the area of oxide solid electrolyte, interface, and anode layer, respectively. The O K-edge fine structures which are typically featured peaks A, B, and C can be clearly distinguished from the spectra, refer to Fig. 48.

The relative peak intensity ratio of peak B and C (I_B/I_C) varies from approximately 0.75 (Fig. 48a, c) to around 0.94 (Fig. 48b). Our EELS spectra analysis clearly indicates that an enhancement of oxygen vacancy ordering is observed in the interfacial zone. Also, similarly, EELS spectra were also observed for the Gd $M_{4,5}$ adsorption

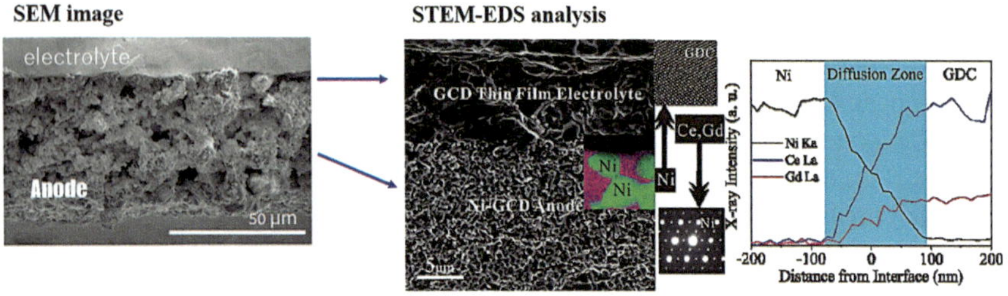

Fig. 47 SEM image and STEM-EDS analysis results observed for the interface of Gd doped CeO_2 thin film and anode layer [adopted from Ref. 277]

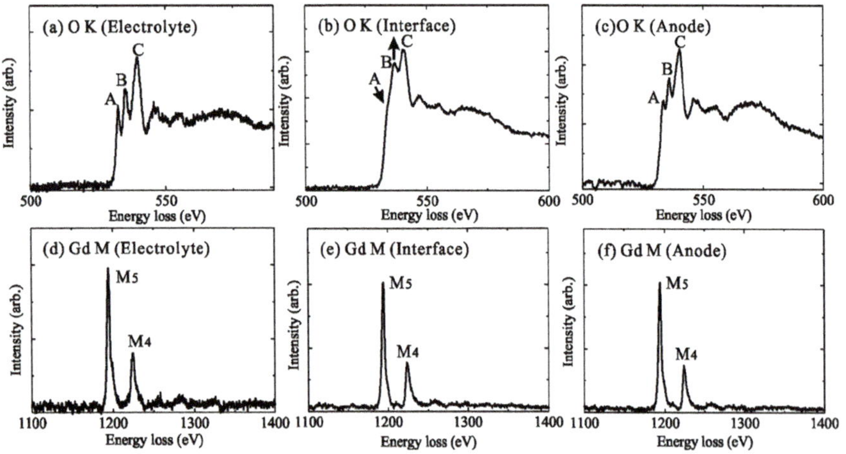

Fig. 48 EELS spectra of O K-edges obtained at the electrolyte: (**a**), electrolyte-anode layer interface: (**b**) and anode layer: (**c**), respectively. Gd $M_{4,5}$ edge spectra obtained at the electrolyte, electrolyte-anode layer interface, and anode layer are demonstrated in (**d**), (**e**), and (**f**), respectively [adopted from Ref. 277]

edges at the area of electrolyte, interface, and anode layer, respectively (Fig. 48d, f). From the comparison of the fine structures of Gd $M_{4,5}$ peaks, there is no obvious variation among observed Gd spectra. Consequently, it indicates that no valence state change occurred for Gd diffusion ions, which is different from that of diffused Ni and Ce ions. On the basis of the STEM-EDS analysis, it is concluded that equal length diffusion mainly occurs at the interfacial region. Different from previous results, not only Ni can diffuse into grains of electrolyte but also rare earth cations such as Ce^{4+} and Gd^{3+} can diffuse into metallic Ni particles that make up the anode. In addition, superstructure with highly ordered oxygen vacancies would be formed at the interfacial region. In terms of bulk atomistic simulation in the interfacial layer, the authors already proposed large defect cluster model whose defect structure is similar to C type rare earth structure [257(a, b)]. In addition, the detailrd defect structure in our cluster model is described in Part 2 chapter "Atomistic Simulation of Subtle Microstructures in CeO_2-Based Electrolyte for Design of the State-of-the-Art Fuel Cells" (by Fei Ye et al.). The authors strongly recommend that the readers check Part 2 chapter "Atomistic Simulation of Subtle Microstructures in CeO_2-Based Electrolyte for Design of the State-of-the-Art Fuel Cells" (by Fei Ye et al.) using this opportunity, and note that similar equal length diffusion and superstructure formation were observed at the interface between $La_{0.6}Sr_{0.4}Co_{0.8}Fe_{0.2}O_3$ cathode and Gd doped CeO_2 electrolyte [281, 282].

As described in this Sect. 3.2.1. 'Defect interface between fluorite related doped CeO_2 and Ni in anode layer', our interactive combination work clearly tells that we must carefully modify the interfacial area with highly ordered oxygen vacancies for fabrication of 'state-of-the-art IT-SOFC' devices and systems.

3.2.2 Defect Interface Among Y_2O_3 Stabilized ZrO_2 and Ni in Anode Layer

Based on all experimental results in our project using doped CeO_2 system, the authors attempted to modify the defect interface using the well tried and researched 8 mol% Y_2O_3 stabilized ZrO_2 (YSZ) system. As described in 'Sect. 1.1.1 Stabilized zirconia', the defect structure change in YSZ was subtle. In addition, it is difficult to expect a co-doping effect. The reason for this is that it is not so easy to dope several dopants at once into the cubic fluorite YSZ lattice as dictated by the phase diagram. Consequently, the defect structure change of YSZ with local ordering of oxygen vacancies was not so clear, while high resolution TEM analysis suggests the formation of Ni–O bond at the interface of Ni and YSZ [283].

The authors assumed that the defect interface modification in the anode which consists of Ni and YSZ has similarity as compared to Ni-doped CeO_2-based oxides. The local ordering would be formed at the Ni-YSZ interface by equal length diffusion, while the diffusion length would be very limited as compared to Ni-doped CeO_2-based oxides. To see co-doping effect at interface between Ni and YSZ based on all experimental data observed for doped CeO_2 series, the authors attempted to dope trace amount of Pt and Pt family metals cation (i.e., Pt^{2+}, Pd^{2+}, Rh^{3+}, Ru^{3+}) whose content is less than 10 ppm (ppm = mg kg^{-1}) in the whole anode composition.

According to the correct measurement of excess heat capacity of YSZ series under Debye temperature region [284–287], the unusual excess excitation mode was observed for stabilized ZrO_2. In contrast, non-stabilized ZrO_2 follows Debye rule as well as all other metal elements. It indicates that lattice vibration in harmonic vibration mode (i.e., in low temperature which is close to absolutely zero K) is excited by dopant of Y cation. The authors concluded that local disordering of oxygen vacancies in YSZ lattice can be created by doping of trace amount of di- or tri valence cation into defect interface area (i.e., into the region of narrow equal length diffusion).

Note that heat capacity consists of lattice specific heat capacity ($C_{ph} = AT^3$) and electron heat capacity ($C_{el} = \gamma T$). In the case of real materials, heat capacity (C_v) can be expressed by Eq. (33).

$$C_v = AT^3 + \gamma T (A, \gamma : \text{constant, under Debye temperature}) \tag{33}$$

Almost all materials follow this rule except YSZ series, while the simple Einstein model cannot explain the temperature dependence of C_v at low temperature.

Consequently, the authors assumed that we have great opportunity to modify local disordering of oxygen vacancies and charge transfer phenomenon at the defect interface using trace amount of di- or tri valence cation doping based on the experimental results of interactive combination of fabrication, microanalysis, and modeling.

In our work, the authors clearly observed improvement of excess overpotentials of anode layer using trace amount of Pt^{2+} cation, refer to Fig. 49 [288].

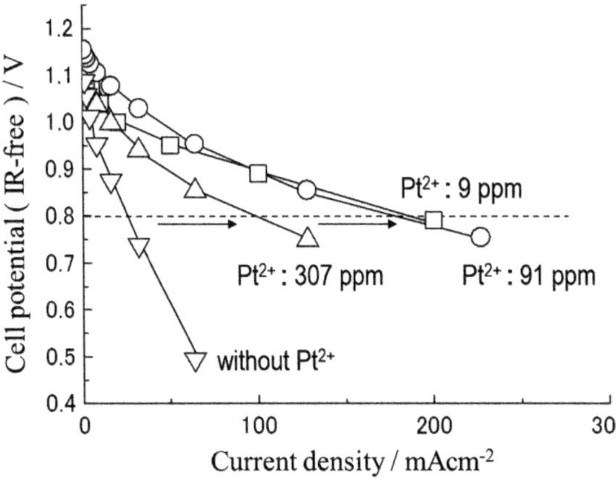

Fig. 49 Comparison of IR-free polarization characteristics (I–V curves) observed for single cells with PtO_x films deposited on the anode. Measurement temperature: 700 °C, Conventional anode (i.e., without Pt^{2+} doping): ▽; Pt^{2+} content: 307 ppm: △; Pt^{2+} content: 91 ppm: ○; Pt^{2+} content: 9 ppm: □. Dashed line indicates the 54% level of power generation efficiency [adopted from Ref. 288]

In this work, we deposited trace amount of PtO_x using the sputtering method. Once deposited, the PtO_x thin film anode layer was reduced at 800°C in a H_2 gas flow. It is a conventional conditioning process of anode layer which consists of NiO and YSZ particles. In this conditioning process of anode, we assumed that Pt^{2+} cation diffused into the anode layer and active Pt^{2+} species fabricated the local disordering area. It is our proposed co-doping effect for the anode layer on YSZ-based IT-SOFC device.

To characterize the defect interface structure, our project team observed the interface structure of the sample with trace amount of Pt^{2+} (**Pt^{2+} content: 9 ppm**) using analytical TEM, refer to Fig. 50 and Table 3.

In outlined green area in Fig. 50a, we can see conventional feature of the Ni lattice. In the FFT pattern obtained from this region shows normal Ni lattice without extra reflection. In contrast, we can see unique defect structural feature from outlined red square region in Fig. 50a which is close to the surface of the Ni. The extra reflections

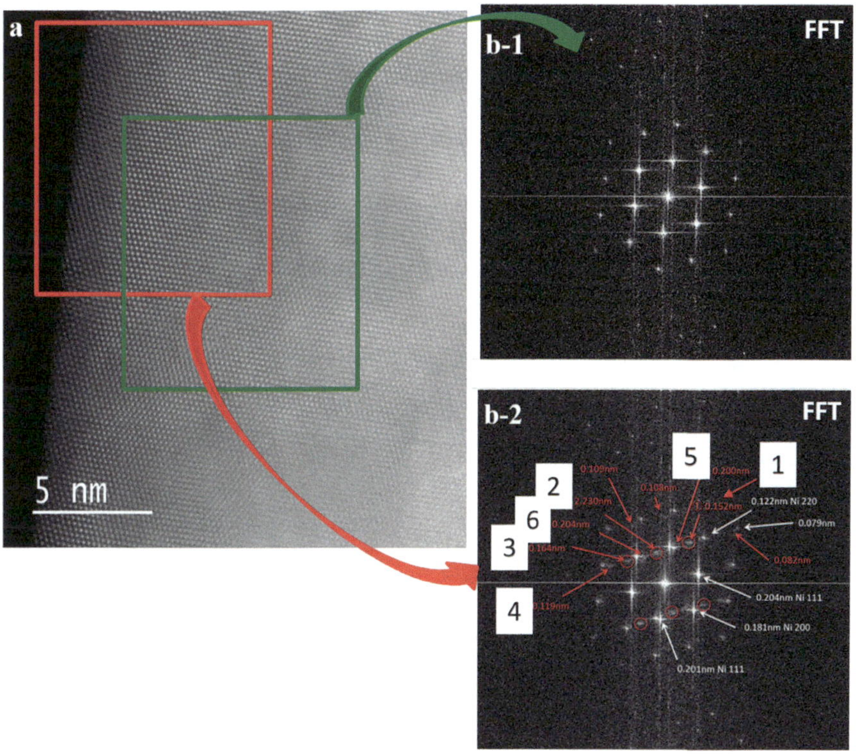

Fig. 50 HAADF-STEM image of Ni particle in reduced anode layer (**a**) and FFT patterns taken from two observation areas: (**b-1**) from green squared area observed in image (**a**), **b-2** from red squared area observed in the image (**a**). The numbers 1 to 5 denote the extra diffractions in FFT recorded from the red squared area of ths figure a. The numbers in this figure b-2 correspond to the number of diffractions in Table 3 [adopted from Ref. 288]

Table 3 Summary of assignments of extra diffraction recorded in Fig. 50b-2 [adopted from Ref. 288]

Diffraction spot	Lattice spacing/nm	Assigned materials and lattice plane
1	0.152	PtO (200)
2	0.230	Pt (200)
3	0.164	Un-known
4	0.199	Pt (311), NiO(222)
5	0.200	Pt(200), Pt(111), Ni(111)
6	0.204	Pt(200), Pt(111), Ni(111), NiO(200)

obtained from the outlined red region in Fig. 50a were clearly observed in the FFT pattern. Also, Pt^{2+} species (i.e., PtO) was detected in the red square area, refer to Table 3.

To conclude the defect interface structure which is fabricated using trace amount of Pt^{2+} cation in Y_2O_3 stabilized ZrO_2 (YSZ), filtering images from extra reflections was observed. (extra spots No.1 to No. 5) as shown in Fig. 51a to e. Also, the filtering image created by a combination of extra spots (No. 1, 2, 4, and 5), refer to Fig. 51f [288].

Figure 51f indicates that the unique defect interface structure was formed on Ni surface by trace amount of Pt^{2+} doping into the YSZ lattice. To conclude why the defect interface modification was useful for a lowering of anode overpotentials in IT-SOFC device, we performed surface atomistic simulation using GULP code, refer to Fig. 52. In our model, we assumed that the interfacial layer on Ni surface consists

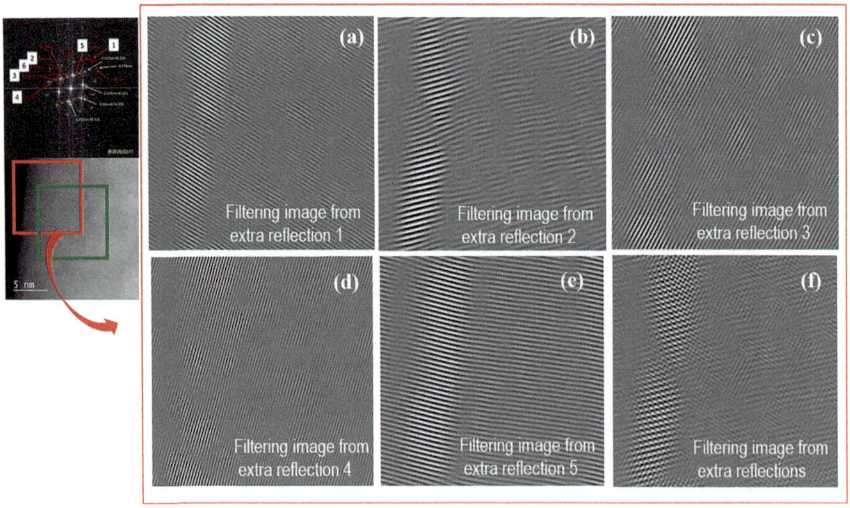

Fig. 51 Filtering images from extra spots No. 1: (**a**); No. 2: (**b**); No. 3: (**c**); No. 4: (**d**); No. 5: (**e**); and combination of diffraction spots No. 1, 2, 4 and 5: (**f**) [adopted from Ref. 288]

of Ni^{2+} (i.e., NiO) as well as from previously published work [283], Pt^{2+}, lattice defects of oxygen and Ni. Based on our defect cluster model (i.e., $V_{Ni}''Pt_i^{\cdot\cdot}V_o^{\cdot\cdot}O_i''$), the authors calculated binding energy (ΔE_b) and got high value such as 1.1 eV [288].

In addition, the authors illustrated conventional active site model [289] and our proposed model based on surface atomistic simulation in Fig. 53.

In the conventional anode active site model, the surface oxygen diffusion on YSZ in the cermet anode is the rate-determining step for the formation of water molecules on the anode below 800°C. In addition, the kinetics of oxygen surface diffusion on the Ni in the anode layer for the formation of water molecules is so slow, and it is almost impossible to create water molecules on the surface of conventional Ni particle, even though the dissociation rate constant of H_2 on Ni is known to be at a high level [288]. On the other hand, the authors concluded that a new active site for maximization of oxygen surface diffusion and promotion of formation reaction of

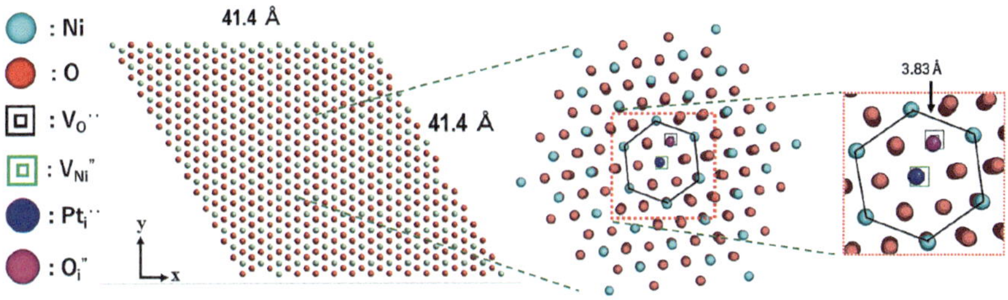

Fig. 52 Schematic diagram of defect cluster model on 7 × 7 NiO (111) surface [adopted from Ref. 288]

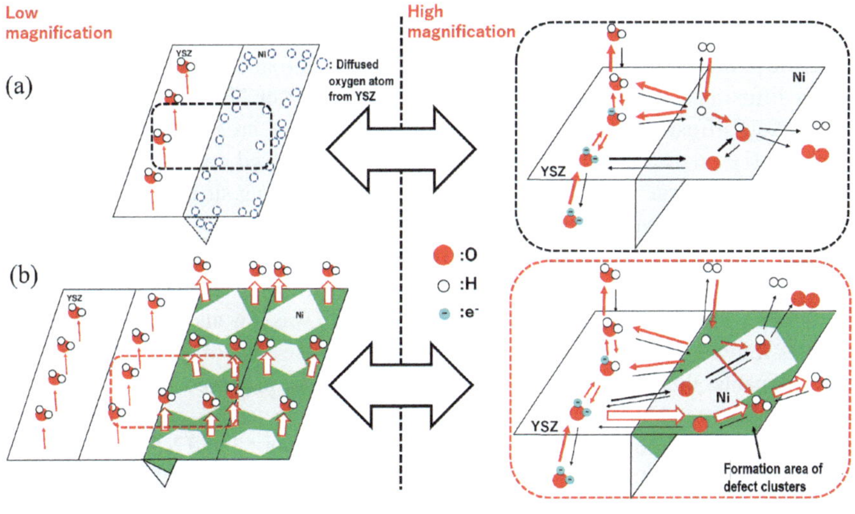

Fig. 53 Conventional active site model: (**a**) and the expected new active site formation (green area) on Ni in anode: (**b**) [adopted from Ref. 288]

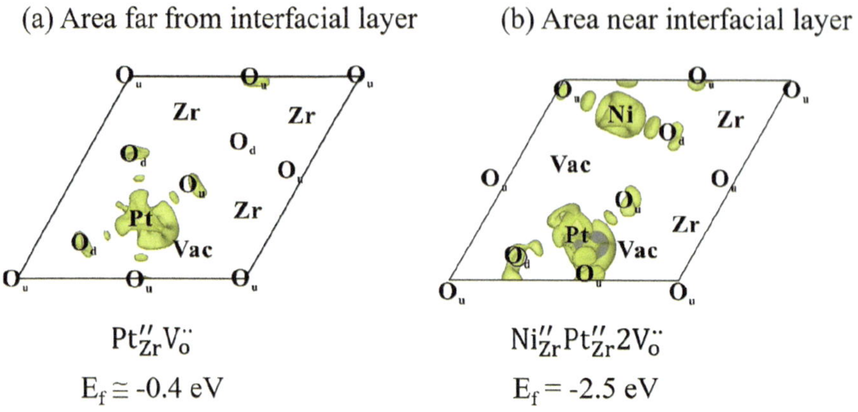

Fig. 54 3D charge density difference maps on fluorite cubic ZrO_2 (111) surface calculated for the area far from interfacial layer: (**a**) and area near interfacial layer: (**b**) [adopted from Ref. 290]

water molecules using the remaining oxygen atoms and lattice defects on the reduced Ni surface using trace amount of Pt^{2+} cation doping was theorized.

In order to conclude co-doping effects of trace amount of Pt^{2+} cations, the authors concluded the defect structure modification on YSZ at the interfacial region in the anode layer using surface first principles calculation. Since this type first-principles calculation work based on microanalysis results is under examination, the authors have just touched on key points of our surface first-principles calculation in this chapter. In this surface first-principles calculation, the authors calculated the formation energy (E_f) of Schottky type defect clusters at the area far from the interfacial layer and area near the interfacial layer (i.e., defect interface between Ni and YSZ in the anode layer), refer to Fig. 54 [290].

As we pointed out that Ni^{2+} cation would diffuse to the surface of YSZ by the equal length diffusion at an area near the defect interface between Ni and YSZ. In contrast, Ni^{2+} cation diffusion would be at a negligible level at an area far from the interfacial layer (i.e., defect interface between Ni and YSZ). Based on this conclusion, we compared 3D charge density difference maps on YSZ using surface first-principles calculation. As shown in Fig. 57b, the calculated E_f value ($= -2.5$ eV) was high enough for formation of defect cluster (i.e., $Ni_{Zr}''Pt_{Zr}''2Vo^{\cdot\cdot}$) and the delocalization of d electron around Pt and Ni was clearly seen at an area near the interfacial layer on the anode, while the calculated E_f value ($\cong -0.4$ eV) was low and the delocalization effect is limited at area far from interfacial layer. In the case of YSZ without additional dopants, the delocalization of d electron was not observed in the simulation at all. It clearly indicates that the delocalization of d electron is promoted by trace amounts of dopant doping at the defect interface between Ni and YSZ. In addition, this delocalization of d electrons would promote 'Schottky anomaly' effect at the three-phase boundary area in the anode layer. Note that the 'Schottky anomaly' is an effect observed in solid-state physics where the specific heat capacity of a solid at low temperature has a peak. Then, the local lattice vibration at defect interface area would be drastically enhanced. It is a similar effect as compared to the local disordering of oxygen vacancies at the defect interface.

In addition, the authors highlighted that the present work was really the first case to make an example of the enhancement of the 'Schottky anomaly' effect by co-doping in fuel cell field. Note that the 'Schottky anomaly' effect in measurement of heat capacity at low temperature was clearly observed in YSZ series by Professor Tooru ATAKE (Emeritus professor of Tokyo Institute of Technology, Japan) et al. [284–287, 291].

Based on the effect of lowering of excess overpotential observed for the anode layer which involves trace amount of Pt^{2+} cation, the authors attempted to dope trace amount of Pt family metal elements such as Pd^{2+}, Rh^{3+}, Ru^{3+} cations, refer to Fig. 55 [292].

In our experiment, the co-doping of trace amount of Rh^{3+} was effective for lowering the overpotentials of the anode layer as well as co-doping of trace amount of Pt^{2+}. Based on this result, we performed the surface first-principles calculation for co-doping of Fe^{2+} and Fe^{3+} using a similar defect cluster model.

Our surface first-principles calculation provided us with high formation energies, refer to Table 4. This table shows one of representative formation energy values in our surface first-principles calculation work. As shown in Table 4, our surface first-principles calculation predicts that Fe^{3+} and Fe^{2+} cations make similar defect clusters (i.e., $Ni_{Zr}''2Fe_{Zr}'2V_o^{\bullet\bullet}$ and $Ni_{Zr}''Fe_{Zr}''2V_o^{\bullet\bullet}$) at areas well away from the interface between Ni and YSZ in the anode.

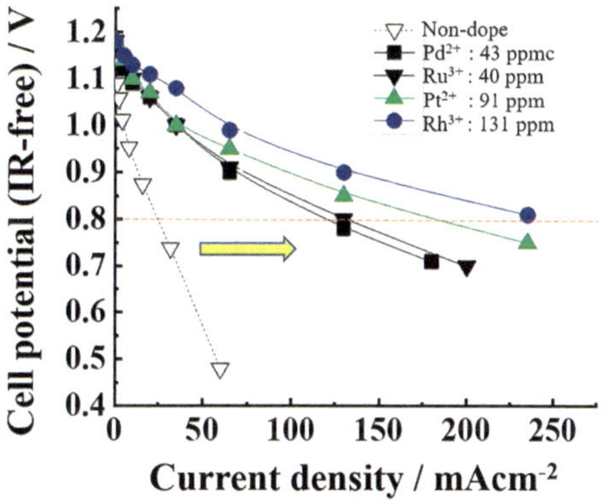

Fig. 55 Comparison of IR-free polarization characteristics (I-V curves) observed for single cells with PtO_x, PdO_x, RuO_x, and RhO_x films deposited on the anode. Measurement temperature: 700 °C, conditioning conditions for anode: 800 °C, 1 h, in a humified H_2 gas flow at room temperature

Table 4 Formation energy (E_f) of defect structure at area far from interface between Ni and YSZ

Defect cluster	Formation energy (E_f)
$2Fe_{Zr}'V_O^{\cdot\cdot}$ (Fe^{3+} doping)	−3.4 eV
$Fe_{Zr}''V_O^{\cdot\cdot}$ (Fe^{2+} doping)	−1.1 eV

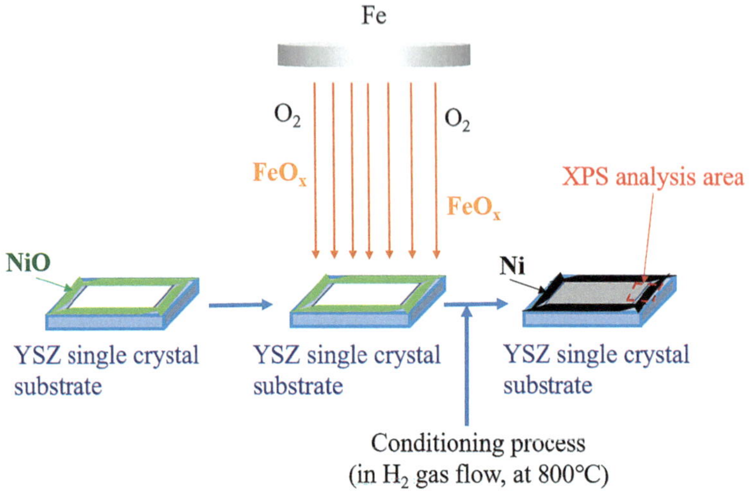

Fig. 56 Schematic diagram of chemical composition analysis on model defect interface. Temperature of YSZ substrate: 110 °C

As we described, we can expect higher formation energy calculated for $Ni_{Zr}''2Fe_{Zr}'2V_o^{\bullet\bullet}$ and $Ni_{Zr}''Fe_{Zr}''2V_o^{\bullet\bullet}$ at the near interface between Ni and YSZ in the anode layer.

Usually, Fe^{3+} species are not stable in a strong reducing atmosphere at conditioning temperatures such as 800 °C. Since our surface first-principles simulation shows us unusual indication for our future challenge, the authors carefully examined the chemical compositions at the model interface of Ni on single crystal YSZ substrate, refer to Fig. 56.

To analyze the chemical composition at defect interface between Ni and YSZ which is doped by small amount of Fe^{2+} and Fe^{3+} cations, FeO_x species was deposited on YSZ single crystal substrate with NiO layer. After deposition of the FeO_x species on the model interface area on YSZ single crystal substrate, the sample was reduced at 800 °C, in H_2 gas flow (i.e., conventional conditioning process). Then, the three Fe species at the interface of Ni and YSZ were observed using XPS, refer to Fig. 57.

Surprisingly, XPS analysis shows us important information about Fe species (i.e., Fe^{3+}, Fe^{2+}, and metallic Fe) from defect interface between Ni and YSZ. After reduction of the interface on single crystal YSZ with Ni at 800°C, Fe^{3+} species remained as major component of the defect interface. We clearly made sure that our surface first-principles calculation data was reasonable while we did not expect the unusual species (i.e., Fe^{3+}) which was seen to remain at the defect interface after strong reduction in H_2 gas flow. In addition, XPS analysis indicates that ($Ni_{Zr}''2Fe_{Zr}'2V_o^{\bullet\bullet}$) defect cluster is formed at the defect interface in the anode layer by deposition of small amount of FeO_x.

Based on all experimental results, we expect that the excess overpotential seen at the anode layer which is modified using multiple dopants such as Fe^{3+} and Fe^{2+} cation

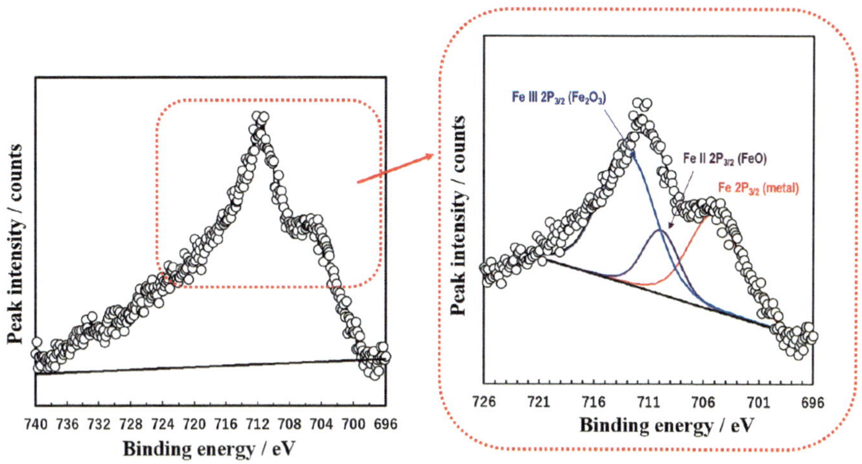

Fig. 57 Fe 2P spectra observed for defect interface area using XPS

will be decreased as well as Pt^{2+} co-doped or Rh^{3+} co-doped Ni-YSZ cermet composition anode. Figure 58 demonstrates IR-free polarization characteristics (I-V curves) observed for the anode which is doped by $Fe^{2+, 3+}$ cations. $Fe^{2+, 3+}$ co-doped Ni-YSZ cermet composition anode showed improved anode performance using single cell of IT-SOFC. The improved performance observed for $Fe^{2+, 3+}$ co-doped anode is comparable to that of Pt^{2+} co-doped or Rh^{3+} co-doped samples at 700°C. Our interactive combination work clearly suggests that the performance (i.e., IR-free value) observed for $Fe^{2+, 3+}$ co-doped Ni-YSZ cermet composition anode will get close to that of Rh^{3+} co-doped Ni-YSZ cermet composition anode, if Fe^{3+}/ Fe^{2+} molar ratio in anode is increased by optimization of fabrication condition.

As described above, the defect interface design for fabrication of 'state-of-the-art IT-SOFC' devices (and systems) can be performed effectively using interactive

Fig. 58 Comparison of IR-free polarization characteristics (I-V curves) observed for single cells with FeO_x films deposited on the anode. Measurement temperature: 700 °C. Conditioning conditions for anode: 800 °C, 1 h, in a humidified H_2 gas flow at room temperature

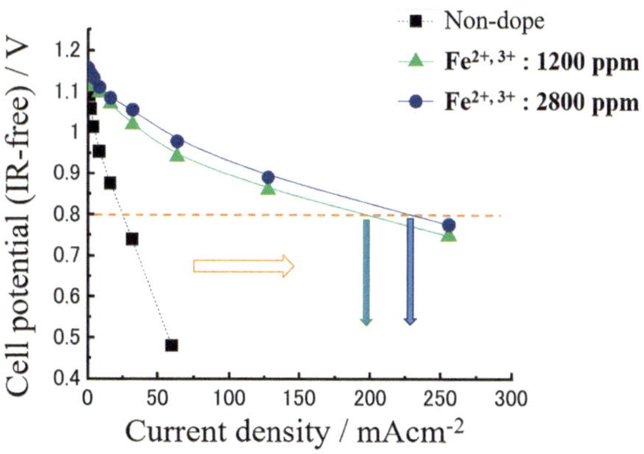

combination of improved and novel fabrication process modification, microanalysis, and surface and bulk modeling.

In Sect. 3.2.3, authors reintroduce another example for improvement of cell performance and its stability using brownmillerite-based oxides.

3.2.3 Defect Interface Between Brownmillerite-Based Oxide, Y_2O_3 Stabilized ZrO_2, and Ni in Anode Layer

As described in Sect. '3.1.2 Brownmillerite type-based system', the electrical conductivity in brownmillerite-based oxide $B_2In_2O_5$ was conspicuously improved by co-doping. In this case, the authors clearly made sure that the defect interface modification is key for enhancement of the electrical conductivity in $Ba_2((Zr_{0.5}, Zn_{0.5})_{0.3}In_{0.7})_2O_5$ system. To decrease the excess anode overpotentials, 'mixed ionic and electronic conducting (MIEC)' nanoparticles are usually used as additives [3]. As a result, based on the experimental results obtained for $Ba_2((Zr_{0.5}, Zn_{0.5})_{0.3}In_{0.7})_2O_5$ system, we attempted to improve the anode performance and long-time stability using $Ba_2((Zr_{0.5}, Zn_{0.5})_{0.3}In_{0.7})_2O_5$ nanoparticle which shows MIEC properties in the anodic condition.

The cell potential (IR-free) values observed for the anode with 0.2 wt% $Ba_2((Zr_{0.5}, Zn_{0.5})_{0.3}In_{0.7})_2O_5$ promoter in the cermet anode layer was compared to that of the conventional cermet anode (i.e., without $Ba_2((Zr_{0.5}, Zn_{0.5})_{0.3}In_{0.7})_2O_5$ additives), refer to Fig. 59a [293]. The cell potential (IR-free) value (125 mA cm^{-2} at 0.8 V) observed for anode with 0.2 wt % $Ba_2((Zr_{0.5}, Zn_{0.5})_{0.3}In_{0.7})_2O_5$ in SOFC single cell was conspicuously higher than that of the conventional anode without $Ba_2((Zr_{0.5}, Zn_{0.5})_{0.3}In_{0.7})_2O_5$ (approximately 25 mA cm^{-2} at 0.8 V) at 700 °C. The excess overpotential on the anode was conspicuously decreased by addition of small amount of $Ba_2((Zr_{0.5}, Zn_{0.5})_{0.3}In_{0.7})_2O_5$ (0.2 wt%) to the cermet anode layer, refer to Fig. 59b [293].

Electrolyte: 8YSZ sintered pellet (thickness, 0.5 mm). Cathode side: O_2 gas flow (80 ml min^{-1}) at room temperature. Anode side: humidified H_2 gas (+3% H_2O) flow (80 ml min^{-1}) at room temperature. Operation temperature: 700 °C. Dashed line in (a) denotes the level of 54% of power generation efficiency. Green arrows show comparison points in activation overpotential region and in concentration overpotential region in this work [adopted from Ref. 293].

In this work, the authors checked the fluctuation of cathode overpotential which is occurred by the gas leak through the cement sealer of performance test equipment of SOFC. In our experiment, the cathode overpotential was negligibly small as compared to the anode overpotential.

Consequently, the authors made sure that small amount of $Ba_2((Zr_{0.5}, Zn_{0.5})_{0.3}In_{0.7})_2O_5$ addition mainly contributed to the lowering of the excess overpotentials of the anode layer of IT-SOFC.

To characterize the anode performance, the authors examined current density vs. power density (I–P) curves using cell potential (IR included) data which are measured at 700 °C, refer to Fig. 60 [293]. The maximum power density (P_{max})

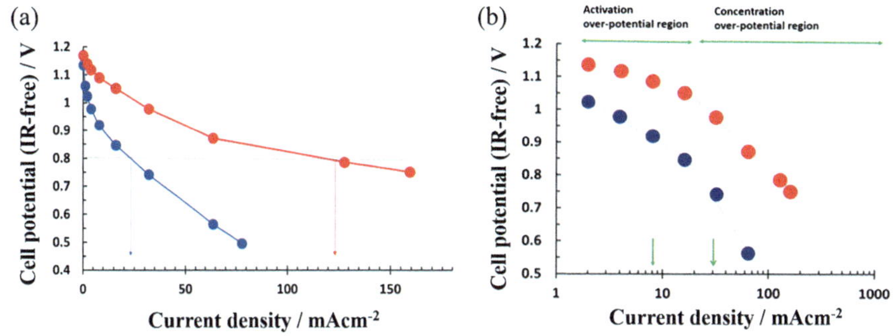

Fig. 59 IR-free polarization characteristics (I-V curves) observed for the conventional cermet anode without $Ba_2((Zr_{0.5}, Zn_{0.5})_{0.3}In_{0.7})_2O_5$ (blue closed circle) and present anode with $Ba_2((Zr_{0.5}, Zn_{0.5})_{0.3}In_{0.7})_2O_5$ (red closed circle) (**a**); Tafel lines derived from IR-free polarization characteristics (I-V curves) observed for the conventional cermet anode without $Ba_2((Zr_{0.5}, Zn_{0.5})_{0.3}In_{0.7})_2O_5$ (blue closed circle) and the present anode with 0.2 wt% $Ba_2((Zr_{0.5}, Zn_{0.5})_{0.3}In_{0.7})_2O_5$ (red closed circle) (**b**). Cermet anode: NiO-8YSZ (mass ratio NiO (4):8YSZ (1)). Cathode: $La_{0.85}Sr_{0.15}MnO_3$

values observed for dense YSZ (thickness, 500 μm) with conventional anode (thickness, 20 μm) at 700 °C was approximately 30 mW cm^{-2}. Also, the P_{max} observed for dense YSZ (thickness, 500 μm) with the present anode including 0.2 wt % $Ba_2((Zr_{0.5}, Zn_{0.5})_{0.3}In_{0.7})_2O_5$ (anode thickness, 20 μm; calcination temperature of anode, 1300 °C) at 700 °C was 62 mW cm^{-2}.

Taking the thickness of dense YSZ solid electrolyte (i.e., 500 μm) into account, maximum power density (P_{max}) value observed for the present anode sample with 0.2 wt% $Ba_2((Zr_{0.5}, Zn_{0.5})_{0.3}In_{0.7})_2O_5$ promoter was at a high level. This comparison of observed P_{max} clearly suggests that addition of small amount of $Ba_2((Zr_{0.5},$

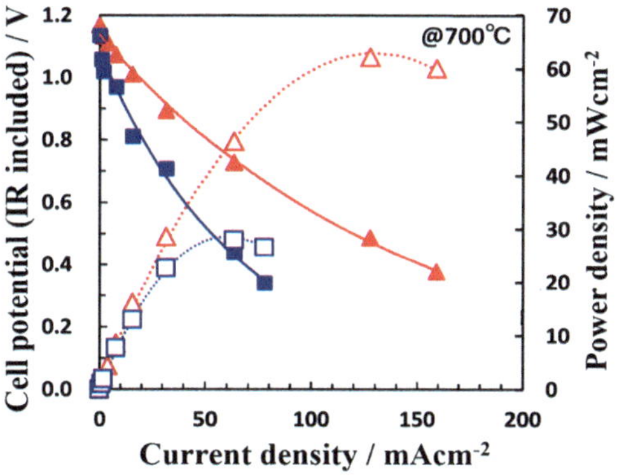

Fig. 60 Cell potential (IR included) and current density vs. power density. Open symbols: current density vs. power density. Closed symbols: cell potential (IR included) vs. current density (i.e., I-V curve). Thickness of dense YSZ solid electrolyte: 500 μm. Operation temperature: 700 °C. Anode gas: humidified H_2 (80 ml min^{-1} at room temperature). Cathode gas: O_2 (80 ml min^{-1} at room temperature) [adopted from Ref. 293]

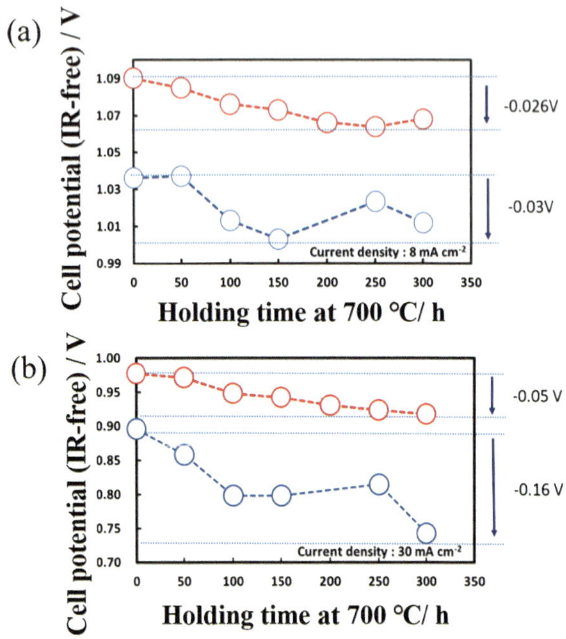

Fig. 61 Comparison of cell potential (IR-free) degradation observed at activation overpotential region (i.e., at 8 $mAcm^{-2}$): (**a**) and at concentration overpotential region (i.e., at 30 mA cm^{-2}): (**b**). Operation temperature: 700 °C. Open red symbols: anode with 0.2 wt% $Ba_2((Zr_{0.5}, Zn_{0.5})_{0.3}In_{0.7})_2O_5$. Open blue symbols: conventional anode without $Ba_2((Zr_{0.5}, Zn_{0.5})_{0.3}In_{0.7})_2O_5$ [adopted from Ref. 293]

$Zn_{0.5})_{0.3}In_{0.7})_2O_5$ is effective in improving the power density at intermediate temperatures such as 700 °C.

Long term stability tests up to 300 h were performed using both the conventional anode without $Ba_2((Zr_{0.5}, Zn_{0.5})_{0.3}In_{0.7})_2O_5$ and the present anode with 0.2 wt%$Ba_2((Zr_{0.5}, Zn_{0.5})_{0.3}In_{0.7})_2O_5$ at 8 $mAcm^{-2}$ (i.e., at activation overpotential region, refer to Fig. 61a and at 30 $mAcm^{-2}$ (i.e., at concentration overpotential region, refer to Fig. 61b [293].

In the activation overpotential region (i.e., at 8 mA cm^{-2}), the cell potential (IR-free) observed for the present anode with 0.2 wt% $Ba_2((Zr_{0.5}, Zn_{0.5})_{0.3}In_{0.7})_2O_5$ slightly decreased during the long-term stability test. However, the observed cell potential (IR-free) after 300 h operation was still higher than the conventional anode without $Ba_2((Zr_{0.5}, Zn_{0.5})_{0.3}In_{0.7})_2O_5$. In the concentration overpotential region, the degradation level observed for the conventional anode (i.e., – 0.16 V) became larger than that of the present anode with 0.2 wt% $Ba_2((Zr_{0.5}, Zn_{0.5})_{0.3}In_{0.7})_2O_5$ promoter (i.e., –0.05 V), refer to Fig. 61b. Since the authors observed large grain growth of Ni in the anode of conventional anode without $Ba_2((Zr_{0.5}, Zn_{0.5})_{0.3}In_{0.7})_2O_5$ promoter using SEM, a large degradation of the anode performance observed for conventional anode would be the result of grain growth of Ni in the anode layer. This result tells us that small amounts of $Ba_2((Zr_{0.5}, Zn_{0.5})_{0.3}In_{0.7})_2O_5$ promoter contributes to both the improvement of the anode performance and suppression of grain growth of Ni in the anode.

To understand why small amount of $Ba_2((Zr_{0.5}, Zn_{0.5})_{0.3}In_{0.7})_2O_5$ promoter contributes to the enhancement of the anode performance, we performed surface atomistic simulation on cubic ZrO_2 <111> using GULP code.

On the cubic ZrO_2 <111> surface, we attempted to assume two kinds of defect clusters such as $\mathbf{2V_{Zr}''''Ba_i^{\bullet\bullet}2In_i^{\bullet\bullet\bullet}4O_i''4V_O^{\bullet\bullet}}$ (i.e., Ba-In cluster) and $\mathbf{2V_{Zr}''''Zn_i^{\bullet\bullet}2In_i^{\bullet\bullet\bullet}4O_i''4V_O^{\bullet\bullet}}$ (i.e., Zn-In cluster), refer to Fig. 62. Our surface atomistic simulation predicts that Ba-In cluster and Zn-In cluster will be formed on surface of YSZ in the anode layer. Also, the formation of those small defect clusters would suppress the large amount of agglomeration of solely dopant oxide such as In_2O_3, BaO, and ZnO on the surface of YSZ and promote the surface oxide ion diffusion which is the rate-determining step of ORR below 800 °C. Since the segregation of solely dopant oxides (i.e., In_2O_3, BaO, and ZnO) was under detectable level at the interface between YSZ and Ni in the anode layer, we expected precipitation of fine promoter particles at the interface between YSZ and Ni particles. Then, the fine promoter particles would suppress the growth of Ni particles under long-term stability test. Consequently, improved stability of the anode was observed using the $Ba_2((Zr_{0.5}, Zn_{0.5})_{0.3}In_{0.7})_2O_5$ as a promoter. Note that the growth of Ni particles is much faster than that of YSZ. In testing the long-term stability of the anode performance under 800 °C, the growth of YSZ particles are negligible.

Moreover, the authors observed additional interesting results during this study. As shown in Fig. 63, we can see a sharp peak which indicates optimum amount of promoter in the relationship between the amount of promoter addition and the cell potential (IR-free) observed for the anode layer.

At the initial stage of our work, it was difficult to find an explanation as to why we can see a sharp peak in Fig. 63. However, our surface atomistic simulation provided us with a useful hint as to a reasonable explanation. The surface atomistic simulation

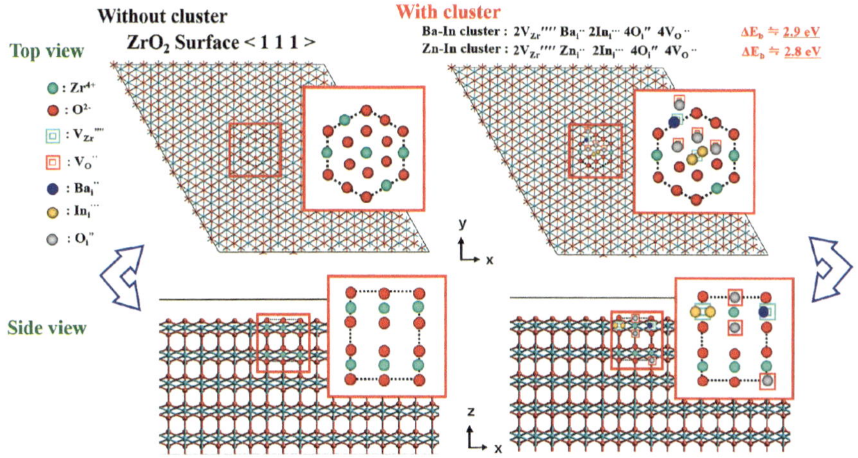

Fig. 62 Illustration of cubic ZrO_2 <111> surface atomic simulation without defect clusterand with defect cluster [adopted from Ref. 293]

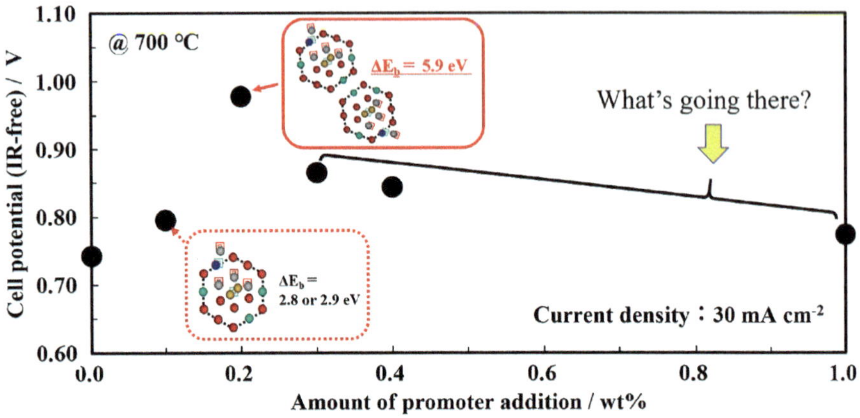

Fig. 63 Relationship between amount of promoter addition and cell potential (IR-free) observed for the anode layer in SOFC single cell. Operation condition of SOFC single cell: refer to Fig. 60

predicts that the formation of small single and twined clusters on the surface of cubic ZrO_2 <111> surface. The calculated binding energies (ΔE_b) for Ba-In cluster and Zn-In cluster are 2.9 eV and 2.8 eV, respectively. Also, ΔE_b values which were calculated for combined defect clusters (Ba-In cluster + Zn-In cluster), (Ba-In cluster + Ba-In cluster), and (Zn-In cluster + Zn-In cluster) were 5.9 eV, 4.0 eV, and 5.8 eV, respectively.

To understand why the cell potential (IR-free) reveals the sharp peak in Fig. 63, we examine the cubic C type rare earth oxide <111> surface (Fig. 64b) based on cubic ZrO_2 <111> surface (Fig. 64a).

On the surface of C type rare earth oxide <111>, the authors attempted to make 4 clusters (i.e., 4 Ba-In clusters, 4 Zn-In clusters, (2Ba-In + 2Zn-In) clusters, (3Ba-In + Zn-In) clusters and (Ba-In + 3Zn-In clusters)) and calculated ΔE_b values as well

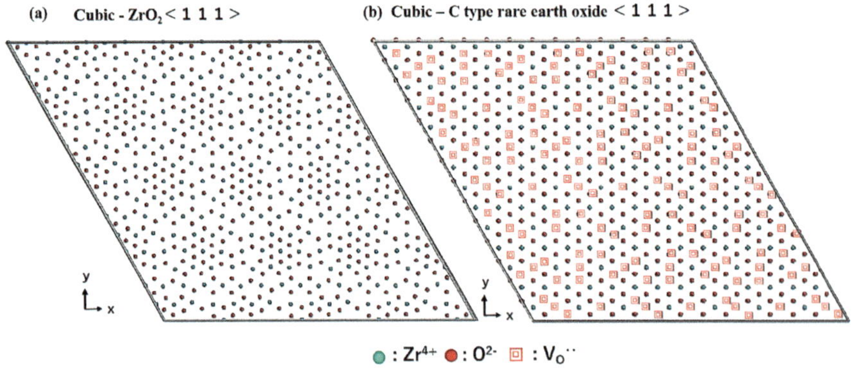

Fig. 64 Model surfaces of cubic ZrO_2 <111> (**a**) and C type rare earth oxide <111> (**b**)

as the combined defect clusters (Ba-In cluster + Zn-In cluster), (Ba-In cluster + Ba-In cluster) and (Zn-In cluster + Zn-In cluster). Unfortunately, ΔE_b values calculated for the large defect clusters which consist of 4 clusters (i.e., 4 Ba-In clusters, 4 Zn-In clusters, (2Ba-In + 2Zn-In) clusters, (3Ba-In + Zn-In) clusters and (Ba-In + 3Zn-In clusters)) were negative values in our surface atomistic simulation.

The authors show one of our cluster models in Fig. 65. When we assumed 2*(Ba-In cluster + Zn-In cluster) clusters which is just a combination of (Ba-In cluster + Zn-In cluster) cluster ($\Delta E_b = 5.9$ eV) on C type rare earth oxide <111> surface, we did not obtain positive ΔE_b calculation for 2*(Ba-In cluster + Zn-In cluster). In contrast, when we assumed the large cluster whose position of oxygen vacancies fit to the oxygen vacancy positions of C type rare earth oxide surface, we can get the positive ΔE_b (= 6.6 eV) as demonstrated in Fig. 66.

This clearly suggests that formation of small defect clusters such as Ba-In cluster, Zn-In cluster, (Ba-In cluster + Zn-In cluster), (Ba-In cluster + Ba-In cluster), and

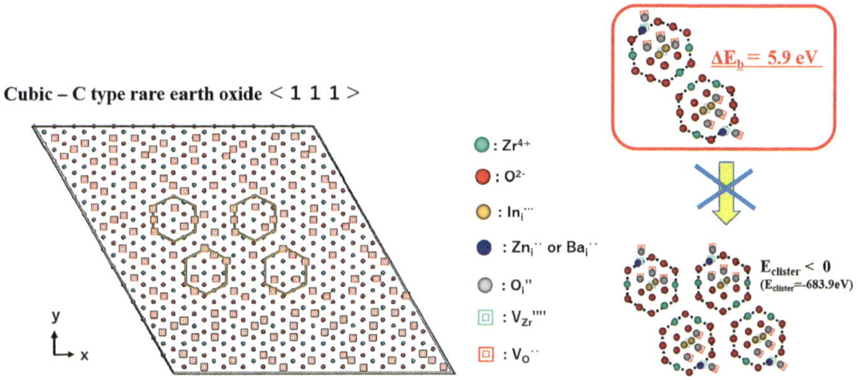

Fig. 65 2*(Ba-In cluster + Zn-In cluster) clusters on C type rare earth oxide <111> surface

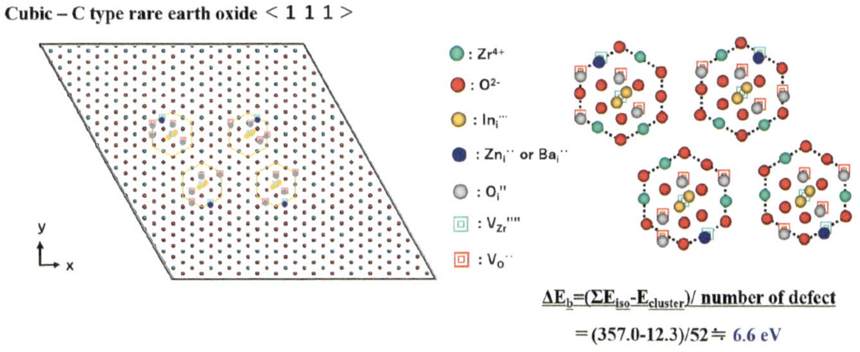

Fig. 66 2*(Ba-In cluster + Zn-In cluster) clusters whose position of oxygen vacancies are same to the oxygen vacancy positions of C type rare earth oxide <111> surface

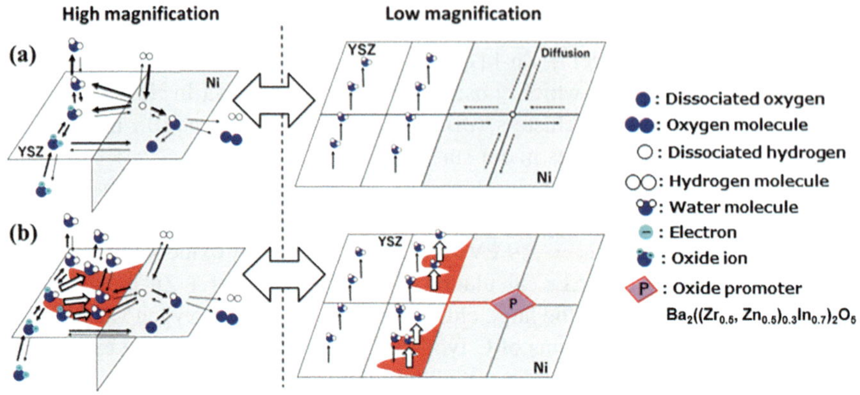

Fig. 67 Conventional active site model: (**a**) and the expected new active site formation (red area) on YSZ in anode: (**b**) [adopted from Ref. 293]

(Zn-In cluster + Zn-In cluster) are all thermodynamically stable on the cubic ZrO_2 <111> surface. However, if the amount of promoter is more than 0.4 wt% in the anode layer, it will be difficult to form the simple combination of active clusters and observe the improvement effect of anode performance. It is because the large combined cluster can be thermodynamically stable when the positions of oxygen vacancies of defect cluster fit to that of C type rare earth oxide surface, refer to Fig. 66. In this case, the surface oxide ion diffusion will be slow by the ordering of oxygen vacancies on YSZ surface. Since the surface diffusion of oxide ion on the surface of YSZ is the rate-determining step in the cathode reaction below 800 °C [289], the cell performance (IR-free) observed for cathode samples, whose content of promoter is more than 0.4 wt%, comes in at a low level as shown in Fig. 63. Consequently, the fabrication condition of anode layer with small amount of promoter should be optimized carefully to maximize the promotion effect in the fuel cell project.

Based on all experimental results, the authors concluded the formation of active site areas which consists of solely cluster and combined defect clusters (i.e., (Ba-In cluster + Zn-In cluster), (Ba-In cluster + Ba-In cluster) or (Zn-In cluster + Zn-In cluster)) on the surface of YSZ, as shown in Fig. 67.

In the fuel cell field, the effect of mixed ionic and electronic conducting (MIEC) nanoparticle additions in improving the anode performance and cathode performance of IT-SOFCs was discussed in previously published work [3]. Since the effect of $Ba_2((Zr_{0.5}, Zn_{0.5})_{0.3}In_{0.7})_2O_5$ promoter is one of MIEC effects, it is concluded that future work using this concept of MIEC effect for development of IT-SOFC devices should follow our research style (i.e., interactive combination of fabrication, micro-analysis, and modeling). Then, we suggest that the readers will have the opportunity to see something new in the material science field for development of 'state-of-the-art of IT-SOFCs'.

3.3 Interactive Combination Work for Lowering of Overpotentials on Electrodes in PEMFC Devices

In this section, the authors reintroduce the representative results for improvement of cathode performance and anode performance in PEMFC devices using interactive combination of fabrication, microanalysis, and modeling similar to the defect interface design for IT-SOFC device. For this challenge, the authors attempted to modify the interface between Pt and cerium oxide (CeO_x) nanoparticle or nanowire in the cathode and anode layer. In the early stages of our work, the authors used CeO_x nanoparticles as oxide support. However, the area of interface of Pt and CeO_x nanoparticles was very narrow. It was hard to expect the fabrication of superior quality Pt-CeO_x nanoparticle support/CB electrodes for cathode side and anode side applications. To maximize the area of key defect interface between Pt and oxide support CeO_x, the authors fabricated CeO_x nanowire in our project. Consequently, the authors re-introduced the experimental results of both Pt-CeO_x nanoparticle/CB and Pt-CeO_x nanowire/CB as anode and cathode in this section.

3.3.1 Lowering of Cathode Overpotentials

The key point in this challenge is the design of the active surface layer of the Pt-O-Ce bonds which is fabricated at the interface between the active metal Pt and the oxide support CeO_x. After we fabricated Pt loaded CeO_x nanoparticle/C (Pt/C weight ratio = 0.2), the authors observed improved oxygen reduction reaction (ORR) activity on Pt-CeO_x nanoparticle/CB (CB: carbon black), refer to Fig. 68a.

Figure 68a clearly suggests that the onset potential and ORR activity observed for Pt loaded CeO_x nanoparticle/CB (Pt/C weight ratio = 0.2) is improved by enhancement of formation area of Pt-O-Ce bond at the Pt-CeO_x interface on the cathode. To compare the cathode performances of Pt-CeO_x(c)/CB which has the highest EASA in the series of Pt-CeO_x nanoparticles/CB and home-made Pt/CB, a single cell was made using the commercially available Nafion® membrane which was sandwiched between cathode (Pt-CeO_x(c)/CB or home-made Pt/CB) and conventional Pt/C anode. The electrochemical polarization curves recorded for the single cell are demonstrated in Fig. 68b. The overpotential recorded for Pt-CeO_x(c)/CB cathode was lower than that of Pt/CB cathode. This reconfirmed that the surface of the Pt is activated by a small amount of CeO_x nanoparticle. This configuration shows better cathode performance as compared with Pt/CB cathode in the single cell (i.e., MEA).

The current density-cell potential curves in Fig. 68b tell us that the excess overpotential observed on the home-made Pt/CB cathode can be decreased by formation of Pt-O-Ce bond at the Pt-CeO_x interface in the cathode layer of the polymer membrane fuel cell device as well as electrochemistry measurement work.

To minimize the excess overpotential observed for Pt surface on the cathode, the authors observed the interface Pt-CeO_x nanoparticle/CB using TEM, refer to Fig. 69. Almost all Pt particles were covered by CeO_x nanoparticle support, refer to Fig. 69a

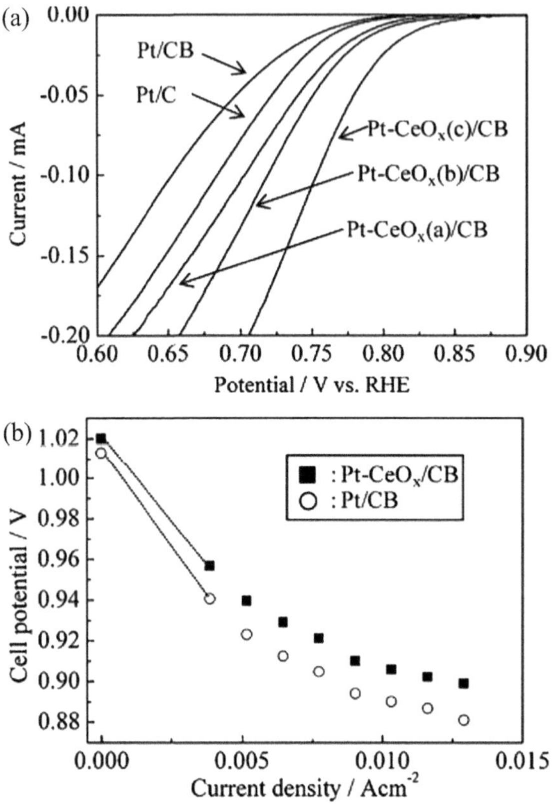

Fig. 68 **a** Voltammograms of ORR on Pt/CB, conventional Pt/C and Pt-CeO_x/CB. Electrochemical Active Surface Area (EASA) of Pt-CeO_x(a)/CB: 14.7 / $m^2 {g_{pt}}^{-1}$, EASA of Pt-CeO_x(b)/CB: 17.8 / $m^2 {g_{pt}}^{-1}$, EASA of Pt-CeO_x(c)/CB: 25.7 / $m^2 {g_{pt}}^{-1}$, EASA of home-made Pt/CB: 9.9 / $m^2 {g_{pt}}^{-1}$, and EASA of conventional Pt/C: 80 / $m^2 {g_{pt}}^{-1}$ [adopted from Ref. 231]. Measurement conditions: refer to section of experiment in this chapter. **b** Electrochemical polarization curves (i.e., I-V curves) for Pt-CeO_x(c)/CB and home-made Pt/CB in the single cell (i.e., MEA); operation temperature: 70 °C, MEA: humidified H_2 (anode side)| Nafion® membrane | humidified O_2 (cathode side), anode: conventional Pt/C provided by Johnson Matthey company, Pt content in anode and cathode: 1 mg cm^{-2} and area of electrode: 3.87 cm^2 [adopted from Ref. 231]

to c. Pt particles which are covered by nano-crystalline CeO_x can be identified by separating the lattices of metal particles and oxide supports by fast Fourier transform (FFT) and inverse FFT (IFFT). As shown in Fig. 69d, the IFFTs of Pt and CeO_x were colorized and superimposed with the HRTEM image. This microanalysis data clearly shows that CeO_x nanoparticle supports could invade the metal surface and form the defect interface which mainly consists of a fluorite structure component. The microanalysis results in Fig. 69 indicate that the metal-support interaction exists in Pt-CeO_x nanoparticle/CB and its influence is increased with an increase in Pt content.

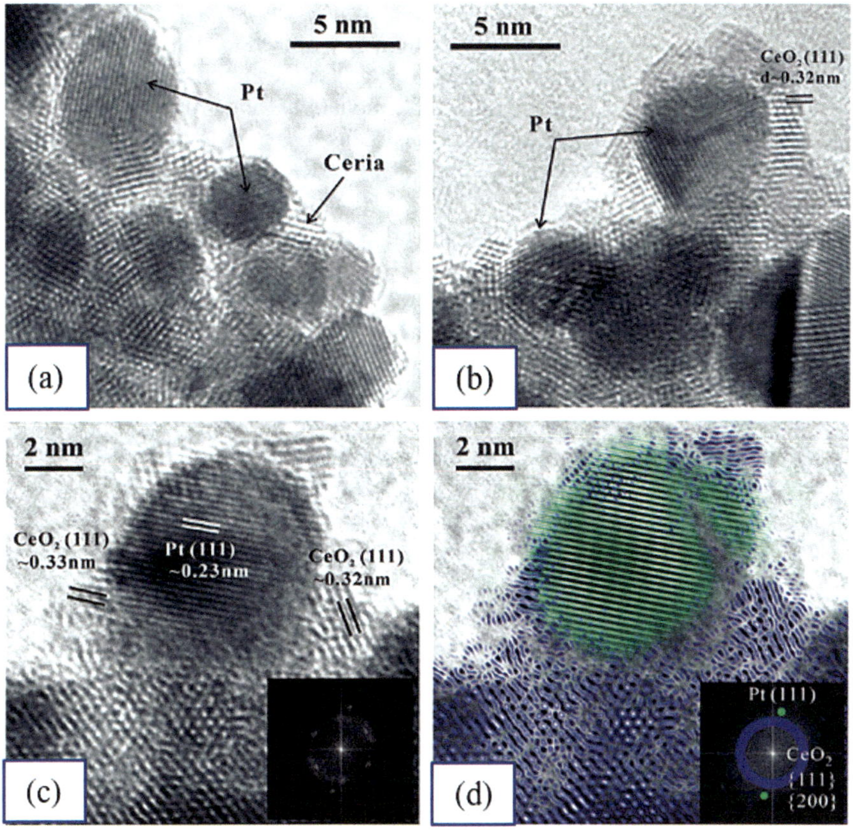

Fig. 69 HRTEM images of samples **a** 20 wt%Pt-CeO_x nanoparticle /CB and **b** and **c** 30 wt%Pt-CeO_x nanoparticle/CB. **c** Lattices of Pt and CeO_x nanoparticle were separated by FFT and IFFT. **d** IFFTs using the reflections of Pt (111), CeO_2 {111}, and CeO_2 {200} (as marked in the inset) were colorized and superimposed with the HRTEM image. Those images were taken from the sample before conditioning treatment [adopted from Ref. 206]

In addition, the authors observed the surface chemical composition of Pt-CeO_x nanoparticle/CB (weight ratio between Pt and C = 0.2) using XPS. Figure 70 shows representative Pt 4f and Ce 3d peaks observed for Pt-CeO_x nanoparticle/CB and Pt 4f peak observed for conventional Pt/C. As shown in Fig. 70b, Pt 4f spectra show two spin —orbital splitting doublets Pt $4f_{7/2} - 4f_{5/2}$. The peaks at 71.1 and 72.8 eV are attributed to metallic Pt (Pt^0) and divalent Pt (Pt^{2+}, PtO). The other extra peak (Pt-O-X (X: oxygen vacancy ($V_o^{\bullet\bullet}$), Ce)) at the binding energy of 72.1 eV which is located at midway position between metallic Pt (Pt^0; $4f_{7/2}$: 71.1 eV) and PtO (Pt^{2+}; $4f_{7/2}$: 72.8 eV) would correspond to the slightly ionized Pt which is attributable to the formation of Pt-O-Ce bond at the interface between Pt and CeO_x nanoparticle. Also, the authors observed small peak of Pt^{4+} species (PtO_2, $4f_{7/2}$: 74.7 eV) on the surface of 'as-prepared' Pt-CeO_x nanoparticle/CB. On the other hand, the peak of

Pt^{4+} species (PtO_2) was under detectable level on conventional Pt/C surface, refer to Fig. 70a. Usually, the electrochemical pre-treatment of the Pt-CeO_x nanoparticle/CB surface was necessary to obtain the stable electrochemical performance. The standard electrochemical pre-treatment for conventional Pt/C and Pt-CeO_x nanoparticle/CB was performed by means of 30 cyclic sweeps in 0.5 M H_2SO_4 aqueous solution using the potential ranging from 0 to 1.0 V (vs. RHE). The influence of another pre-treatment condition (from 0 to 1.5 V (vs. RHE)) on the change of surface composition of conventional Pt/C was demonstrated as reference in same figure. The peak intensity of Pt^{4+} species (PtO_2) was under detectable level by surface cleaning effect of electrochemical pre-treatment (i.e., conditioning process). In contrast, the small extra peak which is assigned to be Pt-O-X (X: oxygen vacancy ($V_o^{\bullet\bullet}$), Ce) remained on the surface of Pt-CeO_x nanoparticle/CB sample. In our work, we concluded that the formation of Pt-O-Ce bond at the interface between Pt and CeO_x nanoparticle is key for enhancement of ORR activity.

Note that the peak intensity of Ce 3d is reduced after the electrochemical pre-treatment (i.e., conditioning process). The reason for this is that free CeO_x without Pt-O-Ce bond is easily dissolved into acidic solution. If the Pt-O-Ce bond was not created at the interface between Pt and CeO_x at all, CeO_x nanoparticle support would

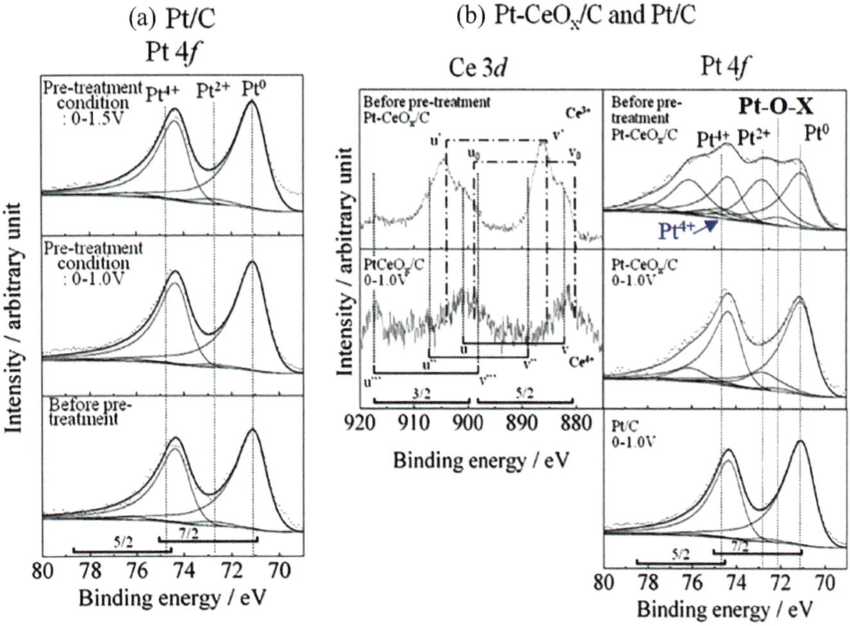

Fig. 70 Pt 4f and Ce 3d peaks observed for Pt-CeO_x nanoparticle/CB and Pt 4f peak observed for conventional Pt/C. Pt-O-X (X: oxygen vacancy ($Vo^{\bullet\bullet}$), Ce) [adopted from Ref. 232]

disappear during the electrochemical measurement and stable observation of electrochemistry in a steady state condition cannot be expected. Based on all these experimental results, it is concluded that we observed the improved ORR activity on Pt-CeO_x nanoparticle/CB in the steady state condition by the formation of Pt-O-Ce bond at the Pt-CeO_x interface on the samples.

To clarify the defect interface structure, the authors fabricated thin films of CeO_x on the conductive Nb-doped $SrTiO_3$ single crystal substrate using PLD method. Also, the interface structural feature was analyzed using high resolution (HR)-TEM analysis (i.e., FFT and IFFT image analyses). In our HR-TEM image analysis, some areas show conventional fluorite reflection (i.e., (2–2)) in FFT patterns. In addition, IFFT pattern which was transformed from the FFT pattern shows a conventional fluorite type lattice contrast, refer to Fig. 71(2–3). On the other hand, we clearly observed extra-reflections from Pt loaded CeO_x thin film on the conductive single crystal (Nb-$SrTiO_3$) in Fig. 71(3–2). IFFT transformed from Fig. 71 (3–2) indicates that the characteristic lattice contrast agrees with C type rare earth oxide pattern, refer to Fig. 71(3–3).

To support this microanalysis data and understand the nature of the Pt-O-Ce bond in detail, the authors performed bulk atomistic simulation using GULP code. Since Pt^{4+} (PtO_2) was under detectable levels in the XPS data and selected area electron diffraction patterns, we assumed that the following defect cluster formation reaction

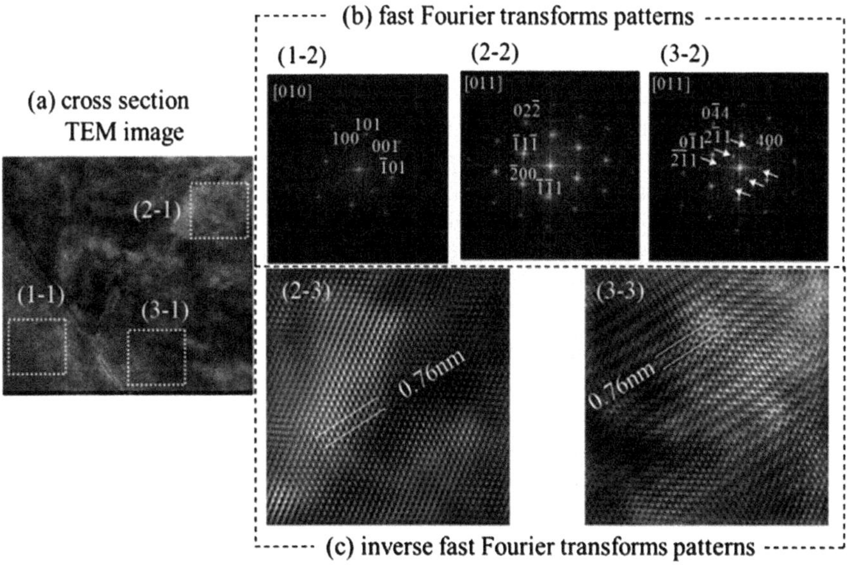

Fig. 71 **a** TEM cross-sectional image taken from electrochemically pre-treated Pt-loaded CeO_x thin film on conductive Nb-doped $SrTiO_3$ (100) single crystal substrate. **b** Fast Fourier transforms patterns: (1–2) transformed from (1–1), (2–2) transformed from (2–1), and (3–2) transformed from (3–1). **c** Inverse fast Fourier transforms patterns: (2–3) transformed from (2–2) and (3–3) transformed from (3–2) [adopted from Ref. 233].

(i.e., Eq. (34)) occurred at the interface between Pt surface and a small amount of remaining CeO_x on Pt.

$$Pt + 2Ce_{ce}{}^{x} \xrightarrow{CeO_2} Pt_{Ce}{}'' 2Vo^{\cdot\cdot} 2Ce_{Ce}{}' \quad (34)$$

where the Kröger-Vink notation is used for the simple explanation of defect cluster formation reaction.

In Eq. (34), the defect cluster $Pt_{Ce}'' 2Vo^{\bullet\bullet} 2Ce_{Ce}'$ consists of two oxygen vacancies, one Pt^{2+} into the Ce^{4+} sub-lattice site, and one Ce^{3+} into Ce^{4+} sub-lattice site. To conclude the relationship between Pt-O-Ce bond and the small defect cluster, the binding energy (ΔE_b) was calculated for small defect clusters. We then attempted to combine those defect clusters to see the feature of the Pt-O-Ce bond at the defect interface on the sample.

Based on Eq. (34), the stable configuration of the $Pt_{Ce}''Vo^{\bullet\bullet}$ cluster in the unit cell of the fluorite-related structure CeO_2 can be estimated using bulk atomistic simulation. The highest binding energy of $Pt_{Ce}''Vo^{\bullet\bullet}$ cluster in all possible configurations in the unit cell was $\Delta E_b = 0.5$ eV (refer to Fig. 74a). In addition, we developed large cluster models by combining the $Pt_{Ce}''Vo^{\bullet\bullet}$ and $Pt_{Ce}''2Vo^{\bullet\bullet}2Ce_{Ce}'$ structures based on the C type rare earth structure, refer to Fig. 74b. Using this model, the calculated ΔE_b became 1.1 eV which was conspicuously higher than the ΔE_b value based on the smallest cluster model (i.e., $Pt_{Ce}''Vo^{\bullet\bullet}$). As described above, our bulk atomistic simulation predicts that Pt-O-Ce bond can be created by formation of large defect clusters on the sample.

To characterize the stability of this Pt-O-Ce bond which was formed at the interface on Pt-CeO_x nanoparticle/CB cathode, we performed a long cycle test of the interface in 0.5 M H_2SO_4 aqueous solution. The ORR activity observed for homemade Pt/CB was almost constant during 1000 cycles test. In contrast, the ORR activity observed for Pt-CeO_x nanoparticle/CB slightly decreased during 1000 cycles test. Despite this, the ORR activity on Pt-CeO_x nanoparticle /CB was still higher than homemade Pt/ CB catalyst, refer to Fig. 73a. To understand why the ORR activity decreased during the 1000 cycle test, we observed the defect interface on the sample that had been subjected to the 1000 cycle test using HR-TEM, refer to Fig. 73b. The HR-TEM microanalysis indicates that the CeO_x found around and between the Pt particles remains in most of the observed areas after 1000 potential sweeps, even though the mass activity was decreased after 1000 cycle sweeps. It is known that the Pt particles acting as electrocatalysts generally grow by the dissolution and subsequent deposition of Pt under the severe electrochemical measurement conditions. Consequently, it can be concluded that the CeO_x layer on the Pt surface can effectively restrain the dissolution and re-deposition of Pt species. Furthermore, the slight increase in the size of Pt particles reveals that the CeO_x layers on the Pt particles could be discontinuous over the whole surface area of Pt. Then, the dissolution and re-deposition of the Pt species could be observed and the mass activity observed for Pt-CeO_x nanoparticle/ CB would slightly decrease during 1000 cycle sweep test, while the CeO_2 layer on the Pt remained intact after the long cycle test.

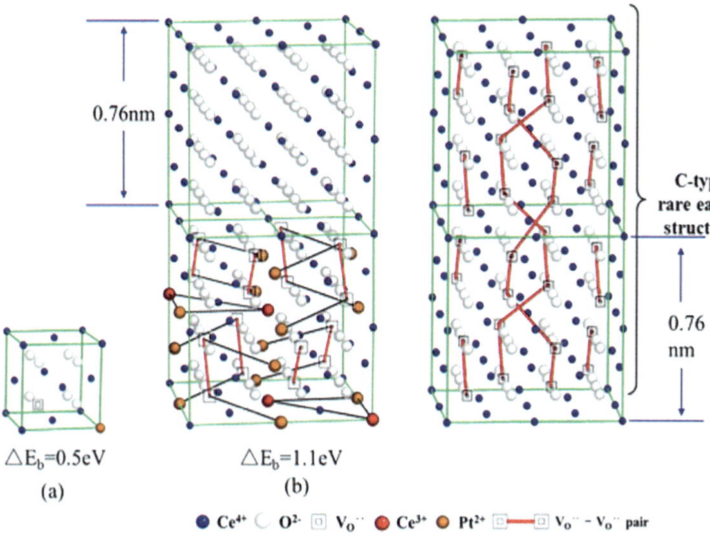

Fig. 72 Calculated binding energy (ΔE_b) and defect structure model based on **a** simple cluster $Pt_{Ce}''Vo^{\bullet\bullet}$ and **b** combined cluster 12 ($Pt_{Ce}''Vo^{\bullet\bullet}$) + 2 ($Pt_{Ce}''2Vo^{\bullet\bullet}2_{Ce}Ce'$). The defect structure of the C type rare earth structure was displayed as reference [adopted from Ref. 233]

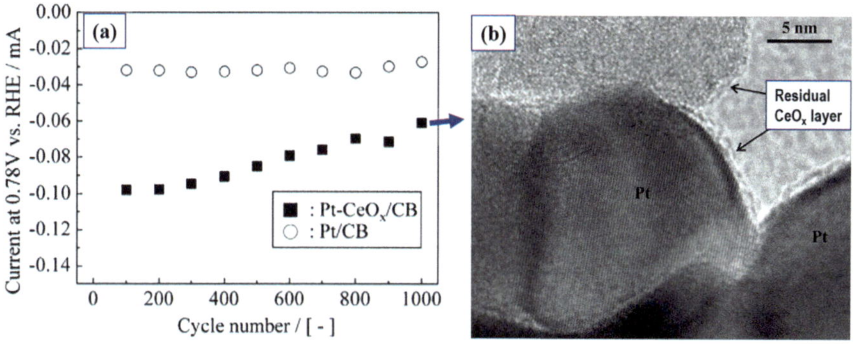

Fig. 73 **a** Stability of mass activity in a long cycle test up to 1000 cycle sweeps in 0.5 M H_2SO_4; Open circle symbols: home-made Pt/CB (Pt/C weight ratio = 0.2), closed square symbols: Pt-CeO_x nanoparticle /CB (Pt/C weight ratio = 0.2), mass activity was observed at 0.81 V vs. RHE. Potential range of cycle test: from 0 to 1.0 V (vs. RHE), Sweep rate: 50 mV s^{-1}. **b** HR-TEM image observed for 1000 cycled sample [adopted from Ref. 207]

To maximize the area of Pt-O-Ce bond on Pt surface and increase both ORR activity and its stability, observed for Pt-CeO_x nanoparticle/CB cathode, the authors fabricated CeO_x nanowire instead of CeO_x nanoparticles. Figure 73 presents SEM and TEM images observed for Pt-CeO_x nanowire/CB.

The products in our fabrication process consisted of only CeO_x nanowires whose average diameter and average length are less than 50 nm and more than 1 μm,

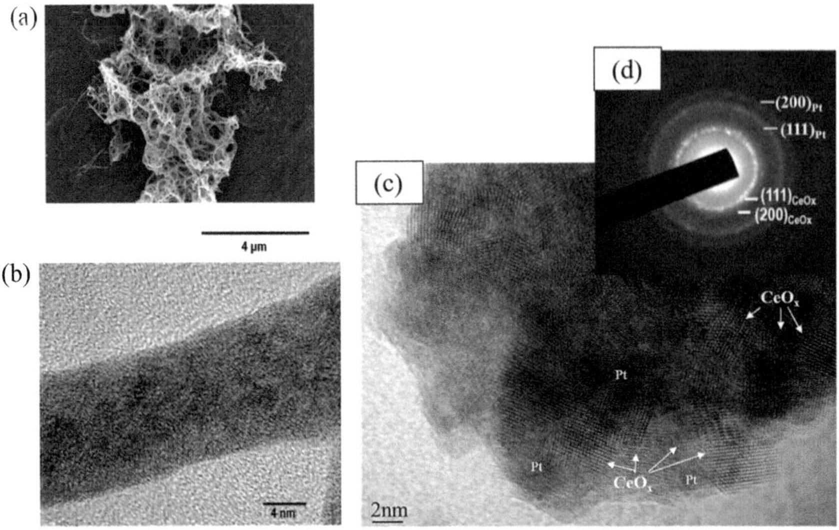

Fig. 74 SEM photograph taken from as prepared CeO_x nanowire (**a**), TEM image observed for as-prepared CeO_x nanowire (**b**), TEM image observed for Pt loaded CeO_x nanowire with selected area electron diffraction pattern (**d**) recorded from corresponding area to (**c**)

respectively. (Fig. 74a). In addition, the fabricated CeO_x nanowires were made up of one-dimensional polycrystalline CeO_x nanoparticles whose average particle size is approximately 3 to 4 nm, refer to Fig. 74b. TEM microanalysis suggests that Pt nano-sized particles were deposited in the cathode sample after Pt particles were loaded on CeO_x nanowires. Nano-sized Pt particles would be in the grain boundary area of CeO_x nanowire, refer to Fig. 74c and d.

To highlight the uniqueness of this Pt-CeO_x nanowire/CB (Pt/C weight ratio = 0.2) cathode, the authors re-introduced the surface chemical composition data observed for Pt-CeO_x nanowire/CB (Pt/C weight ratio = 0.2) using XPS analysis, refer to Fig. 75. The most important difference between Pt-CeO_x nanoparticle/CB surface (i.e., XPS profiles in Fig. 70) and Pt-CeO_x nanowire/CB surface (i.e., XPS profiles in Fig. 75a) is the intensity of Pt-O-X (X: $V_o^{\cdot\cdot}$, Ce) peak observed for Pt-CeO_x nanowire/CB surface. The peak intensity ratio between the Pt-O-X (X: $V_o^{\cdot\cdot}$, Ce) peak and metallic Pt (Pt^0) observed for Pt-CeO_x nanowire/CB surface was approximately 0.3. Since the peak intensity ratio between the Pt-O-X (X: $V_o^{\cdot\cdot}$, Ce) peak and metallic Pt (Pt^0) observed for Pt loaded CeO_x nanoparticle/CB was less than 0.1, the observed peak intensity ratio between the Pt-O-X (X: $V_o^{\cdot\cdot}$, Ce) peak and metallic Pt (Pt^0) observed for the Pt loaded CeO_x nanowire/CB was much higher than that of the Pt loaded CeO_x nanoparticle/CB.

On the other hand, the Ce^{3+} peaks which are labeled as v^0, u^0, v' and u' (where the symbols v^0-u^0, and v'–u' are spin−orbital partners of $3d_{5/2}$ and $3d_{3/2}$, respectively) and Ce^{4+} peaks which are labeled as v, u, v'', u'', v''', and u''' (where the symbols v–u, v''–u'' and v'''–u''' are spin−o rbital partners of $3d_{5/2}$ and $3d_{3/2}$, respectively)

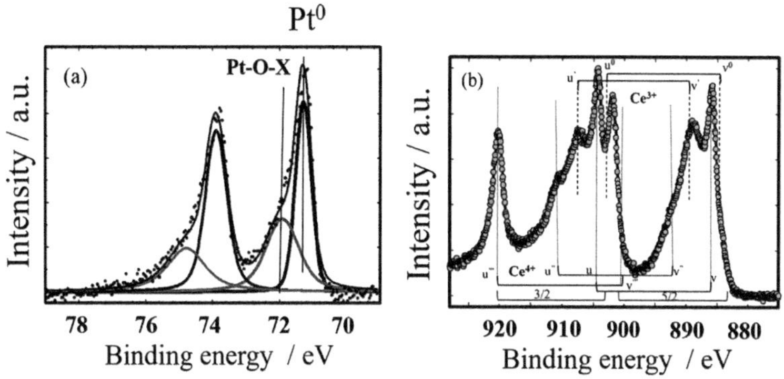

Fig. 75 Pt 4f spectra (Pt $4f_{7/2}-4f_{5/2}$) and Ce 3d spectra (Ce $3d_{5/2}-$Ce $3d_{3/2}$) observed for Pt-CeO_x nanowire/CB (Pt/C weight ratio = 0.2) which was fabricated by conventional chemical process. Extra peak labeled by Pt-O-X (X: $V_o^{\cdot\cdot}$, Ce) indicates that Pt-O-Ce bond was formed by interaction of Pt and CeO_x species (CeO_2 and Ce_2O_3) on the sample surface. Note that XPS profiles of Pt 4f and Ce 3d were measured after the conditioning process. The conditioning process was performed by means of 120 cyclic sweeps in 0.5 M H_2SO_4 aqueous solution using the potential ranging from 0 to 1.0 V (vs. RHE) [adopted from Ref. 235]

were clearly observed in Fig. 75(b). The XPS spectra in Fig. 75b indicate that the surface and interface of the sample consist of CeO_x species (CeO_2 and Ce_2O_3) and $Ce(OH)_3$. Based on XPS analysis data, we concluded that the area of Pt-O-Ce bonds on Pt-CeO_x nanowire/CB was much wider than that of Pt-CeO_x nanoparticle/CB.

After we obtained microanalysis information of the defect interface on Pt-CeO_x nanowire/CB (Pt/C weight ratio = 0.2), the authors examined the electrochemistry including long cycle sweep test of the sample. From our electrochemistry data, the authors re-introduced the data of long cycle sweep test in this chapter, refer to Fig. 76. As demonstrated in this figure, the mass activity observed for Pt-CeO_x nanowire/CB (Pt/C weight ratio = 0.2) was conspicuously increased as compared to conventional Pt/CB (Pt/C weight ratio = 0.2) and Pt-CeO_x nanoparticle/CB (Pt/C weight ratio = 0.2). In addition, the stability of mass activity observed for Pt-CeO_x nanowire/CB (Pt/C weight ratio = 0.2) improves over the Pt-CeO_x nanoparticle/CB (Pt/C weight ratio = 0.2), refer to Figs. 73 and 76. We concluded that activity and its stability was improved by formation of Pt-O-Ce bonds on the sample, as we expected based on the interactive works among fabrication, microanalysis, and modeling.

In order to maximize the activity and decrease the content of Pt in the Pt-CeO_x nanowire/CB, the authors used quantum beam technique for fabrication of Pt-O-Ce bonds on the sample. The reason for this is that the energy of quantum beams such as proton beams or electron beams create active species such as aqueous electron (e_{aq}^-), a hydrogen radical ($H^{\cdot}$), and a hydroxyl radical ($^{\cdot}OH$) radiolysis reaction of water, refer to Eq. (35).

$$H_2O \rightarrow e_{aq}^-, H^+, H^{\cdot}, {}^{\cdot}OH, H_2O_2, H_2 \quad (35)$$

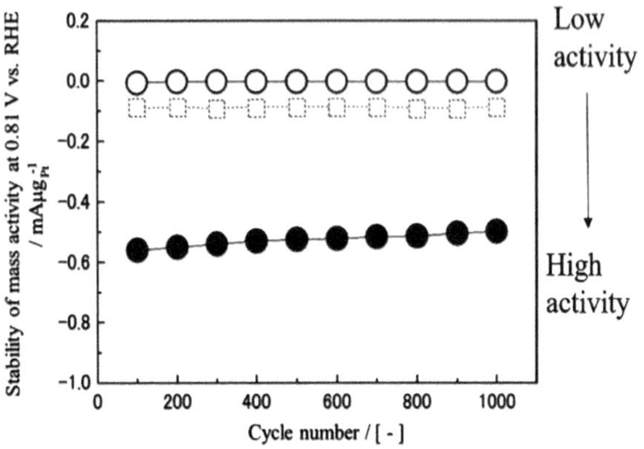

Fig. 76 Stability of mass activity in a long cycle test up to 1000 cycle sweeps in 0.5 M H_2SO_4; Open circle symbols: conventional Pt/C (Pt/C weight ratio = 0.2), open square symbols: Pt-CeO_x nanoparticle /CB (Pt/C weight ratio = 0.2), closed circle symbols: Pt-CeO_x nanowire/CB (Pt/C weight ratio = 0.2), mass activity was observed at 0.81 V vs. RHE. Potential range of cycle test: from 0 to 1.0 V (vs. RHE), Sweep rate: 50 mV s^{-1} [adopted from Ref. 235]

In the radiolysis reaction, e_{aq}^- and $H^·$ create nano-sized metal particles from platinum compounds on oxide supports. Also, we assumed that $^·OH$ reacts with adsorbed OH on CeO_x nanowire which has defect rich surface with a basic nature. We expect that wide area Pt-O-Ce bond will be formed at the interface between Pt and CeO_x nanowire in the Pt-CeO_x nanowire/CB. In addition, we can decrease the Pt content in the Pt-CeO_x nanowire/CB which keeps high ORR activity as well as Pt-CeO_x nanowire/CB (Pt/C weight ratio = 0.2) samples.

As one of the representative examples, the authors re-introduced the experimental results about the defect interface modification using proton beam irradiation method in this chapter. Since the radical density generated by the proton beam irradiation is two orders of magnitude greater than that of the electron beam irradiation, the authors expect that the defect rich CeO_x nanowire surface with Pt nanoparticles will be fully converted to Pt-O-Ce bonds under proton beam irradiation.

As we expected, the top surface of Pt-CeO_x nanowire/CB (Pt/C weight ratio = 0.02) which was irradiated by proton beam shows quite a unique chemical composition. Only Pt-O-Ce bond peaks were observed for proton beam irradiated sample using SXPS analysis, refer to Fig. 77a. Note that the weight ratio between Pt and C of proton beam irradiated sample is 0.02 which is one-tenth as compared to Pt-CeO_x nanowire/CB (Pt/C weight ratio = 0.2) which was fabricated by conventional chemical process. On the other hand, the average surface area consists of Pt^{2+} (i.e., PtO) and Pt-O-Ce bond which is labeled as Pt-O-X (X: $V_o^{··}$, Ce) species, refer to Fig. 77b. Since nano-sized Pt particles are among the grain boundaries of CeO_x NW which is far from the surface area, the peak intensity of metallic Pt would be in trace amount level in SXPS and HXPS analyses.

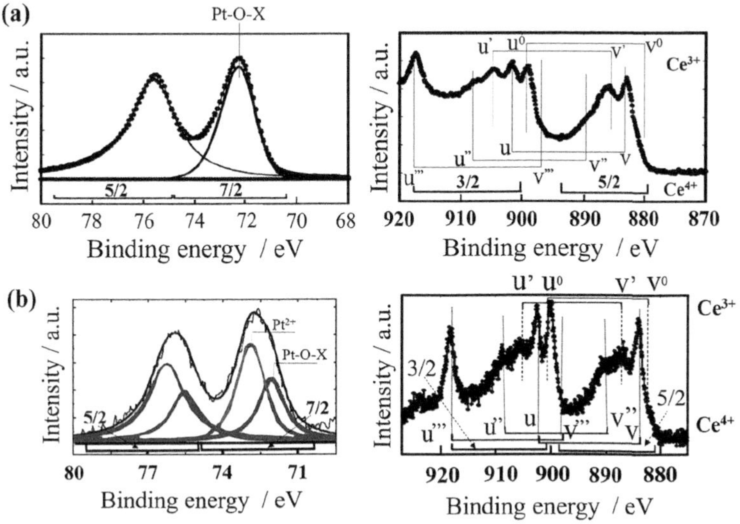

Fig. 77 **a** *SXPS Pt 4f spectra (Pt $4f_{7/2}-4f_{5/2}$) and Ce 3d spectra (Ce $3d_{5/2}-$Ce $3d_{3/2}$) observed for Pt-CeO_x nanowire/CB (Pt/C weight ratio = 0.02) which was fabricated by proton beam irradiation method. **b** *HXPS Pt 4f spectra (Pt $4f_{7/2}-4f_{5/2}$) and Ce 3d spectra (Ce $3d_{5/2}-$Ce $3d_{3/2}$) observed for Pt-CeO_x nanowire/CB (Pt/C weight ratio = 0.02) which was fabricated by proton beam irradiation method. Extra peak labeled by Pt-O-X (X: $V_o^{\cdot\cdot}$, Ce) indicates that Pt-O-Ce bond was formed by interaction of Pt and CeO_x species (CeO_2 and Ce_2O_3) on the sample surface. Note that SXPS and HXPS profiles of Pt 4f and Ce 3d were measured after the conditioning process. The conditioning process was performed by means of 120 cyclic sweeps in 0.5 M H_2SO_4 aqueous solution using the potential ranging from 0 to 1.0 V (vs. RHE) [adopted from Ref. 294]. (*SXPS: soft X-ray photoelectron spectroscopy, *HXPS: hard X-ray (synchrotron X-ray at SPring-8) photoelectron spectroscopy)

To characterize the unique Pt-O-Ce bond on both the top and the average surface area which was formed by proton beam irradiation, we observed the interface structure on Pt-CeO_x nanowire/CB using analytical TEM. (refer to Fig. 78).

Figure 78 clearly indicates that a unique Pt-O-Ce bond layer fully covered the CeO_x nanowire surface. The thickness of Pt-O-Ce bond layer is approximately 10 nm. To support the microanalysis data, the authors performed surface atomistic simulation using GULP code.

Based on SXPS and HXPS analysis data, we assumed that defect cluster consists of Pt^{2+} cation ($Pt_i^{\bullet\bullet}$), Ce^{3+} species ($Ce_i^{\bullet\bullet\bullet}$) and lattice defects such as $V_{Ce}^{''''}$, $V_O^{\bullet\bullet}$, and O_i''. Also, analytical TEM indicates that defect clusters are formed on the surface of CeO_x nanowire. Pt-O-Ce bond which can be active sites on the electrode is formed on the top surface by proton beam irradiation energy. Consequently, our surface defect cluster model should cover all microanalysis possibilities using SXPS, HXPS, and analytical TEM. In our surface atomistic simulation work, two clusters such as Pt^{2+} cluster ($V_{Ce}''''2\,Pt_i^{\bullet\bullet}4\,V_O^{\bullet\bullet}4O_i''$) and Ce^{3+} cluster ($2V_{Ce}''''2Ce_i^{\bullet\bullet\bullet}4\,V_O^{\bullet\bullet}3O_i''$) were assumed on CeO_2 <111> surface. Then, high binding energy (ΔE_b=2.8 eV) for combination of Pt^{2+} cluster and Ce^{3+} cluster was obtained, refer to Fig. 79. Our

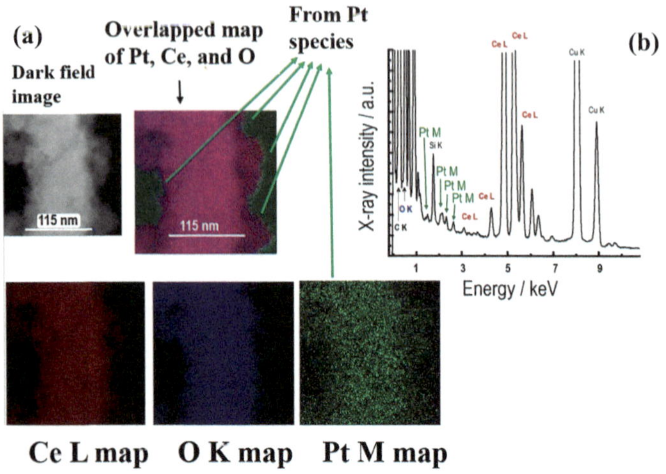

Fig. 78 STEM image–EDS elemental mapping images with dark field image: (**a**) and point analysis profile: (**b**) observed for Pt-CeO_x nanowire/CB (Pt/C weight ratio = 0.02) [adopted from Ref. 294]

surface atomistic simulation predicts that the combination between Pt^{2+} cluster and Ce^{3+} cluster can be stable at 19.9 Å on CeO_2 <111> surface.

To conclude why the distance of two clusters (i.e., Pt^{2+} cluster and Ce^{3+} cluster) is preferable for formation of a stable combination of Pt^{2+} cluster and Ce^{3+} cluster on (111) CeO_2 surface, the positions of the oxygen vacancies on C type rare earth oxide <111> surface with oxygen vacancy distance values were illustrated in Fig. 79a). As

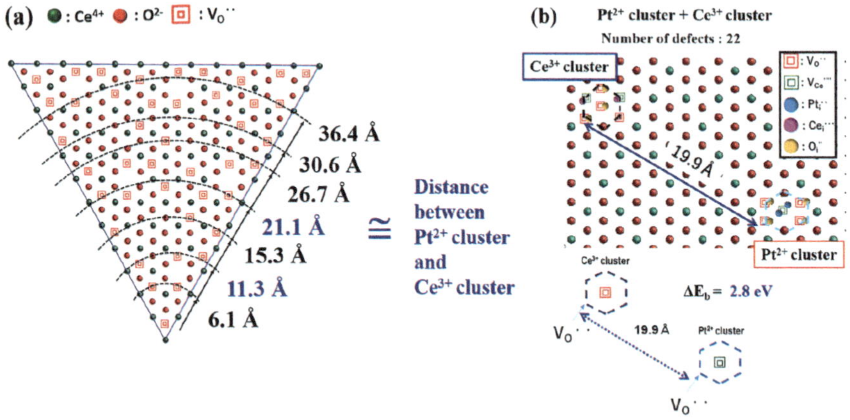

Fig. 79 Schematic diagram <111> surface of C type rare earth oxide structure with distance between oxygen vacancies: (**a**) and Pt^{2+} cluster (V_{Ce}''''2 $Pt_i^{\bullet\bullet}$4 $V_O^{\bullet\bullet}$4O_i'') and Ce^{3+} cluster (2V_{Ce}''''2$Ce_i^{\bullet\bullet\bullet}$4 $V_O^{\bullet\bullet}$3O_i'') with the distance between two clusters (i.e., Pt^{2+} cluster and Ce^{3+} cluster): (**b**) [adopted from Ref. 294]

demonstrated in this figure, our surface atomistic simulation predicts that the combination of two defect clusters (i.e., Pt^{2+} cluster and Ce^{3+} cluster) will be stabilized when the oxygen vacancy position of the combined cluster corresponds to that of the C type rare earth oxide structure which is a related structure of the fluorite type CeO_2. In addition, our surface atomistic simulation tells us that Pt-O-Ce bond consists of Pt^{2+} cluster and Ce^{3+} cluster. The combined clusters can be active sites on Pt-CeO_x nanowire/CB for a lowering of excess overpotentials on the cathode in PEMFC, also it can decrease Pt content in the cathode layer for design of 'state-of-the-art PEMFC' devices. In addition, in-situ XANS analysis using the synchrotron facility demonstrates that the electrochemical redox reaction of CeO_x contributes to minimize the surface oxidation of conventional Pt/C cathode surface in ORR reaction [234]. This in situ XANS analysis supported our ideas.

To characterize the effect of the Pt-O-Ce bond which consists of Pt^{2+} clusters and Ce^{3+} clusters on both fuel cell performance and stability, the authors examined the IR-free cell potential observed for Pt-CeO_x nanowire/CB cathode (Pt/C weight ration = 0.04) whose Pt content is approximately one-fifth of conventional Pt/C cathode (Pt/C weight ratio = 0.2) using MEA at 70°C, refer to Fig. 80. Since sufficient quantities of cathode material cannot be made by the proton beam irradiation method, the authors turned to the electron beam irradiation method as an alternative. Unfortunately, the radical density generated by electron beam irradiation is one hundredth as compared to that of proton beam irradiation. It was hard to fabricate wide Pt-O-Ce bond area on Pt-CeO_x nanowire/CB cathode (Pt/C weight ration = 0.04) using electron beam irradiation method. Consequently, the cell potential (IR-free) observed for Pt-CeO_x nanowire/CB cathode (Pt/C weight ration = 0.04) whose surface was modified by electron beam irradiation was lower than expected. SXPS analysis indicates that the surface of Pt-CeO_x nanowire/CB cathode (Pt/C weight ration = 0.04) was partially covered by Pt-O-Ce bond, while TEM microanalysis shows homogeneous surface layer of Pt on Pt-CeO_x nanowire/CB cathode (Pt/C weight ration = 0.04) as well as Pt-CeO_x nanowire/CB cathode (Pt/C weight ration = 0.04) which was fabricated using proton beam irradiation method. These two microanalysis results tell us that the surface of Pt-CeO_x nanowire/CB cathode (Pt/C weight ration = 0.04) which was modified by electron beam irradiation consists of nano-sized metal Pt particles and Pt-O-Ce bond. As shown in Fig. 80, cell potential (IR-free) observed for the Pt-CeO_x nanowire/CB cathode (Pt/C weight ration = 0.04, Pt content: 0.03 mg_{pt} cm^{-2}) increased with increasing operation time and reached the constant level over a prolonged operational test (up to 84 h).

Usually, the initial cell potential (IR-free) value observed for conventional Pt/C (Pt/C weight ratio = 0.04) in a MEA is a high level in addition to Pt-CeO_x nanowire/C. However, the cell potential (IR-free) value observed for conventional Pt/C (Pt/C weight ratio = 0.04) rapidly decreased with increasing operation time because of the agglomeration of nano-sized metal Pt particles in the cathode layer of the MEA. In the present work, the authors observed conspicuous degradation of cell performance after 40 h operation. Then, a large flooding phenomenon was observed for conventional Pt/C whose Pt content is one-fifth as compared to standard Pt/C in a PEMFC device, (refer to data logger data in Fig. 80).

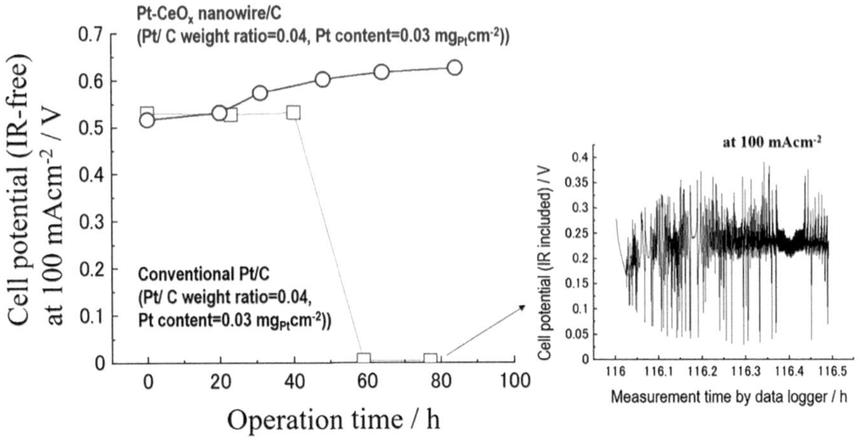

Fig. 80 Electrochemical polarization curves (i.e., I–V curves) observed for Pt-CeO_x nanowire/CB (Pt/C weight ratio = 0.04, Pt content: 0.03 mg_{pt} cm^{-2}, ○) and conventional Pt/C (Pt/C weight ratio = 0.04, Pt content: 0.03 mg_{pt} cm^{-2}, □) in the single cell (i.e., MEA); operation temperature: 70 °C, MEA: humidified H_2 (anode side)| Nafion® membrane | humidified O_2 (cathode side). Anode: conventional Pt/C provided by Johnson Matthey Company. Pt content in anode: 0.15 mg_{pt} cm^{-2}, Pt content in cathode: 0.03 mg_{pt} cm^{-2} [adopted from Ref. 295]

This suggests that we must use Pt-O-Ce interface on the cathode for design of low content Pt cathode in the fuel cell field. In addition, the experimental results in Fig. 80 clearly indicate that the design of Pt-O-Ce bonds on defect rich CeO_x nanowire surface is quite important to keep the favorable balance between high performance of the MEA and its long-term stability for development of 'state-of-the-art PEMFCs'.

As described above, the series of our work about design of active interface on cathode in PEMFC clearly indicates that the interactive combination among fabrication, microanalysis, and modeling is key for development of 'state-of-the-art fuel cells' for our society.

3.3.2 Lowering of Anode Overpotentials

Our team in interactive combination of fabrication methods, microanalysis, modeling work have attempted to improve the carbon monoxide (CO) tolerance on conventional Pt/C anode for development of 'state-of-the-art PEMFCs'. The authors re-introduced several representative data in this chapter as well as design of high-quality cathode materials.

As demonstrated in Fig. 81, Pt-CeO_x nanowire/CB anode showed much better activity in electro-methanol oxidation reaction as compared to conventional Pt/C anode. In addition, we focused on the onset potential of the electro-methanol oxidation reaction for the design of high CO tolerance anode materials for PEMFC devices, refer to Fig. 82. The estimated onset potential of the conventional 20wt% Pt/C was

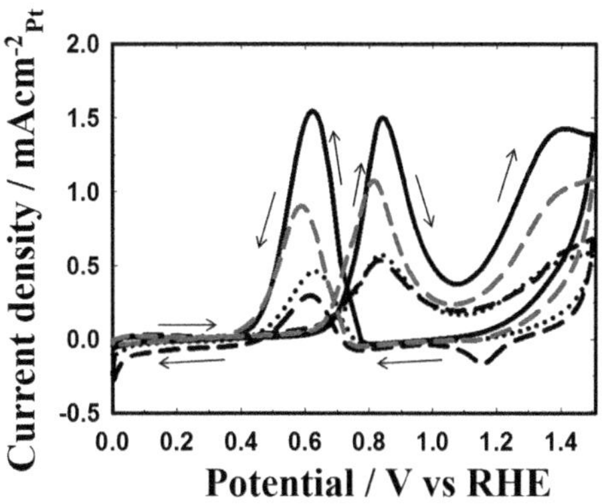

Fig. 81 Cyclic voltammograms of electrochemical methanol oxidation in both forward and back sweeps on 10wt% Pt-CeO_x nanowire/CB anode (Pt/C ratio = 0.4, solid black line), 5wt% Pt-CeO_x nanowire/CB anode (Pt/C ratio = 0.2, dotted black line), 1wt% Pt-CeO_x nanowire/CB anode (Pt/C ratio = 0.1, dashed black line) and conventional 20 wt% Pt/C anode (Pt/C ratio = 0.2, gray dashed line) at 28 °C in the mixed solution of 0.5 M H_2SO_4 aqueous solution and 0.5 M CH_3OH aqueous solution at 10 mV s^{-1} [adopted from Ref. 210]

approximately 0.5 V (vs. RHE). This estimated value corresponds to the previously reported value observed for conventional Pt/C (i.e., 0.5 V vs. RHE) [296]. Moreover, the methanol electro-oxidation on 1wt% Pt-CeO_x nanowire/CB anode commenced from approximately 0.19 V (vs. RHE), which is much lower than that of the conventional Pt/C. The experimental results demonstrated in Fig. 82 suggest that the CO tolerance of nano-sized Pt particles in Pt-CeO_x nanowire/CB, which has Pt-O-X (X: $Vo^{\bullet\bullet}$ or Ce) bond at the interface of Pt and CeO_x nanowire, is much higher than on Pt in conventional Pt/C.

Based on the results of Figs. 81 and 82 the authors assumed that high CO tolerance on Pt which is enhanced by the interaction between Pt and oxide support CeO_x nanowire can be used for improvement of CO tolerance on Pt in the anode layer in a PEMFC device. To examine the CO tolerance on Pt in the anode layer in a MEA (i.e., PEMFC single cell, also known as PEMFC device), the authors added a small amount of CO gas whose concentration is 5 ppm to humidified H_2 during the operation of MEA, refer to Fig. 83. This figure clearly supports that the design of the defect interface between Pt and oxide support CeO_x is quite effective for improvement of CO tolerance on Pt in the anode layer of a MEA. In addition, we clearly observed absorption band shift of CO molecule from low wave number region to high wave number region under the applied potential using in situ IR analysis method. Namely, in-situ IR analysis indicates that adsorption strength of CO on Pt surface in Pt-CeO_x nanoparticle/C became much weaker than that of Pt/C under applied potential [211]. This data clearly supports our idea for design of defect interface structures between Pt and CeO_x support in our project.

The most important point in the results of Fig. 83 is the influence of the CO tolerance on the Pt surface on the area of Pt-O-X bond (X: $V_O^{\bullet\bullet}$ or Ce) on CeO_x

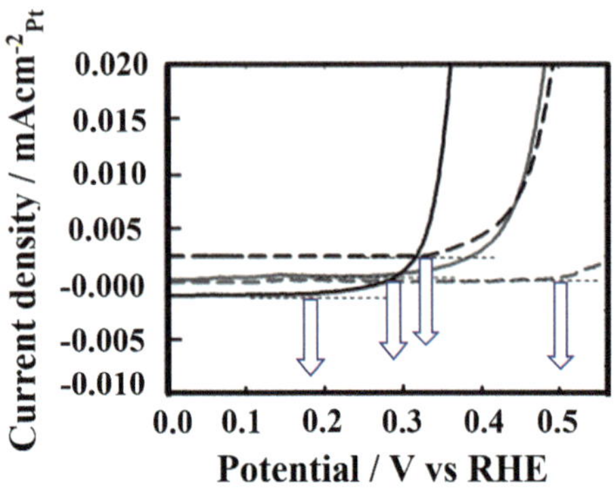

Fig. 82 Onset potential of electrochemical CH_3OH oxidation reaction in forward sweep on 10wt% Pt-CeO_x nanowire/CB anode (Pt/C ratio = 0.4, black dashed line), 5wt% Pt-CeO_x nanowire/CB anode (Pt/C ratio = 0.2, gray solid line), 1wt% Pt-CeO_x nanowire/CB anode (Pt/C ratio = 0.1, black solid line), and conventional 20 wt% Pt/C anode (Pt/C ratio = 0.2, gray dashed line) at 28 ºC in the mixed solution of 0.5 M H_2SO_4 aqueous solution and 0.5 M CH_3OH aqueous solution at 1 mV s^{-1}. Blue arrows show the onset potentials of electrochemical CH_3OH oxidation reaction [adopted from Ref. 210]

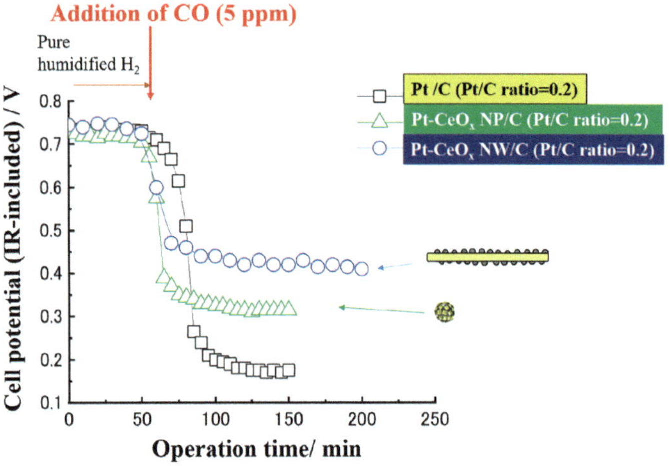

Fig. 83 Schematic diagram of change of cell potential (IR-included) values observed for MEA before and after addition of CO (5 ppm) to humidified H_2 gas. Operation temperature: 70 °C, MEA: humidified H_2 (anode side, 80 mlmin^{-1} at 70 °C) | Nafion® membrane | humidified O_2 (cathode side, 80 mlmin^{-1} at 70 °C). Anode: conventional anode (□),Pt-CeO_x nanoparticle/CB (Pt/C weight ratio = 0.2) (Δ) and Pt-CeO_x nanowire/CB (Pt/C weight ratio = 0.2) (◯). Cathode: conventional Pt/C, Pt content in anode and cathode: 0.15 $mg_{pt}cm^{-2}$

support. As described in ' Sect. 3.3.1 Lowering of the cathode overpotentials', the area of Pt-O-X bond (X: $V_O^{••}$ or Ce) on CeO_x nanowire was much wider than that of CeO_x nanoparticle. In addition, it is well known that conventional CeO_2 surface without Pt-O-X bond (X: $V_O^{••}$ or Ce) is easily corroded in acidic atmosphere under applied potential (i.e., in the electrochemistry measurement condition). In contrast, the authors successfully modified the activity and stability on the defect rich surface of CeO_x at the interface between the anode/ Nafion® membrane (or cathode/Nafion® membrane).

In order to consider the rationale for the formation of Pt-O-X bond (X: $V_O^{••}$ or Ce), an interface layer between Pt and CeO_x, on both cathode and anode, which leads to higher activity, the defect structure formation energy of the defect cluster which consists of oxygen vacancy, Pt^{2+} and Ce cations (i.e., Ce^{3+} and Ce^{4+} species) and the affinity of water molecules to the interfacial layer containing aforementioned defect cluster were calculated using the surface first-principles calculation [295]. In executing the surface first-principles calculations, a simpler model is required to perform the calculations, so instead of the combined defect cluster model shown in Fig. 79, the cluster model consisting of Schottky defects ($2Ce_{Ce}'Pt_{Ce}''2V_O^{••}$) was used to calculate the defect structure formation energy in the interfacial layer including the Schottky defects and affinity for water molecules on that interface layer.

When the Schottky defect ($2Ce_{Ce}'Pt_{Ce}''2V_O^{••}$) model on the (111) CeO_2 surface was assumed using the surface first-principles calculation method, its defect structure formation energy is 2.6 eV. It became close to the binding energy (2.8 eV) which is the defect structure formation energy of the combined defect cluster shown in Fig. 79. This suggests that two possible types of surface defect clusters can be formed at the interfacial layer between Pt and CeO_x NWs. One is the combined defect cluster consisting of Pt^{2+} defect cluster ($V_{Ce}''''2Pt_i^{••}4V_O^{••}4O_i''$) and Ce^{3+} defect clusters ($2V_{Ce}''''2Ce_i^{•••}4V_O^{••}3O_i''$), and another one is Schottky defect type ($2Ce_{Ce}'Pt_{Ce}''2V_O^{••}$) cluster (refer to Fig. 84 for one of examples).

To consider why the formation of an interfacial layer containing aforementioned defect clusters between Pt and CeO_x NW contributed to the improvement of electrocatalytic activity, the authors attempted to consider the reason based on a model with high validity using the surface first-principles calculation method. One of the advantages of using surface first-principles calculation method is that it makes it possible to quantitatively consider the affinity of water molecules on the defect clusters formed in the interfacial layer between Pt and CeO_x nanowire.

Therefore, taking ($2Ce_{Ce}'Pt_{Ce}''2V_O^{••}$) which is a Schottky defect-type cluster as an example, the authors calculated both the adsorption energy (E_a) of water molecule on the metal Pt, $Ce^{(3+)}{}_2O_3$ and ($2Ce_{Ce}'Pt_{Ce}''2V_O^{••}$) Schottky type defect cluster and formation energy (E_f) of them. It was found that there is a trade-off relationship between the formation energy and the adsorption energy of water molecule. Based on this calculation result, the authors further examined the relationship between ($E_a + E_f$) and the Pt^{2+}/Ce^{3+} ratio and found that the ($E_a + E_f$) value is optimal (approximately 3 eV) around the ($2Ce_{Ce}'Pt_{Ce}''2V_O^{••}$) cluster composition (i.e., Pt^{2+}/Ce^{3+} ratio = 0.5). Moreover, it was also found that the calculated value increased as the

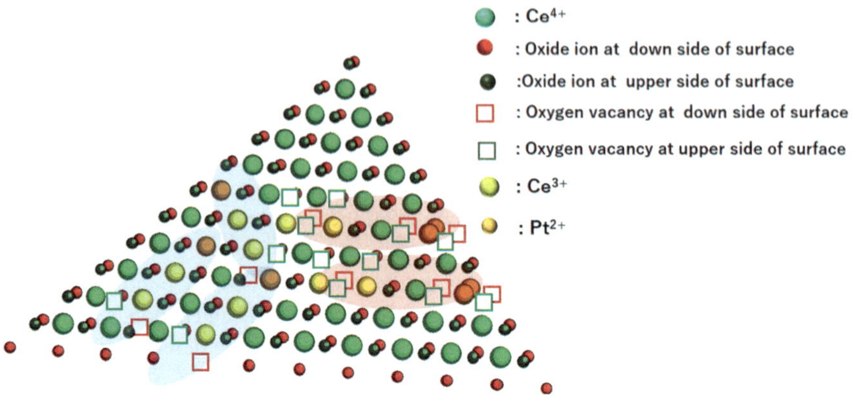

Fig. 84 Schematic diagram of the defect cluster created between Pt and CeO_x NW based on the atomistic simulation and first-principles calculation results on (111) CeO_2 surface. The region painted with a light blue background contains clusters formed by Schottky-type defects, while the region painted with a light red background represents the area where the Pt^{2+} defect cluster and Ce^{3+} defect cluster combine

changes towards ($2E_a + E_f$) (approximately 3.6 eV) and ($3E_a + E_f$) (approximately 4.2 eV). For more details, refer to reference No. [295].

The authors focused on that the length of the water molecule (approximately 11 angstrom) is the same as the length (approx. 11 angstrom) of the Schottky ($2Ce_{Ce}{}'Pt_{Ce}{}''2V_O{}^{\bullet\bullet}$) defect clusters and combined cluster between Pt^{2+} defect cluster and Ce^{3+} defect cluster, and performed the present first-principles calculation. As a result, our first-principles calculations suggest that the affinity of water molecules increases with proceeding defect cluster formation and an electrical double layer (EDL) is more likely to form on the interfacial layer formed between Pt and CeO_x. Since it is generally known that all electrode reactions proceed via EDL, it is noteworthy that the present first-principles calculation also suggests that the continuous formation of EDLs of sufficient length is important for improving electrocatalytic activity on Pt-CeO_x electrodes (i.e., cathode and anode).

Based on the results of the aforementioned calculations and considerations, the key points for activity enhancement on Pt-CeO_x electrodes are summarized using Fig. 85 as follows.

In the case of Pt-loaded CeO_x nanoparticles, the formation of the EDL at the interfacial layer between Pt and CeO_x nanoparticles would be limited and its length would be also very short length because of a remaining CeO_x on the Pt surface. On the other hand, in Pt-loaded CeO_x NWs, defect clusters formed in the interfacial layer between Pt and CeO_x NWs are expected to combine with each other and form over a relatively wide area, so the EDL length on the Pt-loaded CeO_x NWs is expected to be longer. Eventually, the Pt nano-coated layer on top of the CeO_x NWs would have enabled the formation of a continuous and extensive Pt-CeO_x interfacial layer and EDLs on top of Pt-CeO_x electrodes, maximizing the activity of the Pt-CeO_x electrodes, refer to Fig. 85.

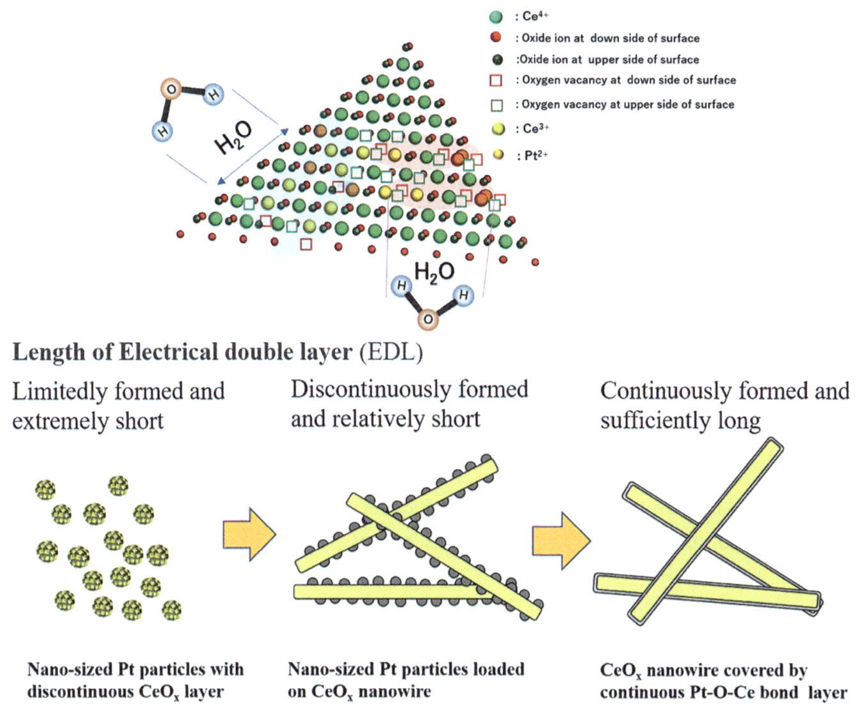

Fig. 85 Schematic illustration of relationship among defect cluster formation, EDL length, and formation of interfacial layer between Pt and CeO_x

As described so far in the present chapter, the design of high-quality Pt-CeO_x electrodes that leads to improved fuel cell performance is thought to be the result of a research style that consistently integrates synthesis, microanalysis, and calculation, and this kind of multidisciplinary research will become increasingly important in fuel cell research and development after the present work.

4 Summary and Future Prospects

The interactive research program which combines novel fabrication, microanalysis, and modeling for the modification of defect interface in/on SOFC components such as electrolyte and anode are introduced using the representative results observed for the rare-earth-doped ceria, stabilized zirconia and brownmillerite based oxides. This is in combination with designing interfaces in the solid electrolyte, solid electrolyte/ anode, and anode layer of single cells. Micro-domains with ordered structure of oxygen vacancies were formed during a slow cooling process in fabrication.

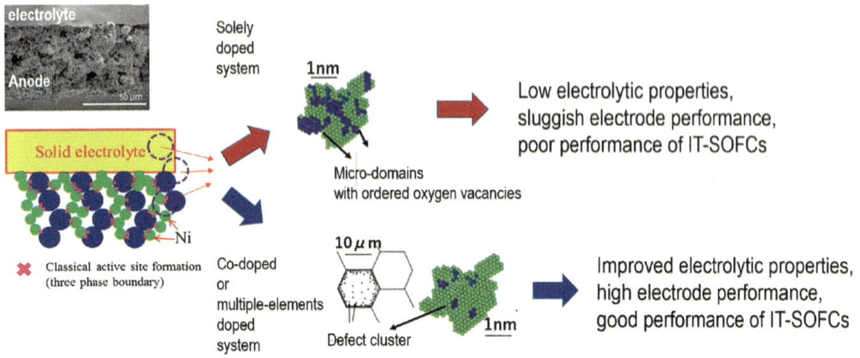

Fig. 86 Schematic diagram of influence of local ordering of oxygen vacancies on properties of fuel cell materials and devices

Namely, the short-range ordered structures of oxygen vacancies are formed as non-equilibrium lattice defects in the interfaces in fuel cell materials. In the solely doped systems, dopant cation is likely to locate in the outer region of defect cluster (i.e., segregation phenomenon). Then, micro-domain regions are easily developed in the interface regions. In contrast, the local ordering of oxygen vacancies is suppressed by formation of small defect clusters. Our first-principles calculation for stabilized ZrO_2 system predicted that small cluster formation by co-doping will promote the delocalization of electron around dopant and disordering of oxygen vacancies at near side of three-phase boundary area of SOFC device. Then, the properties of fuel cell materials and devices are improved in co-doped or multiple-elements doped systems, refer to Fig. 86.

To fabricate the 'state-of-the-art IT-SOFC' devices, the authors recommend the combination of microanalysis and surface/ bulk atomistic simulation with empirical potential at first. It is because the surface/ bulk atomistic simulation can cover the wide region at defect interface of fuel cell materials. This combination work is a first step as a useful support for the conclusion of microanalysis data. Then, we will take the opportunity to modify the fabrication route for tuning the interface structure in the fuel cell materials and devices. After the modified interfaces were fabricated, the combination work of careful microanalysis and first-principles calculation without empirical potential would provide us with useful information about optimized fabrication conditions for final tuning of key interfaces in the fuel cell materials. Let us put it another way, when we observe the defect interface using two eyes such as microanalysis and atomistic simulation (i.e., modeling), we have good opportunity to modify the defect interfaces in fuel cell materials and can improve the performance and its stability. Based on all experimental results, we successfully modified interface defect structure and improved the performance of IT-SOFCs using the interactive combination work of the amalgamation of fabrication, microanalysis, and modeling.

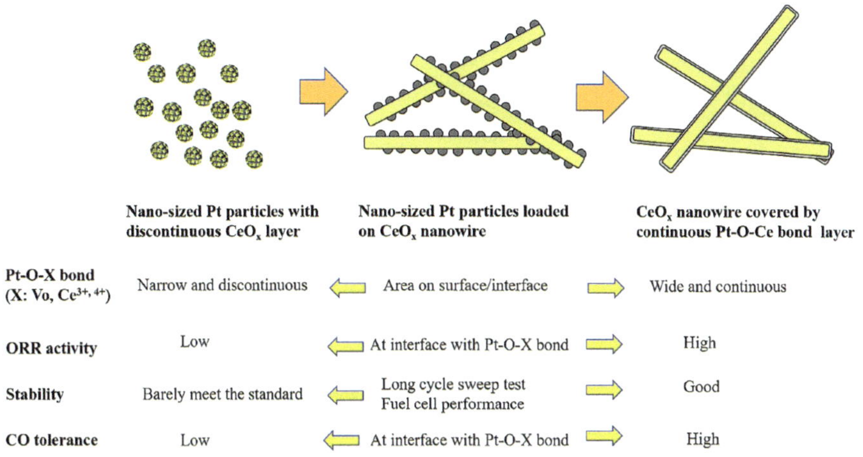

Fig. 87 Schematic diagram of effect of Pt-O-Ce bond at interface between Pt and CeO_x on electrode properties and fuel cell performance

To fabricate the 'state-of-the-art PEMFCs', the authors attempted to maximize the interaction between Pt and CeO_x support. The performance of cathode and anode in PEMFCs was improved by formation of Pt-O-Ce bond on the surface and interface of electrodes in PEMFCs, refer to Fig. 87.

In this challenge, the interactive combination of fabrication, microanalysis and modeling was useful for modifying the interfaces in PEMFCs as well as the modification of interfaces in IT-SOFCs.

Based on all experimental data produced by our team, the authors would like to foresee on future prospect at end of our chapter. As we described in the present chapter, step by step modification of defect interface in fuel cell materials and devices using interactive combination of fabrication, microanalysis, and modeling is required for fabrication of high-quality fuel cell materials and devices which can maintain the favorable balance between performance and durability, refer to Fig. 88. In addition, the authors would like to highlight one more important point for readers. Nowadays, combination work of artificial intelligence (AI) and big databases is a popular R & D field. This combination work minimizes the number of trial-and-error experiments for scientists and engineers in many fields using this deep learning process, refer to Fig. 89.

Certainly, lots of new compositions which show high-quality performance have been reported in the materials science field. This combination work would be useful for R & D in the fuel cell field. However, fuel cell reactions at interfaces in fuel cell devices are complicated as compared to other devices such as power electronics devices, thermoelectric devices, secondary battery, and so on. It is unlikely that the use of conventional combination of AI and big database for the development of high-quality fuel cells will be successful. In R & D of fuel cells, we must be aware of the inherent weakness of combining of AI and database interrogation. That is

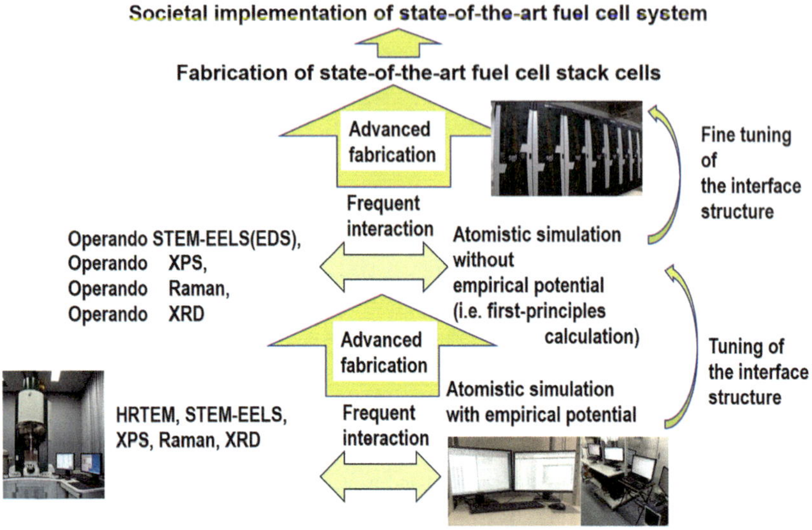

Fig. 88 Schematic diagram about key steps in interactive combination work among fabrication, microanalysis, and modeling for social implementation of high-quality IT-SOFCs

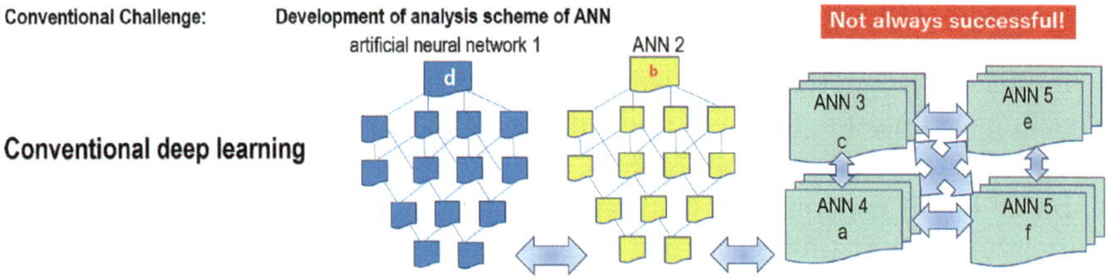

Fig. 89 Schematic diagram of basic concept of deep learning process

there are many null-sets (i.e., empty sets) associated with material science data sets. When important initial data is not in any database, AI cannot start a useful deep learning process at all. In contrast, when we input data of a new design paradigm for fabrication of fuel cells into a null-set of conventional big databases using our proposed interactive combination work, the authors expect that our new data will play the role of an artificial spinal cord function and wake up the artificial brain (i.e., activation of AI) in our fuel cell field. Then, new deep learning process will be developed in R & D field of fuel cells, refer to Fig. 90.

If the readers have the opportunity to access the AI and have relevant information in a database, we foresee those new radical innovations for fabrication of 'state-of-the-art fuel cells' with the quality exceeding the limit of conventional fuel cells will develop in the R & D field. Also, the social implementation of fuel cell will be rapidly accelerated in the near future, refer to Fig. 90.

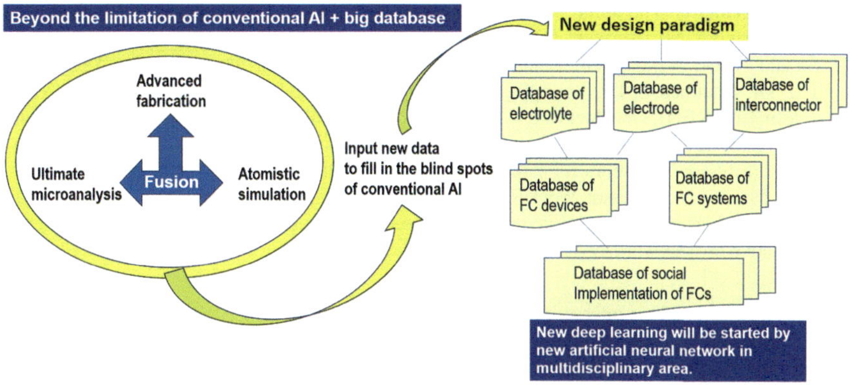

Fig. 90 Schematic diagram of new deep learning process in the fuel cell field

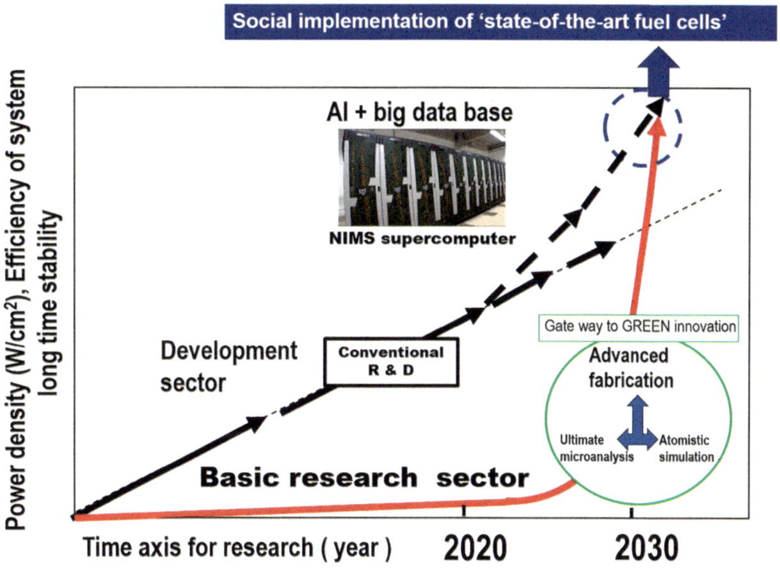

Fig. 91 Future prospect of fuel cell project through our proposed interactive combination

The authors believe that readers of the present chapter have a key to open the door of radical innovation which is demonstrated in Fig. 91. Especially, young readers who have high flexibility research minds and a high level of challenging spirit in the new research, have a great opportunity to progress the radical innovation. We really expect that the young generation of material scientists will attempt to start this challenge after reading of this chapter and obtain significant success in the near future.

Acknowledgements The authors appreciated big support works of Mr. Suga H, Mr. Kobayashi T (Saitama Univ.), Mr. Togasaki H, Ms. Wu YY, Mr. Dendo T, Ms. Isaka N, Mr. Kurashima K, Mr. Suzuki A, Dr. Wang Y, Dr. Takahashi M, Dr. Buchanan R, Dr. Fugane K, Dr. Yang M, Dr. Smid B, Dr. Chauhan S, Dr. Rednyk A, Dr. Tong Ke, Dr. Li JG, Dr. Yan PF, Dr. Li ZM, Dr. Kobayashi T (RIKEN), Dr. Maekawa T, Dr. Ueda S, Dr. Yoshikawa H, Dr. Masuda T, Dr. Noguchi H, Dr. Ikegami T, Professor Kobayashi H, Professor Yamamura H, Professor Atake T, Professor Uosaki K, Professor Nakamura J, Professor Cormack A, Professor Lee JH, Professor Matolinova I, Professor Matolin V, Professor Wepf R and all administration and technical support section staffs of Center for Microscopy and Microanalysis of the University of Queensland for development of our interactive project works.

Funding Supports This interactive project is partially supported by Global Research Center for Environment and Energy based on Nanomaterials Science (GREEN, from FY2009 to FY2018), bilateral joint research projects between Japan and Australia (FY2003-05, FY2005-07, FY2007-09, FY2012-14, and FY2017-19), Grant-in-Aid for Scientific Research (No. 15510084, No. 16079210, No. 19310062, No. 19651038, No. 22310053, No. 25281066, and No. 16H02981) and NIMS Joint Research Hub Program (from FY2018 to FY2020).

References

1. Ota K (2005) Principle of fuel cells for application (review). GS Yuasa Techn Rep 2(1):1–5 (in Japanese)
2. Chang YM, Birnie D III, Kingery WD (1996) Chapter 3 'diffusion as a thermally activated process'. In: Chang YM, Birnie D III, Kingery WD (eds) Physical ceramics, principles for ceramic science and engineering. John Wiley & Sons, Inc. NY, USA, pp 193–195
3. Mori T, Wepf R, Jiang SP (2020) Future prospects for design of 'state-of-the-art' solid oxide fuel cells. J Phys Energy 2:1–9 (article number 031001)
4. Goodenough JB (1978) Solid electrolytes. In: Hagnemuller P, Von Gool W (eds) Academic Press, New York, p 393
5. Kilner JA, Brook RJ (1982) A study of oxygen ion conductivity in doped non-stoichiometric oxides. Solid State Ionics 6:253–259
6. Butler V, Catlow CRA, Fender BEF, Harding JH (1983) Dopant ion radius and ionic conductivity in cerium dioxide. Solid State Ionics 8:109–113
7. Drennan J (1998) The influence of interface structures on the conducting properties of zirconia (ZrO_2)-based solid electrolytes. J Mater Synth Proces 6:181–189
8. (a) Mori T, Drennan J, H.Lee J-H, Li JG, Ikegami T (2002) Oxide ionic conductivity and microstructure of Sm or La doped CeO_2 based system. Solid State Ionics 154: 461–466 (b) Mori T, Drennan J, Wang Y, Lee J H, Li J G, Ikegami T (2003) Electrolytic properties and nanostructural features in the La_2O_3-CeO_2 system. J Electrochem Soc 150:A665–73
9. Mori T, Drennan J, Wang Y, Auchterlonie G, Li JG, Yago A (2003) Influence of nano-structural feature on electrolytic properties in Y_2O_3 doped CeO_2 system. Sci Technol Adv Mat 4:213–220
10. Mori T, Wang Y, Drennan J, Auchterlonie G, Li JG, Ikegami T (2004) Influence of particle morphology on nano-structural feature and conducting property in Sm-doped CeO_2 sintered body. Solid State Ionics 175:641–649
11. Ou DR, Mori T, Ye F, Takahashi M, Zou J, Drennan J (2006) Microstructures and electrolytic properties of yttrium-doped ceria electrolytes: dopant concentration and grain size dependences. Acta Mater 54:3737–3746
12. Mori T, Drennan J (2006) Influence of microstructure on oxide ionic conductivity in doped CeO_2 electrolytes. J Electroceram 17:749–757

13. Catlow CRA (1981) Chapter 2 'Defect clustering in nonstoichiometric oxides. In: Toft Sørensen O (ed) Nonstoichiometric oxides. Academic Press, Cambridge, Massachusetts, USA, pp 61–98
14. Manes L (1981) Chapter 3 'A new method of statistical thermodynamics and its application to oxides of the lanthanide and actinide series. In: Toft Sørensen O (ed) Nonstoichiometric oxides. Academic Press, Cambridge, Massachusetts, USA, pp 99–154
15. Kilner JA, Steele BCH (1981) Chapter 5 'Mass transport in anion-deficient fluorite oxides. In: Toft Sørensen O (ed) Nonstoichiometric oxides. Academic Press, Cambridge, Massachusetts, USA, pp 233–269
16. Kilner JA, Waters CD (1982) The effects of dopant cation–oxygen vacancy complexes on the anion transport properties of non-stoichiometric fluorite oxides. Solid State Ionics 6:253–259
17. Badwal SPS (2006) Ceramic superionic conductor. Mater Sci Tech Ser:568–633
18. Kilner JA (2000) Fast oxygen transport in acceptor doped oxides. Solid State Ionics 129:13–23
19. Badwal SPS, Drennan J (1992) Microstructure/conductivity relationship in the Scandia-zirconia system. Solid State Ionics 53:769–776
20. Badwal SPS, Ciacchi FT, Hannink RHJ (1990) Relationship between phase stability and conductivity of yttria tetragonal zirconia. Solid State Ionics 40(41):882–885
21. Matsui K, Ohgai M (1997) Formation mechanism of hydrous-zirconia particles produced by hydrolysis of $ZrOCl_2$ solutions. J Am Ceram Soc 80:1949–1956
22. Badwal SPS, Drennan J (1987) Grain boundary resistivity and microstructure in zirconia-based oxygen ion conductors. Mater Forum 10:177–183
23. Badwal SPS, Drennan J (1987) Yttria-zirconia: effect of microstructure on conductivity. J Mater Sci 22:3231–3239
24. Badwal SPS, Drennan J, Hughes AE, Sexton BA (1988) A study of impurity phase segregation in fully stabilized yttria-zirconia. Mater Sci Forum 34:195–199
25. Badwal SPS, Drennan J (1989) Grain boundary resistivity in Y-TZP materials as a function of thermal history. J Mater Sci 24:88–96
26. Badwal SPS, Drennan J (1988) The effect of thermal history on the grain boundary resistivity of Y-TZP materials. Solid State Ionics 28–30:1451–1455
27. Butler EP, Drennan J (1982) Microstructure analysis of sintered high-conductivity zirconia with Al_2O_3 additions. J Am Ceram Soc 65:474–478
28. Drennan J, Butler EP (1984) Does alumina act as a grain boundary scavenger in zirconia? Science of Ceramics. In: Proceedings of the 12th international conference, St-Vincent, Italy, vol 12, pp 267–272
29. Rajendran S, Drennan J, Badwal SPS (1987) Effect of alumina additions on the grain boundary and volume resistivity of tetragonal zirconia polycrystals. J Mater Sci Lett 6:1431–1434
30. Badwal SPS, Ciacchi FT, Rajendran S, Drennan J (1998) An investigation of conductivity, microstructure, and stability of electrolyte compositions in the system 9 mol% (Sc_2O_3-Y_2O_3)-ZrO_2 (Al_2O_3). Solid State Ionics 109:167–186
31. Lee J-H, Mori T, Li JG, Ikegami T, Komatsu M, Haneda H (2000) Imaging secondary-ion mass spectroscopy observation of the scavenging of siliceous film from 8-mol%-yttria-stabilized zirconia by the addition of alumina. J Am Ceram Soc 83:1273–1275
32. Lee J-H, Mori T, Li JG, Ikegami T, Komatsu M, Haneda H (2000) The influence of alumina distributions upon scavenging highly resistive grain-boundary phase of 8mol% yttria-stabilized zirconia. Electrochemistry 68:427–432
33. Lee J-H, Mori T, Li JG, Ikegami T, Takenouchi S (2001) Impedance spectroscopic estimation of inter-granular phase distribution in 15mol% calcia-stabilized zirconia/alumina composites. J Eur Ceram Soc 21:13–17
34. Lee J-H, Mori T, Li JG, Ikegami T, Takenouchi S (2001) The influence of alumina addition and its scavenging upon grain-boundary conduction in 15 mol% calcia-stabilized zirconia. Ceram Int 27:269–276
35. Lee J-H, Mori T, Li JG, Ikegami T, Komatsu M, Haneda H (2000) Improvement of grain boundary conductivity of 8mol% yttria stabilized zirconia by precursor scavenging of siliceous phase. J Electrochem Soc 47:2822–2829

36. Lee J-H, Mori T, Li JG, Ikegami T, Drennan J, Kim D-Y (2001) Scavenging of Siliceous Grain-Boundary Phase of 8-mol%-YtterbiaStabilized Zirconia without Additive. J Am Ceram Soc 84:2734–2736
37. (a) Lee J-H, Mori T, Li JG, Ikegami T, Drennan J, Kim DY (2001) Precursor scavenging of the resistive grain-boundary phase in 8 mol% yttria-stabilized zirconia: effect of trace concentration of SiO_2. J Mater Res Soc 16:2377–2383 (b) Mori T, Drennan J, Lee J-H, Li JG, Ikegami T (2002) Improving the ionic conductivity of yttria stabilized zirconia electrolyte materials. Solid State Ionics 154:529–533
38. Lee JH, Mori T, Li JG, Ikegami T, Drennan J, Kim DY (2002) Precursor scavenging of resistive grain boundary phase in 8 mol% ytterbia-stabilized zirconia. J Electrochem Soc 149:J34–J40
39. Steele BCH (2000) Appraisal of $Ce_{1-y}Gd_yO_{2-y/2}$ electrolytes for IT-SOFC operation at 500 °C'. Solid State Ionics 129:95–110
40. Mori T, Drennan J, Wang Y, Auchterlonie G, Li JG (2004) Influence of nano-structural features on electrolytic properties of Gd doped CeO_2 solid electrolytes. J Ceram Soc Jpn 112:S642–S648
41. Mori T, Drennan J (2013) Advanced materials for fuel cells (section No.60). In: Kauffman J, Lee KM (eds) Handbook of sustainable engineering. Springer Publishing Company, NY 10036, pp 1175–87. ISBN: 978-1-4020-8940-4
42. Li F, Ohkubo T, Chen YM, Kodzuka M, Ye F, Ou DR, Mori T, Hono K (2010) Laser-assisted three-dimensional atom probe analysis of dopant distribution in Gd-doped CeO_2. Scr Mater 63:332–335
43. Ye F, Mori T, Ou DR, Takahashi M, Zou J, Drennan J (2007) Ionic conductivities and microstructures of ytterbium doped ceria. J Electrochem Soc 154:B180–B185
44. Blumenthal RN, Brugner FS, Garnier JE (1973) The electrical conductivity of CaO-doped nonstoichiometric cerium dioxide from 700 °C to 1500 °C. J Electrochem Soc 120:1230–1237
45. Tuller HL, Nowick AS (1975) Doped ceria as a solid oxide electrolyte. J Electrochem Soc 122:255–259
46. Kudo T, Obayashi H (1976) Mixed electrical conduction in the fluorite-type $Ce_{1-x}Gd_xO_{2-x/2}$. J Electrochem Soc 123:415–420
47. Mori T, Yamamura H, Sato S (1996) Preparation of alkaline elements doped CeO_2-Sm_2O_3 system and its operation properties as the electrolyte in planar SOFC. J Am Cera Soc 79:3309–3312
48. Mori T, Yamamura H (1998) Preparation of alkali-elements or alkali-earth-element-doped CeO_2-Sm_2O_3 system and its operation properties as the electrolyte in planar solid oxide fuel cells'. J Mater Synth Proces 6:175–179
49. Mori T, Yamamura H, Atake T (1999) Improvement of electrical conductivity in fluorite related Y_2O_3 and fluorite CeO_2 systems based on a unique effective index. J Therm Anal Calorim 57:831–838
50. Mori T, Ikegami T, Yamamura H (1999) Application of a crystallographic index for improvement of the electrolytic properties of CeO_2-Sm_2O_3 system. J Electrochem Soc 146:4380–4385
51. Yamamura H, Katoh E, Ichikawa M, Kakinuma K, Mori T, Haneda H (2000) Multiple doping effect on the electrical conductivity in the $(Ce_{1-x-y}La_xM_y)O_{2-\delta}$ (M=Ca, Sr) system. Electrochemistry 68:455–459
52. Mori T, Lee J-H, Li JG, Ikegami T, Auchterlonie G, Drennan J (2001) Improvement of the electrolytic properties of Y_2O_3 based materials using a crystallographic index. Solid State Ionics 138:277–291
53. Yamamura H, Nishino H, Kakinuma K, Nomura K (2003) Electrical conductivity anomaly around fluorite-pyrochlore phase boundary. Solid State Ionics 158:359–365
54. Yamamura H, Utsunomiya N, Mori T, Atake T (1998) Electrical conductivity in the system ZrO_2-Y_2O_3-Sc_2O_3. Solid State Ionics 107:185–189
55. Murray EP, Tsai T, Barnett SA (1998) Oxygen transfer processes in $(La,Sr)MnO_3/Y_2O_3$-stabilized ZrO_2 cathodes: an impedance spectroscopy study. J Mater Sci:235–243

56. Jiang SP (2006) Activation, microstructure, and polarization of solid oxide fuel cell cathodes. J Solid State Chem 11:93–102
57. Jiang SP (2008) Development of lanthanum strontium manganite perovskite cathode materials of solid oxide fuel cells: a review. J Mater Sci 43:6799–6833
58. Vasylechko L, Matkovski A, Suchocki A, Savytskii D, Syvorotka I (1999) Crystal structure of $LaGaO_3$ and (La, Gd)GaO_3 solid solutions. J Alloy Comp 286:213–218
59. Ishihara T, Matsuda H, Takita Y (1994) Doped $LaGaO_3$ perovskite type oxide as a new oxide ionic conductor. J Am Ceram Soc 116:3801–3803
60. Feng M, Goodenough JB (1994) A superior oxide-ion electrolyte. Eur J Sol State Inor 31:663–672
61. Slater PR, Irvine JTS, Ishihara T, Takita Y (1998) High-temperature powder neutron diffraction study of the oxide ion conductor $La_{0.9}Sr_{0.1}Ga_{0.8}Mg_{0.2}O_{2.85}$ 139:135–143
62. Li G, Morniroli JP, Auchterlonie G, Drennan J, D. Jacob D (2009) An efficient approach to characterize pseudo-merohedral twins by precession electron diffraction: application to the $LaGaO_3$ perovskite. Ultramicroscopy 109:1282–1294
63. Auchterlonie G, Nanoanalysis of dopants in novel oxygen ion conductors. Doctoral thesis submitted for the degree of Doctor of Philosophy at the University of Queensland (School of Mechanical and Mining Engineering) in February 2010. https://espace.library.uq.edu.au/view/UQ:253394
64. Morniroli JP, Auchterlonie G, Drennan J, Zou J (2008) Contribution of electron precession to the study of perovskites displaying small symmetry departures from the ideal cubic ABO_3 perovskite: applications to the $LaGaO_3$ and LSGM perovskites. J Microsc 232:7–26
65. Drennan J, Zelizko V, Hay D, Ciacchi FT, Rajendran S, Badwal SPS (1997) Characterization, conductivity and mechanical properties of the oxygen-ion conductor $La_{0.9}Sr_{0.1}Ga_{0.8}Mg_{0.2}O_{3-x}$. J Mater Chem 7:79–83
66. Kimpton J, Randle TH, Drennan J, Auchterlonie G (2001) An investigation of the microstructure/electrical conductivity relationship in In_2O_3-substituted LSGM. Mater Res Bull 36:639–649
67. Kimpton J (2002) Conductivity and microstructural characterisation of doped zirconia-ceria and lanthanum gallate electrolytes for the intermediate-temperature, solid oxide fuel cell. Doctoral thesis submitted for the degree of Doctor of Philosophy at Swinburne University of Technology (School of Engineering and Science) in December 2002. https://researchbank.swinburne.edu.au/file/8d5c4751-f6aa-4662-b00e-0e60e7664294/1/Justin%20Kimpton%20Thesis.pdf. 4.5 conclusions and future work (v), p 226
68. Goodenough JB, Ruiz-Diaz JE, Zhen YS (1990) Oxide-ion conduction in $Ba_2In_2O_5$ and $Ba_3In_2MO_8$ (M=Ce, Hf, or Zr). Solid State Ionics 44:21–31
69. Goodenough JB, Manthiram A, Kuo JF (1993) Oxygen diffusion in perovskite-related oxides. Mater Chem Phys 35:221–224
70. Adler S, Russek S, Reimer J, Fendorf M, Stacy A, Huang Q, Santoro A, Lynn J, Baltisberger J, Werner U (1994) Local structure and oxide-ion motion in defective perovskites. Solid State Ionics 68:193–211
71. Adler S, Baltisberger J, Werner U, Reimer J (1994) Chemical structure and oxygen dynamics in $Ba_2In_2O_5$. J Am Chem Soc 116:675–681
72. Berastegui P, Hull S, Garcia-Garcia FJ, Eriksson S-G (2002) The crystal structures, microstructure and ionic conductivity of $Ba_2In_2O_5$ and $Ba(In_xZr_{1-x})O_{3-x/2}$. J Solid State Chem 164:119–130
73. Yamamura H, Yamada Y, Mori T, Atake T (1998) Order-disorder transition of oxygen vacancy in the brownmillerite system. Solid State Ionics 108:377–381
74. Yamamura H, Hamazaki H, Kakinuma K, Mori T, Haneda H (1999) Order-disorder transition and electrical conductivity of the brownmillerite solid-solutions system $Ba_2(In, M)_2O_5$ (M=Ga, Al). J Korean Phys Soc 35:S200–S204
75. Yao T, Uchimoto Y, Kinuhata M, Inagaki T, Yoshida H (2000) Crystal structure of Ga-doped $Ba_2In_2O_5$ and its oxide ion conductivity. Solid State Ionics 132:189–198

76. Yoshinaga M, Yamaguchi M, Furuya T, Wang S, Hashimoto T (2004) The electrical conductivity and structural phase transitions of cation-substituted $Ba_2In_2O_5$. Solid State Ionics 169:9–13
77. Kochetova NA, Spesivtseva IV, Animitsa IE (2013) Electrical properties of $Ba_2(In_{1-x} Al_x)_2O_5$ solid solutions. Russ J Electrochem 49:176–180
78. Ta TQ, Tsuji T, Yamamura Y (2006) Thermal and electrical properties of $Ba_2In_2O_5$ substituted for In-site by rare earth elements. J Alloys Compd 408:253–256
79. Noirault S, Quarez E, Piffard Y, Joubert O (2009) Water incorporation into the $Ba_2(In_{1-x}M_x)_2O_5$ ($M = Sc^{3+}$ $0 \leq x < 0.5$ and $M = Y^{3+}$ $0 \leq x < 0.35$) system and protonic conduction. Solid State Ionics 180: 1157–63
80. Saito M, Ito S, Takayama T, Yamamura H (2009) Synthesis of new brownmillerite type compounds $Ba_2(Sc, M')_2O_5$ (M'=Lu, Yb, Tm or Er), $Ba_2(Zn_{1/2}, M'_{1/2})_2O_5$ (M'=Ce, Zr or Hf), $Ba_2(Zn_{2/3}, M'_{1/3})_2O_5$ (M'= Ta or Nb), and $Ba_2(Zn_{3/4}W_{1/4})_2O_5$ and their crystal-chemical properties. Trans Mater Res Soc Jpn 33:233–236
81. Manthiram A, Kuo J-F, Goodenough JB (1993) Characterization of oxygen-deficient perovskites as oxide-ion electrolytes. Solid State Ionics 62:225–234
82. Niwa J, Suehiro T, Kishi K, Ikeda S, Maeda M (2003) Structure and electrical characteristics of Ce^{4+}-doped $Ba_2In_2O_5$. J Mater Sci 38:3791–3795
83. Jankovic J, Wilkinson DP, Hui R (2012) Preparation and characterization of Ce- and La-doped $Ba_2In_2O_5$ as candidates for intermediate temperature (100–500 °C) solid proton conductors. J Power Sources 201:49–58
84. Jayaraman V, Magrez A, Caldes M, Joubert O, Ganne M, Piffard Y, Brohan L (2004) Characterization of perovskite systems derived from $Ba_2In_2O_5$: Part I: The oxygen-deficient $Ba_2In_{2(1-x)}Ti_{2x}O_{5+x}\square_{1-x}$ $(0 \leq x \leq 1)$ compounds. Solid State Ionics 170:17–24
85. Jayaraman V, Magrez A, Caldes M, Joubert O, Taulelle F, Rodriguez-Carvajal J, Piffard Y, Brohan L (2004) Characterization of perovskite systems derived from $Ba_2In_2O_5$: Part II: The proton compounds $Ba_2In_{2(1-x)}Ti_{2x}O_{4+2x}(OH)_y$ $[0 \leq x \leq 1;\ Y \leq 2(1-x)]$. Solid State Ionics 170:25–32
86. Rolle A, Seymour H, Roussel P, Rubbens A, Vannier R-N (2008) Characterisation of $Ba_2In_{2-x}Sn_xO_{5+x/2}$ oxide ion conductors. Ionics 14:477–482
87. Shin JF, Apperley DC, Slater PR (2010) Silicon doping in $Ba_2In_2O_5$: Example of a beneficial effect of silicon incorporation on oxide ion/proton conductivity. Chem Mater 22:5945–5948
88. Rolle A, Vannier RN, Giridharan NV, Abraham F (2005) Structural and electrochemical characterization of new oxide ion conductors for oxygen generating systems and fuel cells. Solid State Ionics 176:2095–2103
89. Shimura T, Yogo T (2004) Electrical properties of the tungsten-doped $Ba_2In_2O_5$. Solid State Ionics 175:345–348
90. Kuramochi H, Mori T, Yamamura H, Kobayashi H, Mitamura T (1994) Preparation and conductivity of $Ba_2In_2O_5$ ceramics. J Ceram Soc Jpn 102:1159–1162
91. Kakinuma K, Yamamura H, Haneda H, Atake T (2001) Oxide-ion conductivity of $(Ba_{1-x}La_x)_2In_2O_{5+x}$ system based on brownmillerite structure. Solid State Ionics 140:301–306
92. Kakinuma K, Yamamura H, Haneda H, Atake T (2002) Oxide-ion conductivity of the perovskite-type solid-solution system, $(Ba_{1-x-y}Sr_xLa_y)_2In_2O_{5+y}$. Solid State Ionics 154–155:571–576
93. Yamamura H, Kutsukake N, Kakinuma K, Mori T, Atake T (1999) Electrical conductivity of oxygen-deficient perovskite solid-solution system (Ba, La) (In, Ga) $O_{3-\delta}$. Electrochemistry 67:765–768
94. Kakinuma K, Takahashi N, Yamamura H, Nomura K, Atake T (2004) Electrical conductivity and local distortion of $(Ba_{0.5}La_{0.5})_2In_2O_{5.5}$ doped with divalent or tetravalent cation in in-site. Solid State Ionics 168: 69–74
95. Tarasova NA, Filinkova YV, Animitsa IE (2013) Electric properties of oxyfluorides $Ba_2In_2O_{5-0.5x}F_x$ with brownmillerite structure. Russ J Electrochem 49:45–51

96. Shin JF, Orera A, Apperley DC, Slater PR (2011) Oxyanion doping strategies to enhance the ionic conductivity in $Ba_2In_2O_5$. J Mater Chem 21:874–879
97. Jiang SP, Badwal SPS (1997) Hydrogen oxidation at the nickel and platinum electrodes on yttria-tetragonal zirconia electrolyte. J Electrochem Soc 144:3777–3784
98. Setoguchi T, Okamoto K, Eguchi K, Arai H (1992) Effects of anode material and fuel on anodic reaction of solid oxide fuel-cells. J Electrochem Soc 139:2875–2880
99. Jiang SP, Chan SH (2004) Development of Ni/Y_2O_3-ZrO_2 cermet anodes for solid oxide fuel cells. Mater Sci Technol 20:1109–1118
100. Wang W, Qu JF, Juliao PSB, Shao ZP (2019) Recent advances in the development of anode materials for solid oxide fuel cells utilizing liquid oxygenated hydrocarbon fuels: a mini review. Energy Technol Ger 7: 33–44
101. Finnerty CM, Coe NJ, Cunningham RH, Ormerod RM (1998) Carbon formation on and deactivation of nickel-based/zirconia anodes in solid oxide fuel cells running on methane. Cat Today 46:137–145
102. Mogensen M, Kammer K (2003) Conversion of hydrocarbons in solid oxide fuel cells. Annu Rev Mater Res 33:321–331
103. Wang W, Su C, Wu Y, Ran R, Shao Z (2013) Progress in solid oxide fuel cells with nickel-based anodes operating on methane and related fuels. Chem Rev 113:8104–8151
104. Yue WX, Li YF, Zheng Y, Wu T, Zhao CH, Zhao J, Geng G, Zhang WQ, Chen J, Zhu JX, Yu B (2019) Enhancing coking resistance of Ni/YSZ electrodes: In situ characterization, mechanism research, and surface engineering. Nano Energy 62:64–78
105. Minh NQ, Takahashi T (1995) Reaction of sulfide impurity. In: Science and technology of ceramic fuel cells, Elsevier, Amsterdam, pp 209–210
106. Kurokawa H, Yang L, Jacobson CP, De Jonghe LC, Visco SJ (2007) Y-doped $SrTiO_3$ based sulfur tolerant anode for solid oxide fuel cells. J Power Sources 164:510–518
107. Lee S, Kim G, Vohs JM, Gorte RJ (2008) SOFC anodes based on infiltration of $La_{0.3}Sr_{0.7}TiO_3$. J Electrochem Soc 155:B1179–B1183
108. Liu Z, Liu B, Ding D, Liu M, Chen F, Xia C (2013) Fabrication and modification of solid oxide fuel cell anodes via wet impregnation/ infiltration technique (review). J Power Sources 237:243–259
109. Rahmanipour M, Pappacena A, Boaro M, Donazzi A (2017) A distributed charge transfer model for IT-SOFCs based on ceria electrolytes. J Electrochem Soc 164:F1249–F1264
110. Liu Y, Shao Z, Mori T, Jiang SP (2020) Development of nickel based cermet anode materials in solid oxide fuel cells–now and future. Mater Rep Energy (in press)
111. Matsui K, Ohgai M (2001) Formation mechanism of hydrous zirconia particles produced by the hydrolysis of $ZrOCl_2$ solutions: III, kinetics study for the nucleation and crystal-growth processes of primary particles. J Am Ceram Soc 84:2303–2312
112. Li JG, Ikegami T, Mori T, Wada T (2001) Reactive $Ce_{0.8}Re_{0.2}O_{1.9}$(Re = La, Nd, Sm, Gd, Dy, Y, Ho, Er, and Y) powders via carbonate precipitation: 1. synthesis and characterization. Chem Mater 13:2913–2920
113. Wang Y, Mori T, Li JG, Yajima Y (2003) Low-temperature fabrication and electrical property of 10 mol% Sm_2O_3-doped CeO_2 ceramics. Sci Technol Adv Mat 4:229–238
114. Wang Y, Mori T, Li JG, Yajima Y, Drennan J (2006) Synthesis, characterization and sinterablity of 10mol% Sm_2O_3-doped CeO_2 nanopowders via carbonate precipitation. J Eur Ceram Soc 26:417–422
115. Li JG, Wang Y, Ikegami T, Mori T, Ishigaki T (2005) Reactive $Ce_{0.8}RE_{0.2}O_{1.9}$ (RE = Gd and Sm) nano-powders: synthesis via homogeneous precipitation, characterization, and low-temperature sintering into dense ceramics. Mat Sci Eng B 121:54–59
116. Li JG, Ikegami T, Wang Y, Mori T (2003) 10mol%-Gd_2O_3-doped CeO_2 solid solutions via carbonate coprecipitation: a comparative study. J Am Ceram Soc 86:915–921
117. Wang Y, Mori T, Drennan J, Li JG, Yajima Y (2004) Low temperature synthesis of 10mol% Gd_2O_3-doped CeO_2 ceramics and its characterization. J Ceram Soc Jpn 112:S41–S45
118. Wang Y, Mori T, Li JG, Ikegami T, Yajima Y (2003) Low-temperature preparation of dense 10mol%-Y_2O_3-doped CeO_2 ceramics using powders synthesized via carbonate coprecipitation. J Mater Res 18:1239–1246

119. Li JG, Ikegami T, Wang Y, Mori T (2002) Nano-crystalline $Ce_{1-x}Y_xO_{2-x/2}$ ($0 \leq x \leq 0.35$) oxides via carbonate precipitation: synthesis and characterization. J Solid State Chem 168:52–59
120. Ye F, Mori T, Ou DR, Zou J, Drennan J (2007) Microstructural characterization of terbium doped ceria. Mater Res Bull 42:943–949
121. Ye F, Mori T, Ou DR, Zou J, Auchterlonie G, Drennan J (2007) Compositional and valent state in-homogeneities and ordering of oxygen vacancies in terbium doped ceria. J Appl Phys 101:1–5 (article number: 113528)
122. Ye F, Mori T, Ou DR (2007) Synthesis and microstructural characterization of $Ce_{1-x}Tb_xO_{2-\delta}$ ($0<x<1$) nano-powders. J Nanosci Nanotechno 7:2521–2525
123. Wang Y, Mori T, Li JG, Drennan J (2005) Synthesis, characterization, and electrical conduction of 10 mol% Dy_2O_3-doped CeO_2 ceramics. J Eur Ceram Soc 25:949–956
124. Ou DR, Mori T, Ye F, Zou J, Auchterlonie G, Drennan J (2007) Microstructural inhomogeneity in holmium-doped ceria and its influence on the ionic conduction. J Electrochem Soc 154:B616–B622
125. Wang Y, Mori T, Li JG (2002) Low temperature synthesis of praseodymium doped ceria nanopowders. J Am Ceram Soc 85:3105–3107
126. Jiang SP, Zhang JP, Zheng XG (2002) A comparative investigation of chromium deposition at air electrodes of solid oxide fuel cells. J Eur Ceram Soc 22:361–373
127. Ou DR, Mori T, Ye F, Miyayama M, Nakayama S, Zou J, Auchterlonie G, Drennan J (2009) Microstructural characteristics of SDC electrolyte film supported by Ni–SDC cermet anode. J Electrochem Soc 156:B825–B830
128. Jiang SP, Callus PJ, Badwal SPS (2000) Fabrication and performance of Ni/ 3 mol% Y_2O_3-ZrO_2 cermet anodes for solid oxide fuel cells. Solid State Ionics 132:1–14
129. Jiang SP, Chan SH (2004) Development of Ni/ Y_2O_3–ZrO_2 cermet anodes for solid oxide fuel cells. Mater Sci Tech Ser 20:1109–1118
130. Souza S, Visco SJ, DeJonghe LC (1997) Thin-film solid oxide fuel cell with high performance at low temperature. Solid State Ionics 98:57–61
131. Kim JW, Virkar AV, Fung KZ, Mehta K, Singhal SC (1999) Polarization effects in intermediate temperature, anode-supported solid oxide fuel cells. J Electrochem Soc 146:69–78
132. Zhao F, Virkar AV (2005) Dependence of polarization in anode-supported solid oxide fuel cells on various cell parameters. J Power Sources 141:79–95
133. Virkar AV, Chen J, Tanner CW, Kim JW (2000) The role of electrode microstructure on activation and concentration polarizations in solid oxide fuel cells. Solid State Ionics 131:189–198
134. Lin Y, Zhan Z, Liu J, Barnett SA (2005) Direct operation of solid oxide fuel cells with methane fuel. Solid State Ionics 176:1827–1835
135. Wilson JR, Kobsiriphat W, Mendoza R, Chen HY, Hiller JM, Miller DJ, Thornton K, Voorhees PW, Adler SB, Barnett SA (2006) Three-dimensional reconstruction of a solid-oxide fuel-cell anode. Nat Mater 5:541–544
136. Wilson JR, Barnett SA (2008) Solid oxide fuel cell Ni-YSZ anodes: effect of composition on microstructure and performance. Electrochem Solid ST 11:B181–B185
137. Wilson JR, Gameiro M, Mischaikow K, Kalies W, Voorhees PW, Barnett SA (2009) Three-dimensional analysis of solid oxide fuel cell Ni-YSZ anode interconnectivity. Microsc Microanal 15:71–77
138. Tanner CW, Fung KZ, Virkar AV (1997) The Effect of porous composite electrode structure on solid oxide fuel cell performance: part I—theoretical analysis. J Electrochem Soc 144:21–30
139. Abdalla AM, Hossain S, Azad AT, Petra PMI, Begum F, Eriksson SG, Azad AK (2018) Nanomaterials for solid oxide fuel cells: a review. Renew Sust Energ Rev 82:353–368
140. Jiang SP (2012) Nanoscale and nano-structured electrodes of solid oxide fuel cells by infiltration: advances and challenges. Int J Hydrog Energy 37:449–470
141. Vohs JM, Gorte RJ (2009) High-performance SOFC cathodes prepared by infiltration. Adv Mater 21:943–956
142. Keyvanfar P, Birss V (2013) Optimization of infiltration techniques used to construct Ni/YSZ anodes. ECS Trans 57:1627–1638

143. Buyukaksoy A, Kammampata SP, Birss VI (2015) Effect of porous YSZ scaffold microstructure on the long-term performance of infiltrated Ni-YSZ anodes. J Power Sources 287:349–358
144. Ideris A, Croiset E, Pritzker M, Amin A (2017) Direct-methane solid oxide fuel cell (SOFC) with Ni-SDC anode-supported cell. Int J Hydrogen Energ 42:23118–23129
145. Wang Z, Li Y, Schwank JW (2014) Evaluation of Ni/SDC as anode material for dry CH_4 fueled solid oxide fuel cells. J Power Sources 248:239–245
146. Yoon D, Manthiram A (2014) Hydrocarbon-fueled solid oxide fuel cells with surface-modified, hydroxylated Sn/Ni–$Ce_{0.8}Gd_{0.2}O_{1.9}$ heterogeneous catalyst anode. J Mater Chem A 2:17041–17406
147. Sumi H, Lee Y-H, Muroyama H, Matsui T, Eguchi K (2010) Comparison between internal steam and CO_2 reforming of methane for Ni-YSZ and Ni-ScSZ SOFC anodes. J Electrochem Soc 157:B1118–B1125
148. Wang W, Jiang SP, Tok AIY, Luo L (2006) GDC-impregnated Ni anodes for direct utilization of methane in solid oxide fuel cells. J Power Sources 159:68–72
149. Chen XJ, Khor KA, Chan SH (2005) Suppression of carbon deposition at CeO_2-modified Ni/YSZ anodes in weakly humidified CH_4 at 850 °C. Electrochem Solid St 8:A79–82
150. Trovarelli A (1996) Catalytic properties of ceria and CeO_2-containing materials. Cataly Rev 38:439–520
151. Springer TE, Zowodzinski TA, Gottesfeld S (1991) Polymer electrolyte fuel cell model. J Electrochem Soc 138:2334–2342
152. Kordesch KV, Simader GR (1995) Environmental impact of fuel cell technology. Chem Rev 95:191–207
153. Ahluwalia RK, Wang X (2005) Direct hydrogen fuel cell systems for hybrid vehicles. J Power Sources 139:152–164
154. Kerres JA (2001) Development of ionomer membranes for fuel cells. J Membrane Sci 185:3–27
155. Zhang H, Shen PK (2012) Recent development of polymer electrolyte membranes for fuel cells. Chem Rev 112:2780–2832
156. Nakajima H, Nomura S, Sugimoto T, Nishikawa S, Honma I (2002) High temperature proton conducting organic-inorganic nanohybrids for polymer electrolyte membrane, part II. J Electrochem Soc 149:A953–A959
157. Honma I, Nakajima H, Nishikawa O, Sugimoto T, Nomura S (2003) Organic/inorganic nanocomposites for high temperature proton conducting polymer electrolytes. Solid State Ionics 162–163:237–245
158. Asano N, Aoki M, Suzuki S, Miyatake K, Uchida H, Watanabe M (2006) Aliphatic/aromatic polyimide ionomers as a proton conductive membrane for fuel cell applications. J Am Chem Soc 128:1762–1769
159. Shao Z-G, Joghee P, Hsing I-M (2004) Preparation and characterization of hybrid Nafion–silica membrane doped with phosphotungstic acid for high temperature operation of proton exchange membrane fuel cells. J Membrane Sci 229:43–51
160. Bae B, Yoda T, Miyatake K, Uchida H, Watanabe M (2010) Proton-conductive aromatic ionomers containing highly sulfonated blocks for high-temperature-operable fuel cells. Angew Chem Int Ed 49:317–320
161. Edwards RL, Demuren A (2019) Interface model of PEM fuel cell membrane steady-state behavior. Int J Energy Environ Eng 10:85–160
162. Yoo E, Okata T, Akita T, Kohyama M, Nakamura J, Honma I (2009) Enhanced electrocatalytic activity of Pt subnanoclusters on graphene nanosheet surface. Nano Lett 9:2255–2259
163. Yano H, Watanabe M, Liyama A, Uchida H (2016) Particle-size effect of Pt cathode catalysts on durability in fuel cells. Nano Energy 29:323–333
164. Watanabe M, Motoo S (1975) Electrocatalysis by ad-atoms part III enhancement of the oxidation of carbon monoxide on platinum by ruthenium ad-atoms. J Electroanal Chem 60:275–283
165. Watanabe M, Uchida M, Motoo S (1987) Preparation of highly dispersed Pt + Ru alloy clusters and the activity for the electrooxidation of methanol. J Electroanal Chem 229:395–406

166. Costamagna P, Srinivasan S (2001) Quantum jumps in the PEMFC science and technology from the 1960s to the year 2000: part I fundamental scientific aspects. J Power Sources 102:242–252
167. Antolini E (2003) Formation of carbon-supported PtM alloys for low temperature fuel cells: a review. Mater Chem Phys 78:563–573
168. Antolini E (2004) Recent developments in polymer electrolyte fuel cell electrodes. J Appl Electrochem 34:563–576
169. Oetjen HF, Schmid VM, Stimming U, Trila F (1996) Performance data of a proton exchange membrane fuel cell using H_2/CO as fuel gas. J Electrochem Soc 143:3838–3842
170. Divisek J, Oetjen HF, Peinecke V, Schmidt VM, Stimming U (1998) Components for PEM fuel cell systems usinghydrogen and CO containing fuels. Electrochim Acta 43:3811–3815
171. Gasteiger HA, Panels JE, Yan SG (2004) Dependence of PEM fuel cell performance on catalyst loading. J Power Sources 127:162–171
172. Watanabe M, Zhu Y, Igarashi H, Uchida H (2000) Mechanism of CO tolerance at Pt-alloy anode catalysts for polymer electrolyte fuel cells. Electrochemistry 68:244–251
173. Wakisaka M, Mitsui S, Hirose Y, Kawashima K, Uchida H, Watanabe M (2006) Electronic structures of Pt-Co and Pt-Ru alloys for CO-tolerant anode catalysts in polymer electrolyte fuel cells studied by EC-XPS. J Phys Chem B 110:23489–23496
174. Shin GY, Yano H, Tryk DA, Watanabe M, Iiyama A, Uchida H (2016) A novel Pt-Co alloy hydrogen anode catalyst with superlative activity, cO-tolerance and robustness. Nanoscale 8:13893–13897
175. Samjeske G, Wang H, Loffler T, Baltruschat H (2002) CO and methanol oxidation at Pt-electrodes modified by Mo. Electrochim Acta 47:3681–3692
176. Götz M, Wendt H (1998) Binary and ternary anode catalyst formulations including the elements W, Sn and Mo for PEMFCs operated on methanol or reformate gas. Electrochim Acta 43:3637–3644
177. Papageorgopoulos DC, Keijzer M, de Bruijn FA (2002) The inclusion of Mo, Nb and Ta in Pt and PtRu carbon supported electrocatalysts in the quest for improved CO tolerant PEMFC anodes. Electrochim Acta 48:197–204
178. Rivera EC, Volpe DJ, Alden L, Lind C, Downie C, Alvarez TV, Angelo ACD, DiSalvo FJ, Abruna HD (2004) Electrocatalytic activity of ordered intermetallic phases for fuel cell applications. J Am Chem Soc 126:4043–4049
179. Liu Y, Li DG, Stamenkovic VR, Soled S, Henao JD, Sun SH (2011) Synthesis of Pt_3Sn alloy nanoparticles and their catalysis for electro-oxidation of CO and methanol. ACS Catal 1:1719–1723
180. Park KW, Choi JH, Kwon BK, Lee SA, Sung YE, Ha HY, Hong SA, Kim H, Wieckowski A (2002) Chemical and electronic effects of Ni in Pt/Ni and Pt/Ru/Ni alloy nanoparticles in methanol electrooxidation. J Phys Chem B 106:1869–1877
181. Deivaraj TC, Chen W, Lee JY (2003) Preparation of PtNi nanoparticles for the electrocatalytic oxidation of methanol. J Mater Chem 13:2555–2560
182. Park KW, Choi JH, Kwon BK, Lee SA, Pak C, Chang H, Sung YE (2004) PtRuRhNi nanoparticle electrocatalyst for methanol electrooxidation in direct methanol fuel cell. J Catal 224:236–242
183. Moghaddam RB, Pickup PG (2012) Support effects on the oxidation of ethanol at Pt nanoparticles. Electrochim Acta 65:210–215
184. Colón-Mercado HR, Kim H, Popov BN (2004) Durability study of Pt_3Ni_1 catalysts as cathode in PEM fuel cell. Electrochem Commun 6:795–799
185. Pietron JI, Pomfret MB, Chervin CN, Long JW, Rolison DR (2012) Direct methanol oxidation at low overpotentials using Pt nanoparticles electrodeposited at ultrathin conductive RuO_2 nanoskins. J Mater Chem 22:5197–5204
186. Kang Y, Murray CB (2010) Synthesis and electrocatalytic properties of cubic Mn−Pt nanocrystals (nanocubes). J Am Chem Soc 132:7568–7569
187. Zhang H, Wang Y, Fachini ER, Cabrera CR (1999) Electrochemically codeposited platinum/molybdenum oxide electrode for catalytic oxidation of methanol in acid solution. Electrochem Solid-State Lett 2:437–439

188. Kang Y, Pyo JB, Ye XC, Gordon TR, Murray CB (2012) Synthesis, shape control, and methanol electro-oxidation properties of Pt–Zn alloy and Pt_3Zn intermetallic nanocrystals. ACS Nano 6:5642–5647
189. Shen P, Chen K, Tseung ACC (1995) CO oxidation on Pt-Ru / WO_3 electrodes. J Electrochem Soc 142:L85–L86
190. Amin RS, Abdel Hameed RM, El-Khatib KM, Elsayed Youssef M, Elzatahry AA (2012) Pt-NiO/C anode electrocatalysts for direct methanol fuel cells. Electrochim Acta 59:499–508
191. Lin CT, Huang HJ, Yang J Jr, Shiao MH (2011) A simple fabrication process of Pt-TiO_2 hybrid electrode for photo-assisted methanol fuel cells. Microelectron Eng 88:2644–2646
192. Konopka DA, Li M, Artyushkova K, Marinkovic N, Sasaki K, Adzic R, Ward TL, Atanasov P (2011) Platinum supported on $NbRu_yO_z$ as electrocatalyst for ethanol oxidation in acid and alkaline fuel cells. J Phys Chem C 115:3043–3056
193. Xu C, Shen PK (2004) Novel Pt/CeO_2/C catalysts for electrooxidation of alcohols in alkaline media. Chem Commun:2238–2239
194. Xu C, Shen PK (2005) Electrochemical oxidation of ethanol on Pt-CeO_2/C catalysts. J Power Sources 142:27–29
195. Xu C, Zeng R, Shen PK, Wei Z (2005) Synergistic effect of CeO_2 modified Pt/C catalysts on the alcohols oxidation. Electrochim Acta 51:1031–1035
196. Takahashi M, Mori T, Kobayashi H, Drennan J, Ou DR, Nishimura C (2005) Preparation and characterization of nano-hetero Pt-pure CeO_2 electrodes supported by carbon materials for direct methanol fuel cells applications. Trans Mater Res Soc Jpn 30:951–954
197. Takahashi M, Mori T, Vinu A, Kobayashi H, Drennan J, Ou DR (2006) Preparation and anode property of Pt-CeO_2 electrodes supported on carbon black for direct methanol fuel cell applications. J Mater Res 21:2314–2322
198. Takahashi M, Mori T, Ye F, Vinu A, Kobayashi H, Drennan J (2007) Design of high quality Pt-CeO_2 composite anodes supported by carbon black for direct methanol fuel cells application. J Am Ceram Soc 90:1291–1294
199. Takahashi M, Mori T, Vinu A, Ou DR, Kobayashi H, Drennan J (2008) Development of high quality Pt-CeO_2 electrodes supported on carbon black for direct methanol fuel cell applications. Adv Appl Ceram 107:57–63
200. Scibioh MA, Kim SK, Cho EA, Lim TH, Hong SA, Ha HY (2008) Pt-CeO_2/C anode catalyst for direct methanol fuel cells. Appl Catal B 84: 773–782
201. Guo DJ, Jing ZH (2010) A novel co-precipitation method for preparation of Pt-CeO_2 composites on multi-walled carbon nanotubes for direct methanol fuel cells. J Power Sources 195:3802–3805
202. Matolin V, Cabala M, Matolinova I, Škoda M, Václavu M, Prince KC, Skála T, Mori T, Yoshikawa H, Yamashita Y, Ueda S, Kobayashi K (2010) Pt and Sn doped sputtered CeO_2 catalysts for fuel cell applications. Fuel Cells 10:139–144
203. Matolin V, Matolinova I, Vaclavu M, Khalakhan I, Vorokhta M, Fiala R, Pis I, Sofer Z, Poltierova-Vejpravova J, Mori T, Potin V, Yoshikawa H, Ueda S, Kobayashi K (2010) Platinum-doped CeO_2 thin film catalysts prepared by magnetron sputtering. Langmuir 26:12824–12831
204. Matolín V, Khalakhan I, Matohínová I, Václavů M, Veltruská K, Vorokhta M (2010) $Pt^{2+,4+}$ ions in CeO_2 rf-sputtered thin films. Surf Interface Anal 42:882–885
205. Zhou Y, Gao Y, Liu Y, Liu J (2010) High efficiency Pt-CeO_2/carbon nanotubes hybrid composite as an anode electrocatalyst for direct methanol fuel cells. J Power Sources 195:1605–1609
206. Ou DR, Mori T, Togasaki H, Takahashi M, Ye F, Drennan J (2011) Microstructures and metal-support interaction of the Pt-CeO_2/C catalysts for direct methanol fuel cell application. Langmuir 27:3859–3866
207. Ou DR, Mori T, Fugane K, Togasaki H, Ye F, Drennan J (2011) Stability of ceria supports in Pt-CeO_x/C catalysts. J Phys Chem C 115:19239–19245
208. Mori T, Ou DR, Zou J, Drennan J (2012) Present status and future prospect of design of Pt-cerium oxide electrodes for fuel cell applications. Prog Nat Sci Mater 22:561–571

209. Tabet-Aoul A, Mohamedi M (2012) Interrelated functionalities of hierarchically CNT/CeO_2/Pt nanostructured layers: Synthesis, characterization, and electroactivity. Phys Chem Chem Phys 14:4463–4474
210. Chauhan S, Richards GJ, Mori T, Yan PF, Hill JP, Ariga K, Zou J, Drennan J (2013) Fabrication of a nano-structured Pt-loaded cerium oxide nanowire and its anode performance in the methanol electro-oxidation reaction. J Mater Chem A 20:6262–6270
211. Mori T, Fugane K, Chauhan S, Ito M, Masuda T, Noguchi H, Uosaki K (2014) Design of Pt-CeO_x hetero-interface on electrodes in polymer electrolyte membrane fuel cells. IOP Conf Ser Mater Sci Eng 1–8; article number: 012010
212. Gong K, Du F, Xia Z, Durstock M, Dai L (2009) Nitrogen-doped carbon nanotube arrays with high electrocatalytic activity for oxygen reduction. Science 323:760–764
213. Qu L, Liu Y, Baek JB, Dai L (2010) Nitrogen-doped graphene as efficient metal-free electrocatalyst for oxygen reduction in fuel cells. ACS Nano 4:1321–1326
214. He F, Chen X, Shen Y, Li Y, Liu A, Liu S, Mori T, Zhang J (2016) Ionic liquid-derived Fe-N/C catalysts for highly efficient oxygen reduction reaction without any supports, templates, and multi-step of pyrolysis. J Mater Chem A 4:6630–6638
215. Guo D, Shibuya R, Akiba C, Saji S, Kondo T, Nakamura J (2016) Active sites of nitrogen-doped carbon materials for oxygen reduction reaction clarified using model catalysts. Science 351:361–365
216. Zhou Z, He F, Shen Y, Chen X, Yang Y, Liu S, Mori T, Zhang YJ (2017) Coupling multiphase-Fe and hierarchical N-doped graphitic carbon as trifunctional electrocatalysts by supramolecular preorganization of precursors. Chem Commun 53:2044–2047
217. Chen X, He F, Shen Y, Yang Y, Mei H, Liu S, Mori T, Zhang YJ (2017) Effect of carbon supports on enhancing mass kinetic current density of Fe-N/C electrocatalysts. Chem Eur J 23:14597–14603
218. Singh SK, Takeyasu K, Nakamura J (2019) Active sites and mechanism of oxygen reduction reaction electrocatalysis on nitrogen-doped carbon materials. Adv Mater 31:1–17 (Article number 1804297)
219. Talapaneni SN, Fugane K, Vinu A, Mori T (2017) Highly efficient electrocatalysis of metal-free, graphitic and sustainable nitrogen doped mesoporous carbon towards oxygen reduction reaction. Adv Porous Mater 5:26–35
220. Yang S, Feng X, Wang X, Mullen K (2011) Graphene-based carbon nitride nanosheets as efficient metal-free electrocatalysts for oxygen reduction reactions. Angew Chem Int Ed 50:5339–5343
221. Yang Z, Yao Z, Li G, Fang G, Nie H, Liu Z, Zhou X, Chen X, Huang S (2012) Sulfur-doped graphene as an efficient metal-free cathode catalyst for oxygen reduction. ACS Nano 6:205–211
222. Zhang YJ, Fugane K, Mori T, Niu L, Ye J (2012) Wet chemical synthesis of nitrogen-doped graphene towards oxygen reduction electrocatalysts without high-temperature pyrolysis. J Mater Chem 22:6575–6580
223. Tian GL, Zhang Q, Zhang B, Jin YG, Huang JQ, Su DS, Wei F (2014) Toward full exposure of “active sites”: nanocarbon electrocatalyst with surface enriched nitrogen for superior oxygen reduction and evolution reactivity. Adv Funct Mater 24:5956–5961
224. Shim J, Lee CR, Lee HK, Lee JS, Cairns EJ (2001) Electrochemical characteristics of Pt-WO_3/C and Pt-TiO_2/C electrocatalysts in a polymer electrolyte fuel cell. J Power Sources 102:172–177
225. Baturina OA, Garsany Y, Zega TJ, Stroud RM, Schull T, Swider-Lyonsa KE (2008) Oxygen reduction reaction on platinum/ tantalum oxide electrocatalysts for PEM fuel cells. J Electrochem Soc 155:B1314–B1321
226. Sasaki K, Zhang L, Adzic RR (2008) Niobium oxide-supported platinum ultra-low amount electrocatalysts for oxygen reduction. Phys Chem Chem Phys 10:159–167
227. Takahashi M, Mori T, Yoshikawa H, Togasaki H, Fugane K, Tada A, Matolin V, Drennan J (2008) Influence of Pt and CeO_2 interaction in Pt-CeO_2 electrode on anode and cathode performance for fuel cell applications. Trans Mater Res Soc Jpn 33:1101–1104

228. Lim DH, Lee WD, Choi DH, Lee HI (2010) Effect of ceria nanoparticles into the Pt/C catalyst as cathode material on the electrocatalytic activity and durability for low-temperature fuel cell. Appl Catal B 94:85–96
229. Yu HB, Kim JH, Lee HI, Scibioh MA, Lee J, Han J, Yoon SP, Ha HY (2005) Development of nanophase CeO_2-Pt/C cathode catalyst for direct methanol fuel cell. J Power Sources 140:59–65
230. Lee KH, Kwon K, Roev V, Yoo DY, Chang H, Seung D (2008) Synthesis and characterization of nanostructured PtCo-CeO_x/C for oxygen reduction reaction. J Power Sources 185:871–875
231. Fugane K, Mori T, Ou DR, Suzuki A, Yoshikawa H, Masuda T, Uosaki K, Yamashita Y, Ueda S, Kobayashi K, Okazaki N, Matolinova I, Matolin V (2011) Activity of oxygen reduction reaction on small amount of amorphous CeO_x promoted Pt cathode for fuel cell application. Electrochim Acta 56:3874–3883
232. Fugane K, Mori T, Ou DR, Yan PF, Ye F, Yoshikawa H, Drennan J (2012) Improvement of cathode performance on Pt-CeO_x by optimization of electrochemical pretreatment condition for PEFC application. Langmuir 28:16692–16700
233. Fugane K, Mori T, Yan PF, Masuda T, Yamamoto S, Ye F, Yoshikawa H, Auchterlonie G, Drennan J (2015) Defect structure analysis of hetero-interface between Pt and CeO_x promoter on Pt electro-catalyst. ACS Appl Mater Inter 7:2698–2707
234. Masuda T, Fukumitsu H, Fugane K, Togasaki H, Matsumura D, Tamura K, Nishihata Y, Yoshikawa H, Kobayashi K, Mori T, Uosaki K (2012) Role of cerium oxide in the enhancement of activity for oxygen reduction reaction at Pt-CeO_x nanocomposite electrocatalyst - an in situ electrochemical X-ray absorption fine structure study. J Phys Chem C 116:10098–10102
235. Chauhan S, Mori T, Masuda T, Ueda S, Richards G, Hill J, Ariga K, Isaka N, Auchterlonie G, Drennan J (2016) Design of low Pt concentration electro-catalyst surfaces with high oxygen reduction reaction activity promoted by formation of heterogeneous interface between Pt and CeO_x nanowire. ACS Appl Mater Inter 8:9059–9070
236. Wang Y, Kageyama H, Mori T, Yoshikawa H, Drennan J (2006) Local structures around Y and Ce cations in 10 mol% Y_2O_3 doped ceria ceramics by XAFS spectroscopy. Solid State Ionics 177:1681–1685
237. Leavitt JA, McIntyre Jr LC, Weller MR (1995) Backscattering spectrometry. In: Tesmer JR, Nastasi M, Barbour JC, Maggiore CJ, Mayer JW (eds) Handbook of modern ion beam materials analysis. Materials Research Society, Pittsburgh, Pennsylvania, Chapter 4, p 40
238. New Energy and Industrial Technology Development Organization (NEDO), PEFC evaluation project, cell evaluation and analysis protocol guideline (electrocatalyst, support, membrane and MEA), edited by Daido University, Ritsumeikan University, Tokyo Institute of Technology, and Japan Automobile Research Institute, January 30th, 2014, pp 12 and 15
239. Williams DB, Carter CB (1996) 17.5. Diffraction from particles. In: Williams DB, Carter CB (eds) Transmission electron microscopy diffraction II. Plenum Publishing Corporation, New York, pp 258–259
240. Li ZP, Mori T, Auchterlonie G, Zou J, Drennan J (2011) Direct evidence of dopant segregation in Gd-doped ceria. Appl Phys Lett 98:1–3 (article number 093104)
241. Li ZP, Mori T, Auchterlonie G, Zou J, Drennan J (2012) Nanodomain formation and distribution in Gd-doped ceria. Mater Res Bull 47:763–767
242. Yan M, Mori T, Ye F, Ou DR, Zou J, Drennan J (2008) Effects of dopant concentration and calcination temperature on the microstructure of Ca-doped ceria nanopowders. J Eur Ceram Soc 28:2709–2716
243. Yan M, Mori T, Zou J, Ye F, Ou DR, Drennan J (2009) TEM and XPS analysis of $Ca_xCe_{1-x}O_{2-y}$ ($x = 0.05$–0.5) as electrolyte materials for solid oxide fuel cells. Acta Mater 57:722–731
244. Yan M, Mori T, Zou J, Drennan J (2009) Effect of grain growth on densification and conductivity of Ca-doped CeO_2 electrolyte. J Am Ceram Soc 92:2745–2750
245. Yan M, Mori T, Zou J, Huang H, Drennan J (2010) Microstructures and mechanical properties of $Ce_{1-x}Ca_xO_{2-y}$ ($x = 0.05, 0.1, 0.2$) with different sintering temperatures. J Eur Ceram Soc 30:669–675

246. Yan M, Mori T, Zou J, Drennan J (2010) Microstructural and chemical aspects of working temperature aged Ca-doped CeO_2. J Eur Ceram Soc 30:2505–2513
247. Ye F, Mori T, Ou DR, Zou J, Drennan J (2008) Compositional dependence of electrical conductivity of $Ce_{1-x}Tb_xO_{2-\delta}$ $(0 \leq x \leq 1)$. Renew Energ Int J 33:331–335
248. Ou DR, Mori T, Ye F, Zou J, Drennan J (2008) Comparison between Y-doped ceria and Ho-doped ceria: electrical conduction and microstructures. Renew Energ Int J 33:197–200
249. Li ZP, Mori T, Ye F, Ou DR, Zou J, Drennan J (2011) Dislocation associated incubational domain formation in lightly gadolinium-doped ceria. Microsc Microanal 1:49–53
250. Li ZP, Mori T, Ye F, Ou DR, Zou J, Drennan J (2011) Ordered structures of defect clusters in gadolinium-doped ceria. J Chem Phys 134:1–7 (artcile number: 224708)
251. Li ZP, Mori T, Ye F, Ou DR, Auchterlonie G, Zou J, Drennan J (2012) Cerium-reduction-induced defects clustering, ordering, and assocaited microstructure evolution in yttrium-doped ceria. J Phys Chem C 116:5435–5443
252. Li ZP, Mori T, Auchterlonie G, Zou J, Drennan J (2012) Incubational domain characterization in lightly doped ceria. J Solid State Chem 192:28–33
253. Li ZM, Mori T, Yan PF, Wu YY, Li ZP (2012) Preparation and performance of intermediate-temperature fuel cells based on Gd-doped ceria electrolytes with different compositions. Mat Sci Eng B 177:1538–1541
254. Li ZP, Mori T, Auchterlonie G, Zou J, Drennan J (2013) Microstructure evolution of yttria-doped ceria in reducing atmosphere. Renew Energ Int J 50:494–497
255. Li ZP, Mori T, Zou J, Drennan J (2013) Defects clustering and ordering in di- and trivalently doped ceria. Mater Res Bull 48:807–812
256. Yan PF, Mori T, Wu YY, Li ZM, Auchterlonie G, Zou J, Drennan J (2013) Microstructural and chemical characterization of ordered structure in yttrium doped ceria. Microsc Microanal 19:102–110
257. (a): Li ZP, Mori T, Zou J, Drennan J (2012) Optimization of ionic conductivity in solid electrolytes through dopant-dependent defect cluster analysis. Phys Chem Chem Phys 14:8369–8375. (b): Li ZP, Mori T, Ye F, Ou D R, Zou J, Drennan J (2011) Structural phase transformation through defect cluster growth in Gd-doped ceria. Phys Rev B 84:1–5 (article number 180201(R))
258. Ikegami T (1996) Contacts and coordination numbers in a compact of polyhedral particles. J Am Ceram Soc 79:148–152
259. Ikegami T (1999) Early-stage sintering in a powder compact of polyhedral particles I models. Ceram Int 25:415–424
260. Ikegami T (1978) Microstructural development during intermediate and final-stage sintering. Acta Metall 35:667–675
261. Ikegami T, Tsutsumi M, Matsuda S, Shirasaki S (1928) A Model of Densification with simultaneous grain growth. J Appl Phys 49:4238–4241
262. Ikegami T (2003) Reconstruction of sintering theories with the aid of statistic method. J Jpn Soc Powder Metall 50:213–219
263. Mori T, Buchanan R, Ou DR, Ye F, Kobayashi T, Kim JD, Zou J, Drennan J (2008) Design of nano-structured ceria based solid electrolytes for development of IT-SOFC. J Solid State Electr 12:841–849
264. Ou DR, Ye F, Mori T (2011) Defect clustering and local ordering in rare earth co-doped ceria. Phys Chem Chem Phys 13:9554–9560
265. Ito S, Saito M, Takayama T, Yamamura H (2010) Synthesis of new brownmillerite-type systems A_2 (M, M')$_2O_5$ (A= Ba, Sr; M= Zn; M'= Zr, Ce, Ti, Hf, Sn). Trans Mater Res Soc Jpn 35:507–509
266. Ito S, Mori T, Yan PF, Auchterlonie G, Drennan J, Ye F, Fugane K, Sato T (2017) High electrcical conductivity in $Ba_2In_2O_5$ brownmillerite based materials induced by design of Frenkel defect structure. RSC Adv 8:4688–4696
267. Kochetova NA, Spesivtseva IV, Animitsa IE (2013) Electrical properties of $Ba_2(In_{1-x}Al_x)_2O_5$ solid solutions. Russ J Electrochem 49:176–180

268. Noirault S, Quarez E, Piffard Y, Joubert O (2009) Water incorporation into the $Ba_2(In_{1-x}M_x)_2O_5$ ($M = Sc^{3+}$ $0 \leq x < 0.5$ and $M = Y^{3+}$ $0 \leq x < 0.35$) system and protonic conduction. Solid State Ionics 180:1157–1163
269. Rolle A, Seymour H, Roussel P, Rubbens A, Vannier RN (2008) Characterization of $Ba_2In_{2e-ex}Sn_xO_{5e+eX/2}$ oxide ion conductors. Ionics 14:477–482
270. Tarasova NA, Filinkova YV, Animitsa IE (2013) Electric properties of oxyfluorides $Ba_2In_2O_{5-0.5x}F_x$ with brownmillerite structure. Russ J Electrochem 49:50–56
271. Rice MJ, Strässier S, G. A. Toombs GA (1974) Superionic conductors: theory of the phase transition to the cation disordered state. Phys Rev Lett 32:596–599
272. Pasternak M, Benczer-Koller N, Yang T, Ruel R, Herber RH (1983) Impurity-induced local disorder in the ordered state of the superionic conductor β-Ag_2Se. Phys Rev B Condens Matter Mater Phys 27:2055–2058
273. Liu Y, Shao ZP, Mori T, Jiang SP (2020) Development of nickel based cermet anode materials in solid oxide fuel cells—now and future. Mater Energy Rep. https://doi.org/10.1016/j.matre.2020.11.002
274. Ou DR, Mori T, Ye F, Kobayashi T, Zou J, Auchterlonie G J, Drennan J (2006) Oxygen vacancy ordering in heavily rare earth-doped ceria. Appl Phys Lett 89:1–3 (article number: 171911)
275. Ou DR, Mori T, Ye F, Zou J, Drennan J (2008) Oxygen-vacancy ordering in lanthanide-doped ceria: dopant-type dependence and structure model. Phys Rev B 77:1–8 (article number: 024108)
276. Ye F, Mori T, Ou DR, Zou J, Drennan J, Nakayama S, Miyayama M (2010) Effect of nickel diffusion on the microstructure of Gd-doped ceria (GDC) electrolyte film supported by Ni-GDC cermet anode. Solid State Ionics 181:646–652
277. Li ZP, Mori T, Auchterlonie G, Guo Y, Zou J, Drennan J, Miyayama M (2011) Mutual diffusion and microstructure evoluation at the electrolyte-anode interface in intermedaite temperature solid oxide fuel cell. J Phys Chem C 115:6877–6885
278. Li ZP, Mori T, Auchterlonie G, Zou J, Drennan J (2011) Superstructure formation and variation in Ni-GDC cermet anodes in SOFC. Phys Chem Chem Phys 13:9685–9690
279. Li ZP, Mori T, Auchterlonie G, Zou J, Drennan J, Miyayama M (2011) Diffusion and segregation along grain boundary at the electrolyte-anode interface in IT-SOFC. Solid State Ionics 191:55–60
280. Li ZP, Mori T, Auchterlonie G, Zou J, Drennan J, Miyayama M (2011) The diffusions and associated interfacial layer formation between thin film electrolyte and cermet anode in IT-SOFC. J Alloy Compd 509:9679–9684
281. Li ZP, Mori T, Auchterlonie G, Zou J, Drennan J (2011) Mutual diffusion occurring at the interface between $La_{0.6}Sr_{0.4}Co_{0.8}Fe_{0.2}O_3$ cathode and Gd-doped ceria electrolyte during IT-SOFC cell preparation. ACS Appl Mater Inter 3:2772–2778
282. Li ZP, Mori T, Auchterlonie G, Zou J, Drennan J (2011) Two types of diffusions at the cathode/electrolyte interface in IT-SOFCs. J Solid State Chem 184:2458–2461
283. Sasaki T, Matsunaga K, Ohta H, Hosono H, Yamamoto T, Ikuhara Y (2004) Atomic and electronic structures of Ni/YSZ(111) interface. Mater Trans 45:2137–2143
284. Shirakami T, Tojo T, Atake T, Mori T, Yamamura H (1995) Low temperature heat capacities of zirconia and yttria-doped zirconia $(ZrO_2)_{1-x}(Y_2O_3)_x$ ($x = 0$,0.02, 0.0396). Thermochim Acta 267: 415–420
285. Tojo T, Atake T, Shirakami T, Mori T, Yamamura H (1996) Low temperature heat capacity of powder and sintered samples of yttria-doped zirconia. Solid State Ionics 86–88:89–92
286. Tojo T, Atake T, Mori T, Yamamura H (1999) Heat capacity and thermodynamic functions of zirconia and yttria stabilized zirconia. J Chem Thermodyn 31:831–845
287. Tojo T, Atake T, Mori T, Yamamura H (1999) Excess heat capacity in yttria stabilized zirconia. J Therm Anal Calorim 57:447–458
288. Rednyk A, Mori T, Yamamoto S, Suzuki A, Yamamoto Y, Tanji T, Isaka N, Kus P, Ito S, Ye F (2018) Design of new active sites on Ni in the anode of intermediate temperature solid oxide fuel cells using trace amount of platinum oxides. ChemPlusChem 83:756–768

289. Liu S, Muhammad A, Mihara K, Ishimori T, Tada T, Koyama M (2017) Predictive micro-kinetic model for solid oxide fuel cell patterned anode: based on an extensive literature survey and exhaustive simulations. J Phys Chem C 121:19069–19079
290. Tong Ke, Ye F, Mori T, unpublished work
291. Shirakami T, Atake T, Mori T, Yamamura H (1995) Low temperature heat capacity of $(ZrO_2)_{1-x}(Y_2O_3)_x$ (x = 0.0776 and 0.0970) and low energy excitations. Solid State Ionics 79:143–146
292. Rednyk A, Mori T, unpublished work
293. Ito S, Mori T, Suzuki A, Okubo H, Yamamoto S, Sato T, Ye F (2019) Design of active site at heterointerface between brownmillerite type oxide promoter and fluorite cubic ZrO_2 in anode of intermediate temperature SOFCs. ACS Appl Energy Mater 2:5183–5197
294. Chauhan S, Mori T, Kobayashi T, Yamamoto S, Ito S, Auchterlonie G, Wepf R, Ueda S, Ye F (2021) Surface layer of Pt-O-Ce bonds on CeO_x nanowire with high ORR activity converted by proton beam irradiation. J Am Ceram Soc 104:1945–1952
295. Mori T, Tong K, Yamamoto S, Chauhan S, Kobayashi T, Isaka N, Auchterlonie G, Wepf R, Suzuki A, Ito S, Ye F (2022) Active Pt-nanocoated layer with Pt–O–Ce bonds on a CeO_x nanowire cathode formed by electron beam irradiation. ACS Omega 7:25822–25836
296. Velázquez-Palenzuela A, Francesc C, Garrido JA, Arias C, Rodriguez RM, Brillas E, Cabot PL (2011) Kinetic analysis of carbon monoxide and methanol oxidation on high performance carbon-supported Pt-Ru electrocatalyst for direct methanol fuel cells. J Power Sources 196:3503

Design of Interface Between Cathode and Electrolyte in SOFCs

San Ping Jiang and Shuai He

Abstract High temperature solid oxide fuel cells (SOFCs) are considered one of the most environmentally friendly and efficient electricity generation technology from fuels such as hydrogen and natural gas with high efficiency and very low greenhouse gas emission. Over the last few decades, the development of SOFC technology has experienced significant progress and much of the recent research have paid great efforts in understanding the processes occurring at the electrode/electrolyte interfaces. As electrochemical reactions mainly proceed at the gas, electrode and electrolyte three phase boundaries (TPBs), the microstructure and properties of the electrode/electrolyte interfaces thus play a crucial role in determining the overall cell performance and durability. This is particularly the case of the O_2 reduction reaction at the cathode/electrolyte interface. Herein, the progress and achievements in the fundamental researches and understanding of the interface evolution behavior and interfacial phenomena are reviewed. The challenges and prospects of the design of cathode and electrolyte interface in SOFCs are discussed.

Keywords Review · Solid oxide fuel cells · Cathode/electrolyte interface · Segregation · Interface formation · Interface design

S. P. Jiang (✉)
National Energy Key Laboratory for New Hydrogen-Ammonia Energy Technologies, Foshan Xianhu Laboratory, Foshan 528216, China
e-mail: s.jiang@curtin.edu.au

WA School of Mines: Mineral, Energy and Chemical Engineering, Curtin University, Perth, WA 6102, Australia

S. He
Petrochemical Research Institute, PetroChina Company Limited, Beijing 102206, China

T. Mori et al. (eds.), *Design of Active Key Interfaces in Fuel Cells*,
https://doi.org/10.1007/978-981-97-5706-0_2

1 Introduction

Solid oxide fuel cell (SOFC) is the most efficient energy-conversion device to convert the chemical energy of fuels like hydrogen, natural gas, etc., to electricity with high efficiency and very low greenhouse emission [1–4]. SOFCs have the significant advantages of high scalability and excellent fuel flexibility, including H_2, methanol, ethanol, natural gas, and hydrocarbons and are an efficient alternative to the conventional combustion technology for electricity generation [5–8]. However, there are major challenges in SOFC technologies, including the material degradation, stack durability, and thermal cycle capability [9–12]. For a given electrode/electrolyte material system, its performance is greatly affected by the electrode and electrolyte processes, such as reactant dissociation, diffusion, mass transport, and charge transfer. The mass transport process is strongly influenced by the micro- and macrostructural features of the electrode, while the charge transfer process, often the rate-determining step at reduced operating temperature, largely depends on the electrocatalytic activity and the nanoscale features of the electrode/electrolyte interface.

The evolution, formation, and change of the interface play a critical role in the performance and durability of SOFCs. For example, the interfacial reaction between $(LaSr)(CoFe)O_3$ (LSCF) cathode and yttria-stabilized zirconia (YSZ) electrolyte would lead to the formation of insulating phases like $SrZrO_3$ and $La_2Zr_2O_7$ and increase the interfacial resistance for the cell [13–16]. The phase and microstructure changes in the interface can be contributed to interfacial reactions [17, 18], micropore formation [19, 20], cation segregation and diffusion [21, 22], delamination [23, 24], and impurity contamination [25, 26]. Among them, the surface segregation at the interface is one of the most important fundamental issues in the activity and performance degradation of cathode materials of SOFCs [27].

Here, the fundamental issues related to the electron and ion transfer processes at the cathode/electrolyte interface will be introduced first. Then the formation and evolution of the cathode/electrolyte interface under high temperature sintering and polarization conditions is reviewed. The inter-relationship between the interface, surface segregation, and interfacial reaction will be highlighted. This is followed by a section of the state-of-the-art techniques employed to investigate the electrode/electrolyte interface evolution behavior under different test environment such as focused ion beam and scanning transmission electron microscopy (FIB-STEM). Finally, the prospect in the design of the effective cathode/electrolyte interface is outlined.

2 Defect Chemistry and Ionic and Electron Transfer Processes at the Interface

At the cathode of SOFCs, the oxygen reduction reaction (ORR) can be illustrated in a H_2-O_2 fuel cell, based on a typical SOFC with a dense oxygen ion conducting YSZ electrolyte, porous lanthanum strontium manganite (LSM) cathode and porous nickel/YSZ (Ni/YSZ) cermet anode. In this fuel cell, oxygen molecules, O_2, are ionized to oxygen ion, $O^{=}$ by taking two electrons and incorporated into oxygen lattice site of the YSZ electrolyte by occupying an oxygen vacancy. The reaction can be expressed as follows in Kröger-Vink notation:

$$\frac{1}{2}O_{2\text{(cathode)}} + 2e^{-} + V_{O}^{\cdot\cdot} = O_{O\text{(electrolyte, cathode)}}^{x} \quad (1)$$

where $V_{O}^{\cdot\cdot}$ is a vacant oxygen lattice site, i.e., oxygen vacancy and O_{O}^{x} is an oxygen ion on a regular oxygen site in the electrolyte lattice. At the same time, oxygen ions flow from the LSM cathode to Ni/YSZ cermet anode through the dense YSZ electrolyte via oxygen vacancies:

$$\mathrm{O}_{\mathrm{O(electrolyte,\ cathode)}}^{\mathrm{x}} \xrightarrow{V_{O}^{\cdot\cdot}} \mathrm{O}_{\mathrm{O(electrolyte,\ anode)}}^{\mathrm{x}} \quad (2)$$

As shown above, ORR requires the supply and release of oxygen ions and electrons simultaneously. Thus, the efficiency of the reaction critically depends on the ionic conductivity of the solid electrolyte and electronic or mixed ionic and electronic conductivity (MIEC) of the cathodes. ORR occurs at the porous electrode/dense electrolyte interface and ionic/electronic transport properties are critically related to microstructure of the cell such as tortuosity and three phase boundary (TPB) areas [28] and surface chemistry such as segregation [29]. Solid-to-solid (StS) contact between the porous electrode and dense electrolyte is in general not necessarily continuous and may be restricted only to a fraction of the interface. The restriction in the interface will cause the misalignment and asymmetric contacts between the working and counter electrodes, and result in the significant change in the electrochemical activity and performance of the cathodes.

Defect chemistry and associated solid state electrochemistry are critical for the fundamental understanding of the electrode/electrolyte interface phenomena in SOFCs. Here only the most common defect chemistry related to the electrode and electrode/electrolyte interface reactions will be introduced, using the well-known cathode materials such as LSM and LSCF. LSM is the best-known electronic conducting electrode materials in SOFCs due to its high electrocatalytic activity for ORR at high temperatures, high electronic conductivity, high structural stability, and chemical compatibility with YSZ electrolyte [30]. The electrical conductivity of LSM varies significantly with the Sr doping and increases with the temperature, a typical *p*-type conductor behavior [31]. Pavone et al showed that Sr substitution of A-site La leads to the formation of electron holes in the LSM electronic structure [32].

At SOFC operating temperatures, the electron holes are localized onto specific Mn ions, forming Mn^{4+} ions, $Mn_{Mn}^{\cdot}$ or electron holes, $\dot{h}$ and causing the local distortion of LSM lattice. The neutrality condition can be simply described as

$$\mathrm{SrO} \xrightarrow{\mathrm{LaMnO_3}} \mathrm{Sr}'_{\mathrm{La}} + Mn^{\cdot}_{\mathrm{Mn}} + \mathrm{O}^{\mathrm{x}}_{\mathrm{O}} \tag{3}$$

LSM is somewhat unusual in comparison to most perovskite oxides in that it can have the oxygen-excess as well as the oxygen-deficient nonstoichiometries. Under normal SOFC operation conditions LSM is in oxygen-excess region and only becomes oxygen-deficient at very low p_{O_2}, e.g., $<10^{-10}$ Pa at 900 °C. In the oxygen-deficient region, the predominant ionic defect is the oxygen vacancy and the charge compensation occurs by the Mn reduction. Since the electronic conduction in LSM is hopping *p*-type mechanism irrespective of the oxygen content, disproportion of Mn ion into Mn^{2+}, Mn^{3+}, and Mn^{4+} occurs [33]. Due to the high oxygen vacancy formation energy [32, 34–36], oxygen vacancy concentration is very low. This would limit the ORR at TPB in the case of LSM cathode.

Very different from LSM, LSCF is the best-known MIEC perovskite electrode in SOFCs and the most representative composition is $La_{0.6}Sr_{0.4}Co_{0.2}Fe_{0.8}O_{3-\delta}$ [37]. LSCF is a typical acceptor-doped cobaltite perovskite-type oxide and is characterized by enhanced lattice oxygen vacancy formation, particularly at increased temperatures. The charge compensation due to the dissolution of SrO in the $La(Fe,Co)O_3$ perovskite lattice can occur by both ionic and electronic defects. However, experimental observation indicates that the charge compensation is mainly maintained by the formation of positively charged oxygen vacancies:

$$2\mathrm{SrO} \xrightarrow{\mathrm{La(Fe,Co)}O_3} 2\mathrm{Sr}'_{\mathrm{La}} + V^{\cdot\cdot}_{O} + 2\mathrm{O}^{\mathrm{x}}_{\mathrm{O}} \tag{4}$$

Since the ionic and electronic compensations occur simultaneously and compete with each other, the charge neutrality condition, assuming only the localized electronic charge carriers to be present and charge compensation also occurs by the valence change of cobalt ions, can be expressed as

$$\left[V^{\cdot\cdot}_{O}\right] + \left[Fe^{\cdot}_{\mathrm{Co}}\right] + \left[Co^{\cdot}_{\mathrm{Co}}\right] = 4\left[Sr'_{\mathrm{La}}\right] \tag{5}$$

The electrochemical activity of the LSCF electrode for ORR decreases significantly with the decrease in operating temperature due to its high activation energy of ORR (e.g., 135–150 kJmol^{-1}). At temperatures below 600 °C, the reaction is increasingly driven to the electrode/electrolyte interface region due to the substantial decrease in the ionic conductivity of LSCF below 600 °C.

3 Interface Formation, Surface Segregation, and ORR

3.1 Interface Formation Induced By High Temperature Sintering

3.1.1 Interface Formation and Cation Interdiffusion

Due to the nature of ceramic components in SOFCs, intimate electrode/electrolyte interface is generally formed by the high temperature sintering process, e.g., ~1150 °C in the case of LSM ($La_{0.8}Sr_{0.2}MnO_3$)/YSZ cell and ~900 °C for $La_{0.6}Sr_{0.4}Co_{0.2}Fe_{0.8}O_{3-\delta}$/gadolinium doped ceria (LSCF/GDC) cell [30, 38, 39]. The cathode/electrolyte interface formed during high temperature sintering can be identified by the formation of convex contact rings on the electrolyte surface [19, 20, 40–43]. In the case of LSM cathode, since it is predominantly an electronic conductor with negligible oxygen ionic conductivity, the ORR primarily occurs at the TPB region. The formation of TPB provides a direct pathway for oxygen ions migration, and thus is critical in determining the overall cell performance and durability [44, 45]. The correlation between TPB and the electrochemical performance of the cell has been extensively studied due to its unique positions in the fundamental understanding of ORR in SOFCs [44, 46].

The interface formation of LSM/YSZ and LSM/GDC induced by high temperature sintering has been investigated systematically by using focus ion beam and scanning transition electron microscopy (FIB-STEM) techniques [47, 48]. Mn diffusion on the YSZ electrolyte surface as well as along its grain boundary during sintering the LSM electrode on the YSZ electrolyte at 1150 °C has been observed, see Fig. 1 [47]. A clear accumulation of Mn cation and in a lesser extent, La cation was detected within the edge of the convex contact ring on the YSZ electrolyte surface, extending as far as 70-90 nm into the YSZ (Fig. 1a). The diffusion of Mn and La into YSZ is attributed to their limited solubility in YSZ, forming MnO-La_2O_3-Y_2O_3-ZrO_2 solid solution. However, at the center region of the convex contact ring on the YSZ electrolyte, the intensity of Mn is qualitatively smaller than that within the rim region of the convex rings, as indicated in Fig. 1b. The formation of convex contact rings at LSM/YSZ electrolyte interface after high temperature sintering has been correlated to cation inter-diffusion, consistent with that reported in the literature [46, 49].

Similar contact convex ring and crystallographic dislocation at the LSM electrode and GDC electrolytes were observed [47]. However, in contrast to LSM/YSZ cells, no cation interdiffusion was detected in the case of GDC electrolyte. This indicates that the interface formation between LSM and YSZ or GDC electrolyte does not necessarily require cation inter-diffusion, i.e., the cation interdiffusion is not an essential condition for the interface formation in the case of LSM-based cathode. The formation of convex rings between LSM electrode and YSZ or GDC electrolyte is essentially due to the minimization of surface energy between these two solid phases during the high temperature sintering processes.

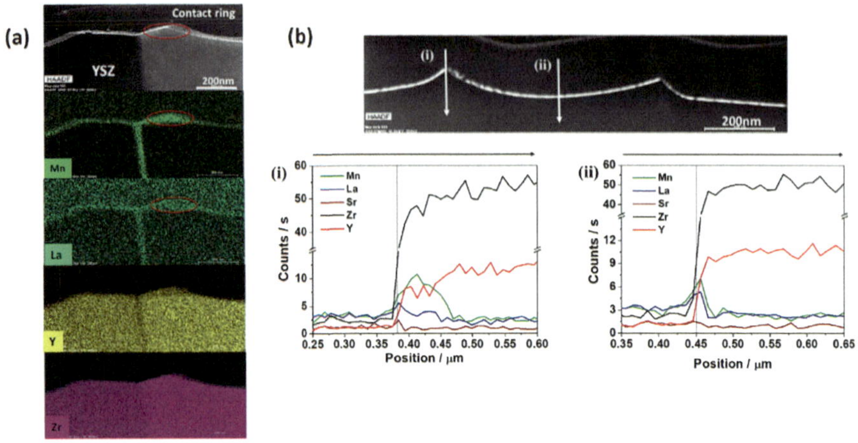

Fig. 1 **a** STEM-EDS element mapping over the cross-section of the YSZ electrolyte in contact with the LSM electrode (removed) after high temperature sintering; **b** STEM-EDS line scan across the convex contact ring ridge and ring center (reproduced from Ref. [47])

Figure 2 shows schematically the formation of LSM/YSZ and LSM/GDC interfaces during the high temperature sintering [47]. Before the sintering, irregular shaped LSM particles are attached to the electrolyte surface with point contacts between LSM particles and electrolyte. After exposed to high temperatures, LSM particles gradually grow and sintering between LSM and YSZ occurs, indicated by the neck formation. The grain growth and neck formation will diminish the curvature of the interface as well as the surface free energy [50, 51]. Cations such as manganese would segregate from the LSM perovskite structure and start to migrate along the YSZ electrolyte surface. In the case of YSZ electrolyte, Mn diffuses into YSZ, particularly at the contact ring edge region, replacing Y in YSZ lattice. On the other hand, Mn would diffuse along the GDC electrolyte surface but cation interdiffusion would not occur at the LSM/GDC interface, most likely due to the very low solubility of Mn in ceria [52]. The high temperature sintering leads to the formation of semi-coherent interface between LSM/YSZ and LSM/GDC crystallographic planes, despite the significant differences in orientation relationship and mismatch factors of these two phases. The electrochemical performance of the LSM electrode indicates that the presence of such crystallographic defects and lattice dislocations does not inhibit the ORR at the interface.

3.2 *Polarization-Induced Interface Formation*

Cathodic polarization promotes charge transfer as well as the migration of oxygen species at the cathode/electrolyte interface; this process impacts greatly on the

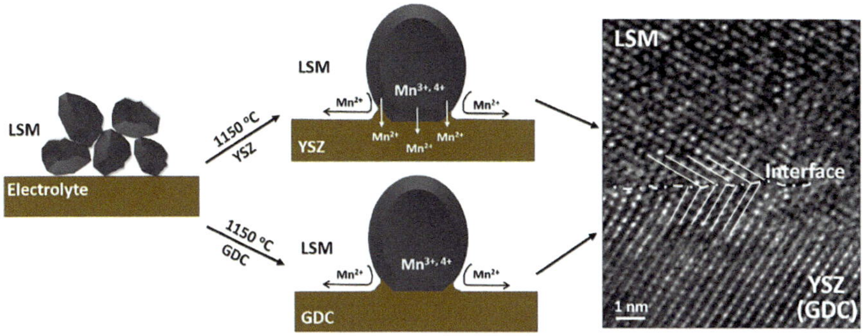

Fig. 2 Schematic diagram showing the Mn migration and interdiffusion and formation of LSM/YSZ and LSM/GDC interfaces under the influence of high temperature sintering (Reproduced from Ref. [47])

microstructure and properties of the interface. Earlier studies show that the polarization has a significant effect on the morphology and topography of the TPB at the interface between LSM electrode and YSZ electrolyte [43, 53]. On the other hand, cathodic polarization can also cause the interface formation. The cathodic polarization induced interface formation has been demonstrated extensively in directly assembled electrodes, including LSM, LSCF, $(La_{0.6}Sr_{0.4})_{0.95}Co_{0.2}Fe_{0.7}Nb_{0.1}O_{3-\delta}$ (LSCFN), $(La_{0.6}Sr_{0.4})_{0.95}(Co_{0.2}Fe_{0.7}Nb_{0.1})_{0.95}Pd_{0.05}O_{3-\delta}$ (LSCFNPd), $La_{0.95}(Co_{0.2}Fe_{0.8})_{0.95}Pd_{0.05}O_{3-\delta}$ (LCFPd), $Sm_{0.95}Co_{0.95}Pd_{0.05}O_{3-\delta}$ (SmCPd), $PrBa_{0.5}Sr_{0.5}Co_{1.5}Fe_{0.5}O_{5+d}$ (PBSCF) electrodes on barrier-layer-free YSZ electrolyte [54–57]. In direct assembly method, electrode coatings are directly applied to the electrolyte surface without pre-sintering at high temperatures. Fig. 3 shows a comparison of conventional electrode fabrication steps and in situ assembly or direct assembly electrode fabrication steps [54]. The direct assembly method significantly reduces the fabrication steps of the electrodes of SOFCs and with direct assembly method, cobaltite-based perovskite (CBP) electrodes can be applied on YSZ electrolyte due to the elimination of high temperature sintering step, thus avoiding the complication and chemical reaction associated with high temperature sintering.

Direct assembly technique offers a new platform for the study of the interface formation and evolution under SOFC operating temperatures. Figure 4 shows typical LSM/YSZ and LSM/GDC interfaces formed via the cathodic polarization treatment. The polarization-induced interface is characterized by particle-shaped contact marks or clusters and the number and distribution of the contact marks greatly depend on the current density as well as the electrode/electrolyte materials. The electrochemical activities of the polarization-induced interface for ORR are very similar to that observed on thermally induced interfaces. The electrochemically induced interface is most likely related to the localized joule heating at the electrode/electrolyte contacts due to the initially high ohmic resistance at the interface. This is supported by the

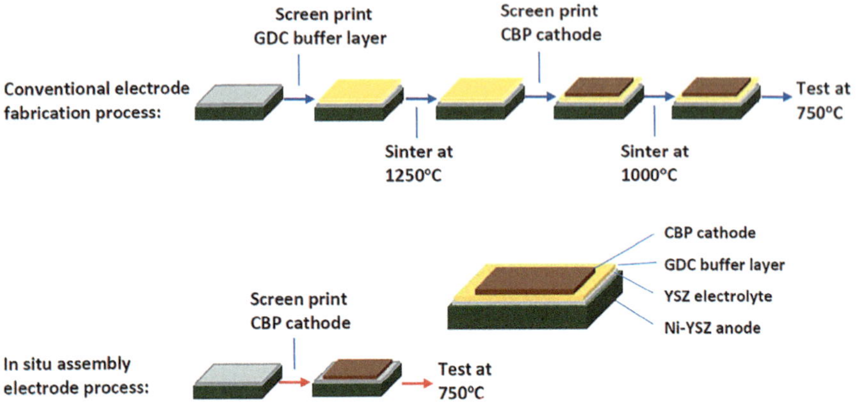

Fig. 3 Comparison of conventional electrode fabrication processes and direct assembly electrode fabrication processes for cobaltite-based perovskite (CBP) electrodes such as LSCF on YSZ electrolyte (Reproduced from Ref. [54])

significant reduction in the cell ohmic resistance when subjecting to the polarization current [43]. Significant reduction in the ohmic resistance of the cell during initial polarization has also been observed for directly assembled LSCFN, LSCFNPd, LCFPd-GDC, and SmCPd-GDC cathodes on YSZ electrolytes [54–57].

3.3 Interface Formation, Segregation, and Interface Reaction

The relationship between the interface formation, segregation, and interfacial reaction is complicated and this is particularly the case for the cobaltite based cathodes such as LSCF and YSZ electrolyte due to the strontium segregation and facile chemical reaction between LSCF and YSZ electrolyte at SOFC fabrication and operating temperatures. The study of the effect of cathodic polarization on strontium segregation and interface formation between LSCF cathode and YSZ electrolyte is made possible by the use of direct assembly technique [58]. In this study, LSCF was directly assembled to barrier-layer-free YSZ electrolyte and the cathode/electrolyte interface evolution was characterized in different stages of cathodic polarization at 750 °C and 1000 mA cm^{-2} using combined FIB-STEM techniques, see Fig. 5 [58]. In the initial stage of cathodic polarization (~1h), the LSCF/YSZ interface forms with substantially reduced cell ohmic resistance. As the directly assembled LSCF electrode particles have very limited contact area to the electrolyte, the high R_{Ω} would result in a much higher current density at the interface, leading to the localized sintering due to the generation of Joule heat and therefore the establishment of interface between LSCF and YSZ electrolyte. Further polarization induces the Sr segregation from the surface of LSCF electrode towards the LSCF/YSZ interface, forming a thin Sr-rich layer (SRL) on the electrolyte surface. The main component

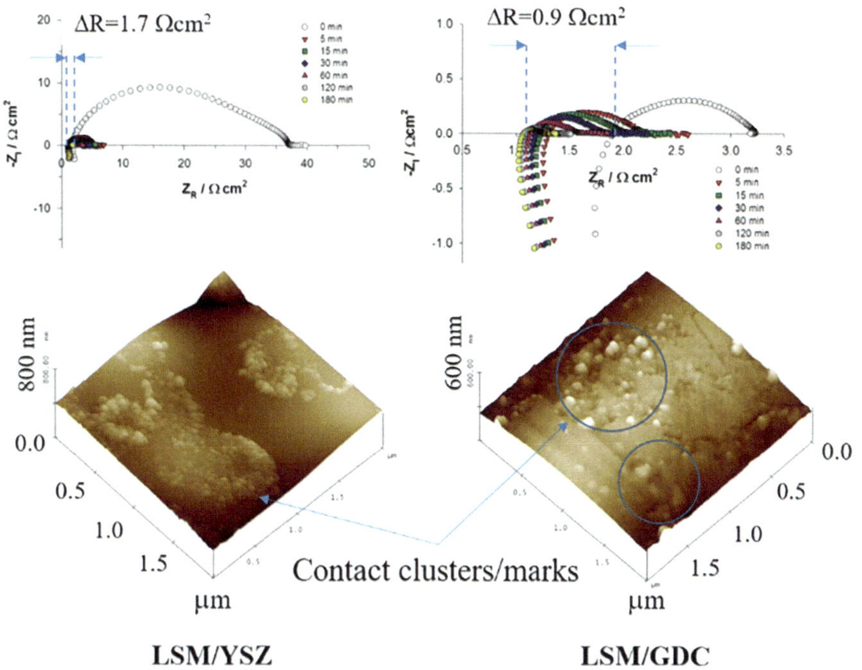

Fig. 4 Impedance curves for ORR on directly assembled LSM cathode on YSZ electrolyte (LSM/YSZ) and directly assembled LSM cathode on GDC (LSM/GDC) electrolyte as a function of cathodic polarization currents of 1000 mA cm^{-2} at 800 °C. AFM micrographs of YSZ and GDC electrolyte surface morphology in contact with directly assembled LSM cathode after polarization at 1000 mA cm^{-2} and 800 °C are also given (Reproduced from Ref. [43])

of the SRL was found to be SrO, which grows epitaxially on the YSZ surface with identical crystalline plane orientation. The segregated Sr subsequently reacts with YSZ, which is kinetically favorable, forming $SrZrO_3$ secondary phase. The thickness of the $SrZrO_3$ layer increases with prolonged polarization. The polarized YSZ electrolyte surface in contact with LSCF particles is characterized by the formation of ring-shaped contact craters.

The relationship between the segregation and interface formation and reaction is also a dynamic one, depending on number of factors such as the activity between electrode and electrolyte, segregation and segregated species of electrode, the presence of catches of the segregated species, etc. This is best illustrated by a comparative study of the interface reaction of LSCF cathode and YSZ or GDC electrolytes under identical conditions [59]. In this study, $La_{0.6}Sr_{0.4}Co_{0.2}Fe_{0.8}O_{3-\delta}$ (LSCF) cathode was directly assembled on YSZ and GDC electrolytes and studied under identical cathodic polarization conditions at 750 °C and 1000 mA cm^{-2}. Sr segregation occurs in both LSCF/YSZ and LSCF/GDC systems. The segregated Sr species migrates to the LSCF electrode/YSZ electrolyte interface, forming $SrZrO_3$ insulating phase. However, in

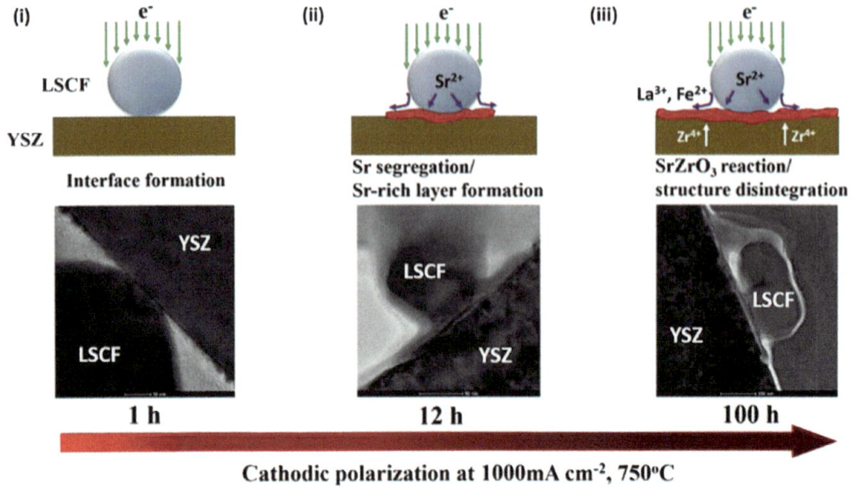

Fig. 5 Schematic diagram showing the influence of cathodic polarization duration on the LSCF/YSZ interface evolution. (i) 1 h: interface formation; (ii) 12 h: Sr segregation and diffusion; (iii) 100 h: Sr layer growth and formation of resistive $SrZrO_3$ phase at the interface (Reproduced from Ref. [58])

the case of LSCF/GDC cells, the inertness of GDC electrolyte suppresses the segregation process of Sr species. Instead, surface segregation B-site Co cation becomes dominant under the cathodic polarization, forming isolated CoO_x particles, see Fig. 6.

It is well known that Sr has a high tendency to segregate at the surface of cobaltite-based perovskites such as LSCF-based electrodes and the surface segregation has a significant detrimental effect on the performance and stability of LSCF-based electrodes for ORR [60–68]. In the case of LSCF/GDC cells, the Sr segregation is substantially retarded, in contrast to the accelerated Sr segregation in the case of LSCF/YSZ cells under identical cathodic polarization conditions (see Fig. 6). This indicates that although the surface segregation of Sr is an inherent property of LSCF based electrode materials [69, 70], the kinetics of segregation is largely determined by the subsequent chemical reaction between the segregated SrO species and the catcher. In the absence of catchers of the segregated species, the segregation rate will be kinetically retarded and slow. Thus, in both LSCF/YSZ and LSCF/GDC cells, the first step is the surface segregation of Sr species at high temperature under the strong oxidizing and/or cathodic polarization conditions:

$$La_{0.6}Sr_{0.4}Co_{0.2}Fe_{0.8}O_{3-\delta} \overset{i_{Cathodic},750^{\circ}C}{\longleftrightarrow} La_{0.6}Sr_{0.4-x}Co_{0.2}Fe_{0.8}O_{3-\delta} + xSrO_s \quad (6)$$

The segregated Sr species will diffuse from electrode surface to the LSCF electrode and electrolyte (i.e., YSZ or GDC) interface:

$$xSrO_s \xrightarrow{\text{diffusion}} xSrO_i \quad (7)$$

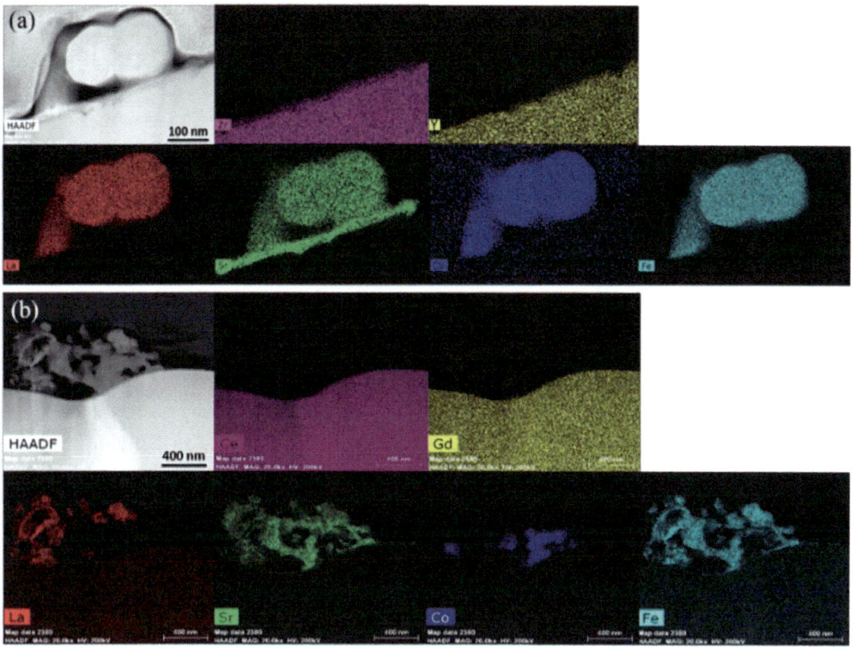

Fig. 6 STEM-EDS mapping of **a** of directly assembled LSCF/YSZ electrolyte interface and **b** of directly assembled LSCF/GDC electrolyte interface after polarization after polarization at 750 °C and 1000 mA cm^{-2} for 100 h (Reproduced and modified from Ref. [59])

In the case of YSZ electrolyte, the chemical catcher of the segregated SrO is Zr, forming $SrZrO_3$, as follows [71]:

$$xSrO_i + Y_{0.16}Zr_{0.84}O_2 \xrightarrow{750^\circ} SrZrO_3 + Y_2O_3 \tag{8}$$

The above reaction accelerates the Sr segregation and leads to the disintegration of LSCF structure and formation of insulating phases at the interface. The formation of insulation zirconate phase is most significant in the degradation of LSCF-based cathodes of SOFCs [71–73]. However, in the case of LSCF/GDC cells, no such catchers for the segregated SrO exist. Thus, there is no driving force for the further and continuous segregation of SrO species. The Sr surface segregation reaction becomes slow and reaches equilibrium under the cathodic polarization conditions:

$$La_{0.6}Sr_{0.4}Co_{0.2}Fe_{0.8}O_{3-\delta} \xleftrightarrow{i_{Cathodic}, 750^\circ C} La_{0.6}Sr_{0.4-x}Co_{0.2}Fe_{0.8}O_{3-\delta} + xSrO_s \tag{9}$$

However, based on the oxygen vacancy mediated surface segregation mechanisms under the cathodic polarization, surface segregation of both the A- and B-site cations can occur in the case of MIEC type electrodes such as LSCF and the degree of the segregation or preferential segregation of A- or B-site cations to the surface is related

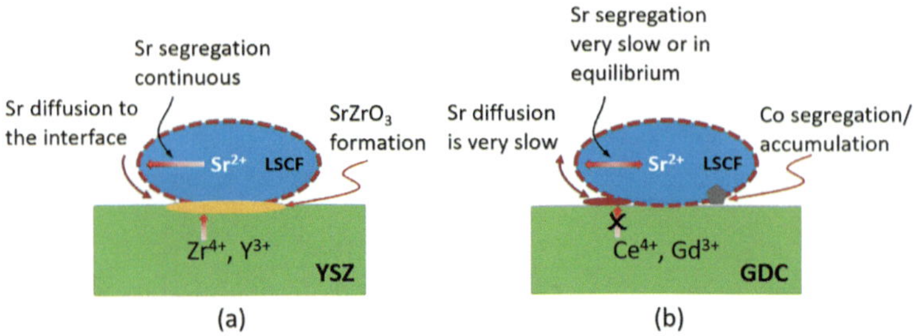

Fig. 7 Schemes of the surface segregation and interface reaction in **a** LSCF/YSZ and **b** LSCF/GDC cells. Red arrow indicates the Sr segregation, diffusion, and reaction direction (Reproduced from Ref. [59])

to the chemical environment or the presence of chemical catchers of system [74]. In the presence of Sr catchers, e.g., Zr in the case of LSCF/YSZ cells, the segregation of A-site cation, Sr, is preferential. On the other hand, the segregation of B-site cations such as Co will take place in the absence of chemical catchers for Sr:

$$La_{0.6}Sr_{0.4}Co_{0.2}Fe_{0.8}O_{3-\delta} \xrightarrow{i_{Cathodic}, 750^{\circ}} La_{0.6}Sr_{0.4-x}Co_{0.2-y}Fe_{0.8}O_{3-\delta} + xSrO_s + CoO_y \quad (10)$$

The segregated Co species can be in the form of CoO or Co_3O_4. The significant segregation of Co species is indicated by the evident accumulation and formation of Co-rich particles in isolated pockets of LSCF electrodes in the case of LSCF/GDC cells (Fig. 6b). Figure 7 shows the schemes of the surface segregation of Sr and/or Co and interface reaction between the segregated species and YSZ or GDC electrolyte [59].

This indicates that the presence of chemical catchers is the driving force for the accelerated Sr segregation of LSCF electrodes under SOFC operation conditions.

4 Experimental Approaches in Interface Investigation

4.1 Focused Ion Beam and Scanning TEM (FIB-STEM)

The instrument has been gaining increasing attention in recent years as an important and effective way to prepare thin lamella for TEM characterization in solid oxide fuel cells [49, 75–80]. Unlike inert gas ion sputtering, FIB ionizes Ga into Ga^+, and then uses electric field acceleration and electrostatic lens aggregation to drive high-energy, high-speed Ga^+ ions to a specified position. The application of combined FIB-STEM technique offers opportunities of more precise analysis of the electrode/electrolyte interface microstructure on a nanoscale. The FIB technique not only allows a TEM lamella to be prepared from a specific area of interest with submicron

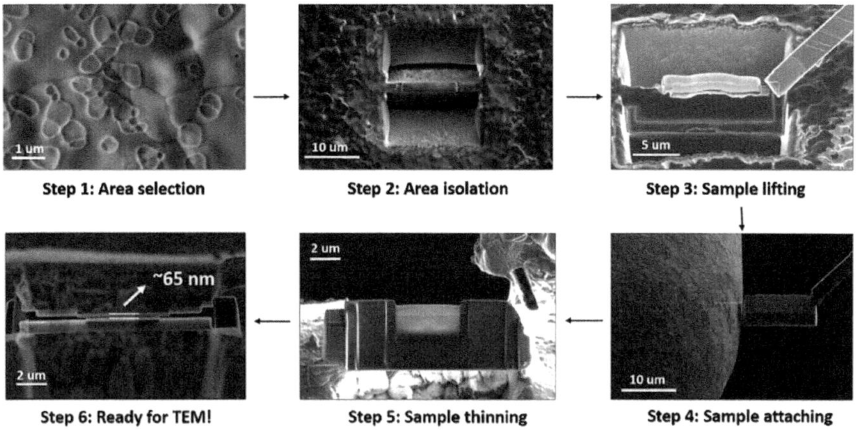

Fig. 8 Flowchart of FIB lamella milling procedures

precision, but also can uniformly mill heterogeneous microstructures that consist of compositionally and structurally distinct regions. The recent development and applications of FIB technique was reviewed by Sezen [81]. Figure 8 shows the typical FIB milling process.

The sample preparation and FIB milling procedures are discussed in detail as follows.

Sample preparation. In order to perform a high-quality FIB milling, sample of interest should be conductive and ideally flat. For non-conductive samples, it is necessary to coat the region of interest with carbon or platinum. Carbon coating is preferred if the crystalline structure of the sample surface is desired. In the case of samples with rough surface or the interface between electrode and electrolyte are of particular interests, resin embedding should be considered. The prepared sample needs to be free from drifting under the microscope working conditions.

FIB milling procedure. First, in order to prevent the damage caused during ion milling, a Pt or C protection layer is deposited on the area of interest using electron beam, then a thick layer (~2 um) is deposited in the same area using ion beam (Step 1). Trenches on both sides of the area with protection layer are milled using ion beam. Typically, large ion beam current is used to make the milling more efficient (Step 2).

Then a "U-cut" is performed with ion beam. It is important to ensure that the sample is cut through, which is indicated by the appearance of a black line below the sample lamella (see Fig. 9 for a typical FIB lamella for HRTEM). After the U-cut, the lamella is welded to the tungsten needle using Pt deposition. The welded lamella is then detached from the bulk (Step 3).

The thick sample lamella is then welded on a FIB sample grid using Pt deposition. To ensure a good attachment, one side of the lamella is welded on the grid, and contact between the needle and the lamella is cut off, then the other side of the lamella is welded on the grid (Step 4).

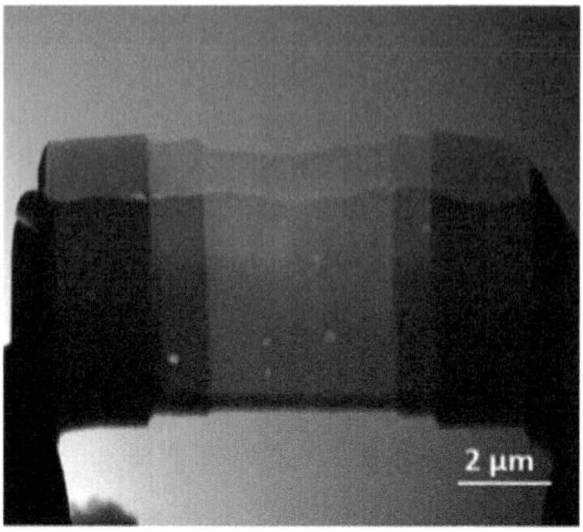

Fig. 9 A typical FIB lamella for HRTEM

Sample thinning. Since the sample is still quite thick (~1 μm), further thinning process is performed by ion milling until the thickness of the lamella is less than 100 nm. The thinning process is critical, as it determines the imaging quality of further TEM analysis (Step 5).

After final low voltage ion beam cleaning, the lamella is ready for TEM examination. The thickness of a FIB lamella should be less than 150nm, and if high resolution TEM (HRTEM) imaging is required, a sample thickness of ~60 nm is desired (see Fig. 9). The FIB instrument adopted here is Dual Beam Focused Ion Beam–Scanning Electron Microscope (FIB-SEM, Helio Nanolab G3 CX, FEI company, US) with Ga^+ source at 30 kV for the ion beam.

FIB has been used to construct 3D microstructural change at the interface under the influence of polarization. For example, Yang et al. [82] to investigated the influence of large cathodic current on the microstructure of A-site deficient LSM/YSZ interface, using FIB 3D reconstruction technique. It was found that after discharge at 1.2 A cm^{-2} for 60 h, the YSZ electrolyte surface was covered by a dense layer of LSM, accompanied with the formation of closed nanopores at the LSM/YSZ interface. The 3D reconstructed results indicate that the total TPB length of the as-prepared cell was reduced from initial 1.08 $\mu m\ \mu m^{-2}$ to 0.49 $\mu m\ \mu m^{-2}$ and 0.26 $\mu m\ \mu m^{-2}$ after the cathodic polarization for 5 h and 40 h, respectively. FIB 3D reconstruction technique can be used for the quantitative analysis of the microstructure and microstructure change of the cathode/electrolyte interface.

FIB in combination with secondary ion mass spectrometry (FIB-SIMS) has been used to analyze the composition of solid surfaces by sputtering and ionizing the surface of a specimen with a focused primary ion beam (usually Ga^+) and collecting and analyzing ejected secondary ions. SIMS is a very sensitive analytical technique that can provide parts per million (ppm) to parts per billion (ppb) sensitivity for most elements when relatively large areas of analysis are available, and has been used widely in examining the evolution of electrode/electrolyte interface in SOC devices [83–86].

4.2 Raman Spectroscopy

Raman spectroscopy is a technique that used to investigate molecule vibration, rotation and other low-frequency modes, and it relies on inelastic scattering of a laser beam in the visible, near infrared, or near ultraviolet range. The laser light interacts with the molecular vibrations in the sample, resulting in the energy of the laser being shifted, and the shift in energy gives information about the vibrational modes in the sample. Due to the nature of the Raman technique, it is widely adopted in SOFCs [52, 87–89].

In combination with atomic force microscope (AFM) technique, Raman microscopy has been effectively used to identify the relationship between the surface chemistry, segregation, and Cr deposition on the LSCF electrode [90]. The microstructure as well as the phase composition and distribution of contaminants on LSCF surface after heat treatment in the absence and presence of Cr_2O_3 at 800 °C in dry and wet air (~3% H_2O) for 96 h were identified. Figure 10 shows typical AFM images and Raman spectrum taking from the LSCF bar surface after heat treatment at 800 °C in dry air for 96 h. Optical and AFM images indicate the presence of two kinds of segregated particles. The Raman spectrum obtained by averaging the recorded spectra from the selected particles on the LSCF surface shows four peaks located at around 672, 600, 506, and 464 cm^{-1} (Fig. 10c). Raman peak of 675 cm^{-1} can be attributed to the Co_3O_4 segregated on the LSCF surface [91]. A mapping of Raman intensity of this Co_3O_4 phase at wavelength 672 cm^{-1} collected from the same area of optical microscope images indicates the distribution of the Co_3O_4 phase (Fig. 10d). The results indicate that the segregated cobalt species have tendencies to agglomerate, forming relatively large Co_3O_4 oxide particles on the LSCF surface. However, Raman mapping of SrO is not applicable because there is no Raman activity for SrO. Nevertheless, the observation of fine particles on the LSCF surface (Fig. 10a and b) indicates SrO segregation from the LSCF perovskite lattice to the surface after heat treatment at 800 °C in dry air for 96 h, consistent with that reported by Oh et al. [60].

4.3 X-ray Photoelectron Spectroscopy (XPS)

X-ray photoelectron spectroscopy (XPS) is a technique for analyzing the surface chemistry of a material, and can measure the elemental composition, empirical formula, chemical state, and electronic state of the elements within a material. XPS spectra are obtained by irradiating a solid surface with a beam of X-rays while simultaneously measuring the kinetic energy and electrons that are emitted from the top 1–10 nm of the material being analyzed. This technique has been widely used to study the surface element composition and chemistry of the electrode/electrolyte interface in SOFCs [73, 92–101].

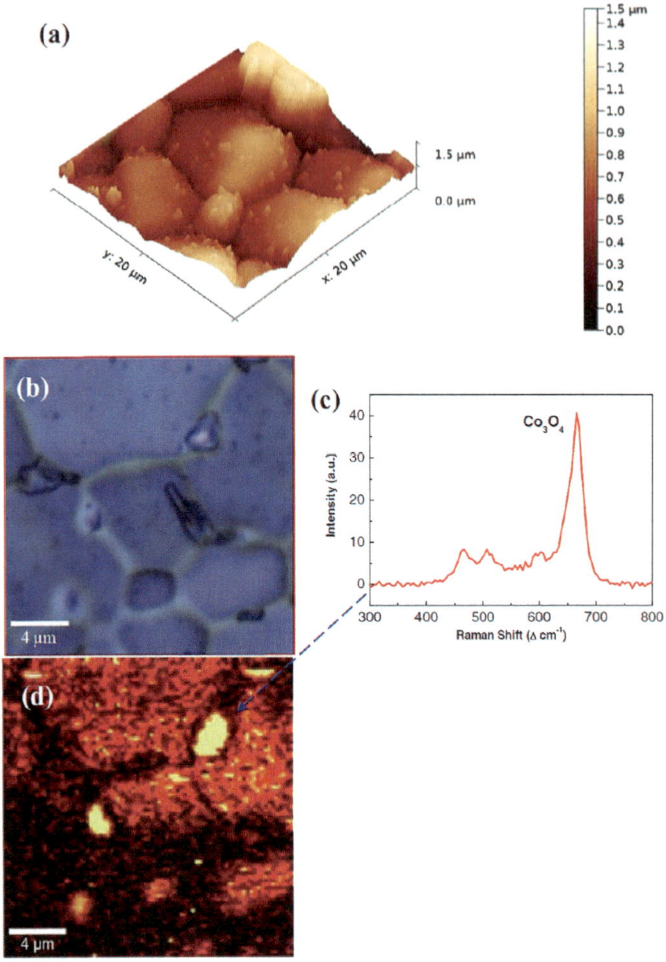

Fig. 10 **a** AFM micrographs and **b** optical microscope images of LSCF surface after heat treatment at 800 °C in the absence of Cr_2O_3 and dry air for 96 h, **c** Raman spectra from a selected particle on the LSCF surface, and **d** Raman mapping at wavelength of 675 cm^{-1}, corresponding to Co_3O_4 phase collected from the same area of optical microscope image in (**b**) (Reproduced from Ref. [90])

Backhaus-Ricoult et al. [95] conducted an in situ study using XPS to investigate the LSM electrode/YSZ electrolyte interface chemistry after cathodic polarization. Under increasingly negative bias, Mn was found to exhibit a continuous reduction in its average surface oxidation state. Under anodic polarization or open circuit condition, Mn adopted an oxidation state of 4^+ at the LSM surface, and a mixed oxidation state of 3^+–4^+ was found under low cathodic polarization. Under strong cathodic polarization, Mn reached an average surface oxidation state close to 2^+. Manganese was observed to spread in response to an increase in cathodic bias from the TPB over the electrolyte surface and was also found to retreat slowly when the bias

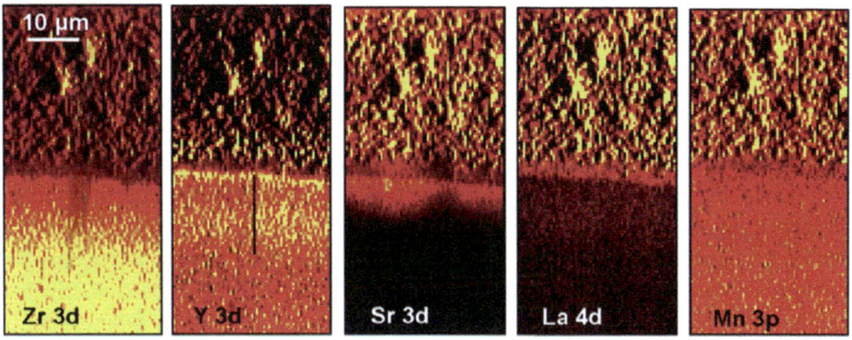

Fig. 11 XPS spectro-images of Zr 3d, Y 3d, Sr 3d, La 4d, and Mn 3p peaks in the TPB region at bias of −0.6 V. The upper portion of the image is LSM particles, while the bottom is the YSZ electrolyte. The yellow contrast corresponds to high concentration (Reproduced from Ref. [95])

was released or faster when an anodic bias was applied (Fig. 11). It has been proposed that the electrochemical activation of LSM cathode under cathodic polarization may be related to the rapid spread of the partially reduced manganese oxide species from the three-phase boundary over the electrolyte surface, thus providing high electronic conductivity in the electrolyte surface and promoting the direct incorporation of oxygen into the electrolyte.

5 Future Prospect

Interface formation between two solid phases, i.e., porous cathode and dense electrolyte in the case of SOFCs, is essentially driven by the minimization of the contact or surface energy of the two phases under high temperature sintering or under cathodic polarization. In the case of cathode/electrolyte interface, there are no visible differences in the electrochemical activity for ORR at the interfaces induced by high temperature or polarization, although the morphology of the thermally and polarization-induced interface is very different. However, in addition to the initial stage of the interface formation, the cathode/electrolyte interface stability is probably the most critical for the long-term durability operation of SOFCs. This is in turn closely related to the interface reaction and cation segregation, as discussed in this Chapter. Therefore, from the design point of view, the interface design can be approached from segregation mitigation and *in situ* active or protective interface formation.

5.1 Surface Segregation Mitigation

Surface segregation of cations in particular Sr of perovskite oxides can be supressed and mitigated by introducing A-site deficiencies in LSCF [102, 103] and A-site or B-site doping [104–111]. For example, Lu et al. [103] reported that a 5% A-site deficient LSCF oxygen electrode, $(La_{0.6}Sr_{0.4})_{0.95}Co_{0.2}Fe_{0.8}O_3$, can significantly reduce the formation of $SrZrO_3$ at the SDC/YSZ interface as compared with stoichiometric LSCF. A-site non-stoichiometry can also effectively reduce the reactivity between LSM electrodes and YSZ electrolytes [112, 113]. Introducing dopant is effective to enhance the activity and to increase the structural stability, thus suppressing Sr segregation of cobaltite base perovskites. It has been shown that B-site doping with high valence cations such as Nb [106–111], Sb [114, 115], Mo [116], Ta [117], Y [118–121], Sc [122] and W [123, 124] can be effective in enhancing the structural and compositional stability of cobalt-containing perovskite oxygen electrode materials such as LSCF, BSCF, $SrCoO_{3-\delta}$ and $BaCoO_{3-\delta}$. For example, we studied a Nb-doped $La_{0.24}Sr_{0.16}Ba_{0.6}Co_{0.5}Fe_{0.44}Nb_{0.06}O_{3-\delta}$ (LSBCFN) electrode prepared by a new direct mixing synthesis of $(La_{0.6}Sr_{0.4})(Co_{0.2}Fe_{0.8})O_{3-\delta}$ (LSCF) and $Ba(Co_{0.7}Fe_{0.2}Nb_{0.1})O_{3-\delta}$ (BSCN) [125]. The new LSBCFN electrode showed high electrical conductivities (124 S cm^{-1} at 600 °C) and better electrochemical activities for ORR as compared with those of LSCF or BCFN in temperatures ranging from 600 to 900 °C. Most important, the LSBCFN electrodes show an excellent tolerance towards Cr deposition as a result of the significantly reduced surface segregation of Sr and Ba.

However, a balance exists between structural stability and electrocatalytic activity in the doping of high valence cations in perovskite oxygen electrode B-sites. For example, Fan et al. [126] reported that in the case of Nb-doped BSCF, Nb-doping can enhance the structural stability of BSCF but at the expense of reduced oxygen permeation activity. The doping of 10% Nb in BSCF B-sites led to a 10% lowering in oxygen flux whereas the doping of 20% Nb drastically increased performance losses to 65%. Similarly, Chen et al. [127] reported that Nb doping in LSCF B-sites significantly reduced SrO surface segregation and enhanced the stability of oxygen electrodes but also decreased electrode activity, whereas Pd doping promoted ORR activities due to the exsolution of catalytically active Pd/PdO nanoparticles under the influence of cathodic polarization. As a result, the synergistic effects of Nb and Pd co-doping can lead to increased electrocatalytic activities and operating stabilities for LSCF electrodes.

5.2 In situ Active Interface Formation

More recently, we have found that by introducing doped Bi_2O_3 materials such as $Er_{0.4}Bi_{1.6}O_3$ (ESB) into LSM, $Sm_{0.95}Co_{0.95}Pd_{0.05}O_{3-\delta}$ (SmCPd) and $La_{0.6}Sr_{0.4}Co_{0.2}Fe_{0.7}Nb_{0.1}O_{3-\delta}$ (LSCFNb) via a decoration method, ESB-decorated composite electrodes provided significantly enhanced cathode performances for ORR in SOFCs [128–130]. In the case of LSCFNb electrodes directly assembled on barrier-layer-free YSZ electrolyte cells, ESB decoration substantially increased the power performance and stability of the cell, achieving a peak power density of 1.32 Wcm^{-2} and excellent stability at 750 °C and 250 $mAcm^{-2}$ for 100 h [130]. The study indicates the redistribution and migration of the decorated ESB phase in the ESB-LSCFNb composite towards the YSZ electrolyte under the influence of cathodic polarization and the formation of a thin ESB layer at the cathode/YSZ electrolyte interface (Fig. 12) [130]. The *in situ* formed ESB layer can not only prevent direct contact and subsequent reaction between segregated SrO and YSZ electrolyte, but also significantly promote oxygen migration/diffusion at the interface for O_2 reduction reactions, resulting in remarkable increases in power output and decreases in activation energy.

The electrochemical activity of the cathode can also be enhanced by engineering of the defect concentration and distribution at the interface between the electrolyte and the cathode. Kim et al showed recently that by modifying the interfaces between YSZ electrolyte and LSM cathode by introducing an infiltrated YSZ layer with various Y_2O_3 content, the enriched oxygen vacancies at the interfaces significantly promoted the ORR kinetics with the extended and active reaction sites of LSM cathodes [131].

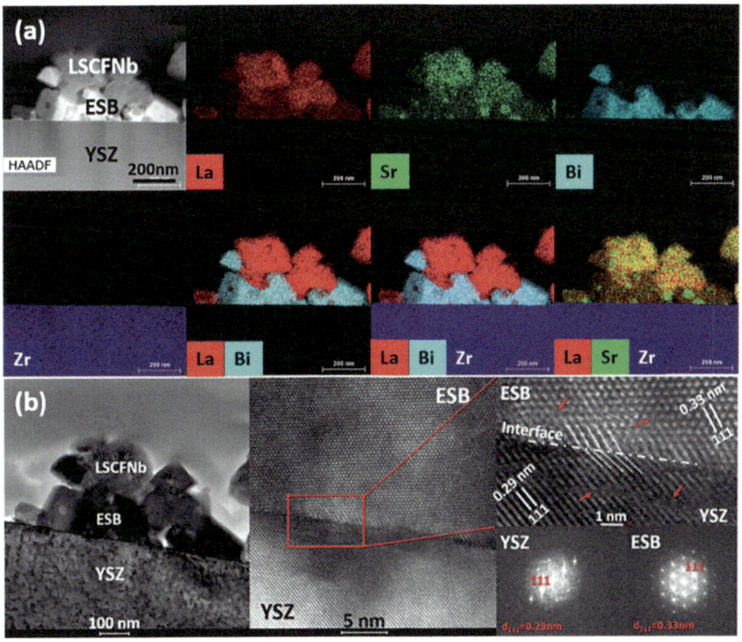

Fig. 12 **a** STEM-EDS element mapping and **b** HRTEM and FFT images of the ESB-LSCFNb/YSZ interface of an anode supported YSZ electrolyte cell with a directly assembled ESB-LSCFNb electrode after polarization at 250 $mAcm^{-2}$ and 750 °C for 100 hrs. Red arrows indicate lattice distortion (Reproduced from Ref. [130])

This implies that the defect chemistry at the interface can play an important role in the electrocatalytic activity of the electrodes for ORR.

Long-term operational durability for 30,000 or higher is most critical for the commercial viability of SOFCs technologies. One of the main factors affecting the long-term stability of a SOFC is the stability of the cathode/electrolyte interface [132]. The phase migration and in situ deposition and formation of highly active and protective layers on electrode/electrolyte interfaces under the influence of polarization potential have shown particular potential in the mitigation and inhibition of the detrimental effects of surface segregation in cobaltite-based electrodes of SOFCs [57]. However, very little is known for the microstructure and phase changes at the interface and their implications for the performance stability. More concentrated and significant efforts are required to fundamental understanding of the cathode/electrolyte interface and interface stability in particular during the long-term stability operation of SOFCs.

Acknowledgement This project is supported by the Australian Research Council under Discovery Project scheme (DP180100731, DP180100568).

References

1. Fabbri E, Bi L, Pergolesi D, Traversa E (2012) Towards the next generation of solid oxide fuel cells operating below 600 degrees c with chemically stable proton-conducting electrolytes. Adv Mater 24:195–208
2. Laguna-Bercero MA (2012) Recent advances in high temperature electrolysis using solid oxide fuel cells: a review. J Power Sources 203:4–16
3. Lee KT, Yoon HS, Wachsman ED (2012) The evolution of low temperature solid oxide fuel cells. J Mater Res 27:2063–2078
4. Mori T, Wepf R, Jiang SP (2020) Future prospects for the design of "state-of-the-art" solid oxide fuel cells. J Phys Energy 2:031001
5. Wang W, Su C, Wu Y, Ran R, Shao Z (2013) Progress in solid oxide fuel cells with nickel-based anodes operating on methane and related fuels. Chem Rev 113:8104–8151
6. Jeong J, Baek S-W, Bae J (2014) A diesel-driven, metal-based solid oxide fuel cell. J Power Sources 250:98–104
7. Gür TM (2016) Comprehensive review of methane conversion in solid oxide fuel cells: prospects for efficient electricity generation from natural gas. Prog Energy Combust Sci 54:1–64
8. Papurello D, Borchiellini R, Bareschino P, Chiodo V, Freni S, Lanzini A, Pepe F, Ortigoza GA, Santarelli M (2014) Performance of a solid oxide fuel cell short-stack with biogas feeding. Appl Energy 125:254–263
9. Wang S, Jiang SP (2017) Prospects of fuel cell technologies. Natl Sci Rev 4:163–166
10. Singh B, Ghosh S, Aich S, Roy B (2017) Low temperature solid oxide electrolytes (LT-SOE): a review. J Power Sources 339:103–135
11. Bucher E, Sitte W (2011) Long-term stability of the oxygen exchange properties of (La, Sr)$_{1-z}$(Co, Fe)$O_{3-\delta}$ in dry and wet atmospheres. Solid State Ionics 192:480–482
12. Tu H, Stimming U (2004) Advances, aging mechanisms and lifetime in solid-oxide fuel cells. J Power Sources 127:284–293

13. Mai A, Becker M, Assenmacher W, Tietz F, Hathiramani D, Iverstiffee E, Stover D, Mader W (2006) Time-dependent performance of mixed-conducting SOFC cathodes. Solid State Ionics 177:1965–1968
14. Understanding YH, Compatibility M (2003) Annu Rev Mater Res 33:581–610
15. Yokokawa H (1991) Thermodynamic analysis of reaction profiles between $LaMO_3$ (M=Ni,Co,Mn) and ZrO_2. J Electrochem Soc 138
16. Petric A (2000) Evaluation of La–Sr–Co–Fe–O perovskites for solid oxide fuel cells and gas separation membranes. Solid State Ionics 135:719–725
17. Figueiredo F (1997) Reactions between a zirconia-based electrolyte and $LaCoO_3$-based electrode materials. Solid State Ionics 101–103:343–349
18. Chen M, Liu YL, Hagen A, Hendriksen PV, Poulsen FW (2009) LSM-YSZ reactions in different atmospheres. Fuel Cells 9:833–840
19. Matsui T, Komoto M, Muroyama H, Eguchi K (2014) Interfacial stability between air electrode and ceria-based electrolyte under cathodic polarization in solid oxide fuel cells. Fuel Cells 14:1022–1027
20. Kuznecov M (2003) Diffusion controlled oxygen transport and stability at the perovskite/electrolyte interface. Solid State Ionics 157:371–378
21. Daio T, Bishop SR, Yildiz B, Tuller HL, Lyth SM, Hyodo JK, Kaneko SM, Perry NH (2017) Specific cation diffusion across the $(La,Sr)MnO_3$ $(Ce,Gd)O_2$ interface in SOFCs
22. Wang F, Brito ME, Yamaji K, Cho D-H, Nishi M, Kishimoto H, Horita T, Yokokawa H (2014) Effect of polarization on Sr and Zr diffusion behavior in LSCF/GDC/YSZ system. Solid State Ionics 262:454–459
23. Simner SP, Anderson MD, Engelhard MH, Stevenson JW (2006) Degradation mechanisms of La–Sr–Co–Fe–O_3 SOFC cathodes. Electrochem Solid-State Lett 9:A478
24. Kim Y-M, Kim-Lohsoontorn P, Bae J (2010) Effect of unsintered gadolinium-doped ceria buffer layer on performance of metal-supported solid oxide fuel cells using unsintered barium strontium cobalt ferrite cathode. J Power Sources 195:6420–6427
25. Kim JY, Sprenkle VL, Canfield NL, Meinhardt KD, Chick LA (2006) Effects of chrome contamination on the performance of $La_{0.6}Sr_{0.4}Co_{0.2}Fe_{0.8}O_3$ cathode used in solid oxide fuel cells. J Electrochem Soc 153
26. Chen K, Liu SS, Guagliardo P, Kilburn MR, Koyama M, Jiang SP (2015) A fundamental study of boron deposition and poisoning of $La_{0.8}Sr_{0.2}MnO_3$ cathode of solid oxide fuel cells under accelerated conditions. J Electrochem Soc 162:F1282–F1291
27. Chen K, Jiang SP (2020) Surface segregation in solid oxide cell oxygen electrodes: phenomena, mitigation strategies and electrochemical properties. Electrochem Energy Rev
28. Wilson JR, Kobsiriphat W, Mendoza R, Chen HY, Hiller JM, Miller DJ, Thornton K, Voorhees PW, Adler SB, Barnett SA (2006) Three-dimensional reconstruction of a solid-oxide fuel-cell anode. Nat Mater 5:541–544
29. Jiang SP (2007) Activation, microstructure, and polarization of solid oxide fuel cell cathodes. J Solid State Electrochem 11:93–102
30. Jiang SP (2008) Development of lanthanum strontium manganite perovskite cathode materials of solid oxide fuel cells: a review. J Mater Sci 43:6799–6833
31. Mizusaki J, Yonemura Y, Kamata H, Ohyama K, Mori N, Takai H, Tagawa H, Dokiya M, Naraya K, Sasamoto T, Inaba H, Hashimoto T (2000) Electronic conductivity, seebeck coefficient, defect and electronic structure of nonstoichiometric $La_{1-x}Sr_xMnO_3$. Solid State Ionics 132:167–180
32. Pavone M, Munoz-Garcia AB, Ritzmann AM, Carter EA (2014) First-principles study of lanthanum strontium manganite: insights into electronic structure and oxygen vacancy formation. J Phys Chem C 118:13346–13356
33. Poulsen FW (2000) Defect chemistry modelling of oxygen-stoichiometry, vacancy concentrations, and conductivity of $(La_{1-x}Srx)(y)MnO_3$+/-delta. Solid State Ionics 129:145–162
34. Nowotny J, Rekas M (1998) Defect chemistry of $(La, Sr)MnO_3$. J Am Ceram Soc 81:67–80
35. Badwal SPS, Jiang SP, Love J, Nowotny J, Rekas M, Vance ER (2001) Chemical diffusion in perovskite cathodes of solid oxide fuel cells: the Sr doped $LaMn_{1-x}M_xO_3$ (M=Co, Fe) systems. Ceram Int 27:419–429

36. Lee YL, Kleis J, Rossmeisl J, Morgan D (2009) Ab initio energetics of LaBO3(001) (B=Mn, Fe, Co, and Ni) for solid oxide fuel cell cathodes. Phys Rev B 80
37. Jiang SP (2019) Development of lanthanum strontium cobalt ferrite perovskite electrodes of solid oxide fuel cells—a review. Int J Hydrog Energy 44:7448–7493
38. Moon H, Kim S, Hyun S, Kim H (2008) Development of IT-SOFC unit cells with anode-supported thin electrolytes via tape casting and co-firing. Int J Hydrogen Energy 33:1758–1768
39. Fukui T, Ohara S, Naito M, Nogi K (2002) Performance and stability of SOFC anode fabricated from NiO–YSZ composite particles. J Power Sources 110:91–95
40. Matsui T, Mikami Y, Muroyama H, Eguchi K (2013) Influence of (La, Sr)$MnO_{3+\delta}$ cathode composition on cathode/electrolyte interfacial structure during long-term operation of solid oxide fuel cells. J Power Sources 242:790–796
41. Liu YL, Hagen A, Barfod R, Chen M, Wang HJ, Poulsen FW, Hendriksen PV (2009) Microstructural studies on degradation of interface between LSM–YSZ cathode and YSZ electrolyte in SOFCs. Solid State Ionics 180:1298–1304
42. Matsui T, Mikami Y, Muroyama H, Eguchi K (2010) Quantitative analysis of microstructural change at the interface between (La, Sr)MnO_3 cathode and YSZ electrolyte upon discharge operation. J Electrochem Soc 157:B1790
43. Jiang SP (2015) Thermally and electrochemically induced electrode/electrolyte interfaces in solid oxide fuel cells: an AFM and EIS study. J Electrochem Soc 162:F1119–F1128
44. Fukunaga H (1996) The relationship between overpotential and the three phase boundary length. Solid State Ionics 86–88:1179–1185
45. Mukhopadhyay M, Mukhopadhyay J, Sharma AD, Basu RN (2011) In-situ patterned intra-anode triple phase boundary in SOFC electroless anode: an enhancement of electrochemical performance. Int J Hydrogen Energy 36:7677–7682
46. Horita T, Tsunoda T, Yamaji K, Sakai N, Kato T, Yokokawa AH (2002) Microstructures and oxygen diffusion at the $LaMnO_3$ film/yttria-stabilized zirconia interface. Solid State Ionics 152–153:439–446
47. He S, Chen K, Saunders M, Li J, Cui CQ, Jiang SP (2017) A FIB-STEM study of $La_{0.8}Sr_{0.2}MnO_3$ cathode and Y_2O_3-ZrO_2/Gd_2O_3-CeO_2 electrolyte interfaces of solid oxide fuel cells. J Electrochem Soc 164:F1437–F1447
48. Li N, Ai N, Chen K, Cheng Y, He S, Saunders M, Dodd A, Suvorova A, Jiang SP (2016) In situ assembled $La_{0.8}Sr_{0.2}MnO_3$ cathodes on a Y_2O_3–ZrO_2 electrolyte of solid oxide fuel cells—interface and electrochemical activity. RSC Adv 6:99211–99219
49. Grosjean A, Sanseau O, Radmilovic V, Thorel A (2006) Reactivity and diffusion between $La_{0.8}Sr_{0.2}MnO_3$ and ZrO_2 at interfaces in SOFC cores by TEM analyses on FIB samples. Solid State Ionics 177:1977–1980
50. Cologna M, Raj R (2011) Surface diffusion-controlled neck growth kinetics in early stage sintering of Zirconia, with and without applied DC electrical field. J Am Ceram Soc 94:391–395
51. Kingery WD (1976) Introduction to ceramics. In: Kingery WD, Bowen HK, Uhlmann DR (eds) Wiley, New York
52. Zhao L, Hyodo J, Ishihara T, Sasaki K, Bishop SR (2013) XRD and Raman spectroscopy study of Mn solubility in cerium oxide. ECS Trans 57:1607–1612
53. Jiang SP, Wang W (2005) Effect of polarization on the interface between (La, Sr)MnO_3 electrode and Y_2O_3-ZrO_2 electrolyte. Electrochem Solid-State Lett 8:A115
54. Chen K, Li N, Ai N, Li M, Cheng Y, Rickard WDA, Li J, Jiang SP (2016) Direct application of cobaltite-based perovskite cathodes on the yttria-stabilized zirconia electrolyte for intermediate temperature solid oxide fuel cells. J Mater Chem A 4:17678–17685
55. Li N, Ai N, He S, Cheng Y, Rickard WDA, Chen K, Zhang T, Jiang SP (2018) Effect of Pd doping on the activity and stability of directly assembled $La_{0.95}Co_{0.19}Fe_{0.76}Pd_{0.05}O_{3-\delta}$ cathodes of solid oxide fuel cells. Solid State Ionics 316:38–46
56. Chen K, He S, Li N, Cheng Y, Ai N, Chen M, Rickard WDA, Zhang T, Jiang SP (2018) Nb and Pd co-doped $La_{0.57}Sr_{0.38}Co_{0.19}Fe_{0.665}Nb_{0.095}Pd_{0.05}O_{3-\delta}$ as a stable, high performance electrode for barrier-layer-free Y_2O_3-ZrO_2 electrolyte of solid oxide fuel cells. J Power Sources 378:433–442

57. Chen ZY, Jiang WJ, Lu Z, Wang ZH, Chen ZX, Jiang SP, Lin TC, Shao YQ, Tang D, Chen KF, Ai N (2020) Accelerating effect of polarization on electrode/electrolyte interface generation and electrocatalytic performance of $Er_{0.4}Bi_{1.6}O_3$ decorated $Sm_{0.95}CoO_3-$delta cathodes. J Power Sources:465
58. He S, Saunders M, Chen K, Gao H, Suvorova A, Rickard WDA, Quadir Z, Cui CQ, Jiang SP (2018) A FIB-STEM study of strontium segregation and interface formation of directly assembled $La_{0.6}Sr_{0.4}Co_{0.2}Fe_{0.8}O_{3-\delta}$Cathode on Y_2O_3-ZrO_2 electrolyte of solid oxide fuel cells. J Electrochem Soc 165:F417–F429
59. Sun Y, He S, Saunders M, Chen K, Shao Z, Jiang SP (2020) A comparative study of surface segregation and interface of $La_{0.6}Sr_{0.4}Co_{0.2}Fe_{0.8}O_{3-\delta}$ electrode on GDC and YSZ electrolytes of solid oxide fuel cells. Int J Hydrog Energy
60. Oh D, Gostovic D, Wachsman ED (2012) Mechanism of $La_{0.6}Sr_{0.4}Co_{0.2}Fe_{0.8}O_3$ cathode degradation. J Mater Res 27:1992–1999
61. Yu Y, Ludwig KF, Woicik JC, Gopalan S, Pal UB, Kaspar TC, Basu SN (2016) Effect of Sr content and strain on Sr surface segregation of $La_{1-x}Sr_xCo_{0.2}Fe_{0.8}O_{3-\delta}$ as cathode material for solid oxide fuel cells. ACS Appl Mater Interfaces 8:26704–26711
62. Liu YH, Chen KF, Zhao L, Chi B, Pu J, Jiang SP, Jian L (2014) Performance stability and degradation mechanism of $La_{0.6}Sr_{0.4}Co_{0.2}Fe_{0.8}O_3$-delta cathodes under solid oxide fuel cells operation conditions. Int J Hydrog Energy 39:15868–15876
63. Zhao L, Drennan J, Kong C, Amarasinghe S, Jiang SP (2014) Insight into surface segregation and chromium deposition on $La_{0.6}Sr_{0.4}Co_{0.2}Fe_{0.8}O_3$-delta cathodes of solid oxide fuel cells. J Mater Chem A 2:11114–11123
64. Pan ZH, Liu QL, Zhang L, Zhang XW, Chan SH (2015) Effect of Sr surface segregation of $La_{0.6}Sr_{0.4}Co_{0.2}Fe_{0.8}O_3$-delta electrode on its electrochemical performance in SOC. J Electrochem Soc 162:F1316–F1323
65. Simner SP, Anderson MD, Engelhard MH, Stevenson JW (2006) Degradation mechanisms of La-Sr-Co-Fe-O_3SOFC cathodes. Electrochem Solid State Lett 9:A478–A481
66. Ding HP, Virkar AV, Liu ML, Liu F (2013) Suppression of Sr surface segregation in $La_{1-x}SrxCo_{1-y}FeyO_3$-delta: a first principles study. Phys Chem Chem Phys 15:489–496
67. Araki W, Yamaguchi T, Arai Y, Malzbender J (2014) Strontium surface segregation in $La_{0.58}Sr_{0.4}Co_{0.2}Fe_{0.8}O_3$-delta annealed under compression. Solid State Ionics 268:1–6
68. Niania M, Podor R, Ben Britton T, Li C, Cooper SJ, Svetkov N, Skinner S, Kilner J (2018) In situ study of strontium segregation in $La_{0.6}Sr_{0.4}Co_{0.2}Fe_{0.8}O_3$-delta in ambient atmospheres using high-temperature environmental scanning electron microscopy. J Mater Chem A 6:14120–14135
69. Li YF, Zhang WQ, Zheng Y, Chen J, Yu B, Chen Y, Liu ML (2017) Controlling cation segregation in perovskite-based electrodes for high electro-catalytic activity and durability. Chem Soc Rev 46:6345–6378
70. Druce J, Tellez H, Burriel M, Sharp MD, Fawcett LJ, Cook SN, McPhail DS, Ishihara T, Brongersma HH, Kilner JA (2014) Surface termination and subsurface restructuring of perovskite-based solid oxide electrode materials. Energy Environ Sci 7:3593–3599
71. He S, Saunders M, Chen KF, Gao HF, Suvorova A, Rickard WDA, Quadir Z, Cui CQ, Jiang SP (2018) A FIB-STEM study of strontium segregation and interface formation of directly assembled $La_{0.6}Sr_{0.4}Co_{0.2}Fe_{0.8}O_3$ delta cathode on Y_2O_3-ZrO_2 electrolyte of solid oxide fuel cells. J Electrochem Soc 165:F417–F429
72. Wang F, Nishi M, Brito ME, Kishimoto H, Yamaji K, Yokokawa H, Horita T (2014) Sr and Zr diffusion in LSCF/10GDC/8YSZ triplets for solid oxide fuel cells (SOFCs). J Power Sources 258:281–289
73. Vovk G, Chen XH, Mims CA (2005) In situ XPS studies of perovskite oxide surfaces under electrochemical polarization. J Phys Chem B 109:2445–2454
74. Chen K, Jiang SP (2020) Surface segregation in solid oxide cell oxygen electrodes: phenomena, mitigation strategies and electrochemical properties. Electrochem Energy Rev. https://doi.org/10.1007/s41918-020-00078-z

75. Baram M, Kaplan WD (2008) Quantitative HRTEM analysis of FIB prepared specimens. J Microsc 232:395–405
76. Soldati AL, Baqué L, Troiani H, Cotaro C, Schreiber A, Caneiro A, Serquis A (2011) High resolution FIB-TEM and FIB-SEM characterization of electrode/electrolyte interfaces in solid oxide fuel cells materials. Int J Hydrogen Energy 36:9180–9188
77. Liu SS, Matsumura S, Koyama M (2015) Boundary observation and contrast tuning of Ni/YSZ anode by TEM and FIB-SEM. ECS Trans 68:1275–1279
78. Liu SS, Takayama A, Matsumura S, Koyama M (2016) Image contrast enhancement of Ni/YSZ anode during the slice-and-view process in FIB-SEM. J Microsc 261:326–332
79. Shearing PR, Gelb J, Brandon NP (2010) X-ray nano computerised tomography of SOFC electrodes using a focused ion beam sample-preparation technique. J Eur Ceram Soc 30:1809–1814
80. Jublot M, Texier M (2014) Sample preparation by focused ion beam micromachining for transmission electron microscopy imaging in front-view. Micron 56:63–67
81. Sezen M (2016) Modern electron microscopy in physical and life sciences
82. Yang J, Muroyama H, Matsui T, Eguchi K (2012) The influence of discharge operation on the microstructure of strontium-doped lanthanum manganite cathode for solid oxide fuel cells. J Power Sources 204:25–33
83. Nagasawa T, Shimura T, Shikazono N, Hanamura K (2019) Characterization of active sites in solid oxide fuel cell composite cathode through oxygen isotope labeling combined with three-dimensional microstructural analysis by focused ion beam—scanning electron microscopy. ECS Trans 91:1371–1375
84. Nagasawa T, Hanamura K (2019) Investigation of oxide ion flux at cathode/electrolyte interface in solid oxide fuel cell. J Power Sources 412:695–700
85. Develos-Bagarinao K, Budiman RA, Liu S-S, Ishiyama T, Kishimoto H, Yamaji K (2020) Evolution of cathode-interlayer interfaces and its effect on long-term degradation. J Power Sources:453
86. Chanthanumataporn M, Nagasawa T, Hanamura K (2019) Direct visualization of oxygen incorporation for oxygen reduction on porous mixed-Conducting LaSrCoO3 cathode. J Electrochem Soc 166:F581–F586
87. Zhao L, Zhang J, Becker T, Jiang SP (2014) Raman spectroscopy study of chromium deposition on $La_{0.6}Sr_{0.4}Co_{0.2}Fe_{0.8}O_3$-cathode of solid oxide fuel cells. J Electrochem Soc 161:F687–F693
88. Agarkov DA, Burmistrov IN, Tsybrov FM, Tartakovskii II, Kharton VV, Bredikhin SI (2017) In-situ Raman spectroscopy analysis of the interfaces between Ni-based SOFC anodes and stabilized zirconia electrolyte. Solid State Ionics 302:133–137
89. Agarkov DA, Burmistrov IN, Tsybrov FM, Tartakovskii II, Kharton VV, Bredikhin SI (2018) In-situ Raman spectroscopy analysis of the interface between ceria-containing SOFC anode and stabilized zirconia electrolyte. Solid State Ionics 319:125–129
90. Zhao L, Zhang J, Becker T, Jiang SP (2014) Raman spectroscopy study of chromium deposition on $La_{0.6}Sr_{0.4}Co_{0.2}Fe_{0.8}Co_3$ cathodes of solid oxide fuel cells. J Electrochem Soc 161:F687–F693
91. Liu M, Ding D, Blinn K, Li X, Nie L, Liu M (2012) Enhanced performance of LSCF cathode through surface modification. Int J Hydrogen Energy 37:8613–8620
92. la O' GJ, Savinell RF, Shao-Horn Y (2009) Activity enhancement of dense strontium-doped lanthanum manganite thin films under cathodic polarization: a combined AES and XPS study. J Electrochem Soc 156:B771–B781
93. Baumann FS, Fleig Jr, Konuma M, Starke U, Habermeier HU, Maier J (2005) Strong performance improvement of $La_{0.6}Sr_{0.4}Co_{0.8}Fe_{0.2}O_{3-\delta}$ SOFC cathodes by electrochemical activation. J Electrochem Soc 152:A2074
94. Wu QH, Liu M, Jaegermann W (2005) X-ray photoelectron spectroscopy of $La_{0.5}Sr_{0.5}MnO_3$. Mater Lett 59:1980–1983
95. Backhaus-Ricoult M, Adib K, St Clair T, Luerssen B, Gregoratti L, Barinov A (2008) In-situ study of operating SOFC LSM/YSZ cathodes under polarization by photoelectron microscopy. Solid State Ionics 179:891–895

96. Jalili H, Han JW, Kuru Y, Cai Z, Yildiz B (2011) New insights into the strain coupling to surface chemistry, electronic structure, and reactivity of $La_{0.7}Sr_{0.3}MnO_3$. J Phys Chem Lett 2:801–807
97. Cai Z, Kubicek M, Fleig J, Yildiz B (2012) Chemical heterogeneities on $La_{0.6}Sr_{0.4}CoO_{3-\delta}$ thin films—correlations to cathode surface activity and stability. Chem Mater 24:1116–1127
98. Wei T, Zhang Q, Huang Y-H, Goodenough JB (2012) Cobalt-based double-perovskite symmetrical electrodes with low thermal expansion for solid oxidefuel cells. J Mater Chem 22:225–231
99. Chen K, Hyodo J, O'Donnell KM, Rickard W, Ishihara T, Jiang SP (2014) Effect of volatile boron species on the electrocatalytic activity of cathodes of solid oxide fuel cells: III. $Ba_{0.5}Sr_{0.5}Co_{0.8}Fe_{0.2}O_3$-electrodes. J Electrochem Soc 161:F1163–F1170
100. Wang CC, Chen K, Jiang SP (2016) Mechanism and kinetics of SO2Poisoning on the electrochemical activity of $La_{0.8}Sr_{0.2}MnO_3$ cathodes of solid oxide fuel cells. J Electrochem Soc 163:F771–F780
101. Götsch T, Praty C, Grünbacher M, Schlicker L, Bekheet MF, Doran A, Gurlo A, Tada M, Matsui H, Ishiguro N, Kloetzer B, Penner S (2017) Iron exsolution phenomena in lanthanum strontium ferrite SOFC anodes. ECS Trans 78:1327–1341
102. Ding H, Virkar AV, Liu M, Liu F (2013) Suppression of Sr surface segregation in $La_{1\text{-}x}Sr_xCo_{1\text{-}y}FeyO_{3-\delta}$: a first principles study. Phys Chem Chem Phys 15:489–496
103. Lu Z, Darvish S, Hardy J, Templeton J, Stevenson J, Zhong Y (2017) SrZrO3 formation at the interlayer/electrolyte interface during $(La_{1\text{-}x}Sr_x)1\text{-}\delta Co_{1\text{-}y}FeyO_3$ cathode sintering. J Electrochem Soc 164:F3097–F3103
104. Yoo S, Jun A, Ju Y-W, Odkhuu D, Hyodo J, Jeong HY, Park N, Shin J, Ishihara T, Kim G (2014) Development of double-perovskite compounds as cathode materials for low-temperature solid oxide fuel cells. Angew Chem Int Ed 53:13064–13067
105. Choi S, Park S, Shin J, Kim G (2015) The effect of calcium doping on the improvement of performance and durability in a layered perovskite cathode for intermediate-temperature solid oxide fuel cells. J Mater Chem A 3:6088–6095
106. Ito W, Nagai T, Sakon T (2007) Oxygen separation from compressed air using a mixed conducting perovskite-type oxide membrane. Solid State Ionics 178:809–816
107. Nagai T, Ito W, Sakon T (2007) Relationship between cation substitution and stability of perovskite structure in $SrCoO_{3-\delta}$-based mixed conductors. Solid State Ionics 177:3433–3444
108. Chen XB, Jiang SP (2013) Highly active and stable $(La_{0.24}Sr_{0.16}Ba_{0.6})(Co_{0.5}Fe_{0.44}Nb_{0.06})O_{3-\delta}$ (LSBCFN) cathodes for solid oxide fuel cells prepared by a novel mixing synthesis method. J Mater Chem A 1:4871-4878
109. Zhang P, Guan G, Khaerudini DS, Hao X, Han M, Kasai Y, Sasagawa K, Abudula A (2014) Properties of A-site nonstoichiometry $(Pr_{0.4})xSr_{0.6}Co_{0.2}Fe_{0.7}Nb_{0.1}O3-_{\sigma}$ $(0.9 \leq x \leq 1.1)$ as symmetrical electrode material for solid oxide fuel cells. J Power Sources 248:163–171
110. Li J, Yang C, Liu M (2016) High performance intermediate temperature solid oxide fuel cells with $Ba_{0.5}Sr_{0.5}Co_{0.8}Fe_{0.1}Nb_{0.1}O-\delta$ as cathode. Ceram Int 42:19397–19401
111. Wang F, Nakamura T, Yashiro K, Mizusaki J, Arnezawa K (2014) Effect of Nb doping on the chemical stability of BSCF-based solid solutions. Solid State Ionics 262:719–723
112. Jiang SP, Love JG, Zhang JP, Hoang M, Ramprakash Y, Hughes AE, Badwal SPS (1999) The electrochemical performance of LSM/zirconia-yttria interface as a function of a-site non-stoichiometry and cathodic current treatment. Solid State Ionics 121:1–10
113. Jiang SP, Zhang JP, Ramprakash Y, Milosevic D, Wilshier K (2000) An investigation of shelf-life of strontium doped $LaMnO_3$ materials. J Mater Sci 35:2735–2741
114. Aguadero A, Perez-Coll D, de la Calle C, Alonso JA, Escudero MJ, Daza L (2009) $SrCo_{1\text{-}x}Sb_xO_{3-\delta}$ perovskite oxides as cathode materials in solid oxide fuel cells. J Power Sources 192:132–137
115. Wang F, Igarashi K, Nakamura T, Yashiro K, Mizusaki J, Amezawa K (2016) Tailoring the chemical stability of cobalt-rich perovskite mixed conductor. Solid State Ionics 288:2–5
116. Aguadero A, Pérez-Coll D, Alonso JA, Skinner SJ, Kilner J (2012) A new family of Mo-doped $SrCoO_{3-\delta}$ perovskites for application in reversible solid state electrochemical cells. Chem Mater 24:2655–2663

117. Li M, Zhao M, Li F, Zhou W, Peterson VK, Xu X, Shao Z, Gentle I, Zhu Z (2017) A niobium and tantalum co-doped perovskite cathode for solid oxide fuel cells operating below 500 °C. Nat Commun 8:13990
118. Meffert M, Unger LS, Grünewald L, Störmer H, Wagner SF, Ivers-Tiffée E, Gerthsen D (2017) The impact of grain size, A/B-cation ratio, and Y-doping on secondary phase formation in $(Ba_{0.5}Sr_{0.5})(Co_{0.8}Fe_{0.2})O_{3-\delta}$. J Mater Sci 52:2705–2719
119. Foit SR, Vinke IC, de Haart LGJ, Eichel R-A (2017) Power-to-syngas: an enabling technology for the transition of the energy system? Angew Chemie-Int Edn 56:5402–5411
120. Duan C, Hook D, Chen Y, Tong J, O'Hayre R (2017) Zr and Y co-doped perovskite as a stable, high performance cathode for solid oxide fuel cells operating below 500 °C. Energy Environ Sci 10:176–182
121. He W, Wu X, Yang G, Shi H, Dong F, Ni M (2017) $BaCo_{0.7}Fe_{0.22}Y_{0.08}O_{3-\delta}$ as an active oxygen reduction electrocatalyst for low-temperature solid oxide fuel cells below 600° C. ACS Energy Lett 2:301–305
122. Zhou W, Shao Z, Ran R, Cai R (2008) Novel $SrSc_{0.2}Co_{0.8}O_{3-\delta}$ as a cathode material for low temperature solid-oxide fuel cell. Electrochem Commun 10:1647–1651
123. Popov M, Bychkov S, Nemudry A (2016) Modification of mixed conducting $Ba_{0.5}Sr_{0.5}Co_{0.8}Fe_{0.2}O_{3-\delta}$ by partial substitution of cobalt with tungsten. Russ J Electrochem 52:648–654
124. Shin JF, Xu W, Zanella M, Dawson K, Savvin SN, Claridge JB, Rosseinsky MJ (2017) Self-assembled dynamic perovskite composite cathodes for intermediate temperature solid oxide fuel cells. Nat Energy 2:16214
125. Chen XB, Jiang SP (2013) Highly active and stable $(La_{0.24}Sr_{0.16}Ba_{0.6})(Co_{0.5}Fe_{0.44}Nb_{0.06})O_3$-delta (LSBCFN) cathodes for solid oxide fuel cells prepared by a novel mixing synthesis method. J Mater Chem A 1:4871–4878
126. Fang SM, Yoo CY, Bouwmeester HJM (2011) Performance and stability of niobium-substituted $Ba_{0.5}Sr_{0.5}Co_{0.8}Fe_{0.2}O_{3-\delta}$ membranes. Solid State Ionics 195:1–6
127. Chen Y, Yoo S, Li X, Ding D, Pei K, Chen D, Ding Y, Zhao B, Murphy R, Deglee B, Liu J, Liu M (2018) An effective strategy to enhancing tolerance to contaminants poisoning of solid oxide fuel cell cathodes. Nano Energy 47:474–480
128. Ai N, Li N, He S, Cheng Y, Saunders M, Chen K, Zhang T, Jiang SP (2017) Highly active and stable Er0.4Bi1.6O3 decorated $La_{0.76}Sr_{0.19}MnO_{3+\delta}$ nanostructured oxygen electrodes for reversible solid oxide cells. J Mater Chem A 5:12149–12157
129. Ai N, Chen M, He S, Chen K, Zhang T, Jiang SP (2018) High performance nanostructured bismuth oxide–cobaltite as a durable oxygen electrode for reversible solid oxide cells. J Mater Chem A 6:6510–6520
130. He S, Zhang Q, Maurizio G, Catellani L, Chen KF, Chang QB, Santarelli M, Jiang SP (2018) In situ formation of $Er_{0.4}Bi_{1.6}O_3$ protective layer at cobaltite cathode/Y_2O_3-ZrO_2 electrolyte interface under solid oxide fuel cell operation conditions. Acs Appl Mater Interfaces 10:40549–40559
131. Kim SJ, Koo JY, Mun T, Choi M, Lee W (2020) Tailoring defect chemistry at interfaces for promoted oxygen reduction reaction kinetics. J Mater Chem A 8:23313–23322
132. Matsui T, Yoshida N, Muroyama H, Eguchi K (2020) Electrocatalytic activity compensation mechanism upon long-term operation of solid oxide fuel cells; LSCM cathode/SDC interlayer/YSZ electrolyte system. J Mater Chem A 8:11867–11873

Design of HT-PEMFC Devices

Jiayi Tang, Chao Su, and Zongping Shao

Abstract Polymer electrolyte membrane fuel cells (HT-PEMFCs) that operate at higher temperatures (>100 °C) have emerged as one branch of next-generation power sources for efficient energy conversion applications. The operating environment of HT-PEMFCs brings great opportunities for faster electrode reaction kinetics, alleviates the poisoning effect of electrocatalysts while implementing fuels with a higher level of impurities, and allows for simpler water and heat management systems. Attracted by the exceptional merits, research efforts have been devoted to solve the technical challenges in the design and fabrication of HT-PEMFC devices to make this fuel cell technology compelling for the future progress in clean energy market. This chapter mainly focuses on the research and development efforts in recent years towards the design of novel HT-PEMFC devices, involving the development of key materials, such as electrocatalysts and membranes, the design of reaction interface and electrode structure, as well as the fabrication and integration of the stack and system. The cutting-edge electrode designs and fundamental issues associated with the HT-PEMFC technology are highlighted and identified to provide an extensive view of the current research and development status, and shed some light on the future research directions from components design to device application level.

Keywords High-temperature PEMFCs · PBI membrane · Membrane electrode assembly · Electrocatalysts

J. Tang · Z. Shao (✉)
WA School of Mines: Minerals, Energy and Chemical Engineering, Curtin University, Perth, WA 6102, Australia
e-mail: zonping.shao@curtin.edu.au

C. Su
School of Energy and Power, Jiangsu University of Science and Technology, Zhenjiang 212100, China

T. Mori et al. (eds.), *Design of Active Key Interfaces in Fuel Cells*,
https://doi.org/10.1007/978-981-97-5706-0_3

1 Introduction

For centuries, with serious shortage of energy resources, researchers are keen for seeking stable supplies of clean energy to produce electricity. Fuel cell technology thus emerged as it could extract electricity with high efficiency from clean fuels, for example, hydrogen. Proton-exchange membrane fuel cell (PEMFC) is gaining significant momentum with a large amount of publications appearing in recent years, and their applications have penetrated to many commercial scenarios, especially automobiles. By applying gases as fuels, PEMFCs are commonly operated in a range of temperature from 80 to 200 °C based on the capability and stability of proton exchange membrane (PEM). A standard H_2/AIR PEMFC prototype is illustrated in Fig. 1, in which PEM is sandwiched by two catalyst layers, namely, anode catalyst layer (ACL) and cathode catalyst layer (CCL) [1].

CLs need to be coupled with a complex gas diffusion layer (GDL), which aims to realize proper mass transportation during electrochemical reactions. The CL design is a long-time research hotspot in the development of PEMFC devices. As shown in the figure, it is a complicated issue to construct proper triple phase boundaries (TPBs), with desirable electrochemical reaction kinetics, efficient proton and electron transport, fast reactant and product transfer, water and heat exchange simultaneously [2]. Besides the core electrode region in PEMFC devices, the optimization of flow geometry in flow field plate (FFP) also plays an important role to facilitate the transformation of fuels, water, and heat during electrode reactions [3]. PEMFCs that operate under 100 °C are classified as low-temperature PEMFCs (LT-PEMFCs), and those operating above 100 °C usually require specific design of PEM to withstand high temperature, are called high-temperature PEMFCs (HT-PEMFCs). Several compelling reasons why researchers want to operating the PEMFC at high temperature are as follows: (1) Electrochemical kinetics with the electrodes are greatly enhanced; (2) Water management in cell and the coupled cooling system could be simplified; (3) The operating condition of HT-PEMFCs allows for fuels with impurities, like CO. Owing to these apparent merits, HT-PEMFCs stand out for

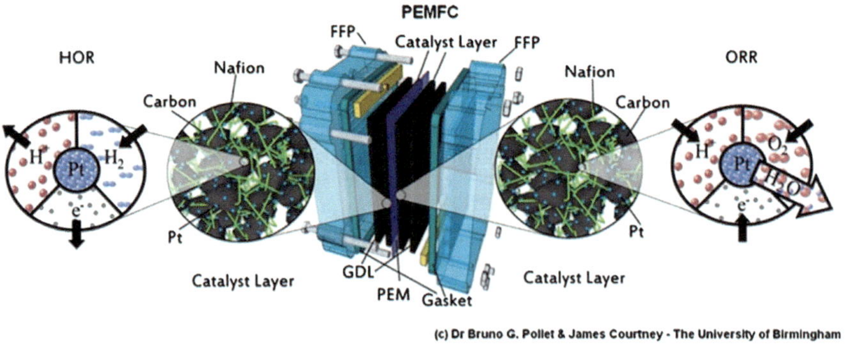

Fig. 1 Layout view of a single PEMFC. (Reproduced from Pollet et al. [1])

a wider range of application scenarios, compared with LT-PEMFCs, which makes them more promising power sources for vehicles, spacecraft, and combined heat and power (CHP) systems [4].

Though high-temperature operation brings lots of advantages, there exist some challenges to the components in electrode under high temperature. Conventional perfluorosulfonic acid (PFSA)-based materials, for example, Nafion® membrane requires humidification to aid proton conductivity during PEMFC operation. High temperature of >100 °C could largely limit the humification of the membrane with liquid water, thus puts great threat to the conductivity and stability of the membrane, as shown in Fig. 2 [5]. Most of the research works in developing HT-PEMFCs are dedicated to the development of efficient and stable PEMs. Phosphoric acid (PBI)-doped membranes are expected to be the most promising brunch of material that could be operated without humidification under high temperature in HT-PEMFCs, if that acid leaching issue could get rid of. One typical membrane electrode assembly (MEA) fabricated for HT-PEMFC is Celtec®-P 1000 is available from BASF Fuel Cell Inc., which have been demonstrated with long-term durability under operation at 160 °C with no humidification, showing a voltage drop of less than 6 $\mu V\ h^{-1}$ and a lifetime of at least 20, 000 h (Fig. 3) [6].

This chapter mainly focuses on the research and development efforts in recent years towards the fabrication of novel HT-PEMFC devices, from the perspectives of key components design, electrodes construction, stack and system integration, as well as the commercial application status. Advances based on the following issues in developing HT-PEMFC technology are discussed with emphasis in this chapter.

- Development of active and stable electrocatalysts for HT-PEMFCs;
- Progress for improving chemical and mechanical properties of PBI-membranes;

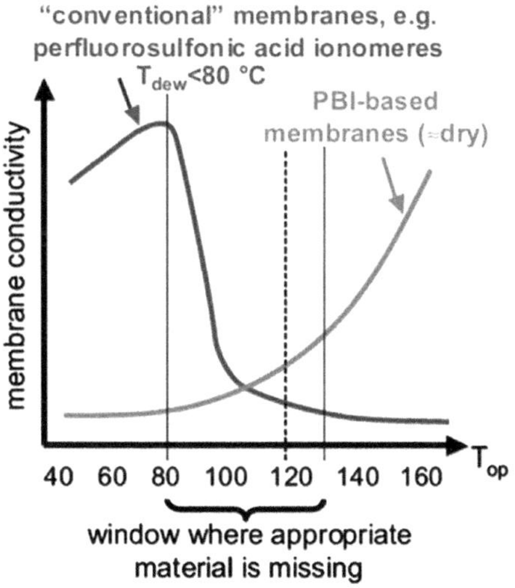

Fig. 2 Schematic of the conductivity change with the operating temperature. Difference between conventional PFSA-based material and PBI-based material. (Reproduced from Wieser [5])

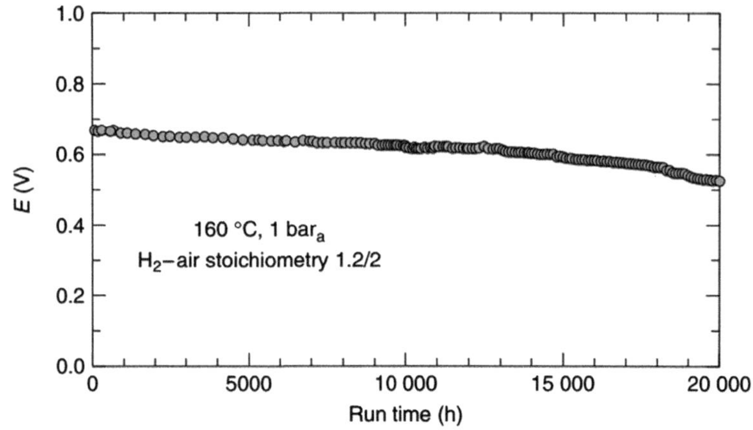

Fig. 3 Long-term durability test of Celtec®-P1000 MEA operating at 160 °C, 1 bar$_a$, H_2/air stoichiometry of 1.2/2. (Reproduced from Seel et al. [6])

- Optimization of MEA architecture and TPBs to improve mass transportation and electrode kinetics;
- Integrated stack design fabricated with simplified system.

Finally, the remaining technical challenges and future perspectives will be proposed for further investigation in this area.

2 Design of Key Components in HT-PEMFCs

2.1 Electrocatalysts

2.1.1 Electrode Reaction Kinetics in HT-PEMFCs

The cathodic reaction during PEMFC operation has long been more concerned by researchers, since the overpotential for hydrogen oxidation reaction (HOR) at the anode is negligibly small in the absence of mass transport compared with that of oxygen reduction reaction (ORR) at the cathode. Electrochemical reactions on the electrodes take place at the TPBs, as shown in Fig. 4, [2] consisting of active sites for catalytic reactions (e.g., platinum nanoparticles), support materials for electron transfer, ionomer for proton conduction, as well as primary and secondary pores for efficient mass transportation. The efficiency of electrode reactions is closely linked to the porous structure of CCLs, besides the intrinsic electrocatalytic kinetics of the active sites.

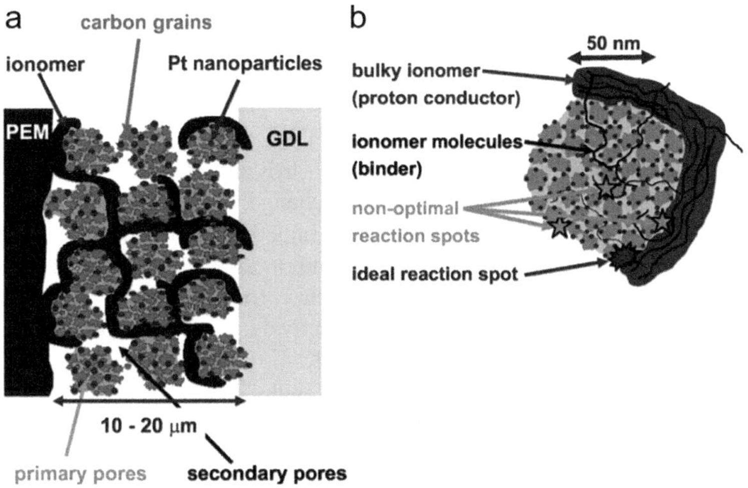

Fig. 4 Illustration of fine CL structure in the electrode. **a** Macro-scale components and porous structures; **b** Micro-scale of the composition and reaction spots of the single agglomerate. (Reproduced from Eikerling [2])

The electrode reactions in a typical PBI/H_3PO_4-based MEA are as follows:

- Anode:

$$H_2 + 2e^- \rightarrow 2H^+$$

$$H_2PO_4^- + H^+ = H_3PO_4$$

- Cathode:

$$O_2 + 4H^+ + 4e^- \rightarrow 2H_2O$$

$$PBI \cdot H^+ = PBI + H^+$$

- Membrane:

$$H_3PO_4 + PBI = H_2PO_4^- + PBI \cdot H^+$$

Take the most widely implemented platinum (Pt)-based electrocatalyst as an example. The electrode reaction kinetics for the cathodic ORR and are anodic HOR are greatly enhanced at elevated temperatures in HT-PEMFCs. According to Butler-Volmer equation, the exchange current densities for ORR and HOR on Pt catalyst as a function of temperature are expressed as follows [7].

$$I_c = i^0_{O_2}\left(e^{\frac{n_{\alpha O}\alpha_c F\eta_c}{RT}} - e^{\frac{n_{\alpha O}(1-\alpha_c)F\eta_c}{RT}}\right)(\text{ORR on Pt/PtO})$$

$$I_a = i^0_{H_2}\left(e^{\frac{n_{\alpha H}\alpha_H F\eta_a}{RT}} - e^{\frac{n_{\alpha H}(1-\alpha_H)F\eta_a}{RT}}\right)(\text{HOR on Pt})$$

I_c and I_a are the current densities for cathodic ORR and anodic HOR, respectively. $i^0_{O_2}$ and $i^0_{H_2}$ are the corresponding apparent exchange current densities for the two electrodes, which can be measured through AC impedance method [8]. η_c and η_a are the overpotentials for ORR and HOR [7]; α_c and α_H are the corresponding electron transfer coefficients. $n_{\alpha O}$ and $n_{\alpha H}$ are the electron transfer numbers for ORR and HOR of the rate-determining step (RDS), respectively. T is the temperature (K), and R is the universal gas constant (8.314 J mol^{-1} K^{-1}).

The anodic HOR has not been a big concern in the development of PEMFC. The HOR at the anode proceeds through the *Tafel-Volmer*, and/or the *Heyrovsky-Volmer routes,* as presented with the following equations [9–11].

$$2Pt + H_2 \leftrightarrow 2Pt - 2H_{ads}(\text{Tafel reaction, 1})$$

$$Pt + H_2 \leftrightarrow Pt - H_{ads} + H^+ + e^-(\text{Heyrovshi reaction, 2})$$

$$2Pt - H_{ads} \leftrightarrow 2\,Pt + 2H^+ + 2e^-(\text{Volmer reaction, 3})$$

The reaction rate of HOR on Pt is dependent on the Pt crystal orientation following the trend of Pt (110) > Pt (100) > Pt (111). The *Tafel* reaction is considered to be the RDS in the case of Pt (110) [12, 13]. Under low-temperature operation, CO molecules are prone to adsorb onto Pt surface, thereby cause serious poisoning of the active sites. The poisoning effect of CO on the Pt surface over a temperature range of 80–200 °C with PBI membranes was investigated by Li et al. [14] It was found that the CO tolerance can be sufficiently enhanced at elevated temperature, from 10 to 20 ppm CO at 80 °C to 1000 ppm at 130 °C, even up to 30,000 ppm when operating temperature reaches 200 °C, thus the design of anode catalyst with HT-PEMFC is much simpler, compared with LT-PEMFC.

In terms of the electrode reactions in PEMFCs, the exchange current density value for the ORR at the cathode is approximately ~10^{-8} to 10^{-9} A cm^{-2}, which is much smaller than that of the HOR (~10^{-3} to 10^{-4} A cm^{-2}) at the anode, hence the anticipated cell potential ($E_{iR\text{-}free}$) during operation primarily relies on the reversible cell voltage (E_{rev}) in subtraction of the voltage loss of ORR [15].

$$E_{iR-\text{free}} = E_{\text{rev}} - \eta_{\text{ORR}}$$

The E_{rev} of the cell operating at ambient pressure under H_2/O_2 and constant relative humidity (RH) is dependent on the cell temperature (T) as calculated with the following equation[16]:

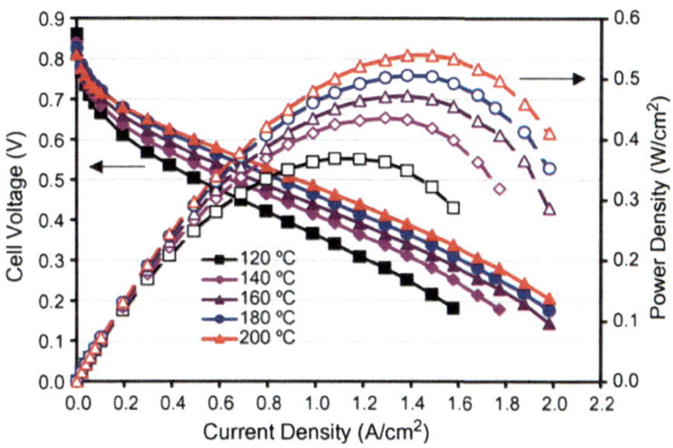

Fig. 5 Polarization curves obtained at elevated temperatures. MEA: PA-PBI-membrane fabricated with Pt-based electrocatalysts in both electrodes with a total Pt loading of 1.7 mg cm^{-2}; Operating conditions: dry H_2/O_2 gases, An/Ca stoichiometry 1.5/2, ambient backpressure. (Reproduced from Zhang et al. [19])

$$E_{\text{rev}} = 1.23 - 0.9 \times 10^{-3}(T - 298) - \frac{RT}{2F}\ln\left(\frac{RT}{100}\right)$$

From the equation, an increase in the operating temperature from 80 to 160 °C could increase the reaction rate of the ORR on Pt by two orders of magnitude. The voltage loss for the ORR at the cathode can be calculated through *Tafel* equation:

$$\eta_{\text{ORR}} = b \log \frac{i + i_x}{10 L_{\text{ca}} A_{\text{Pt,el}} i^*_{O,S} \exp\left[-E_c^{\text{rev}}/RT(1 - T/T^*)\right]}$$
$$b = 2.303\,\text{RT}/\alpha_c \text{F}$$

where i_x is the current caused by H_2 crossover through the membrane from anode to cathode, L_{ca} (mg_{Pt} cm^{-2}) is the Pt loading within the cathode, $A_{Pt,el}$ ($m^2{}_{Pt}$ g^{-1}) is the effective Pt surface area, $i^*_{O,S}$ is the specific exchange current density normalized to the reference oxygen partial pressure of 101.3 kPa and reference temperature (T^*) of 333 K, and E_c^{rev} is the activation energy of the ORR at the E_{rev}, which can be obtained from electrolyte experiments [17]. In HT-PEMFCs, the Tafel slope (b) of the ORR is found to increase with the temperature at low current density regions, while it is less dependent on temperature in high current density regions [18].

Figure 5 shows the enhanced cell performance with elevated temperature [19]. The maximum output power density is also as a function of temperature, indicating the benefit of high-temperature operation. The temperature effect is more pronounced at lower temperature (120–160 °C) than at higher temperature (>160 °C) in terms of the cell performance growth with the temperature.

2.1.2 Pt/Pt Alloy Electrocatalysts

From 1960s to present, great progress has been made towards developing efficient and stable catalytic materials for PEMFCs. The principles of catalyst design for LT- and HT-PEM fuel cells are essentially similar, mainly include (1) morphology and composition control to provide enough redox sites, (2) electronic structure optimization to achieve highly active sites, (3) support material development to favor the catalytic properties of the metal nanoparticles and to promote the electronic conductivity of the catalyst. Pt group metal and metal alloys are still considered as the most promising class of electrocatalysts under acidic environment. In recent years, precious-metal (PM) free electrocatalysts attract significant interests considering the scarcity and high price of Pt. In terms of the increased electrode kinetics with higher temperatures in HT-PEMFC, it is possible to be fitted with PM-free electrocatalysts, for example, iron (Fe) and cobalt (Co) have been demonstrated to have high activity, [20–22] but few reports give the practical applications in HT-PEMFCs. Possibly because acid leaching under high-temperature operating conditions would like to cause dissolution of these transition metals, thus lead to serious degradation of electrode components, making these materials hard to apply in real circumstances.

In terms of the operating conditions in HT-PEMFCs, the electrode reaction kinetics are expected to be enhanced, in addition, CO poisoning is likely to be avoided, thus making catalyst design much easier. However, the major challenge for the electrocatalysts showing high activity and stability in HT-PEMFC is the leaching of phosphate anions from H_3PO_4-doped PBI membranes. The existence of phosphoric acid (PA) could block some of the active sites in the electrode. Usually a higher Pt loading within the electrode is required for HT-PEMFC, compared with that for LT-PEMFC.

Pt-alloy catalysts referring to modified Pt electronic structure were developed for PEMFCs, however, the corresponding electrodes under high temperature with PBI based-membrane have been reported to be influenced by phosphoric acid leaching. Rotating disk electrode (RDE) methods have been implemented to study the influence of adsorbed phosphate ions on Pt single crystal, and it was found that these anions can strongly adsorb onto the Pt surface, thus cause great barrier for the adsorption of the reactants [23]. Neyerlin et al. studied the different catalytic performance of Pt-Ni alloy/C with PBI-membrane and Nafion® 112 membrane [15]. According to the exchange current density and activation energy at zero overpotential for ORR with Pt-Ni/C under typical LT-PEMFC conditions, the as calculated theoretical performance with the same catalyst under HT-PEMFC conditions is presented in Fig. 6a. However, the real performance based on PBI-membrane is approximately two orders of magnitude reduction as for the exchange current density, though taking humidity affect into account. Besides the influence of proton concentration decreased at low RH, the inferior performance for ORR is also likely to stem from the poisoning effect of H_2PO_4 as has been widely proposed [24, 25].

Polymer-coated Pt electrocatalysts were designed to improve the durability [26, 27]. Pyridine-containing polybenzimidazole (PyPBI) was applied to strengthen the Pt interaction with carbon nanotubes (CNTs) [28]. Zhang et al. [29] reported the synthesis of PBI-coated Pt/CNT, as illustrated in Fig. 7. The PBI polymer could bind with the pre-treated ox-CNT substrate tightly due to the formation of acid–base interaction between the –COOH groups on the surface of ox-CNT and –NH groups of the PBI. More importantly, the homogenous dispersion of PBI on the ox-CNT, with the assistance of the leached PA from the PA-doped PBI membrane, could make

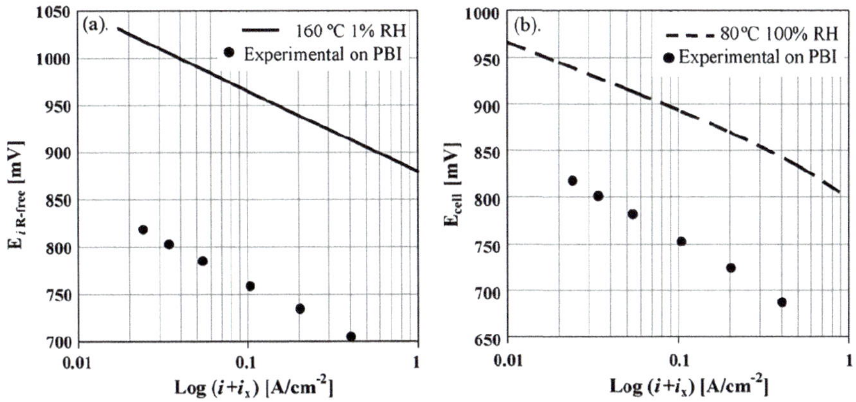

Fig. 6 Experimental and theoretical *iR*-free polarization curve achieved with Pt-Ni cathode based on PBI membrane: **a** 160 °C and 1%RH, **b** 80 °C and 100%RH. (Reproduced from Neyerlin et al. [15])

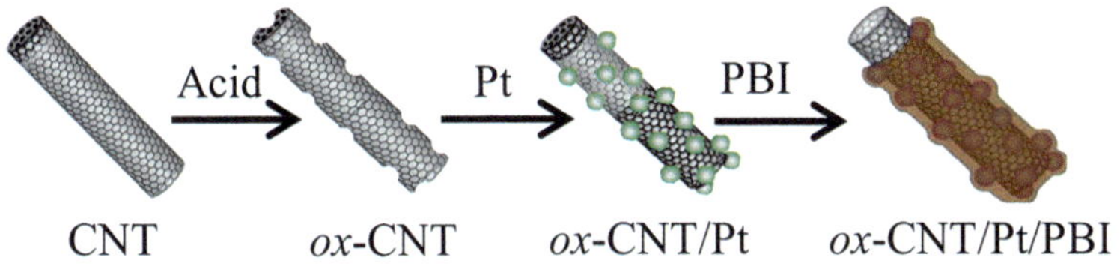

Fig. 7 Schematic illustration of the preparation procedure of ox-CNT/Pt/PBI. (Reproduced from Zhang et al. [29])

a more efficient proton transport network in the catalyst layer due to the PA bind with PBI, thus enhance the performance of the electrode.

2.1.3 Carbon Supports

In the state-of-art PEMFCs, carbon black is commonly used as support material for active nanoparticles, for example, Vulcan, Ketjen, and acetylene black (AB) are the most widely used commercial supports for Pt nanoparticles in electrodes [30]. Thermodynamically, the equilibrium potential of carbon corrosion is 0.207 V (vs. RHE) [31]. Carbon corrosion in the electrodes of PEMFC is likely to cause the detachment of Pt nanoparticles and further aggregation into larger particles, thereby resulting in a loss of active surface area [32–40]. Efforts have been devoted to improve the stability of this kind of material for HT-PEMFC via various methods, which mainly focus on two aspects: (1) Enhancement of metal-substrate interaction (SMSI) between metal nanoparticles and the carbon substrates. (2) Improvement of reactants accessibility on the surface of catalysts.

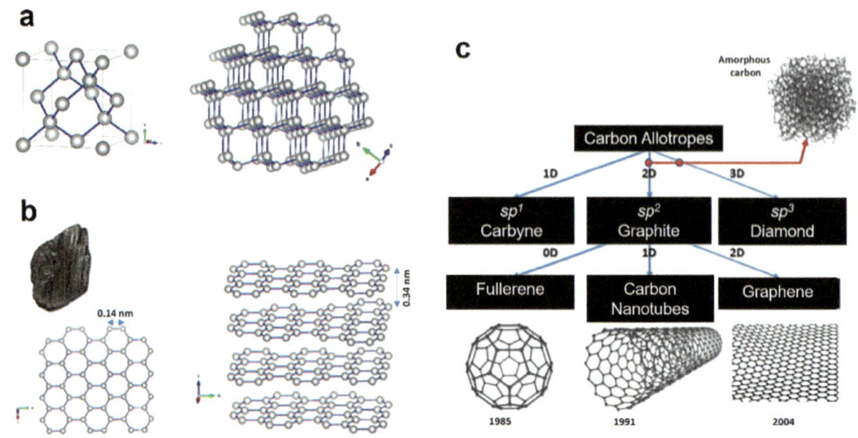

Fig. 8 Some carbon allotropes used as catalyst supports in electrocatalysis: **a** diamond, **b** graphite, **c** fullerene, CNTs, and graphene

Depending on the synthesis routes of carbon material, different allotropes, like diamond, graphite, fullerene, as well as three-dimensional (3D) structures have been obtained and applied for electrocatalysis study,[41] as depicted in Fig. 8. In contrast to diamond, graphite (Fig. 8b) is soft and brittle, with individual carbon atom sp^2 hybridized, and bonded to three other carbon atoms, forming a trigonal planar geometry [42]. 3D carbon supports with high electron conductivity and mesoporous volumes (Fig. 8c), such as carbon nanotubes (CNTs) and graphene/graphene oxide (GO/RGO), [43–47] were highly valued to provide large surface area for the dispersion of catalytic nanoparticles and benefit the diffusion of the reactant species during the electrode reactions.

The oxidation of carbon monoxide was reported to show a favorable interaction of Pt NPs with graphitic carbon supports, due to the decreased ratio of the disordered band intensity with the ordered graphitic band intensity [45, 48]. Another design principle for improving the catalytic performance with carbon supports is by surface modification. Oxygen functional groups, such as carboxylic acids, anhydrides, lactones, and phenols, could increase the surface acidity of carbon material to favor the adsorption of reactant molecules [49]. These groups can be further modified into nitrogen groups via reactions with nitrogen reagents (e.g., NH_3 and amines), which promote basicity and improve interaction with acid molecules [50]. Polymer wrapping is considered as another effective way for carbon support modification [21, 51–53]. Polymer wrapped surface was found to expedite the anchoring of metal precursors during synthesis procedures, PBI and PyPBI modified CNTs have shown improved catalytic durability and catalyst utilization in HT-PEMFCs, as these polymers could coordinated with catalyst nanoparticles and support with attractive forces such as van der Waals forces, electrostatic forces, or hydrogen bonding [54].

2.1.4 Oxide Supports and Oxide-Carbon Composites

Semiconductor oxides, such as SnO_2, TiO_2, and WO_3, [55–60] have been explored as promising candidates as support materials for Pt NPs in electrocatalysis. On one hand, the corrosion resistance of these oxides and the strong interaction with active metal NPs brings enhanced stability of the catalyst, with respect to the carbon support materials. On the other hand, the special valence bands of the semiconductor oxides could modify the electronic structures of metal active sites, thus improve the catalytic activity. Taking Pt/TiO_2 catalyst as an example, the strong interfacial interaction between Pt NPs and the oxide substrate could make Pt and Ti atoms to form a metal–metal bond [61]. The modification of Pt electronic structure can be associated with the charge transfer at the Pt-Ti interface [62].

Compared with carbon, the inferior conductivity of oxide materials may limit their application as substrate materials for electrode reactions in fuel cells. Therefore, oxide–carbon composites have emerged as an attractive brunch of support materials [63, 64]. Pt NPs as the catalytic active centers could be selectively deposited onto these substrates via UV reduction deposition method [45, 65] or chemical reduction method, [66, 67] to form Pt-based oxide–carbon composites ($Pt/M_xO_y/C$). For example, carbon supported Pt combining with zirconium oxides (Pt_4ZrO_2/C) [68] was developed as an alternative for cathode reaction in HT-PEMFCs. The catalyst presented higher sintering resistance after potential sweep test compared with Pt/C. In addition, through proper control of the mass ratio of oxide and carbon, the composite material could exhibit a similar conductivity to that of carbon [69]. In the presence of carbon, oxides are served as anchoring points of the catalytic nanoparticle centers [70–72]. In terms of other oxide support systems such as $Pt/CeO_x/C$, refer to Chap. 1 (Mori T et al.) as well.

2.1.5 Degradation Mechanisms of Electrocatalyst

Although PBI-based HT-PEMFC brings great opportunity with enhanced electrode kinetics, the leaching of acid especially under high-temperature operations poses serious threats to the stability of metal-based electrocatalysts, thus hindering the large-scale implementation of HT-PEMFCs [73, 74]. Studies have shown that the morphology and size distribution of Pt/C nanoparticles at both electrodes in high-temperature PBI-based PEMFC have changed a lot after only 600 h operation [75]. Rapid growth of the Pt NPs is observed in the presence of H_3PO_4 under high operating temperature in cell, as shown in Fig. 9. The particle size distribution of Pt at the initial stage is relatively narrow in a range of 2–5 nm. After 600 h life test, the size distribution of Pt particles is quite broad, and the agglomeration of the Pt particles on the cathode was found more severe than that of anode with a size distribution between 4 and 14 nm.

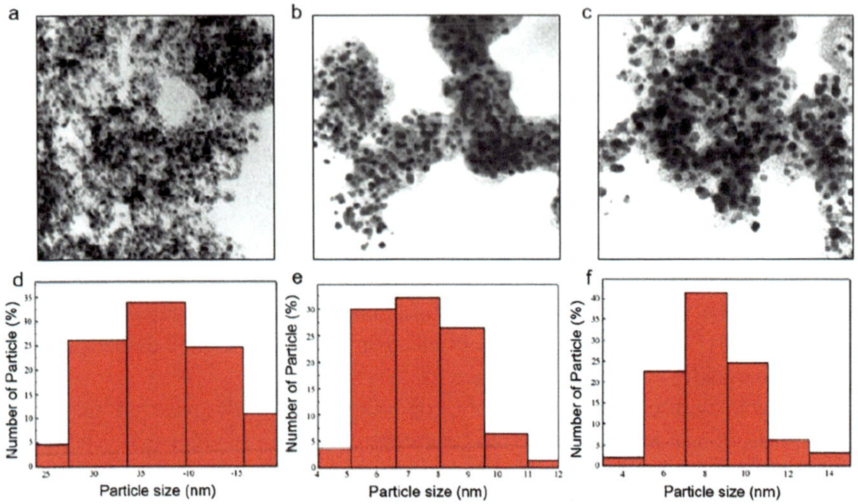

Fig. 9 TEM images of the Pt/C catalysts: **a** Pt/C catalyst at the beginning of life; **b** Anode Pt/C catalyst after 600 h life test; **c** Cathode Pt/C catalyst after 600 h life test; **d**–**f** Corresponding particle size distribution histograms. (Reproduced from Liu et al. [75])

The proposed mechanisms that contribute to the degradation of cathode catalyst are as follows:

(1) *Catalyst particle sintering caused by crystallite migration.*

This degradation mechanism could happen especially when the Pt nanoparticles disperse in a wide range of size distribution. Simonsen et al. studied the *Ostwald* ripening effect [76] of Pt nanoparticles with the *in-situ* transmission electron microscopy (TEM) method [77]. The sintering of supported Pt nanoparticles is typically attributed to the mass-transport mechanisms involving crystallite or atomic migration, which is likely to be mediated by Pt-oxygen species. Bett et al. found that the existence of H_3PO_4 could cause the transformation of Pt crystallite, and the decrease of the crystallite density, thus lead to an increase in Pt particle size [78].

(2) *Metal dissolution under specific operating potentials.*

The Pt dissolution mechanism was proposed by Meyers and Darling following the formation and decomposition of surface *PtO* during cell operation at intermediate potentials [79].

$$\mathrm{Pt} \rightarrow \mathrm{Pt}^{2+} + 2\mathrm{e}^{-}$$

$$\mathrm{Pt} + \mathrm{H_2O} \rightarrow \mathrm{PtO} + 2\mathrm{H}^{+} + 2\mathrm{e}^{-}$$

$$\mathrm{PtO} + 2\mathrm{H}^{+} \rightarrow \mathrm{Pt}^{2+} + \mathrm{H_2O}$$

Recent work has demonstrated the strong effect of humidification on the stability of Pt catalyst [80]. The loss of Pt surface area was found less severe in an unhumidified environment which can be explained by the protective effect of dehydration at high temperature with phosphoric acid. Pt dissolution mechanism suggests the necessity of water to provide ion conductivity. Furthermore, the lower mobility of Pt due to the higher viscosity of dehydrated solvent also restricts the Pt dissolution. Thus, from this point of view, the dissolution of Pt may not be the major cause of cell performance degradation under operating conditions in HT-PEMFC compared with LT-PEMFCs.

(3) *Carbon support corrosion.*

Carbon support corrosion in PEMFC under some special operating conditions, such as *start-up* and *shut-down* is considered as a major cause of cell performance degradation. Early in the 1970s, the stability of carbon-supported catalysts at high temperature (165 °C) was concerned in the development of PAFCs [81–83]. Under the operating conditions of HT-PEMFCs, carbon corrosion is accelerated by reacting with oxygen atoms generated by catalyst particles, as expressed with the following equations, leading to a time-resolving oxidation of carbon support in the catalyst layer [84, 85].

$$C_S \rightarrow C_S^+ + e^-$$

$$C_S^+ + \frac{1}{2}H_2O \rightarrow C_SO + H^+$$

$$C_SO + H_2O \rightarrow CO_{2(g)} + 2H^+ + 2e^-$$

2.2 Membrane

2.2.1 Proton Exchange and Physicochemical Property of PBI Membrane

Grotthus mechanism, as proposed by Von Grotthus, is a proton jump mechanism. It is commonly accepted that the proton transfer in PA-doped PBI membrane mainly goes through the Grotthus-type hopping mechanism [86, 87], where the protons conduct from an imide site to a neighboring vacant site. The migration of proton is assisted by a counter anion. With the assistance of water, a proton arrives at one H_2O molecule, making it temporary a hydronium ion (H_3O^+), which can be transferred with the diffusion of H_2O molecule (vehicle mechanism as illustrated in Fig. 10a). Through Grotthus mechanism, proton moves with the detachment of H_3O^+ directly for the next hop, and this process is additionally accompanied with H_2O molecule (Fig. 10b) [7]. In terms of the phosphoric acid monitored proton transfer, the similar mechanism

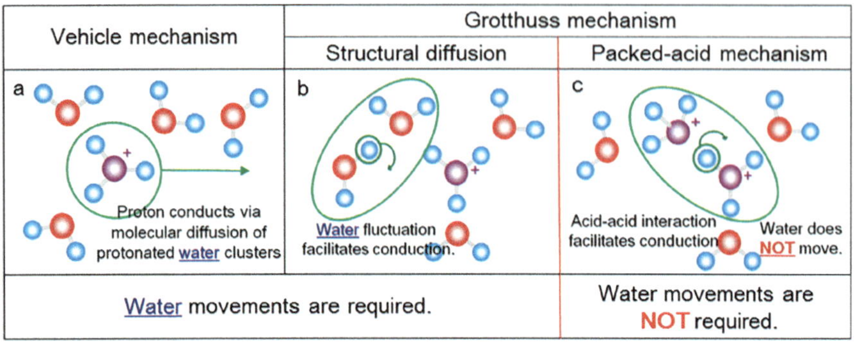

Fig. 10 Mechanisms of proton transfer: **a** vehicle mechanism, **b** structural diffusion, and **c** packed-acid mechanism. (Reproduced from Ogawa et al. [89])

is observed with the Grotthus chain between PA and the ion pair $H_4PO_4^+$/ $H_2PO_4^-$ [86, 88]. Ogawa et al. [89] proposed the packed-acid mechanism (Fig. 10c) without water accompanied in this situation, and reorientation is avoided by the acid-acid interaction, which is typically observed in the media of low humidity and consisting of highly concentrated acids.

Proton conductivity in H_3PO_4/PBI material mainly depends on the acid doping level and the relative humidity [87, 90]. PA-doped level of PBI membranes was reported to attribute to different mechanisms of proton transfer [91–93]. Under the operating conditions of HT-PEMFCs, PA undergoes a series of transforms and changes in composition via interactions with water and the PBI membrane, resulting in discriminate ionic conductivity and viscosity as studied by Korte et al. [94] It was demonstrated that under the operating conditions of HT-PEMFC, PA and water phases would undergo a thermodynamic equilibrium with a preferential evaporation of water. At lower acid-doping level, protons will be conducted by the hopping mechanism mainly between H_3O^+ and H_2O species on neighboring polymer chains, thus resulting in low proton conductivity. At high PA concentrations, an increasing fraction of protons will be transferred directly between $H_4PO_4^+$ and H_3PO_4 molecules, indicating potentially high proton conductivity.

As the polymer electrolyte membrane also serves as the separator between anode and cathode in the fuel cell, the gas permeability of the membrane is of particular importance. As hydrogen generally has a higher permeation rate than that of oxygen, due to its smaller molecular size. At elevated temperature from 120 to 180 °C, an increased hydrogen permeability was reported with acid doped-PBI by He et al. of around 120–400 × 10^{-7} mol cm^{-1} s^{-1} bar^{-1} compared with that of pristine PBI 1.6–4.3 × 10^{-17} mol cm^{-1} s^{-1} bar^{-1} [95]. When doped with acid, the PBI membrane showed a volume swelling by 100–110%, resulting in a significant increase of permeability by 2–3 orders of magnitude as shown in Fig. 11. The solubility coefficient (*C/p*, mol cm^{-3} bar^{-1}, where *p* is the applied gas pressure, and C is the gas concentration in the polymer) is used as a parameter for gas diffusion. In the low range of gas solubility or pressure, according to Henry's law, the value is assumed to be

Fig. 11 Hydrogen (*open*) and oxygen (*solid*) permeability of PBI membrane at different temperatures before (*cycles*) and after (*squares*) being doped with PA. (Reproduced from He et al. [95])

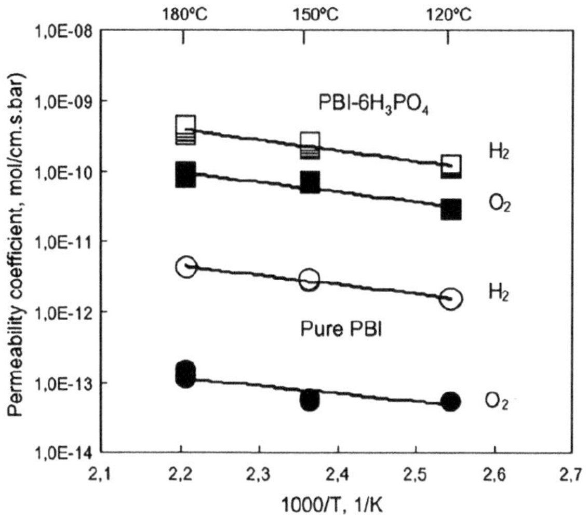

a constant over the pressure range, the diffusion coefficient can then be calculated from the permeability. For PBI membrane doped with 6 H_3PO_4 per repeat unit at 150 °C the diffusion coefficient was, much higher than pure PBI membrane.

2.2.2 Synthesis and Modification of Membranes Used in HT-PEMFCs

Numerous efforts have been devoted to the development of effective and stable proton exchange materials for HT-PEMFC. To date, many issues of HT-PEMFC technology especially in durability and performance still involves the ability of proton exchange membrane (PEM). The PEM consists of the polymeric backbone, side chains, and any fillers or support materials which are added for some desirable functions. The exchange capability of the PEM is relied on the proton carriers in membrane, such as phosphoric acid (H_3PO_4, PA). A range of PEMs have been developed aiming to the high-temperature operating conditions in HT-PEMFC, including PBI membranes, inorganic–organic composite membranes, sulfonated hydrocarbon polymer membranes, etc. which are classified by the polymer composition.

The application of traditional perfluorosulfonic acid (PFSA) membranes, such as Nafion®, which are widely used in LT-PEMFCs, is largely limited in HT-PEMFCs because of the high gas permeability and sharp decline in conductivity of the membrane at high operating temperature and low humidity, as we illustrated at the beginning of this chapter. So far, PBI-based membranes have been considered as the most promising candidates for HT-PEMFCs, owing to the merits of high conductivity, [87, 91–93, 96] excellent mechanical properties, [97] and thermal stability [98]. PBI materials in a broad sense refer to a family of aromatic heterocylic polymers containing benzimidazole units [99]. Li et al. reviewed the structure modification and synthesis routes of PBI membranes [99]. With the progress made in material science,

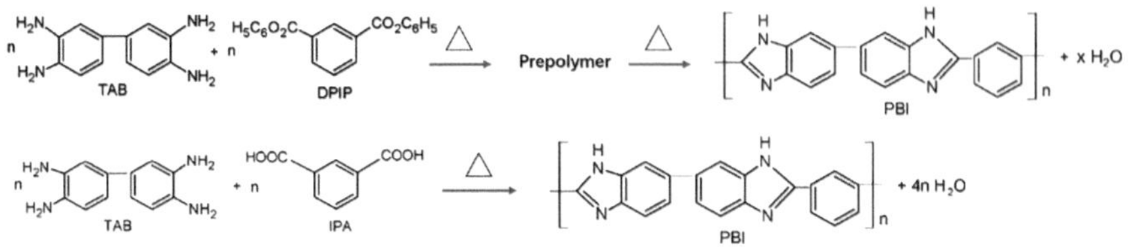

Fig. 12 The two-step synthesis of PBI proposed by Vogel and Marvel (*top*), and the one-step process proposed by Choe (*bottom*)

acid–base chemistry method plays a significant role in forming solid membranes with good homogeneity. Polymers bearing basic sites, such as alcohol, ether, imine, or amide groups could react with medium or strong acids. The basic polymer serves as the proton acceptor in a typical acid–base reaction to form ion pair. In 1961, Vogel and Marvel [100] firstly reported a two-step synthesis of PBI by process heating of an equimolar mixture of tetraaminobiphenyl (TAB) and phenyl isophthalate (DPIP) monomers, as shown in Fig. 12. By replacing DPIP with isophthalic acid (IPA) in the presence of organo phosphorus and silicon compounds as the catalyst, a simplified one-step strategy was proposed by Choe (Fig. 12) [101].

With the development of acid–base chemistry method, polymer membranes containing a high concentration of PA with excellent homogeneity and strength were developed by researchers. At the initial stage, electrolyte based on concentrated PA was found to be capable to operate at a very low humidity without significant loss of the proton conductivity. PA is served as the proton carrier which could realize a high proton conductivity over a wide range of temperature up to more than 200 °C, [102] thus making PA-PBI membranes one of the most promising candidates for HT-PEMFCs. Homogeneous solution synthesis by applying solvents such as polyphosphoric acid (PPA) has been widely used, [103–105] and highly valued for the preparation of small-scale batches of linear, high molecular weight polymers especially in laboratory for membrane casting. In this process, PA-doped membranes can be made by directly casting the solution mixture to form a thin film. Then the sol–gel transition subsequently takes place by hydrolyzing the PPA or other solvents under the environment with carefully controlled temperature and humidity to tune the acid content in the membrane, [106] as illustrated in Fig. 13.

Acid doping chemistry could enhance the proton conductivity of PBI membranes, however, high-level acid doping is found to decrease the mechanical stability of the membrane, thereby great efforts have been made towards the modification of the polymer structures to solve this problem. Commonly used modification includes crosslinking, [107–109] organic–inorganic compositing, [110, 111] and other functionalization strategies[112] mainly aim to bring favorable membrane properties, such as high molecular weight and good solubility to enhance the mechanical stability and proton conductivity of PBI membranes [113].

Cross-linking is considered as an effective way to enhance mechanical strength. In general, there are two widely used methods of crosslinking: ionic crosslinking [114–117] and covalent crosslinking [118–120]. Hasiotis et al. [121] were the first

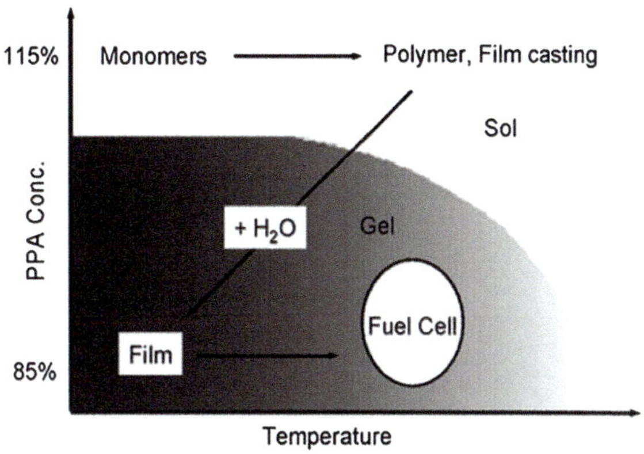

Fig. 13 State diagram of the PPA sol–gel process. (Reproduced from Xiao et al. [106])

to transform the crosslinking concept to high-temperature fuel cell membranes by mixing a molar excess of polybenzimidazole with sulfonated chloromethylated (PSU), followed by doping with PA. The acidic polymer works as the macromolecular cross-linker in the blend membrane, and the phosphoric acid provides H^+ conductivity.

Ionic crosslinking occurs through the interaction of polymeric blends with polymeric acids, usually providing sulphonic group and polymeric bases with imidazole group (Fig. 14) [122]. The introduction of ionic cross-links into the PBI membranes can prevent the PA dissolution during the doping procedure. In addition, the water uptake of the membrane could be reduced [123, 124]. Ionic crosslinking between the acid and base blend polymers thus is likely to improve the stability and conductivity of the membrane.

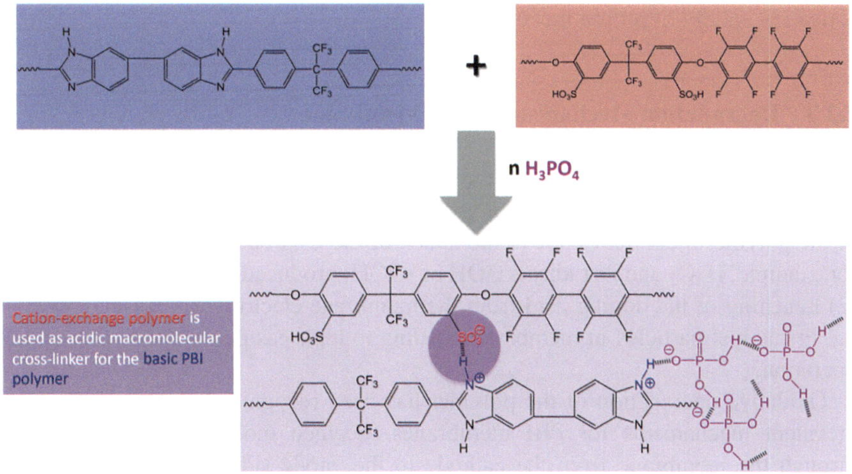

Fig. 14 Ionical crosslinking and the formation of the hydrogen bridge network by PA doping of the membrane MJK1845 as an example [125]

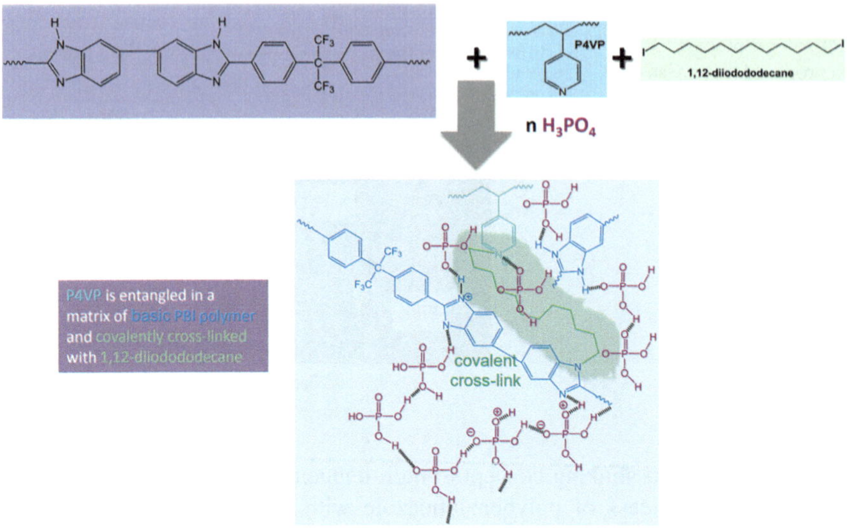

Fig. 15 Covalent crosslinking of the membrane MJK1844as an example [125]

Covalent crosslinking in PBI membranes is thought to be stable with acid-doping and oxidative environment, thus makes them promising candidates for fuel cell applications [109]. The imidazole group in the PBI membrane could be cross-linked by an organic acid or its halide with two or more functional groups per molecule to form covalent linkages [126]. Figure 15 shows an example of covalent crosslinking. The degree of covalent crosslinking was studied to show an influence on the proton conductivity of the membrane [127]. A too-dense cross-linked network structure within the membrane would possibly inhibit the water absorption and increase the activation energy of proton movement.

2.2.3 Degradation Mechanism of PBI Membrane

The degradation mechanism of PBI membrane has been widely concerned during the development of HT-PEMFC technology, mainly from three aspects: (1) Degradation of the polymer membranes due to the attack of the C-H bond by oxidative species, for example, H_2O_2 and its radicals (•OH or •OOH) produced during electrocatalysis. (2) Leaching of the doping acid from the membrane electrolyte. (3) Deposition of the Pt catalyst particles in membrane, leading to high gas permeability during cell operation.

Oxidative degradation of the polymer has been recognized as one of the most prevalent mechanisms for PBI membranes. Oxygen molecules that permeated through the membrane from the cathode to the anode side could possibly cause the formation of oxidative species H_2O_2, as proposed by Laconti et al. [128] The degradation of PBI membrane is also studied by Liao et al. [129] when exposed to the Fenton solution as an accelerated parameter. During the Fenton test, severe weight loss was observed with PBI membrane, which is significantly higher than that of Nafion 117 membrane, as shown in Fig. 16a. Changes of the two monomers, 3,3′–4,4′-tetraaminobiphenyl (TAB) and isophthalic acid (IPA) in PBI, have been investigated by FTIR spectra before and after the Fenton test (Fig. 16b), indicating that the PBI degradation is due to the attacks of oxidative radicals at the amino N–H and carboxylic acid (–COOH) groups with the network. Operating under pure oxygen was reported to accelerate the membrane oxidative degradation [130]. The acid–base interaction between the PA molecules and amino N–H groups was investigated in their later work, [131] and was found that the acid–base interaction could alleviate the attack of hydroxyl radicals of the polymer, thus slow down the membrane degradation.

Increase in PA-doping level is generally thought to give rise to the proton conductivity of PBI membranes, nevertheless, this advantage has to make a sacrifice of the

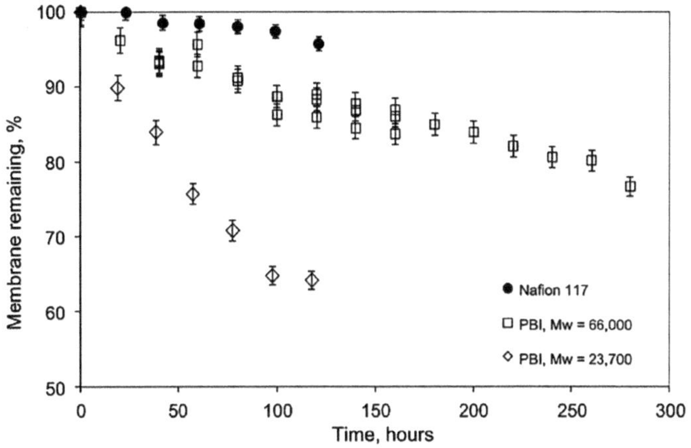

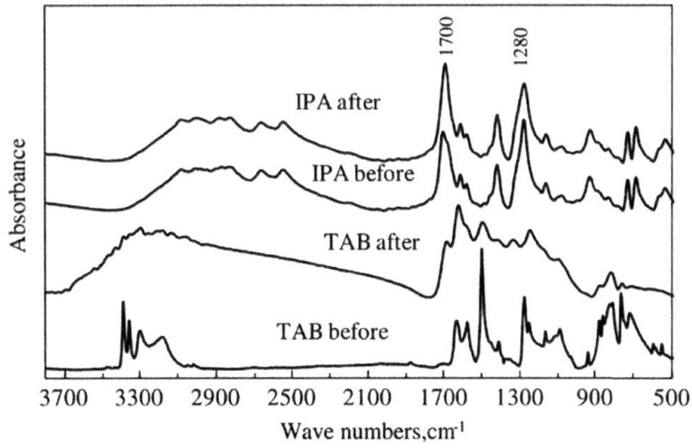

Fig.16 **a** Weight losses of PBI and Nafion membranes immersed in Fenton solution, **b** FTIR spectra of TAB and IPA before and after Fenton test. (Reproduced from Liao et al. [131])

mechanical strength of PBI membrane with a large amount of PA doping. During the operation of HT-PEMFC, when temperature falls below 100 °C and liquid water is produced, The PA-doped polymer membranes can bleed out a part of the PA molecules [132]. The released PA can cause blocking of the active sites with the catalysts, as well as the damage of fuel cell system. The above-mentioned PA leaching process off the membrane can be expressed by the following equations. The PA in the membrane is gradually dehydrated into polyphosphoric acid, resulting in a continuous acid loss in membrane and reducing the stability of the polymer structure [23, 133].

$$H_3PO_4 + PBI \leftrightarrow H_2PO_4^- + PBI \cdot H^+$$

$$H_3PO_4 + H_2O \leftrightarrow H_2PO_4^- + H_3O^+$$

Both the above reactions could increase the concentration of $H_2PO_4^-$ anions, which assists the proton transfer via the Grotthus chain. Recently, some work dedicated to solve the leaking out problem of acid doped membrane. Chen et al. [134] reported the synthesis of long alkyl side chains with quaternary ammonium groups grafted PBI membranes (PBI-Sc-X). The strong interaction between quaternary ammonium and biphosphate ion-pairs effectively retains PA in the base-acid doped membrane, achieving a better proton conductivity of 0.104 S cm^{-1}, which is comparable to that of Nafion membrane [135] and much better PA retention than pure PBI membrane.

Oono et al. [136] studied the in-situ decline of the PA-doped membrane during operation of MEA in HT-PEMFC under temperature of 150 °C at a current density of 0.2 A cm^{-2} for periods of time. PA distributions across sections of MEAs were characterized and analyzed by using transmission electron microscopy and electron probe micro-analysis (EPMA), as shown in Fig. 17. The initial PA concentration was very low in both cathode and anode catalyst layers, however, it increased substantially over operating time, indicating that migration of PA from the membrane. The PA appears to be uniformly distributed in both catalyst layers after 15,000 h of operation. After 17,860 h test, a further increase of the depleted region occurs, accompanied by the rapid thinning of the membrane.

Besides the chemical degradation, the polymer membrane is also sensitive to some physical parameters during real cell operation. The fabrication manner of the HT-PEMFC stack usually makes the membrane under a compressive force between the bipolar plates. Moreover, the swelling and shrinking of the membrane during load cycle or relative humidity cycle is likely to cause mechanical stress to the membrane, [137] thus leading to time-dependent deformation of the membrane framework. Pristine PBI membranes possess a good mechanical tensile strength of ~60–70 MPa under dry environment, and ~100–160 MPa when saturated with water [138]. However, the backbones of the polymer in the membrane can be separated by the free acid in the case of PA-doping membrane, resulting in a dramatic decrease of the mechanical strength of polymer, especially at high temperature [97, 138].

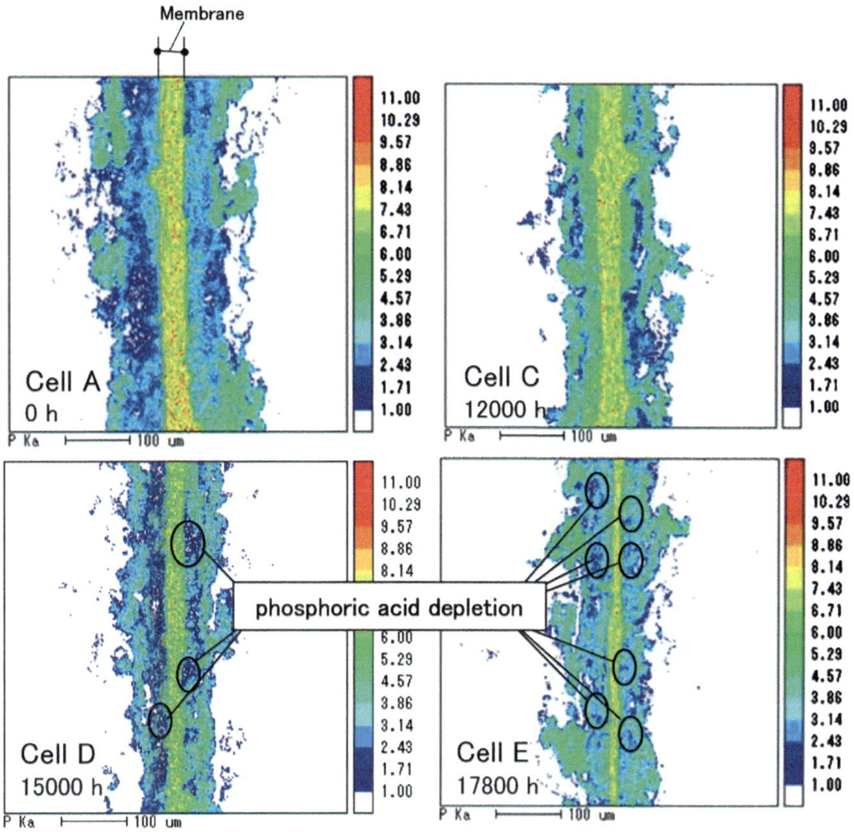

Fig. 17 EPMA analysis of PA leaching in cross sections of the MEA operating at 150 °C and at 0.2 A cm^{-2} for **a** 0 h, **b** 12,000 h, **c** 15,000 h, and **d** 17,860 h. (Reproduced from Oono et al. [136])

3 Fabrication and Structure Control of Membrane Electrode Assembly (MEA)

3.1 Selection of "Binder" in Catalyst Layer (CL)

MEA design (Fig. 18) is a complicated task, involves both the development of advanced electrode materials, as well as the proper design of geometry structures to guarantee the best trade-off between catalyst utilization and mass transportation. Electrocatalysts in CLs need to combine with proton-conducting ionomer, which served as binders to form continuous proton transport network within the MEA. The polymeric binders are also considered as bridges among catalyst particles to avoid the large agglomeration of active sites, thus favor the formation of more TPBs for

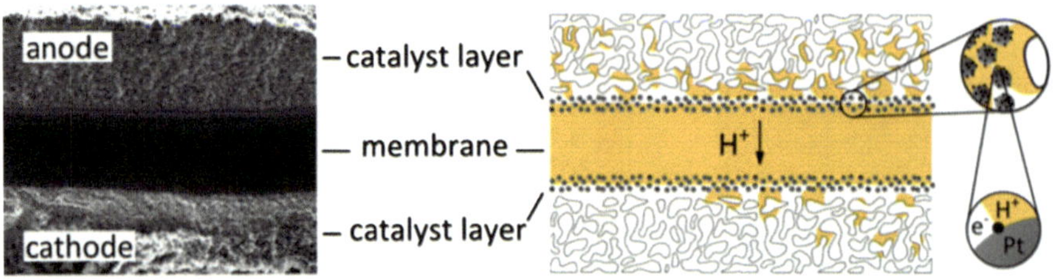

Fig. 18 Cross-section image of an HT-PEMFC MEA (*captured* via *secondary electron detection, left*) and sketch (*right*) [203]

reaction [139, 140]. The mass ratio of ionomer and carbon support (I/C) is an important parameter in CL design used as manifestation of correlation between reaction and transportation.

The fuel cell performance is heavily determined by the composition and structure parameters of CL, as well as the electrochemical reactions efficiency [141]. At the TPBs where the electrochemical reactions happen, the electrocatalyst, electrolyte binder, and reactants/products are in simultaneous contact [142]. The binders in the CL play a vital role to keep the integrity of the electrode, meanwhile provide proton transport pathways. In PBI-based HT-PEMFCs, different polymer binders, such as Nafion, PBI, polytetrafluoroethylene (PTFE), polyvinylidene difluoride (PVDF), and PBI-PVDF have been evaluated by Su et al. [143] to provide an overview of the optimization of gas diffusion electrode (GDE). Conventional sulphonated and PBI polymers used as proton conductors for HT-PEMFC were demonstrated to show inferior cell performance, [144] possibly because of the blocking effect with Nafion and PBI polymers, leading to high ohmic resistance. PTFE is more attractive polymer binder, as it could facilitate the formation of CL featuring with high catalyst utilization and proper hydrophobicity [145, 146]. The concentration of the binders in CL is also an important parameter in the CL design. Mack et al. [147] investigated the optimization of the polytetrafluoroethylene (PTFE) content (ranging from 5 to 40 wt%) in the catalyst layers in terms of the HT-PEMFC performance. Lower PTFE contents significantly reduced the charge-transfer resistance and hydrophobicity, owing to higher porosity and better wettability of the surface with PA. For PVDF GDE and PBI/PVDF GDE, the hydrophobicity and porosity of their CLs are relatively low, leading to a uniform and fine GDEs surface.

The transport of reactants (H_2 and air) and charge carriers (proton and electron) within the CLs are of the trade-off relationship to achieve high performance in HT-PEMFC. In PBI-based HT-PEMFCs, PA is commonly employed to facilitate the proton transfer within the membrane. An increase in the amount of PA in the CL would increase proton conductivity but also makes it difficult for oxygen to approach the reaction sites, as shown in Fig. 19, thus leading to a lower proton resistance (R_p) but higher mass transportation resistance (R_m). PA leaching has been considered as a serious cause to the degradation of HT-PEMFC performance. Well-defined CLs with

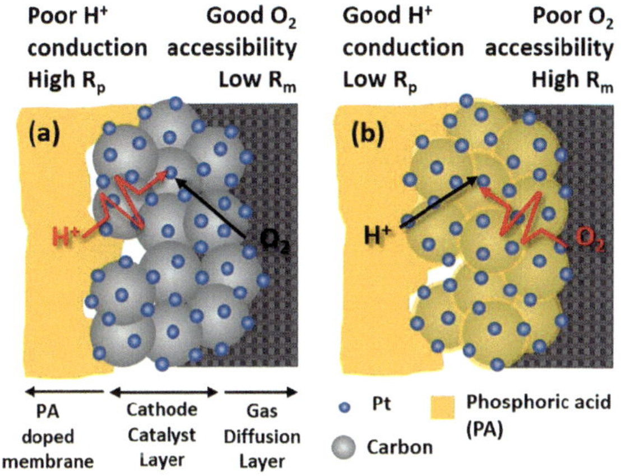

Fig. 19 Schematic illustrations of catalyst layers with **a** a small amount of PA, **b** a large amount of PA. (Reproduced from Lee et al. [149])

hydrophobic binders and crack structures have been reported as critical parameters to alleviate the effect of large-scope distribution of PA in MEAs [148, 149].

3.2 *Optimization of Pt Loading and CL Thickness*

As mentioned in the foregoing, based on the same binder, the I/C ratio is an important parameter in CL design, to construct proper TPBs for electrode reactions. It is worth mentioning that recent progress has been made towards overcoming the high Pt content of the electrodes in HT-PEMFCs [150–157]. The Pt loading for HT-PEMFCs in early research work was around 1 $mg_{Pt}cm^{-2}$, which had been far from reaching the 2020 set target (<0.125 $mg_{Pt}cm^{-2}$ per MEA) by U.S. Department of Energy (DOE). The Pt loading, as well as the CL thickness in the electrodes, are recognized as critical factors to influence the cell performance.

In the case of Pt/C electrocatalyst, the Pt percentage in the electrodes has an apparent relation with the thickness of the CLs. High Pt percentage of the catalyst could mean a thinner catalyst layer in the electrode, with a definite Pt loading. On one hand, a thinner CL makes the transport of reactants and products less tortuous. On the other hand, a thinner CL with proper Pt loading was reported to feature with larger amount of mesopores in the electrode, which is more likely to give the best combination of mass transfer characteristics and electrode reaction kinetics within the CL, leading to the best fuel cell performance. Lobato et al. [158] in their work studied the potential influence of Pt percentage on the porous structure and performance of the electrode in PBI-based HT-PEMFC. As shown in Fig. 20a, logically, the lower Pt percentage brings larger amount of mesopores in the CL, also the corresponding higher carbon content could mean thicker CL in the electrode. In general, the thinner CL may reduce the reactant's transport tortuosity, while lower Pt percentage may benefit the accessibility of the active sites. Thus, the MEA with the best combination

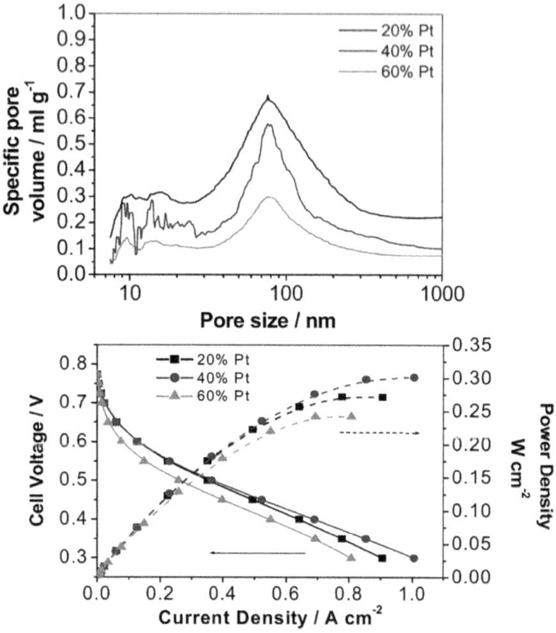

Fig. 20 **a** Porosity of the electrodes with different Pt/C ratio, **b** Polarization curves with different Pt/C content in cathode catalyst layer. (Reproduced from Lobato et al. [158])

of catalytic activity and mass transfer characteristic gives the best cell performance (40 wt.% Pt/C in this work as shown in Fig. 20b).

In addition, with low Pt loading, the number of active catalytic sites is limited, thus the distribution of polymeric binders needs to be taken special consideration to build suitable TPBs and improve the catalyst utilization. The Pt/C ratio of the catalyst can be designed to realize the desirable thickness of the CL, as well as maximize the active sites utilization with low Pt loading [154]. Martin et al. in their work explored the possibility of reducing the Pt loading with both electrodes to as low as 0.1 $mg_{Pt}cm^{-2}$, and achieved an overall Pt utilization of 2.5 kW mg_{Pt}^{-1} under H_2/O_2 operating conditions in the HT-PEMFC. With the fixed Pt loading of 0.1 $mg_{Pt}cm^{-2}$ in the electrodes, the Pt/C ratio (wt.%) of the catalyst ranging from 10 to 60% was reported to result in a ~15 um thickness difference of the catalyst layer. In this case, a relatively thicker CL was found to benefit the fuel cell performance, mainly because the proper distribution of PA in the CL reduced the ohmic resistance within the electrode. In the recent work of the same group,[152] they reported an ultra-low Pt loading achieved in the HT-PEMFC with 0.05 $mg_{Pt}cm^{-2}$ at anode and 0.1 $mg_{Pt}cm^{-2}$ at cathode [153]. It was found that the binder-less MEA with ultra-low Pt loading showed a comparable performance to the reported electrodes with 10-times-higher Pt loading [159, 160].

3.3 Gas Diffusion Layer (GDL) and Microporous Layer (MPL)

GDL is an integral part of MEA in PEMFCs which is commonly impregnated with a microporous layer (MPL), aiming to realize the gas fuels transport and diffusion, heat exchange, as well as water management especially for LT-PEMFCs [161, 162]. Carbon cloth or carbon paper (Fig. 21) is widely used as porous substrate material in GDL due to its high gas permeability, excellent electronic and heat conductivity [163]. MPLs are characterized by a microporous carbon structure and a hydrophobic polymer impregnated surface, which is designed to increase the limiting current of the electrode and reduce the mass transport resistance [164].

Fick's law has been applied to describe the gas diffusive behavior in GDLs involved in fuel cells, by using the following expression [165]:

$$\frac{\partial C_i}{\partial t} = D_i^{\mathrm{eq}} \frac{\partial^2 C_i}{\partial z^2}$$

where C_i is the gas concentration, D_i^{eq} is the diffusion coefficient, which need to be modified with different pore geometry. Z is the length of the GDL sample. Mass transfer in GDLs is also taken place by Knudsen diffusion (K_n) within the micropores, [166] where λ_g is the mean molecular free path and d_p is the pore diameter, which requires comprehensive consideration to predict the gas diffusion pathways in the GDL. For O_2 and N_2 gas flow in the GDL, the Knudsen number is around 0.001, thereby enabling the use of Darcy's law for calculation of gas permeability, however, in the case of the MPL, the use of Darcy's law might be erroneous [167].

$$K_n = \lambda_g / d_p$$

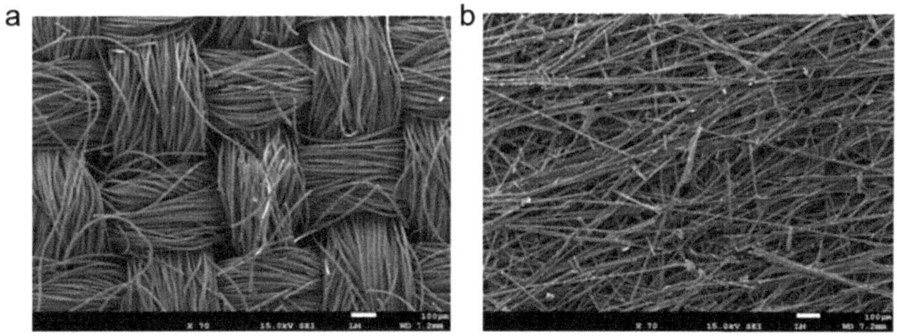

Fig. 21 SEM images of **a** carbon cloth and **b** carbon paper. (Reproduced from Jayakumar et al. [163])

Lobato et al. demonstrated the optimized cell performance in HT-PEMFC through control of MPL thickness with different carbon contents (from 0.5 to 4 mg cm^{-2}) [168].

It was found that the meso and micropores regions could be tuned through control of carbon content in MPLs [169]. Figure 22a shows the AC impedance spectra obtained from MPLs with different carbon contents ranging from 0.5 to 4.0 mg cm^{-2}. The AC impedance analysis confirmed that the improvement of cell performance (Fig. 22b) is originated from an optimal MPL structure that allows for the improvement of the reactant/product gases transport, as well as the reduction of the fuel cell ohmic resistance in the electrode. A thicker MPL with high carbon content could be beneficial at the activation polarization region (low current densities), because the accessible active surface area increases with the thickness of MPL. The performance loss takes place at mass transportation region (high current densities) and is dominated by the resistance related to the mass transfer of gas reactants to catalyst sites and the proton transfer within the catalyst layer (O_2/H^+ transportation resistance, R_{mt}).

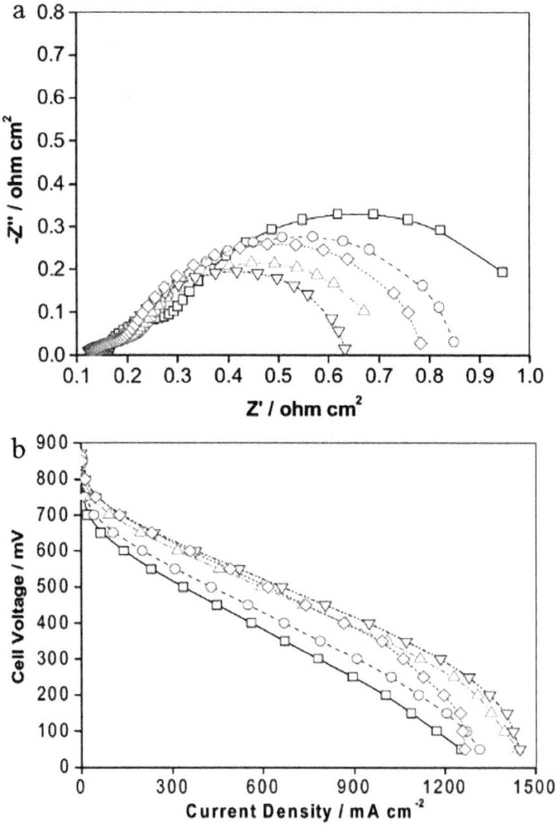

Fig. 22 **a** Impedance spectra and **b** Polarization curve under H_2/air obtained with cathode MPLs containing different carbon contents. (□) without MPL, (○) with 0.5 mg cm^{-2} carbon content MPL, (Δ) with 1 mg cm^{-2} carbon content MPL, (∇) with 2 mg cm^{-2} carbon content MPL, (◇) 4 mg cm^{-2} carbon content MPL. (Reproduced from Lobato et al. [168])

3.4 Fabrication Methods of MEA

As described at the beginning of this chapter, MEA is served as the key part in fuel cell devices, with an integrated structure consisting of membrane, catalyst layers, and gas diffusion layers. The assembly strategies play a significant role on the structure and morphology of the electrodes, which would greatly influence the properties and performance of the MEA. Preparing the catalyst coated substrate (CCS) or the catalyst coated membrane (CCM) are two commonly used methods for electrode fabrication, with the difference in the substrates for direct catalyst ink coating. The CCS method involves coating the catalyst ink onto the substrate, usually the gas diffusion layer, while in the CCM method, the catalyst ink is directly deposited onto the membrane. The chemical and physical properties of the catalyst ink have great impact on the morphology of catalyst layer [170]. Huang et al. [171] studied the effects of dispersed solvents in catalyst ink, including water, methanol, ethylene glycol, glycerin, 1,2-propylene glycol, and 1,3-propylene glycol and on the structure and the structure-related performance of the catalyst layer. The CLs formed from different ink dispersion solvents showed diverse morphology, as shown in Fig. 23. Catalyst layers which originated from higher boiling point solvents present less cracks than those from low boiling point solvents, such as water and methanol. In other words, the faster evaporation of the low boiling point solvents is likely to make more cracks in CLs. The best uniformity of CL was achieved with ethylene glycol as the dispersion agent, which shows a major impact on the MEA performance.

Several depositing techniques have been developed to form uniform catalyst layer, such as spraying, casting, and slot-die coating [172]. Ultrasonic spraying is

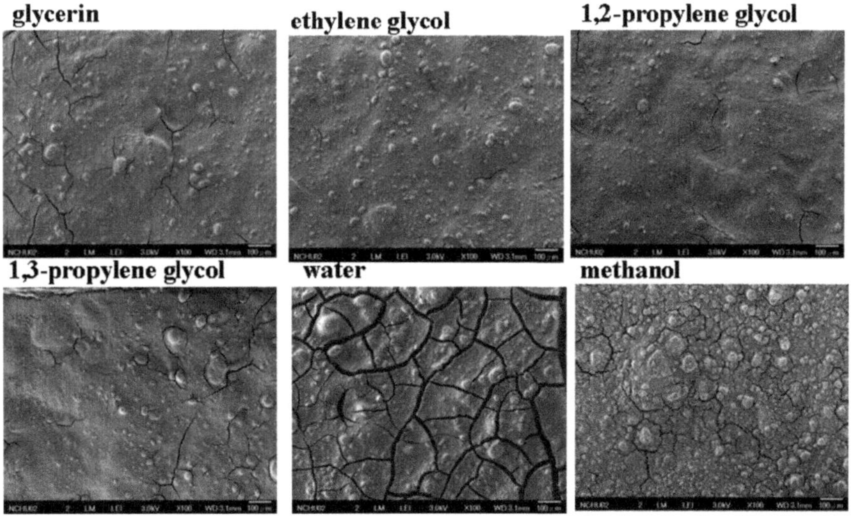

Fig. 23 Surface morphology of CLs formed from various catalyst dispersion inks. (Reproduced from Huang et al. [171])

more commonly applied in laboratory. The solid content of the catalyst ink has to be controlled at a low threshold for insurance of viscosity during spraying, which largely limit the large-scale production efficiency, while coating methods, such as tape coating and slot die coating have been developed to solve this problem, which could adapt to higher solid content with the ink and meet the catalyst loading requirement by single-step operation, making it one of the most promising techniques for large-scale manufacturing of fuel cells [141, 173]. Different electrode preparation crafts have been investigated and compared by Bodner et al. [173] Through spraying method (Fig. 24a), catalyst is dispersed in a volatile liquid medium and then sprayed onto the substrates, namely, GDL or Membrane by the carrier gas, usually air or nitrogen. In order to achieve a homogeneous catalyst layer, this process needs to be repeated several times, with precise control of catalyst loading being deposited to the substrate. The disadvantage of this method is apparent that the whole manufacturing process is time-consuming [174].

Tape casting and slot-die coating techniques (Fig. 24b, c respectively) are similar, which commonly involve catalyst ink with higher viscosity. In the coating process, a wet catalyst film is formed on the substrate with a well-defined droplet size and nozzle moving speed. The loading of the catalyst is determined by the solid content of the ink and the thickness of wet-film formed during coating process. The final catalyst layer homogeneity is largely dependent on ink stability, blade moving speed, drying rate, as well as the substrate properties. Different techniques used for preparing the electrode would directly affect the structure of the catalyst layer. Many of the research works by applying spraying method demonstrated that the catalyst layers produced are featured with similar structure, mostly dense, and with relatively low porosity [152, 154–156, 175, 176]. The in-plane and through-plane X-ray microtomography images of the catalyst layers made from spraying and coating methods are shown in Fig. 25 [177]. The catalyst layer can be easily distinguished from the microporous layer (MPL) with variation in brightness. Compared with the coated catalyst layer, the sprayed one appears much more heterogeneous, which contains voids and void clusters of different shapes and sizes (Fig. 25a–c). The drying mode is also different from the coating process. It usually takes longer time for the paste to settle on the substrate and the coated layer to get completely dry out. Therefore, a relatively tense CL with less voids and fewer cracks was witnessed with coating method (Fig. 25d–f).

The prepared electrode CCM or CCS need to be further assembled by hot pressing with GDL or membrane, respectively to comprise MEA. The pre-manufactured catalyst-loaded GDE via CCS process is then pressed together with a membrane to form an MEA, while the CCM pre-prepared by directly deposited the catalyst onto the polymeric membrane is then fitted with the GDLs to form the MEA. The purpose of this procedure is to reduce the contact resistance between different layers. The compression temperature and pressure should be well controlled to make close interaction between layers, at the same time keep the pore structures in CL and GDL.

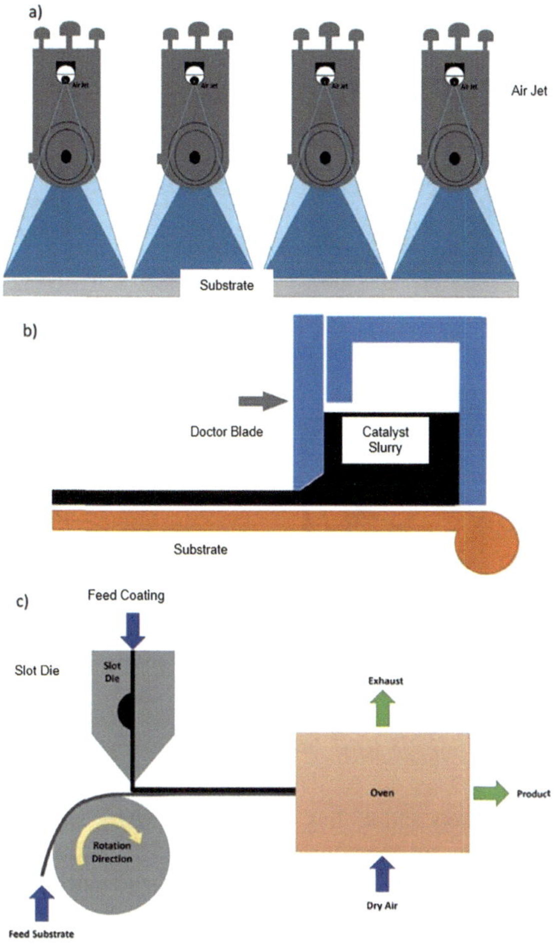

Fig. 24 Scheme of different electrode preparation methods: **a** spray coating, **b** tape casting, and **c** slot-die coating. (Reproduced from Bodner et al. [173])

4 HT-PEMFC Stack and System R&D Progress

Besides the design of the core electrode part, other components such as gaskets, sealings, and bipolar plates with multiple functions must be assembled to build a complete HT-PEMFC stack. In addition, the operation of the HT-PEMFC needs the assistance of an integrated system to realize the transport and distribution of fuel gases to the electrodes, apply load, collect and diagnose the electrical signals, as well as provide thermal management and mechanical support to the HT-PEMFC device or stack.

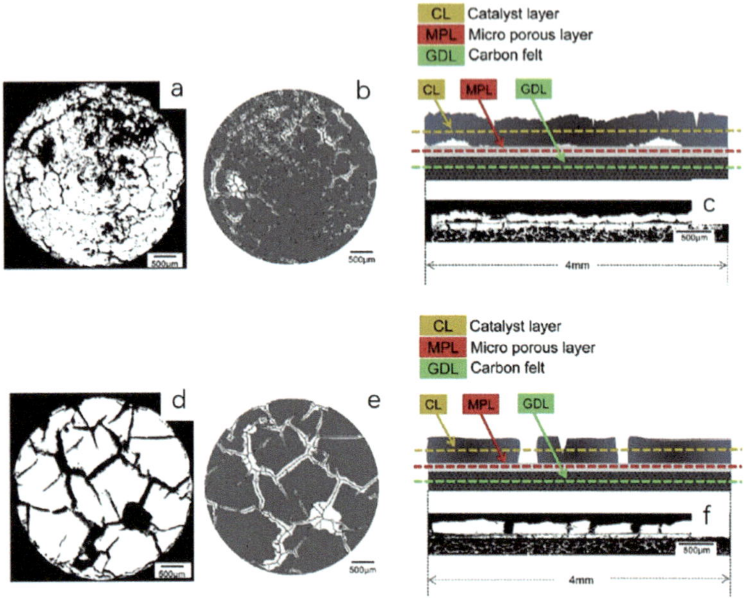

Fig. 25 In-plane and through-plane X-ray microtomography images of the electrode prepared with different techniques: **a–c** spraying, and **d–f** coating. (Reproduced from Mack et al. [177])

4.1 Gaskets and Sealing

The gaskets and sealing used in HT-PEMFCs also need to fulfill multiple requirements. In an operating PEM fuel cell, they are used to prevent the leakage of gases and coolant, and compensate tolerance for dimensional changes within the temperature range. They should be electrical insulated and provide adequate contact pressure between bipolar plates and GDLs over a long period of time, [178] to realize long-term stability under operating conditions in HT-PEMFC. Compressible materials, such as silicon rubber, perfluoroelastomer (FFKM), fluorocarbon (FKM), tetrafluoroethylene-propylene (AFLAS ®), ethylene propylene rubber (EPM, EPDM), and PEFE have been selected and modified to fulfill the sealing and supporting requirements used as gaskets and seals in HT-PEMFCs.

4.2 Bipolar Plates and Flow Field

Among the components of the HT-PEMFC device, the bipolar plate is a multifunctional key component that contributes to gas fuels delivery, high electrical, and mechanical supports of the stack. High electrical conductivity is required to minimize the ohmic loss in the stack, while high mechanical property is needed to withstand the compaction pressure during stack assembly.

Various materials have been selected for bipolar plates in HT-PEMFCs, for example, stainless steel (SS), [179, 180] graphite, [181] as well as other carbon composites [182, 183]. SS bipolar plates have emerged as promising candidates mainly because of their high electrical conductivity, and mechanical properties available in thin thicknesses, which making them commercially conducive to obtain a higher volume specific power output. Graphite is developed as a reference material for bipolar plates, in terms of the high corrosion resistance, but the brittleness of graphite often implies the thick bulk plates, thus makes them incompatible with mass production. Solutions have been investigated by developing carbon composite bipolar plate as the substitute for the brittle graphite, for example, carbon/cyanate ester-modified epoxy, [183] carbon/silicone elastomer, [184] carbon fiber composites, [185, 186] and so on.

Different from LT-PEMFCs, higher operating temperature of the cell can eliminate the need for the designing of the flow field to cater for liquid water removal, since water exists in vapor form, so low-pressure drop should be adapted to enhance the mass transport. Specific considerations should be given to the operating temperature range and the presence of PA in bipolar plate design. PA residue on the plate can compromise the gas transport in the GDL, cause oxidation process, and accelerate the material corrosion. In some cases, PA leaching from the MEA can be excluded via the flow field of the bipolar plate made from the well-defined material of plate which does not absorb PA [187].

4.3 Stack and System

According to different application scenario, multiple cells are normally assembled in series to form a fuel cell stack in order to provide required power density. Barreras et al. assembled a 5-cell HT-PEMFC stack by using commercially available Celtec P-1000 MEAs from BASF [188]. This stack features with several novel aspects, such as the new supplement way of reactant gases to the flow field, the design of the flow field geometry in terms of both the cathode and anode plates, as well as the new concept with assembly of heating system to realize the rapid start-up of the stack. The 5-cell HT-PEMFC stack yields a maximum power of 650 W. Budak et al. [189] recently developed a 12-cell stack feeding with reformat gas mixture containing H_2, CO_2, and CO. The prepared HT-PEMFC stack is assembled with cooling plates to prevent coolant leakage between the cells, and coupled with a thermo-oil-based liquid cooling system.

Compared with LT-PEMFC, HT-PEMFC system requires a less complex module with simplified water management and better tolerance of impurities, which eliminates the need for gas pre-purification in reformate operated system. In addition, there is less need for gas humidifiers in the system design. As the HT-PEMFC provides the possibility to use fuels from a wide variety of sources, for example, hydrocarbon and

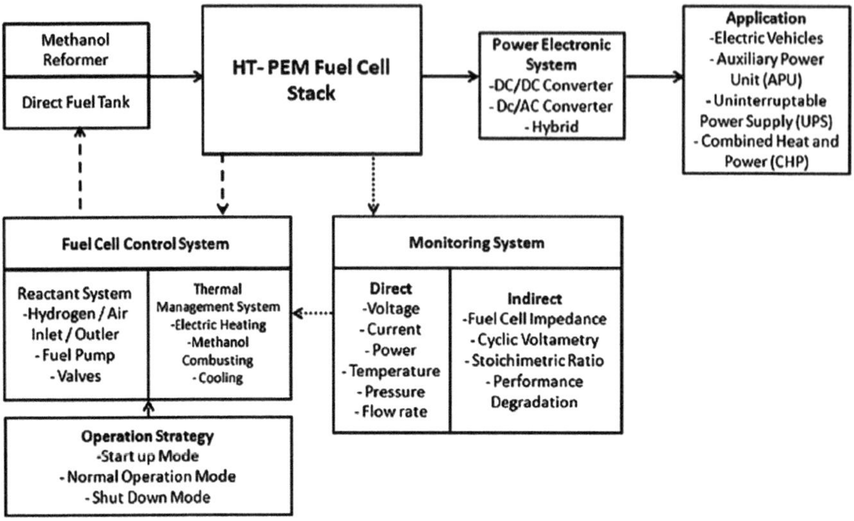

Fig. 26 Typical components in a HT-PEMFC system. (Reproduced from Rosli et al. [196])

alcohol fuels, some efforts have been demonstrated to the use of a fuel processing unit converting the raw fuel into a hydrogen-rich gas to increase efficiency [190–193].

A comprehensive model of a HT-PEMFC system is presented in Fig. 26, comprising a steam-reforming reactor to provide gas fuels, an integrated fuel cell stack, cooling system and heat reservoir, DC/AC module, and other auxiliary equipment, typically includes the monitor system. Two most widely-used cooling techniques for HT-PEMFC stacks are air cooling and liquid cooling. In the case of liquid cooling by means of thermo-oils or water, a separate cooling circuit plays a significant role to isolate the coolant flow from the electrode and the reactant gases [194]. Air cooling is achieved by increasing the air stoichiometric ratio in the cathode, [195, 196] thus is limited by the total cell volume and power output [178, 197].

Arsalis analyzed a micro-CHP HT-PEMFC system in terms of the operating parameters for maximizing total system efficiency without violating the requirements of the system [198]. The designed system is able to provide maximum loads of 100 kW_e with the highest calculated combined cogeneration system efficiency of 83.08%, while the fuel cell stack efficiency is 37.74%.

5 Conclusion and Challenges Towards HT-PEMFC Technology

Research and development progress in the last two decades with HT-PEMFC devices has made them promising candidates for next-generation power sources in auxiliary power units (APU) and combined heat-and-power (CHP) residential systems applications. With the commercial promotion from companies, like BASF and Danish Power Systems. HT-PEMFC devices have been reported to realize a long continuous operation lifetime exceeding 15,000 h.

In this chapter, R&D progress involving the design and fabrication of the HT-PEMFC is presented, from the perspectives of key components synthesis, electrodes construction, stack and system integration, as well as the commercial application status. Advances based on the PBI-membranes allow for the high proton exchange capability of the membrane, maintaining high proton conductivity over 0.01 S/cm with no water under high operating temperatures. In addition, the high operating temperature of 150–200 °C largely enhance the electrode kinetics in HT-PEMFCs. Besides, there are many other advantages to this technology. The high tolerance of fuel impurities brings larger flexibility for using a variety of fuels, including alcohols, biodiesel, and other liquid fuels, [199–201] which can be efficiently used in an HT-PEMFC system with a steam reformer. Higher operating temperature also enables simpler water management and easier flow field design. The increased heat transfer rate with larger temperature gradient between the fuel cell and the external environment could allow for the recycle of high-quality heat for cogeneration purposes.

Several remaining technical challenges associated with the HT-PEMFC technology need to be overcome to realize the large-scale commercialization:

- The components and materials used in this technology need to be carefully selected to eliminate the degradation caused by acid leaching. Novel research is required in order to ensure stringent durability of HT-PEMFC devices for various applications.
- The high cost of HT-PEMFCs is still the major concern for commercial application. Even though the PBI membranes cost around half the price of Nafion® membranes, the higher Pt loading still makes them more expensive than LT-PEMFCs. Research efforts towards developing effective and stable Pt-free electrocatalysts under HT-PEMFC operating conditions are still in a scarcity. The electrocatalysts account for around 45% of the total stack cost with HT-PEMFC devices, which makes the cost of HT-PEMFC far from the 2020 DOE target of $ 1000 kW^{-1} [202].
- By switching to high temperature operation, it usually takes HT-PEMFC devices much longer time to start up and reach the acceptable operating status, which could limit their application scenarios, especially in portable devices. In addition, when PA-doped membrane is used, an efficient and complex heat management strategy is badly required to precisely monitor the temperature of fuel cell to avoid the acid leaching that may cause harmful degradation of cell components and system.

References

1. Pollet BG, Staffell I, Shang JL (2012) Current status of hybrid, battery and fuel cell electric vehicles: from electrochemistry to market prospects. Electrochim Acta 84:235–249
2. Eikerling M (2006) Water management in cathode catalyst layers of PEM fuel cells: a structure-based model. J Electrochem Soc 153(3):E58
3. Vargas J, Ordonez J, Bejan A (2004) Constructal flow structure for a PEM fuel cell. Int J Heat Mass Transf 47(19–20):4177–4193
4. Pollet BG, Franco AA, Su H, Liang H, Pasupathi S (2016) Proton exchange membrane fuel cells. In: Barbir F, Basile A, Veziroğlu TN (eds) Compendium of hydrogen energy. Woodhead Publishing, Oxford, pp 3–56
5. Wieser C (2004) Novel polymer electrolyte membranes for automotive applications–requirements and benefits. Fuel Cells 4(4):245–250
6. Seel D, Benicewicz B, Xiao L, Schmidt T (2010) High-temperature polybenzimidazol-based membranes. In: Handbook of fuel cells
7. Bard A, Faulkner L (2001) Electrochemical methods: fundamentals and applications. John Wiley & Sons, Inc., New York
8. Zhang J, Tang Y, Song C, Zhang J, Wang H (2006) PEM fuel cell open circuit voltage (OCV) in the temperature range of 23 C to 120 C. J Power Sources 163(1):532–537
9. Tafel J (1905) The polarisation of cathodic hydrogen development. Zeitschrift Fur Physikalische Chemie-Stochiometrie Und Verwandtschaftslehre 50(6):641–712
10. Heyrovský J (1927) A theory of overpotential. Recl Trav Chim Pays-Bas 46(8):582–585
11. Erdey-Gruz T, Volmer M (1930) The theory of hydrogen high tension. Z Physik Chem 150:203–213
12. Marković N, Grgur B, Ross PN (1997) Temperature-dependent hydrogen electrochemistry on platinum low-index single-crystal surfaces in acid solutions. J Phys Chem B 101(27):5405–5413
13. Marković NM, Ross PN Jr (2002) Surface science studies of model fuel cell electrocatalysts. Surf Sci Rep 45(4–6):117–229
14. Li Q, He R, Gao J-A, Jensen JO, Bjerrum NJ (2003) The CO poisoning effect in PEMFCs operational at temperatures up to 200 C. J Electrochem Soc 150(12):A1599–A1605
15. Neyerlin K, Singh A, Chu D (2008) Kinetic characterization of a Pt-Ni/C catalyst with a phosphoric acid doped PBI membrane in a proton exchange membrane fuel cell. J Power Sources 176(1):112–117
16. Bernardi DM, Verbrugge MW (1992) A mathematical model of the solid-polymer-electrolyte fuel cell. J Electrochem Soc 139(9):2477
17. Stamenković V, Schmidt T, Ross P, Marković N (2003) Surface segregation effects in electrocatalysis: kinetics of oxygen reduction reaction on polycrystalline Pt_3Ni alloy surfaces. J Electroanal Chem 554:191–199
18. Parthasarathy A, Srinivasan S, Appleby AJ, Martin CR (1992) Temperature dependence of the electrode kinetics of oxygen reduction at the platinum/Nafion® interface-a microelectrode investigation. J Electrochem Soc 139(9):2530
19. Zhang J, Tang Y, Song C, Zhang J (2007) Polybenzimidazole-membrane-based PEM fuel cell in the temperature range of 120–200 C. J Power Sources 172(1):163–171
20. Villers D, Jacques-Bédard X, Dodelet J-P (2004) Fe-based catalysts for oxygen reduction in PEM fuel cells pretreatment of the carbon support. J Electrochem Soc 151(9):A1507–A1515
21. Oh H, Oh JG, Roh B, Hwang I, Kim H (2011) Development of highly active and stable non-precious oxygen reduction catalysts for PEM fuel cells using polypyrrole and a chelating agent. Electrochem Commun 13(8):879–881
22. Cheng Y, He S, Lu S, Veder J, Johannessen B, Thomsen L, Saunders M, Becker T, De Marco R, Li Q (2019) Iron single atoms on graphene as nonprecious metal catalysts for high-temperature polymer electrolyte membrane fuel cells. Adv Sci 6(10):1802066

23. Lobato J, Canizares P, Rodrigo MA, Linares JJ (2007) PBI-based polymer electrolyte membranes fuel cells: temperature effects on cell performance and catalyst stability. Electrochim Acta 52(12):3910–3920
24. Ross P, Andricacos P (1983) The effect of H_2PO_4-anion on the kinetics of oxygen reduction on Pt. J Electroanal Chem 154(1–2):205–215
25. Floriano J, Ticianelli E, Gonzalez E (1994) Influence of the supporting electrolyte on the oxygen reduction reaction at the platinum/proton exchange membrane interface. J Electroanal Chem 367(1–2):157–164
26. Yang Z, Moriguchi I, Nakashima N (2016) A highly-durable CO-tolerant poly (vinylphosphonic acid)-coated electrocatalyst supported on a nanoporous carbon. ACS Appl Mater Inter 8(14):9030–9036
27. Xiao L, Zhang H, Jana T, Scanlon E, Chen R, Choe EW, Ramanathan L, Yu S, Benicewicz B (2005) Synthesis and characterization of pyridine-based polybenzimidazoles for high temperature polymer electrolyte membrane fuel cell applications. Fuel Cells 5(2):287–295
28. Fujigaya T, Okamoto M, Nakashima N (2009) Design of an assembly of pyridine-containing polybenzimidazole, carbon nanotubes and Pt nanoparticles for a fuel cell electrocatalyst with a high electrochemically active surface area. Carbon 47(14):3227–3232
29. Zhang Q, Ling Y, Cai W, Yu X, Yang Z (2017) High performance and durability of polymer-coated Pt electrocatalyst supported on oxidized multi-walled in high-temperature polymer electrolyte fuel cells. Int J Hydrog Energy 42(26):16714–16721
30. Luo Y, Alonso-Vante N (2015) The effect of support on advanced Pt-based cathodes towards the oxygen reduction reaction state of the art. Electrochim Acta 179:108–118
31. Takeuchi N, Fuller TF (2008) Modeling and investigation of design factors and their impact on carbon corrosion of PEMFC electrodes. J Electrochem Soc 155(7):B770
32. Ferreira PJ, Shao-Horn Y, Morgan D, Makharia R, Kocha S, Gasteiger HA (2005) Instability of Pt/C electrocatalysts in proton exchange membrane fuel cells a mechanistic investigation. J Electrochem Soc 152(11):A2256–A2271
33. Debe MK (2012) Electrocatalyst approaches and challenges for automotive fuel cells. Nat Chem 486(7401):43–51
34. Ge J, St-Pierre J, Zhai Y (2014) PEMFC cathode catalyst contamination evaluation with a RRDE-methyl methacrylate. Int J Hydrog Energy 39(32):18351–18361
35. Ma J, Habrioux A, Alonso-Vante N (2014) The effect of substrates at cathodes in low-temperature fuel cells. ChemElectroChem 1(1):37–46
36. Mayrhofer KJ, Meier JC, Ashton SJ, Wiberg GK, Kraus F, Hanzlik M, Arenz M (2008) Fuel cell catalyst degradation on the nanoscale. Electrochem Commun 10(8):1144–1147
37. Shao Y, Yin G, Gao Y (2007) Understanding and approaches for the durability issues of Pt-based catalysts for PEM fuel cell. J Power Sources 171(2):558–566
38. Shao-Horn Y, Sheng W, Chen S, Ferreira PJ, Holby E, Morgan D (2007) Instability of supported platinum nanoparticles in low-temperature fuel cells. Top Catal 46(3–4):285–305
39. Speder J, Zana A, Spanos I, Kirkensgaard JJ, Mortensen K, Arenz M (2013) On the influence of the Pt to carbon ratio on the degradation of high surface area carbon supported PEM fuel cell electrocatalysts. Electrochem Commun 34:153–156
40. Speder J, Zana A, Spanos I, Kirkensgaard JJ, Mortensen K, Hanzlik M, Arenz M (2014) Comparative degradation study of carbon supported proton exchange membrane fuel cell electrocatalysts–the influence of the platinum to carbon ratio on the degradation rate. J Power Sources 261:14–22
41. Alonso-Vante N (2018) Precious versus non-precious electrocatalyst centers. In: Chalcogenide materials for energy conversion. Springer, pp 101–168
42. Yahya N (2013) Carbon and oxide nanostructures. Springer
43. Cheng Y (2013) Highly effective and CO-tolerant PtRu electrocatalysts supported on poly (ethyleneimine) functionalized carbon nanotubes for direct methanol fuel cells. Electrochim Acta 99:124–132
44. Kou R, Shao Y, Wang D, Engelhard MH, Kwak JH, Wang J, Viswanathan VV, Wang C, Lin Y, Wang Y (2009) Enhanced activity and stability of Pt catalysts on functionalized graphene sheets for electrocatalytic oxygen reduction. Electrochem Commun 11(5):954–957

45. Ma J, Habrioux A, Morais C, Lewera A, Vogel W, Verde-Gomez Y, Ramos-Sanchez G, Balbuena PB, Alonso-Vante N (2013) Spectroelectrochemical probing of the strong interaction between platinum nanoparticles and graphitic domains of carbon. ACS Catal 3(9):1940–1950
46. Sebastián D, Calderón J, González-Expósito J, Pastor E, Martínez-Huerta M, Suelves I, Moliner R, Lázaro M (2010) Influence of carbon nanofiber properties as electrocatalyst support on the electrochemical performance for PEM fuel cells. Int J Hydrog Energy 35(18):9934–9942
47. Wang X, Li W, Chen Z, Waje M, Yan Y (2006) Durability investigation of carbon nanotube as catalyst support for proton exchange membrane fuel cell. J Power Sources 158(1):154–159
48. Hasche F, Oezaslan M, Strasser P (2010) Activity, stability and degradation of multi walled carbon nanotube (MWCNT) supported Pt fuel cell electrocatalysts. Phys Chem Chem Phys 12(46):15251–15258
49. Szymanski GS, Karpinski Z, Biniak S, Światkowski A (2002) The effect of the gradual thermal decomposition of surface oxygen species on the chemical and catalytic properties of oxidized activated carbon. Carbon 40(14):2627–2639
50. Mangun CL, Benak KR, Economy J, Foster KL (2001) Surface chemistry, pore sizes and adsorption properties of activated carbon fibers and precursors treated with ammonia. Carbon 39(12):1809–1820
51. Zhang S, Shao Y, Yin G, Lin Y (2010) Carbon nanotubes decorated with Pt nanoparticles via electrostatic self-assembly: a highly active oxygen reduction electrocatalyst. J Mater Chem A 20(14):2826–2830
52. Yoo SJ, Kim S, Jeon T, Hwang SJ, Lee J, Lee S, Lee K, Cho Y, Sung Y, Lim T (2011) Enhanced stability and activity of Pt-Y alloy catalysts for electrocatalytic oxygen reduction. Chem Commun 47(41):11414–11416
53. Zhang S, Shao Y, Yin G, Lin Y (2011) Self-assembly of Pt nanoparticles on highly graphitized carbon nanotubes as an excellent oxygen-reduction catalyst. Appl Catal B-Environ 102(3):372–377
54. Berber MR, Hafez IH, Fujigaya T, Nakashima N (2014) Durability analysis of polymer-coated pristine carbon nanotube-based fuel cell electrocatalysts under non-humidified conditions. J Mater Chem A 2(44):19053–19059
55. Bauer A, Lee K, Song C, Xie Y, Zhang J, Hui R (2010) Pt nanoparticles deposited on TiO_2 based nanofibers: electrochemical stability and oxygen reduction activity. J Power Sources 195(10):3105–3110
56. Huang SY, Ganesan P, Popov BN (2011) Titania supported platinum catalyst with high electrocatalytic activity and stability for polymer electrolyte membrane fuel cell. Appl Catal B-Environ 102(1–2):71–77
57. Kang E, An S, Yoon S, Kim JK, Lee J (2010) Ordered mesoporous WO_3-X possessing electronically conductive framework comparable to carbon framework toward long-term stable cathode supports for fuel cells. J Mater Chem A 20(35):7416–7421
58. Takasaki F, Matsuie S, Takabatake Y, Noda Z, Hayashi A, Shiratori Y, Ito K, Sasaki K (2011) Carbon-free Pt electrocatalysts supported on SnO_2 for polymer electrolyte fuel cells: electrocatalytic activity and durability. J Electrochem Soc 158(10):B1270
59. Timperman L, Lewera A, Vogel W, Alonso-Vante N (2010) Nanostructured platinum becomes alloyed at oxide-composite substrate. Electrochem Commun 12(12):1772–1775
60. Yan Z, Wei W, Xie J, Meng S, Lü X, Zhu J (2013) An ion exchange route to produce WO_3 nanobars as Pt electrocatalyst promoter for oxygen reduction reaction. J Power Sources 222:218–224
61. Jiang D, Overbury S, Dai S (2012) Structures and energetics of Pt clusters on TiO_2: interplay between metal-metal bonds and metal-oxygen bonds. J Phys Chem C 116(41):21880–21885
62. Lewera A, Timperman L, Roguska A, Alonsovante N (2011) Metal-support interactions between nanosized Pt and metal oxides (WO_3 and TiO_2) studied using X-ray photoelectron spectroscopy. J Phys Chem C 115(41):20153–20159

63. Huang SY, Ganesan P, Popov BN (2010) Electrocatalytic activity and stability of niobium-doped titanium oxide supported platinum catalyst for polymer electrolyte membrane fuel cells. Appl Catal B-Environ 96(1–2):224–231
64. Camacho BR, Morais C, Valenzuela MA, Alonsovante N (2013) Enhancing oxygen reduction reaction activity and stability of platinum via oxide-carbon composites. Catal Today 202:36–43
65. Zhang F, Chen J, Zhang X, Gao W, Jin R, Guan N, Li Y (2004) Synthesis of titania-supported platinum catalyst: the effect of pH on morphology control and valence state during photodeposition. Langmuir 20(21):9329–9334
66. Beak S, Jung D, Nahm KS, Kim P (2010) Preparation of highly dispersed Pt on TiO_2-modified carbon for the application to oxygen reduction reaction. Catal Lett 134(3):288–294
67. Shim J, Lee C, Lee H, Lee J, Cairns EJ (2001) Electrochemical characteristics of Pt-WO_3/C and Pt-TiO_2/C electrocatalysts in a polymer electrolyte fuel cell. J Power Sources 102(1):172–177
68. Liu G, Zhang H, Zhai Y, Zhang Y, Xu D, Shao ZG (2007) Pt_4ZrO_2/C cathode catalyst for improved durability in high temperature PEMFC based on H_3PO_4 doped PBI. Electrochem Commun 9(1):135–141
69. Ma J, Valenzuela E, Gago AS, Rousseau J, Habrioux Al, Alonso-Vante N (2014) Photohole trapping induced platinum cluster nucleation on the surface of TiO_2 nanoparticles. J Phys Chem C 118(2):1111–1117
70. Liu R, Wu D, Feng X, Mullen K (2010) Nitrogen-doped ordered mesoporous graphitic arrays with high electrocatalytic activity for oxygen reduction. Angew Chem Int Edit 49(14):2565–2569
71. Masuda T, Fukumitsu H, Fugane K, Togasaki H, Matsumura D, Tamura K, Nishihata Y, Yoshikawa H, Kobayashi K, Mori T (2012) Role of cerium oxide in the enhancement of activity for the oxygen reduction reaction at Pt-CeO_x nanocomposite electrocatalyst- an in situ electrochemical X-ray absorption fine structure study. J Phys Chem C 116(18):10098–10102
72. Ou DR, Mori T, Fugane K, Togasaki H, Ye F, Drennan J (2011) Stability of ceria supports in Pt-CeO_x/C catalysts. J Phys Chem C 115(39):19239–19245
73. Tseung A, Dhara S (1975) Loss of surface area by platinum and supported platinum black electrocatalyst. Electrochim Acta 20(9):681–683
74. Honji A, Mori T, Tamura K, Hishinuma Y (1988) Agglomeration of platinum particles supported on carbon in phosphoric acid. J Electrochem Soc 135(2):355
75. Liu G, Zhang H, Hu J, Zhai Y, Xu D, Shao ZG (2006) Studies of performance degradation of a high temperature PEMFC based on H_3PO_4-doped PBI. J Power Sources 162(1):547–552
76. Wilson MS, Garzon FH, Sickafus KE, Gottesfeld S (1993) Surface area loss of supported platinum in polymer electrolyte fuel cells. J Electrochem Soc 140(10):2872
77. Simonsen SB, Chorkendorff I, Dahl S, Skoglundh M, Sehested J, Helveg S (2010) Direct observations of oxygen-induced platinum nanoparticle ripening studied by in situ TEM. J Am Chem Soc 132(23):7968–7975
78. Bett J, Kinoshita K, Stonehart P (1974) Crystallite growth of platinum dispersed on graphitized carbon black. J Catal 35(2):307–316
79. Darling RM, Meyers JP (2003) Kinetic model of platinum dissolution in PEMFCs. J Electrochem Soc 150(11):A1523–A1527
80. Sondergaard T, Cleemann LN, Zhong L, Becker H, Steenberg T, Hjuler HA, Seerup L, Li Q, Jensen JO (2018) Catalyst degradation under potential cycling as an accelerated stress test for PBI-based high-temperature PEM fuel cells-effect of humidification. Electrocatalysis 9(3):302–313
81. Giordano N, Antonucci P, Passalacqua E, Pino L, Arico A, Kinoshita K (1991) Relationship between physicochemical properties and electrooxidation behaviour of carbon materials. Electrochim Acta 36(13):1931–1935
82. Passalacqua E, Antonucci P, Vivaldi M, Patti A, Antonucci V, Giordano N, Kinoshita K (1992) The influence of Pt on the electrooxidation behaviour of carbon in phosphoric acid. Electrochim Acta 37(15):2725–2730

83. Tedman CS (ed) (1974) Corrosion problems in energy conversion and generation. Corrosion Division, Electrochemical Society
84. Yu X, Ye S (2007) Recent advances in activity and durability enhancement of Pt/C catalytic cathode in PEMFC: part ii: degradation mechanism and durability enhancement of carbon supported platinum catalyst. J Power Sources 172(1):145–154
85. Zhang S, Yuan X, Wang H, Mérida W, Zhu H, Shen J, Wu S, Zhang J (2009) A review of accelerated stress tests of MEA durability in PEM fuel cells. Int J Hydrog Energy 34(1):388–404
86. Liu Z, Tsou Y, Calundann G, De Castro ES (2011) New process for high temperature polybenzimidazole membrane production and its impact on the membrane and the membrane electrode assembly. J Power Sources 196(3):1055–1060
87. Ma Y, Wainright JS, Litt MH, Savinell RF (2004) Conductivity of PBI membranes for high-temperature polymer electrolyte fuel cells. J Electrochem Soc 151(1)
88. Vilciauskas L, Tuckerman ME, Bester G, Paddison SJ, Kreuer KD (2012) The mechanism of proton conduction in phosphoric acid. Nat Chem 4(6):461–466
89. Ogawa T, Kamiguchi K, Tamaki T, Imai H, Yamaguchi T (2014) Differentiating Grotthuss proton conduction mechanisms by nuclear magnetic resonance spectroscopic analysis of frozen samples. Anal Chem 86(19):9362–9366
90. Daletou MK, Kallitsis JK, Voyiatzis G, Neophytides SG (2009) The interaction of water vapors with H_3PO_4 imbibed electrolyte based on PBI/polysulfone copolymer blends. J Membr Sci 326(1):76–83
91. Bouchet R, Siebert E (1999) Proton conduction in acid doped polybenzimidazole. Solid State Ion 118(3):287–299
92. Glipa X, Bonnet B, Mula B, Jones DJ, Roziere J (1999) Investigation of the conduction properties of phosphoric and sulfuric acid doped polybenzimidazole. J Mater Chem A 9(12):3045–3049
93. Kawahara M, Morita J, Rikukawa M, Sanui K, Ogata N (2000) Synthesis and proton conductivity of thermally stable polymer electrolyte: poly(benzimidazole) complexes with strong acid molecules. Electrochim Acta 45(8):1395–1398
94. Korte C, Conti F, Wackerl J, Lehnert W (2016) Phosphoric acid and its interactions with polybenzimidazole-type polymers. In: Li Q, Aili D, Hjuler HA, Jensen JO (eds) High temperature polymer electrolyte membrane fuel cells: approaches, status, and perspectives. Springer International Publishing, Cham, pp 169–194
95. He R, Li Q, Bach A, Jensen JO, Bjerrum NJ (2006) Physicochemical properties of phosphoric acid doped polybenzimidazole membranes for fuel cells. J Membr Sci 277(1):38–45
96. He R, Li Q, Xiao G, Bjerrum NJ (2003) Proton conductivity of phosphoric acid doped polybenzimidazole and its composites with inorganic proton conductors. J Membr Sci 226(1):169–184
97. Litt M, Ameri R, Wang Y, Savinell R, Wainwright J (1998) Polybenzimidazoles/phosphoric acid solid polymer electrolytes: mechanical and electrical properties. MRS Online Proc Libr Archive:548
98. Samms S, Wasmus S, Savinell RF (1996) Thermal stability of proton conducting acid doped polybenzimidazole in simulated fuel cell environments. J Electrochem Soc 143(4):1225–1232
99. Li Q, Jensen JO, Savinell RF, Bjerrum NJ (2009) High temperature proton exchange membranes based on polybenzimidazoles for fuel cells. Prog Polym Sci 34(5):449–477
100. Vogel H, Marvel C (1961) Polybenzimidazoles, new thermally stable polymers. J Polym Sci Pol Chem 50(154):511–539
101. Choe EW (1994) Catalysts for the preparation of polybenzimidazoles. J Appl Polym Sci 53(5):497–506
102. Kreuer K (1996) Proton conductivity: materials and applications. Chem Mat 8(3):610–641
103. Eaton PE, Carlson GR, Lee JT (1973) Phosphorus pentoxide-methanesulfonic acid: convenient alternative to polyphosphoric acid. J Org Chem 38(23):4071–4073
104. Hedberg F, Marvel C (1974) A new single-step process for polybenzimidazole synthesis. J Polym Sci Pol Chem 12(8):1823–1828

105. Kim H, Cho SY, An SJ, Eun YC, Kim J, Yoon H, Kweon H, Yew KH (2004) Synthesis of poly(2,5-benzimidazole) for use as a fuel-cell membrane. Macromol Rapid Commun 25(8):894–897
106. Xiao L, Zhang H, Scanlon E, Ramanathan L, Choe E-W, Rogers D, Apple T, Benicewicz BC (2005) High-temperature polybenzimidazole fuel cell membranes via a sol-gel process. Chem Mat 17(21):5328–5333
107. Yang JS, Cleemann LN, Steenberg T, Terkelsen C, Li QF, Jensen JO, Hjuler HA, Bjerrum NJ, He RH (2014) High molecular weight polybenzimidazole membranes for high temperature PEMFC. Fuel Cells 14(1):7–15
108. Song M, Lu X, Li Z, Liu G, Yin X, Wang Y (2016) Compatible ionic crosslinking composite membranes based on SPEEK and PBI for high temperature proton exchange membranes. Int J Hydrog Energy 41(28):12069–12081
109. Li Q, Pan C, Jensen JO, Noyé P, Bjerrum NJ (2007) Cross-linked polybenzimidazole membranes for fuel cells. Chem Mat 19(3):350–352
110. Liu YL (2009) Preparation and properties of nanocomposite membranes of polybenzimidazole/sulfonated silica nanoparticles for proton exchange membranes. J Membr Sci 332(1–2):121–128
111. Chang YN, Lai JY, Liu YL (2012) Polybenzimidazole (PBI)-functionalized silica nanoparticles modified PBI nanocomposite membranes for proton exchange membranes fuel cells. J Membr Sci 403:1–7
112. Wang KY, Chung TS, Qin JJ (2007) Polybenzimidazole (PBI) nanofiltration hollow fiber membranes applied in forward osmosis process. J Membr Sci 300(1–2):6–12
113. Aili D, Jensen JO, Li Q (2016) Polybenzimidazole membranes by post acid doping. Springer, High temperature polymer electrolyte membrane fuel cells, pp 195–215
114. Deimede V, Voyiatzis GA, Kallitsis JK, Qingfeng L, Bjerrum NJ (2000) Miscibility behavior of polybenzimidazole/sulfonated polysulfone blends for use in fuel cell applications. Macromol 33(20):7609–7617
115. Hasiotis C, Deimede V, Kontoyannis CG (2001) New polymer electrolytes based on blends of sulfonated polysulfones with polybenzimidazole. Electrochim Acta 46(15):2401–2406
116. Kerres J, Ullrich A, Meier F, Haring T (1999) Synthesis and characterization of novel acid–base polymer blends for application in membrane fuel cells. Solid State Ion 125(1):243–249
117. Schonberger F, Hein M, Kerres J (2007) Preparation and characterisation of sulfonated partially fluorinated statistical poly(arylene ether sulfone)s and their blends with PBI. Solid State Ion 178(7):547–554
118. Gillham J (1963) Polymer structure: cross-linking of a polybenzimidazole. Science 139(3554):494–495
119. Sukumar PR, Wu W, Markova D, Ünsal Ö, Klapper M, Müllen K (2007) Functionalized poly(benzimidazole)s as membrane materials for fuel cells. Macromol Chem Phys 208(19–20):2258–2267
120. Wang S, Zhang G, Han M, Li H, Zhang Y, Ni J, Ma W, Li M, Wang J, Liu Z (2011) Novel epoxy-based cross-linked polybenzimidazole for high temperature proton exchange membrane fuel cells. Int J Hydrog Energy 36(14):8412–8421
121. Hasiotis C, Qingfeng L, Deimede V, Kallitsis JK, Kontoyannis CG, Bjerrum NJ (2001) Development and characterization of acid-doped polybenzimidazole/sulfonated polysulfone blend polymer electrolytes for fuel cells. J Electrochem Soc 148(5)
122. Kerres J (2016) Applications of acid–base blend concepts to intermediate temperature membranes. Springer, High temperature polymer electrolyte membrane fuel cells, pp 59–89
123. Kerres J, Tang C-M, Graf C (2004) Improvement of properties of poly (ether ketone) ionomer membranes by blending and cross-linking. Ind Eng Chem Res 43(16):4571–4579
124. Kerres J, Ullrich A, Meier F, Häring T (1999) Synthesis and characterization of novel acid–base polymer blends for application in membrane fuel cells. Solid State Ion 125(1–4):243–249
125. Mack F, Aniol K, Ellwein C, Kerres J, Zeis R (2015) Novel phosphoric acid-doped PBI-blends as membranes for high-temperature PEM fuel cells. J Mater Chem A 3(20):10864–10874

126. Davis HJ, Thomas NW (inventors) (1977) Chemical modification of polybenzimidazole semipermeable (Celanese Corporation). US patent 4,020,142
127. Wang C, Li Z, Sun P, Pei H, Yin X (2020) Preparation and properties of covalently crosslinked polybenzimidazole high temperature proton exchange membranes doped with high sulfonated polyphosphazene. J Electrochem Soc 167(10):104517
128. Laconti A, Liu H, Mittelsteadt C, McDonald R (2006) Polymer electrolyte membrane degradation mechanisms in fuel cells-findings over the past 30 years and comparison with electrolyzers. ECS Trans 1(8):199
129. Liao JH, Li QF, Rudbeck HC, Jensen JO, Chromik A, Bjerrum NJ, Kerres J, Xing W (2011) Oxidative degradation of polybenzimidazole membranes as electrolytes for high temperature proton exchange membrane fuel cells. Fuel Cells 11(6):745–755
130. Schmidt TJ, Baurmeister J (2006) Durability and reliability in high-temperature reformed hydrogen PEFCs. ECS Trans 3(1):861
131. Liao J, Yang J, Li Q, Cleemann LN, Jensen JO, Bjerrum NJ, He R, Xing W (2013) Oxidative degradation of acid doped polybenzimidazole membranes and fuel cell durability in the presence of ferrous ions. J Power Sources 238:516–522
132. Yu S, Xiao L, Benicewicz BC (2008) Durability studies of PBI-based high temperature PEMFCs. Fuel Cells 8:165–174
133. Orfanidi A, Daletou MK, Sygellou L, Neophytides SG (2013) The role of phosphoric acid in the anodic electrocatalytic layer in high temperature PEM fuel cells. J Appl Electrochem 43(11):1101–1116
134. Chen H, Wang S, Liu F, Wang D, Li J, Mao T, Liu G, Wang X, Xu J, Wang Z (2020) Base-acid doped polybenzimidazole with high phosphoric acid retention for HT-PEMFC applications. J Membr Sci 596
135. Kreuer KD, Portale G (2013) A critical revision of the nano-morphology of proton conducting ionomers and polyelectrolytes for fuel cell applications. Adv Funct Mater 23(43):5390–5397
136. Oono Y, Sounai A, Hori M (2012) Long-term cell degradation mechanism in high-temperature proton exchange membrane fuel cells. J Power Sources 210:366–373
137. Kerres J (2005) Blended and cross-linked ionomer membranes for application in membrane fuel cells. Fuel Cells 5(2):230–247
138. Li Q, Rudbeck HC, Chromik A, Jensen JO, Pan C, Steenberg T, Calverley M, Bjerrum NJ, Kerres J (2010) Properties, degradation and high temperature fuel cell test of different types of PBI and PBI blend membranes 347(1):260–270
139. Litster S, McLean G (2004) PEM fuel cell electrodes. J Power Sources 130(1–2):61–76
140. Seland F, Berning T, Børresen B, Tunold R (2006) Improving the performance of high-temperature PEM fuel cells based on PBI electrolyte. J Power Sources 160(1):27–36
141. Siddique N, Liu F (2010) Process based reconstruction and simulation of a three-dimensional fuel cell catalyst layer. Electrochim Acta 55(19):5357–5366
142. Kazeminasab B, Rowshanzamir S, Ghadamian H (2017) Nitrogen doped graphene/cobalt-based catalyst layers of a PEM fuel cell: performance evaluation and multi-objective optimization. Korean J Chem Eng 34(11):2978–2983
143. Su H, Pasupathi S, Bladergroen B, Linkov V, Pollet BG (2013) Optimization of gas diffusion electrode for polybenzimidazole-based high temperature proton exchange membrane fuel cell: Evaluation of polymer binders in catalyst layer. Int J Hydrog Energy 38(26):11370–11378
144. Modestov A, Tarasevich M, Filimonov VY, Leykin AY (2009) Influence of catalyst layer binder on catalyst utilization and performance of fuel cell with polybenzimidazole-H_3PO_4 membrane. J Electrochem Soc 156(5):B650
145. Park JO, Kwon K, Cho MD, Hong SG, Kim T, Yoo DY (2011) Role of binders in high temperature PEMFC electrode. J Electrochem Soc 158(6):B675
146. Mazúr P, Soukup J, Paidar M, Bouzek K (2011) Gas diffusion electrodes for high temperature PEM-type fuel cells: role of a polymer binder and method of the catalyst layer deposition. J Appl Electrochem 41(9):1013–1019
147. Mack F, Morawietz T, Hiesgen R, Kramer D, Zeis R (2013) PTFE distribution in high-temperature PEM electrodes and its effect on the cell performance. ECS Trans 58(1):881–888

148. Gurau V, De Castro ES (2020) Prediction of performance variation caused by manufacturing tolerances and defects in gas diffusion electrodes of phosphoric acid (PA)-doped polybenzimidazole (PBI)-based high-temperature proton exchange membrane fuel cells. Energies 13(6):1345
149. Lee E, Kim D, Pak C (2020) Effects of cathode catalyst layer fabrication parameters on the performance of high-temperature polymer electrolyte membrane fuel cells. Appl Surf Sci 510:145461
150. Du H, Wang C, Yang C, Hsu H, Chang S, Huang H, Lai S, Chen J, Yu TL, Chen L (2014) A high performance polybenzimidazole-CNT hybrid electrode for high-temperature proton exchange membrane fuel cells. J. Mater. Chem. A. 2(19):7015–7019
151. Kim S, Myles TD, Kunz HR, Kwak D, Wang Y, Maric R (2015) The effect of binder content on the performance of a high temperature polymer electrolyte membrane fuel cell produced with reactive spray deposition technology. Electrochim Acta 177:190–200
152. Martin S, Garciaybarra PL, Castillo JL (2018) Ten-fold reduction from the state-of-the-art platinum loading of electrodes prepared by electrospraying for high temperature proton exchange membrane fuel cells. Electrochem Commun 93:57–61
153. Martin S, Jensen JO, Li Q, Garcia-Ybarra PL, Castillo JL (2019) Feasibility of ultra-low Pt loading electrodes for high temperature proton exchange membrane fuel cells based in phosphoric acid-doped membrane. Int J Hydrog Energy 44(52):28273–28282
154. Martin S, Li Q, Jensen JO (2015) Lowering the platinum loading of high temperature polymer electrolyte membrane fuel cells with acid doped polybenzimidazole membranes. J Power Sources 293:51–56
155. Martin S, Li Q, Steenberg T, Jensen JO (2014) Binderless electrodes for high-temperature polymer electrolyte membrane fuel cells. J Power Sources 272:559–566
156. Su H, Jao TC, Barron O, Pollet BG, Pasupathi S (2014) Low platinum loading for high temperature proton exchange membrane fuel cell developed by ultrasonic spray coating technique. J Power Sources 267:155–159
157. Yao D, Zhang W, Ma Q, Xu Q, Pasupathi S, Su H (2019) Achieving high Pt utilization and superior performance of high temperature polymer electrolyte membrane fuel cell by employing low-Pt-content catalyst and microporous layer free electrode design. J Power Sources 426:124–133
158. Lobato J, Canizares P, Rodrigo MA, Linares JJ, Ubeda D, Pinar FJ (2010) Study of the catalytic layer in polybenzimidazole-based high temperature PEMFC: effect of platinum content on the carbon support. Fuel Cells 10(2):312–319
159. Galbiati S, Baricci A, Casalegno A, Marchesi R (2013) Degradation in phosphoric acid doped polymer fuel cells: a 6000 h parametric investigation. Int J Hydrog Energy 38(15):6469–6480
160. Kannan A, Li Q, Cleemann L, Jensen J (2018) Acid distribution and durability of HT-PEM fuel cells with different electrode supports. Fuel Cells 18(2):103–112
161. Ramasamy RP, Kumbur EC, Mench MM, Liu W, Moore D, Murthy M (2008) Investigation of macro-and micro-porous layer interaction in polymer electrolyte fuel cells. Int J Hydrog Energy 33(13):3351–3367
162. Gostick JT, Fowler MW, Pritzker MD, Ioannidis MA, Behra LM (2006) In-plane and through-plane gas permeability of carbon fiber electrode backing layers. J Power Sources 162(1):228–238
163. Jayakumar A, Sethu SP, Ramos M, Robertson J, Al-Jumaily A (2015) A technical review on gas diffusion, mechanism and medium of PEM fuel cell. Ionics 21(1):1–18
164. Mathias M, Roth J, Fleming J, Lehnert W, Vielstich W (2003) Handbook of fuel cells: fundamentals, technology, and applications, vol 3, pp 517–37
165. Webb SW, Pruess K (2003) The use of Fick's law for modeling trace gas diffusion in porous media. Transp Porous Media 51(3):327–341
166. Malek K, Coppens M-O (2003) Knudsen self-and Fickian diffusion in rough nanoporous media. J Chem Phys 119(5):2801–2811
167. Pant LM, Mitra SK, Secanell M (2012) Absolute permeability and Knudsen diffusivity measurements in PEMFC gas diffusion layers and micro porous layers. J Power Sources 206:153–160

168. Lobato J, Cañizares P, Rodrigo MA, Úbeda D, Pinar FJ, Linares JJ (2010) Optimisation of the microporous layer for a polybenzimidazole-based high temperature PEMFC-effect of carbon content. Fuel Cells 10(5):770–777
169. Park S, Lee J-W, Popov BN (2006) Effect of carbon loading in microporous layer on PEM fuel cell performance. J Power Sources 163(1):357–363
170. Fernandez R, Ferreira-Aparicio P, Daza L (2005) PEMFC electrode preparation: influence of the solvent composition and evaporation rate on the catalytic layer microstructure. J Power Sources 151:18–24
171. Huang DC, Yu PJ, Liu FJ, Huang SL, Hsueh KL, Chen YC, Wu CH, Chang WC, Tsau FH (2011) Effect of dispersion solvent in catalyst ink on proton exchange membrane fuel cell performance. Int J Electrochem Sci 6(7):2551–2565
172. Aziz AA, Bakar SA, Rusop M (2010) Carbon nanostructured materials. Springer, Carbon and oxide nanostructures, pp 165–193
173. Bodner M, García HR, Steenberg T, Terkelsen C, Alfaro SM, Avcioglu GS, Vassiliev A, Primdahl S, Hjuler HA (2019) Enabling industrial production of electrodes by use of slot-die coating for HT-PEM fuel cells. Int J Hydrog Energy 44(25):12793–12801
174. Mauger SA, Neyerlin KC, Yangneyerlin AC, More KL, Ulsh M (2018) Gravure coating for roll-to-roll manufacturing of proton-exchange-membrane fuel cell catalyst layers. J Electrochem Soc 165(11)
175. Liang H, Su H, Pollet BG, Linkov V, Pasupathi S (2014) Membrane electrode assembly with enhanced platinum utilization for high temperature proton exchange membrane fuel cell prepared by catalyst coating membrane method. J Power Sources 266:107–113
176. Devrim Y, Arıca ED (2019) Multi-walled carbon nanotubes decorated by platinum catalyst for high temperature PEM fuel cell. Int J Hydrog Energy 44(34):18951–18966
177. Mack F, Klages M, Scholta J, Jörissen L, Morawietz T, Hiesgen R, Kramer D, Zeis R (2014) Morphology studies on high-temperature polymer electrolyte membrane fuel cell electrodes. J Power Sources 255:431–438
178. Janßen H, Supra J, Lüke L, Lehnert W, Stolten D (2013) Development of HT-PEFC stacks in the kW range. Int J Hydrog Energy 38(11):4705–4713
179. André J, Antoni L, Petit J-P (2010) Corrosion resistance of stainless steel bipolar plates in a PEFC environment: a comprehensive study. Int J Hydrog Energy 35(8):3684–3697
180. Lee SJ, Lai JJ, Huang CH (2005) Stainless steel bipolar plates. J Power Sources 145(2):362–368
181. Derieth T, Bandlamudi G, Beckhaus P, Kreuz C, Mahlendorf F, Heinzel A (2008) Development of highly filled graphite compounds as bipolar plate materials for low and high temperature PEM fuel cells. J New Mat Electr Sys 11(1):21
182. Lee D, Lim JW (2017) Cathode/anode integrated composite bipolar plate for high-temperature PEMFC. Compos Struct 167:144–151
183. Lee D (2016) Carbon composite bipolar plate for high-temperature proton exchange membrane fuel cells (HT-PEMFCs). J Power Sources 327:119–126
184. Lee D, Lim JW, Nam S, Choi I (2015) Gasket-integrated carbon/silicone elastomer composite bipolar plate for high-temperature PEMFC. Compos Struct 128:284–290
185. Kim M, Lim JW (2015) Surface modification of carbon fiber phenolic bipolar plate for the HT-PEMFC with nano-carbon black and carbon felts. Compos Struct 119:630–637
186. Lim JW, Kim M, Yu YH (2014) Development of carbon/PEEK composite bipolar plates with nano-conductive particles for High-Temperature PEM fuel cells (HT-PEMFCs). Compos Struct 118:519–527
187. Li Q, Aili D, Hjuler HA, Jensen JO (2016) High temperature polymer electrolyte membrane fuel cells. Springer, Switzerland
188. Barreras F, Lozano A, Roda V, Barroso J, Martin J (2014) Optimal design and operational tests of a high-temperature PEM fuel cell for a combined heat and power unit. Int J Hydrog Energy 39(10):5388–5398
189. Budak Y, Devrim Y (2019) Micro-cogeneration application of a high-temperature PEM fuel cell stack operated with polybenzimidazole based membranes. Int J Hydrog Energy 45(60):35198–35207

190. Avgouropoulos G, Paxinou A, Neophytides SG (2014) In situ hydrogen utilization in an internal reforming methanol fuel cell. Int J Hydrog Energy 39(31):18103–18108
191. Weng F, Cheng C, Chen K (2013) Hydrogen production of two-stage temperature steam reformer integrated with PBI membrane fuel cells to optimize thermal management. Int J Hydrog Energy 38(14):6059–6064
192. Korsgaard AR, Nielsen MP, Kaer SK (2008) Part two: control of a novel HTPEM-based micro combined heat and power fuel cell system. Int J Hydrog Energy 33(7):1921–1931
193. Korsgaard AR, Nielsen MP, Kaer SK (2008) Part one: a novel model of HTPEM-based micro-combined heat and power fuel cell system. Int J Hydrog Energy 33(7):1909–1920
194. Scholta J, Messerschmidt M, Jorissen L, Hartnig C (2009) Externally cooled high temperature polymer electrolyte membrane fuel cell stack. J Power Sources 190(1):83–85
195. Liu Y, Lehnert W, Janßen H, Samsun RC, Stolten D (2016) A review of high-temperature polymer electrolyte membrane fuel-cell (HT-PEMFC)-based auxiliary power units for diesel-powered road vehicles. J Power Sources 311:91–102
196. Rosli R, Sulong AB, Daud WRW, Zulkifley MA, Husaini T, Rosli MI, Majlan EH, Haque MA (2017) A review of high-temperature proton exchange membrane fuel cell (HT-PEMFC) system. Int J Hydrog Energy 42(14):9293–9314
197. Andreasen SJ, Kær SK, Sahlin SL, Justesen KK (2013) Design and control of high temperature PEM fuel cell systems using methanol reformers with air or liquid heat integration. Paper presented at 5th international conference FDFC2013
198. Arsalis A (2012) Modeling and simulation of a 100 kWe HT-PEMFC subsystem integrated with an absorption chiller subsystem. Int J Hydrog Energy 37(18):13484–13490
199. Araya SS, Zhou F, Liso V, Sahlin SL, Vang JR, Thomas S, Gao X, Jeppesen C, Kær SK (2016) A comprehensive review of PBI-based high temperature PEM fuel cells. Int J Hydrog Energy 41(46):21310–21344
200. Mazur P, Soukup J, Paidar M, Bouzek K (2011) Gas diffusion electrodes for high temperature PEM-type fuel cells: role of a polymer binder and method of the catalyst layer deposition. J Appl Electrochem 41(9):1013–1019
201. Iulianelli A, Ribeirinha P, Mendes A, Basile A (2014) Methanol steam reforming for hydrogen generation via conventional and membrane reactors: a review. Renew Sust Energ Rev 29:355–368
202. Park JO, Hong S-G (2016) Design and optimization of HT-PEMFC MEAs. In: Li Q, Aili D, Hjuler HA, Jensen JO (eds) High temperature polymer electrolyte membrane fuel cells: approaches, status, and perspectives. Springer International Publishing, Cham, pp 331–352
203. Engl T, Gubler L, & Schmidt T J (2016) Catalysts and catalyst-layers in HT-PEMFCs. High Temperature Polymer Electrolyte Membrane Fuel Cells: Approaches, Status, and Perspectives, 297–313.

Modeling

Atomistic Simulation of Subtle Microstructures in CeO_2-Based Electrolyte for Design of the State-of-the-Art Fuel Cells

Fei Ye, Ding Rong Ou, Zhipeng Li, and Toshiyuki Mori

Abstract The research work of the microstructure of rare-earth-doped ceria as electrolyte in solid oxide fuel cell is introduced in this chapter. The results of subtle microstructures at atomic and nano scales including defect clusters and nano-sized domains studied by atomistic computer simulation combined with microstructure observation by transmission electron microscopy are focused on. These inhomogeneities are formed by aggregation of vacancies and dopants ions. The phase transformation from F-type structure to C-type structure in doped ceria is not a sudden change, but gradually takes place from clusters to domains, and then to C-type precipitates with increasing the doping concentration. Several stable cluster structures can be considered as unit structures for larger clusters. A unique dumbbell structure formed with six oxygen vacancies has been identified as the building block for defect cluster growth and sequentially for the evolution from F- to C-type structure. Furthermore, the results of various types of doped ceria suggest that the subtle microstructure change is a common phenomenon in rare-earth-doped ceria. The capability of the growth of defect cluster is affected by dopant type, which can explain the dopant type dependency of domain development with increasing doping concentration in

F. Ye (✉)
Department of Materials Science and Engineering, Southern University of Scence and Technology, Nanshan District, 1088 Xueyuan Road, Shenzhen 518055, China
e-mail: yef3@sustech.edu.cn

D. R. Ou
Hydrogen Energy Research Center, Guangzhou Power Supply Bureau, Guangdong Power Grid Co. Ltd., 2 Keyun North Road, Tianhe District, Guangzhou 510645, China
e-mail: oudingrong0626@guangzhou.csg.cn

Z. Li
Institute of Flexible Electronics, Northwestern Polytechnical University, 1 Dongxiang Road, Chang'an District, Xi'an 710072, China
e-mail: iamzpli@nwpu.edu.cn

T. Mori
Global Research Center for Environmental and Energy Based on Nanomaterials (GREEN), Center for Green Research on Energy and Environmental Materials, National Institute for Materials Science, 1-1 Namiki, Tsukuba 305-0044, Ibaraki, Japan
e-mail: mori.toshiyuki@nims.go.jp

T. Mori et al. (eds.), *Design of Active Key Interfaces in Fuel Cells*,
https://doi.org/10.1007/978-981-97-5706-0_4

heavily doped ceria. Then, an example of doped ceria sintered at different temperature shows that the ionic conductivity can be improved by controlling these subtle microstructures.

Keywords Microstructure · Ordering · Cluster · Ceria · Fuel cell

1 Introduction

1.1 Limitation of Ionic Conductivity of Ceria-Based Electrolytes

Fuel cells are electrochemical devices, by which the chemical energy of a fuel is converted directly into electricity by electrochemical reaction. The structure of a typical fuel cell resembles a typical battery that consists of an electrolyte layer in contact with an anode and a cathode on either side. When a gaseous fuel and an oxidant are respectively fed to the anode and cathode, the electrochemical reactions take place at the electrodes to produce an electric current. Since individual fuel cells produce voltage about 1 V, they must be combined by interconnects to produce appreciable voltage. The interconnects provide an electrical series connection between adjacent fuel cells to form stacks. They also provide a gas barrier that separates the fuel and oxidant of adjacent cells.

The components in fuel cell stacks involve a variety of materials. Great efforts have been made to develop new materials or improve the properties of existing materials for the fuel cell application. In the case of solid oxide fuel cells (SOFCs), traditionally they operate in a temperature range from 800 to 1000 °C. Such high temperature is conductive to overcome the high ohmic loss from a thick electrolyte. However, such high temperature results in high cost of materials and operation of SOFCs. If the operating temperature is decreased, a broader set of materials can be used as the components due to the decrease in the requirement on the material properties. Moreover, the thermal management is simplified, and the degradation of the cell and stack components is less. Because of these advantages, the development of SOFCs operating in the intermediate temperature (IT) range of 600–800 °C has attracted broad attention.

However, IT-SOFCs also accompany with other problems. Especially, the electrolyte conductivity and the catalytic activity of electrodes decrease with the decrease in temperature. For the most commonly used electrolyte material, yttrium stabilized ZrO_2 (YSZ), a typical approach to minimize ohmic loss is to decrease the thickness to satisfy the requirement of electrical conduction in the IT range.

New candidate materials for electrolyte application have been investigated. Rare-earth-doped (RE-doped) ceria is an important candidate material, that has relatively high ionic conductivity even at temperature below 600 °C [1]. The conductivity of RE-doped ceria is originated from oxygen vacancies created by the substitution of trivalent rare-earth cations. For instance, the addition of Gd_2O_3 into CeO_2 forms vacancies according to the equation (in Kröger-Vink notation).

$$Gd_2O_3 \xrightarrow{CeO_2} 2\,Y'_{Ce} + V_{\ddot{O}} + 3\,O_O^{\times} \quad (1)$$

Obviously, the vacancies are formed to maintain electrical neutrality. The general formula of Gd-doped ceria is expressed as $Gd_xCe_{1-x}O_{2-x/2}$.

From Eq. (1), it could be expected that the conductivity of doped ceria should increase with increasing doping concentration due to the increase in the concentration of oxygen vacancies. Generally, precipitates (M_2O_3) with cubic structure (C-type structure) form in ceria with fluorite structure (F-type structure) at a doping concentration about 40–60 at.% for rare-earth dopants [2]. Hence, the high solubility of rare earth dopant was traditionally regarded as an advantage. However, numerous studies have demonstrated that the doping concentration corresponding to the maximum conductivity was much lower than the solubility [3].

As shown in Fig. 1a, an example of Gd-doped ceria, the maximum conductivity at 500 °C corresponds to a doping concentration about 10–20 at.%. Figure 1b is another example of Y-doped ceria [4]. The maximum conductivity at 500 °C also occurs at a concentration about 15 at.%. In addition, the conductivity follows the Arrhenius behavior, which is expressed using the equation

$$\sigma T = A\exp\left(-\frac{E_a}{RT}\right), \quad (2)$$

where A is a pre-exponential constant, σ is the conductivity, E_a is the activation energy, and R is the gas constant (8.314 J·mol^{-1}·K^{-1}). The activation energies determined using this equation are also shown in Fig. 1. They increase with increasing

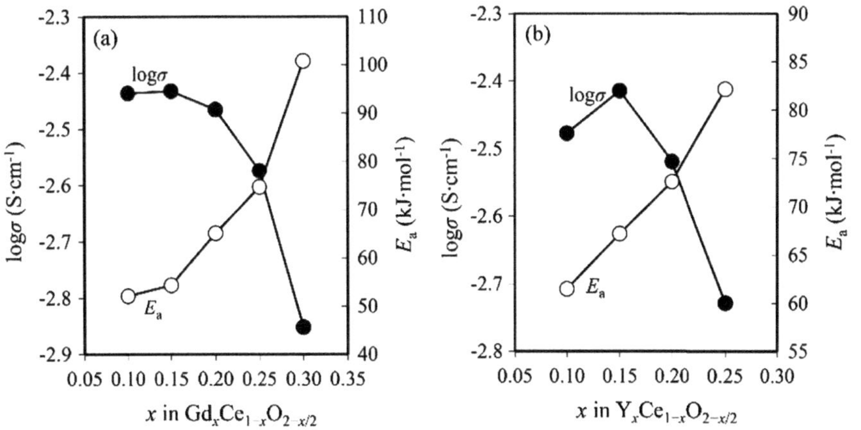

Fig. 1 The conductivity σ and the activation energy E_a of **a** $Gd_xCe_{1-x}O_{2-x/2}$ and **b** $Y_xCe_{1-x}O_{2-x/2}$ at 500 °C as a function of the doping concentration. (**b** is reproduced from Ref. [4])

the doping concentration although the conductivity increases initially and decreases afterward.

1.2 Subtle Microstructures in RE-Doped Ceria

As a common saying, the properties of materials are closely related to their microstructure. The microstructure features of materials can be studied from atomic scale to micron or even millimeter scale. With the development of material characterization and computer simulation technologies, material research and development has been able to be carried out in atomic and nano scales, so that the ability of material optimization and design is greatly improved.

For ceria-based electrolytes, the limitation of the conductivity is closed related to the subtle microstructures. One of the subtle microstructure changes is the formation of defect clusters due to the attraction between the dopant cations and vacancies [5, 6]. The local ordering of the clusters has been investigated by atomistic computer simulation [5, 7–10]. The effect of the dopant-vacancy cluster on the ionic conductivity was extensively studied in various types of doped ceria [11–14]. The increase in the amount of the clusters with increasing doping concentration results in a decrease in the conductivity, since these clusters lower the concentration of freely mobile oxygen vacancies.

Besides the clusters, nano-sized domains were extensively observed in heavily doped ceria (doping concentration > 10 at.%) by high resolution transmission electron microscopy (HRTEM), such as in Y, La, Sm, Gd, Dy, Ho, Yb doped ceria [4, 15–19]. In the selected area electron diffraction (SAED) patterns, besides the reflections of F-type structure, diffuse scattering and extra spots corresponding to the formation of nanosized domains were observed. Moreover, locally ordered oxygen vacancies in the domains were confirmed by electron energy loss spectroscopy (EELS) [18, 20–24]. Since the domains are also formed by the aggregation of dopant cations and oxygen vacancies, they result in a decrease in the concentration of freely mobile oxygen vacancies. Consequently, the domains also contribute to the decrease in the conductivity of heavily doped ceria.

These subtle microstructures in doped ceria were also extensively studied by spectroscopic techniques. X-ray absorption fine structure spectroscopy (XAFS) is a synchrotron-based technique that provides information of local atomic and electronic structure of materials. The extended XAFS (EXAFS), which is a portion of the XAFS spectrum extended to about 100–1500 eV above the absorption edge, provides information about nearest neighbor bonding environments, such as the coordination numbers and interatomic distances. It has been used to measure the distance from Ce or dopant cations to their neighboring anions and other cations [25]. The average nearest neighbor distances of doped ceria show that the lattice strongly deviates from fluorite symmetry at the unit cell level [26, 27]. In Y-doped ceria, the results show that the local structures around the Ce cation are similar to that in the F-type structure though the coordination numbers of Ce–O tend to be smaller than 8, while the local structures around Y cation are significantly different from those around Ce cation, and show more resemblance to that in the C-type Y_2O_3 [28]. In Gd-doped ceria, it is also confirmed that the Gd-O distance undergoes a monotonic decrease towards the average value in Gd_2O_3 [29].

Raman spectroscopy is a readily accessible experimental technique comparing with XAFS, which is another technique that can be used to investigate bonding and short-range order in oxide materials by detecting lattice vibration. Replacing Ce^{4+} ions with cations having an ionic radius larger than the critical radius for zero lattice shifts the Raman frequency towards lower Raman wavenumbers, whereas the opposite is true for smaller cations [30, 31]. In agreement with the EXAFS conclusions, Raman spectra demonstrate additional defect related Raman modes, indicating that the fluorite symmetry is no longer present on the unit cell level.

Similarly, nuclear magnetic resonance (NMR) spectroscopy can give insights into local structural configurations. Evaluation of ^{17}O spectra reveals that the dopant ions are randomly distributed at least for low doping to moderate doping levels for larger dopants (Gd, Sm, Nd, La) [32, 33]. In the case of Y-doped ceria, ^{89}Y spectra show a strong preference of vacancy association with Y dopants [34].

These observations of subtle microstructures seem in contradiction with the X-ray diffraction (XRD) studies that indicate fluorite symmetry for doped ceria under solution limit [35–38]. This proves that the global average does not provide the comprehensive description of structure. However, the spectroscopic techniques are still in-direct analysis since the detailed structure features have to be determined from deep analysis of spectra.

It has been suggested that the clusters and domains are preliminary states or symptoms of a transition from F-type structured ceria to C-type trivalent oxide based on the comprehensive observations [22]. Therefore, it is important to systematically study the features of the clusters and the domains, as well as the microstructure evolution from F-type structure to C-type structure with increasing doping concentration. Insights gained from such studies can facilitate our understanding of the phase transformation at atomic scale. Consequently, the electrical conductivity of doped ceria for SOFC applications can be improved by controlling the microstructures.

In this chapter, the investigations of atomistic simulation combined with TEM microstructure analysis on short-range ordered structures of clusters and domains in RE-doped ceria as electrolyte in SOFCs are reviewed. The subtle microstructures at atomic and nano scales and their effect on electrical conductivity are elucidated. The results of Gd-doped ceria are mainly described, and some results of Y-doped ceria and other types of doped ceria are involved to show the applicable of the results for RE-doped ceria and dopant type dependency of the microstructure development. Then, the method to control the formation of the subtle microstructures is discussed.

2 Experimental and Theoretical Methods

2.1 Experimental Methods

The microstructure features in sintered rare-earth (RE) doped ceria electrolyte were studied experimentally. The ammonium carbonate co-precipitation method was used to synthesize the powders [39]. The powders were calcined under flowing oxygen (about 200 mL/min) at 600–800 °C for 2 h followed by isostatically pressing under 200 MPa pressure. Then the compacted bodies were sintered in a temperature range from 900 to 1500 °C for 6 h.

The crystal phases were identified by XRD (RINT2200HF). The average grain size of sintered samples was measured from the micrographs taken by scanning electron microscopy (SEM, Hitachi S-5000) using the linear intercept method [40].

The detailed microstructures were characterized by transmission electron microscopy (TEM, JEOL JEM-2000EX, or Tecnai F30). The TEM samples were prepared by ultrasonically cutting from the sintered samples followed by mechanical grinding, dimpling, and ion milling. The areas for taking the SAED patterns were about 0.2 mm in diameter. To obtain the weak extra spots and diffuse scattering in the patterns, a relatively faint beam was applied and the exposure time for taking the patterns was 45 s. The EELS analysis was performed on TEM equipped with Gatan imaging filter (GIF). Because the subtle microstructures are at atomic or nanoscale levels, it is relatively difficult to perform an EELS analysis on them separately. Therefore, the EELS spectra were acquired from larger regions. According to the collection aperture size of 2 mm and the magnification, 10 k or 20 k, the area for taking the EELS spectra was about 0.1–0.2 mm. The regions selected for the EELS study were deviated from the strong Bragg reflections and chosen to have a thickness of 0.1–0.3 inelastic mean free-paths (about 10–30 nm). To avoid the influence of irradiation damage [41], a spread beam and a short acquirement time of no longer than 5 s were applied. The power-law technique was used to subtract the background of the EELS spectra [42].

2.2 Theoretical Methods

The atomistic computer simulation of short-range ordered structure in RE-doped ceria was based on energy minimization calculation. The interactions between ions consist of long-range Coulombic forces and short-range forces. The general formula of Coulombic interaction was accepted to describe the long-range force. The pair potential of Buckingham potential form was accepted to describe the short-range force, which was expressed as.

$$S_{ij} = Aexp\left(-r_{ij}/\rho\right) - Cr_{ij}^{-6}, \qquad (3)$$

Table 1 Parameters for the computer simulation of RE-doped ceria

Short-range potential parameters			
Species/Refs.	A (eV)	ρ (Å)	C (eVÅ^6)
Ce^{4+}–O^{2-}/[43]	1809.68	0.3547	20.40
Y^{3+}–O^{2-}/[44]	1766.40	0.33849	19.43
Sm^{3+}–O^{2-}/[45]	1944.44	0.3414	21.49
Gd^{3+}–O^{2-}/[7]	1885.75	0.3399	20.34
Dy^{3+}–O^{2-}/[46]	1807.84	0.3393	18.77
Yb^{3+}–O^{2-}/[7]	1649.80	0.3386	16.57
O^{2-}–O^{2-}/[47]	9547.96	0.2192	32.0
Shell parameters			
Species/Refs	X (|e|)	Y (|e|)	k (eVÅ^{-2})
Ce^{4+}/[43]	4.20	−0.20	177.84
O^{2-}/[47]	0.04	−2.04	6.3

where the distance between two atoms, and A, ρ, and C are adjustable parameters. The parameters for the calculation of RE-doped ceria are listed in Table 1, which were determined empirically from some known properties of related oxides, such as lattice parameters and elastic modulus [7, 43–47]. The short-range interaction was set to zero beyond 20 Å, since it was negligible beyond a few lattice spacings.

The electronic polarizability of ions was described by the shell model [48]. The shell model of an ion consists of a massive core and a massless shell. Their charges are $X|e|$ and $Y|e|$, respectively, and the overall charge state of each ion is equal to $(X + Y)|e|$. The core and shell are connected by a harmonic spring with force constant k. For the results reviewed in this chapter, O^{2-} and Ce^{4+} ions were treated as polarizable as listed in Table 1.

The Mott-Littleton two-region approach [49] was used for the simulation, which is coded in the GULP program [50]. In this approach, the lattice is partitioned into two regions as shown in Fig. 2. They are a spherical inner region I and an outer region II. The defects are introduced at the center of region I. Region II extends to infinity. An interfacial region IIa is introduced to ensure a smooth transition between regions I and IIb. Atomic interactions in region I are calculated explicitly to relax the positions of all ions to zero strain, while the ion displacements in region IIa are determined via the Mott-Littleton approximation. To ensure no appreciable change in the defect energy if the region sizes were increased further, an inner region I with a radius of $3.0a_0$ ($a_0 = 5.411$ Å, the lattice constant of ceria) and an interfacial region IIa with a radius of $6.0a_0$ were used.

The binding energies of the clusters were used to compare the stability of the clusters. They were calculated by

$$E_b = \Sigma E_{isolated} - E_{cluster}, \quad (4)$$

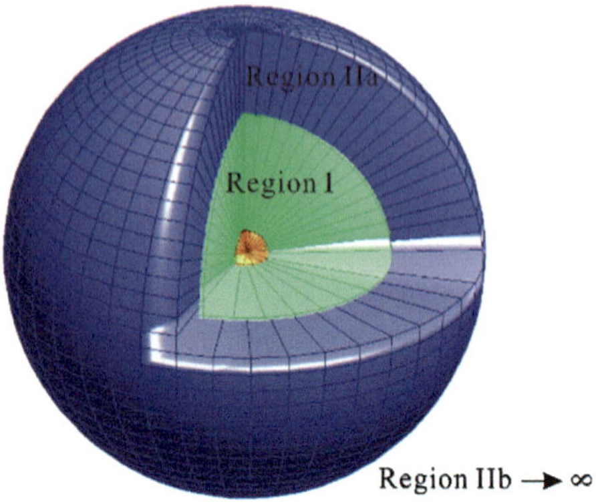

Fig. 2 Definition of the two regions in the Mott-Littleton two-region approach

where E_{cluster} is the formation energy of a defect cluster, and $\Sigma E_{\text{isolated}}$ is the sum of the defect formation energies of the corresponding isolated defects. A positive binding energy indicates a preference for the cluster over the isolated point defects. That is, higher binding energy denotes greater stability. Electrically neutral clusters were examined, in which each oxygen vacancies was associated with two trivalent cations since the substitution of dopant cations for Ce^{4+} was accompanied by the formation of an oxygen vacancy for every two dopant cations. For each cluster size, various arrangements of oxygen vacancies and dopant cations were examined to find the cluster with the highest binding energy.

3 Results and Discussion

3.1 *Crystal Phases in RE-Doped Ceria*

The precipitation of C-type oxide in F-type ceria matrix with increasing the doping concentration can be readily confirmed by XRD. In Fig. 3, the $Gd_xCe_{1-x}O_{2-x/2}$ samples solely contain fluorite structure when $x \leq 0.3$. When $4 \leq x \leq 0.7$, extra peaks of the C-type precipitates occur as indicated by arrows. The intensity of the extra peaks increases with increasing Gd concentration. The sample becomes single-phase oxide with C-type structure when $x = 0.8$.

As shown in Fig. 4, the structure of the C-type precipitates (space group $Ia\overline{3}$) is closely related to the fluorite structure (F-type structure) of ceria (space group $Fm\overline{3}m$). A unit cell of the C-type structure can be constructed using eight unit cells of F-type structure by removing 25% of oxygen ions and introducing atomic shifts.

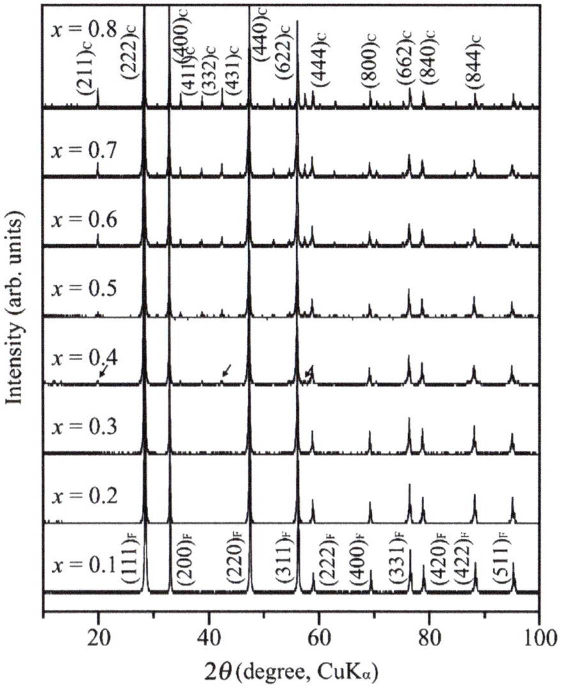

Fig. 3 XRD profiles of the $Gd_xCe_{1-x}O_{2-x/2}$ sintered samples. The profiles with $x = 0.1$ and 0.8 are indexed based on F-type and C-type structures (denoted by subscripts F and C), respectively. (Reproduced from Ref. [15])

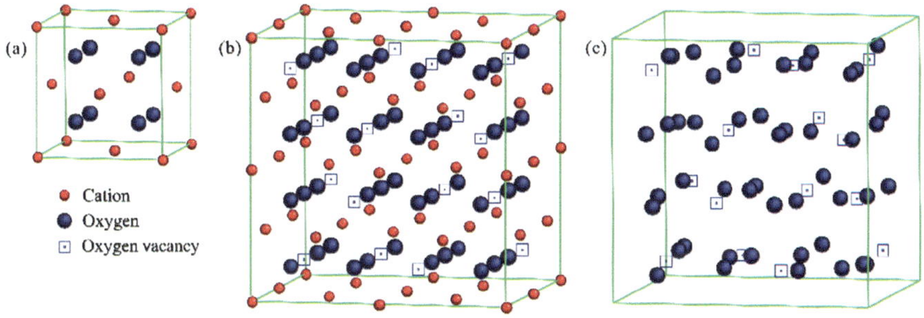

Fig. 4 Schematic illustration of constructing a unit cell of C-type structure from **a** F-type structure by **b** removing 25% oxygen ions and **c** introducing atomic shifts. The shifted cations are omitted in (**c**) for clarity. (Reproduced from Ref. [15])

3.2 Structure of Defect Clusters

Defect clusters originating from aggregation of dopants and charge-compensating oxygen vacancies have been widely investigated in doped ceria due to their appreciable impact on ionic conductivity of electrolytes in SOFC. Theoretical models have been developed for distributions and interactions of defects in RE-doped ceria. However, in most previous investigations by computer simulation [5, 7–10], the clusters that have been examined only contain one or two oxygen vacancies, which are not large enough to analyze the ordered structure of the clusters. In addition, the lack of the knowledge of larger cluster also hinders the study of the domain structure, since the domains are formed from the development of defect clusters with increasing doping concentration.

To address these issues, the authors of this chapter studied larger cluster in Gd-doped ceria containing up to 6 vacancies and 12 dopant cations. A helical chain model with up to four vacancies was first proposed [51, 52]. However, restricted by the low symmetry and nonconvergence trait of such helical chain structure, it is difficult to explain the structure change in the phase transformation from F-type structure to C-type structure [53]. Therefore, it is essential to further explore the interactions, ordered structures, and the growth pathway of defect clusters.

Various configurations of defect clusters have been considered to assess their stability. An oxygen vacancy occupies the first, second, or third nearest-neighbor site with respect to the dopant cation. In other words, they are separated by <111>/4, <113>/4, and <133>/4, respectively. For the smallest defect cluster $1V_{\ddot{O}}2Gd'_{Ce}$ that is not charge neutral [Fig. 5a], the defect formation energies with oxygen vacancy located at the three nearest-neighbor sites are 48.04, 48.26, and 48.46 eV, respectively, indicating the first neighbor is the preferred site.

The structure of the simplest neutralized cluster $1V_{\ddot{O}}2Gd'_{Ce}$ with the highest binding energy is shown in Fig. 5b. For defect clusters containing more than two vacancies, the adjacent vacancies usually occur along either <110>/2 or <111>/2. Figure 5c represents the most stable $2V_{\ddot{O}}4Gd'_{Ce}$ defect cluster. All dopants prefer to occupy the sites that are the first neighbor to one vacancy, and the first or second neighbor to the other vacancy to ensure the lowest overall defect energy. The much higher binding energy of $2V_{\ddot{O}}4Gd'_{Ce}$ than $1V_{\ddot{O}}2Gd'_{Ce}$ also indicates an intrinsic trend for cluster growth.

Figure 5d represents the most stable cluster constructed by three vacancies, which have a shape of isosceles triangle with two equal edges along <110>/2, forming an interior angle of 120°, and the lattice vector of the other edge is <112> . As the cluster grows to $4V_{\ddot{O}}8Gd'_{Ce}$, a symmetric tetrahedron is formed [Fig. 5e]. The three faces of the tetrahedron are constructed by three isosceles triangles that are the same as that in Fig. 5d. This suggests a possible growth mechanism for defect clusters that the stable clusters, regardless of sizes, may be constructed with the same basic structure of isosceles triangle units.

The development of the clusters was further examined by defect energy calculations of larger clusters. For $5V_{\ddot{O}}10Gd'_{Ce}$ [Fig. 5f], the newly evolved vacancies will

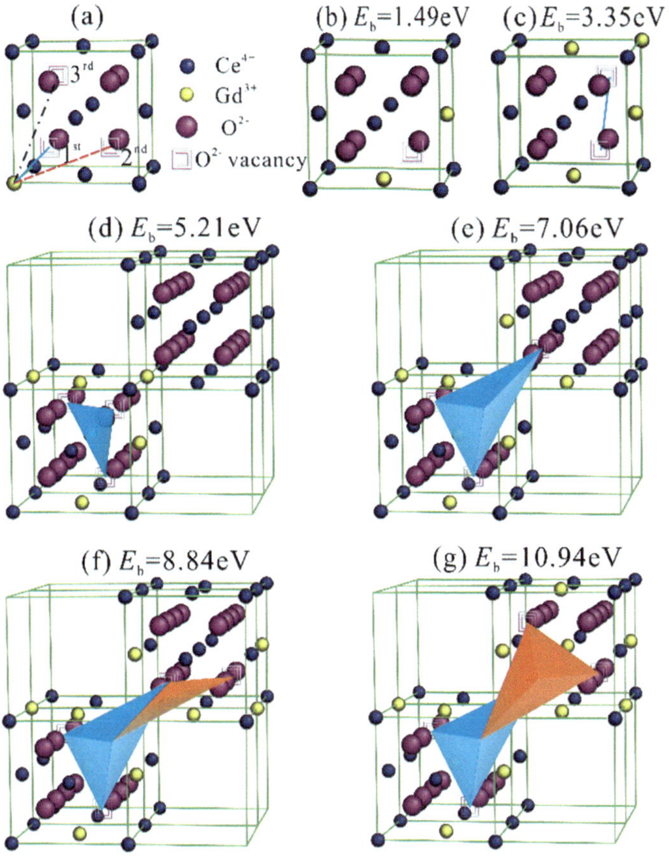

Fig. 5 **a** $1V_{\ddot{O}}Gd'_{Ce}$ clusters with oxygen vacancies at the first, second, and third neighbor site with respect to the dopant cation Gd^{3+}. The structures of defect clusters **b** $1V_{\ddot{O}}2Gd'_{Ce}$, **c** $2V_{\ddot{O}}4Gd'_{Ce}$, **d** $3V_{\ddot{O}}6Gd'_{Ce}$, **e** $4V_{\ddot{O}}8Gd'_{Ce}$, **f** $5V_{\ddot{O}}10Gd'_{Ce}$, **g** $6V_{\ddot{O}}12Gd'_{Ce}$ with highest binding energy comparing with other configurations. (Reproduced from Ref. [54])

occupy the site along <110>/2 with respect to its nearest vacancy in the existing tetrahedron structure, and are involved to form the isosceles triangle structure. When the cluster grows up to $6V_{\ddot{O}}12Gd'_{Ce}$, a distinct configuration named as a dumbbell structure is formed [Fig. 5g]. The dumbbell structure can be considered as a combination of six isosceles triangles, or two tetrahedrons connected by sharing one tetrahedral edge. The defect cluster can further develop in this way.

The defect clusters containing one to six vacancies are compared in terms of the total binding energy as shown in Fig. 6. It can be seen that the binding energy tends to increase with cluster size, indicating the higher stability of larger clusters and the intrinsically attractive interaction between the nearby defects.

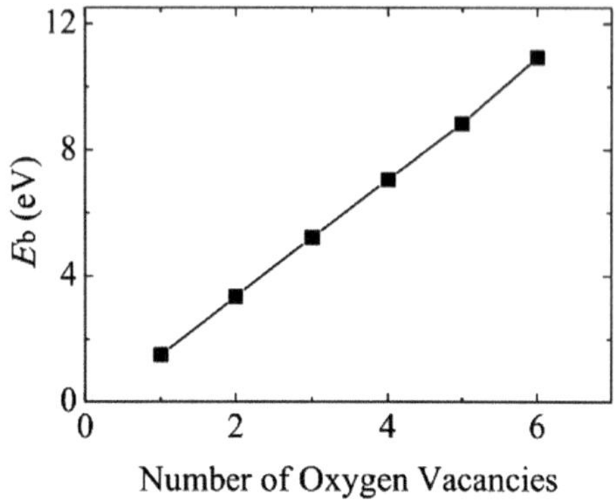

Fig. 6 Binding energy E_b as a function of the number of oxygen vacancies in defect clusters. (Reproduced from Ref. [54])

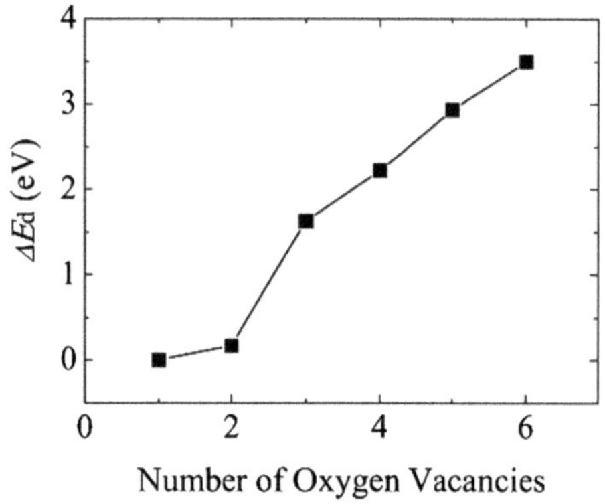

Fig. 7 Difference of defect energy between the clusters with the lowest and the second lowest defect energy ΔE_d as a function of the number of oxygen vacancies in defect clusters. (Reproduced from Ref. [54])

The stability of the defect clusters is also confirmed by calculating the defect energy difference between cluster with the lowest and the second lowest defect energy. It is calculated by.

$$\Delta E_d = E_{2ndlowest} - E_{lowest} \tag{5}$$

As shown in Fig. 7, when defect clusters are relatively small, containing one or two vacancies, the defect energy difference is not significant, suggesting that the clusters with different structure have similar formation probability in the matrix. When the cluster grows larger, there is a jump in the energy difference for $3V_{\ddot{O}}6Gd'_{Ce}$. Beyond this cluster size, the defect energy difference linearly increases with cluster size. Therefore, the cluster with three vacancies is a critical size for the formation of the unique cluster structures.

Interestingly, there is a close correspondence between the cluster structure and the C-type structure. Figure 8 shows the distribution of the oxygen vacancies in the C-type structure corresponding to the F-type structure. The cations and oxygen atoms are omitted. It is note that any three adjacent vacancies can form an isosceles triangle. More nearby vacancies with tetrahedrons and dumbbells as shown in Fig. 5 can be

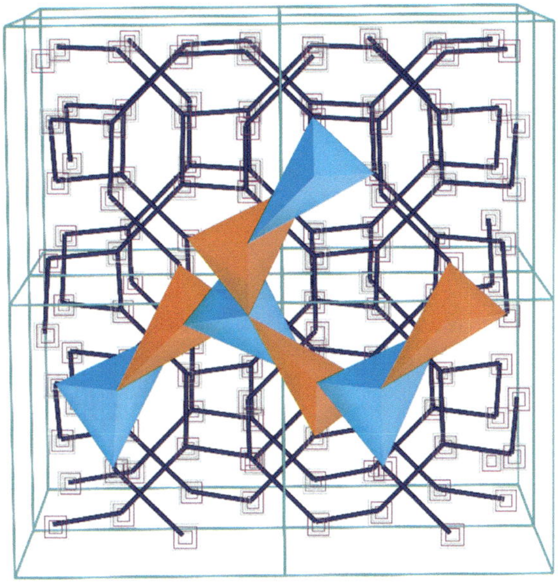

Fig. 8 Oxygen vacancies in the C-type structure, with other ions omitted for clarity. The neighboring vacancies are connected by blue lines. (Reproduced from Ref. [54])

determined in the C-type structure. The dumbbells are connected with each other simply by coordinated spatial rotation due to the nonplanar separation of <110>/2 lattice vectors. These structures connect and occupy all vacancy sites in the C-type structure. Therefore, they may be regarded as the basic building block to construct the vacancy sublattice in the C-type structure. The phase transformation from F-type structure to the C-type structure with increasing the doping concentration can be further understood based on the structure evolution of cluster growth.

3.3 *Microstructural Features Nano-sized Domains*

The detailed microstructure features at nanoscale were examined by TEM. Figure 9 shows the HRTEM images of the GDC samples. The $Gd_{0.1}Ce_{0.9}O_{1.95}$ sample only shows fluorite structure. Some brighter regions that only cover several atoms as indicated by arrows have small lattice distortion. It is possible that defect clusters are embedded in these regions. In the $Gd_{0.2}Ce_{0.8}O_{1.9}$ and $Gd_{0.3}Ce_{0.7}O_{1.85}$ samples shown in Fig. 9b and c, nano-sized domains can be observed as those roughly enclosed by dashed lines. These domains are homogeneously distributed in the samples. Moreover, the distribution density and size of the domains slightly increases with increasing the doping concentration.

The formation of the domains is associated with lattice distortion and dislocations. The dislocations are marked by "T" in and around these domains. In the [112] HRTEM image [Fig. 9d], the lattice distortion can be seen more clearly. The domains

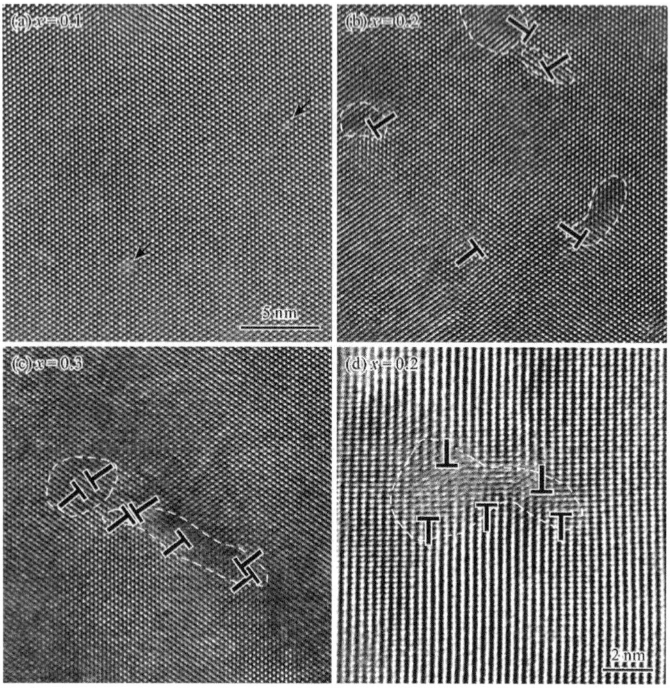

Fig. 9 [110] HRTEM images of $Gd_xCe_{1-x}O_{2-x/2}$ samples with **a** $x = 0.1$, **b** $x = 0.2$, and **c** $x = 0.3$. **d** [112] HTREM image of $Gd_{0.2}Ce_{0.8}O_{1.9}$. (**a**, **b**, and **c**) is reproduced from Ref. [15], and **d** is reproduced from Ref. [23])

can also be distinguished from the matrix by fast Fourier transform (FFT) and inverse FFT (IFFT). Figure 10 shows the FFT and IFFT images from the HRTEM image of the sample $Gd_{0.2}Ce_{0.8}O_{1.9}$. The spots of F-type structure and very weak diffuse scattering can be seen in the FFT image. After excluding the spots of F-type structure, an IFFT image is obtained in which the lattice of the domains remains clear while the lattice of the matrix is blurred. Therefore, the lattice in the domains should slightly deviate from the ideal F-type structure.

Figure 11 shows the SAED patterns of the Gd-doped ceria samples. Both diffuse scatterings and extra reflections are shown in the patterns due to the formation of the domains. Based on the SAED patterns, the three-dimensional distributions of the extra reflections and diffuse scatterings in a unit cell in reciprocal space are constructed. As shown in Fig. 12a the satellite extra reflections around a reflection of fluorite structure can be described as $G_F \pm \{110\} * /2$, where $\boldsymbol{G}_F$ is a reciprocal lattice vector of the fluorite structure. It is noted that the extra spots are absent in the [100] SAED pattern, while they can be seen on the (100) reciprocal plane in Fig. 12a. It is possible because part of the extra spots in the [110], [111], and [112] SAED patterns are formed by double diffraction [22]. However, double diffraction generally takes place under certain conditions. Another possible reason is that the

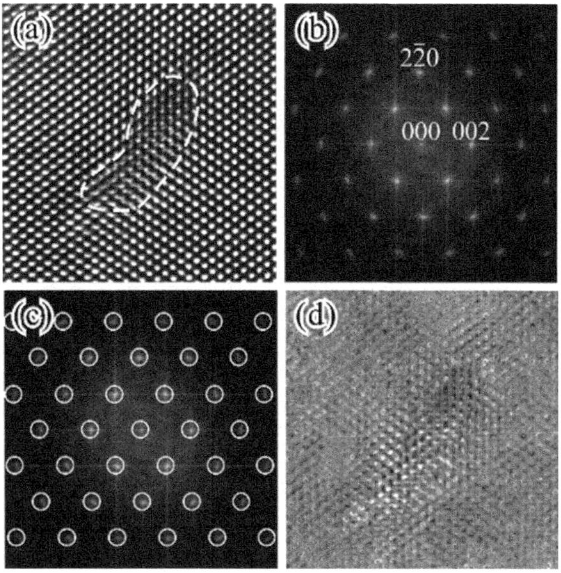

Fig. 10 FFT of HRTEM images of the $Gd_{0.2}Ce_{0.8}O_{1.9}$ sample. **a** [110] HRTEM image. **b** The corresponding FFT image. **c** FFT image with circular mask filters. **d** IFFT image corresponding to the HRTEM image obtained after excluding the spots of fluorite structure. (Reproduced from Ref. [53])

extra spots deviate slightly from the (100) reciprocal plane due to the lattice distortion in the domains [53]. In the reciprocal unit cell (Fig. 12b), the distribution of diffuse scatterings is composed of eight cylindroids with their rotational axes parallel to the <111> directions of the bcc reciprocal lattice. The overlaps between the cylindroids are adjusted as small planes. The arc for building The rotational cylindroids is built using an arc, which is a part of the circle with the center at the point of 3/4 (011)* and crossing the reflections of (000)* and (111)*.

The compositional feature of the domains was investigated qualitatively using EFTEM elemental mapping. Figure 13 shows the EFTEM elemental maps of Ce and Gd elements of $Gd_{0.1}Ce_{0.9}O_{1.95}$ and $Gd_{0.3}Ce_{0.7}O_{1.85}$. The contrast in Fig. 13b suggests the inhomogeneous composition distribution in the $Gd_{0.3}Ce_{0.7}O_{1.85}$ sample. Dark regions in the Ce map with a size about 10–20 nm are brighter in the Gd map. They have a higher Gd concentration than their surrounding matrix. We believe these regions are the nano-sized domains since the size and distribution of these regions are similar to the domains shown in the HRTEM images.

Since the dopant cations tend to segregate in the domains as confirmed by the EFTEM maps, the oxygen vacancies may be trapped by the segregated dopant cations due to the Coulombic interaction. Thus, the ordered structure of the domains involves an ordering of these oxygen vacancies, which can be confirmed by EELS. Figure 14 shows oxygen K-edge taken from the $Gd_xCe_{1-x}O_{2-x/2}$ samples. The area for the analysis was about 0.2 μm, which was much larger than the domain size to ensure the spectra corresponding to the average results. There are three peaks marked by A, B, and C in the figure. Peak A is from the hybridization between oxygen and Ce^{4+} cations [55, 56], which is obvious in the sample with $x = 0.1$ and becomes weak in the samples with $x = 0.2$ and 0.3 due to the decrease in Ce concentration.

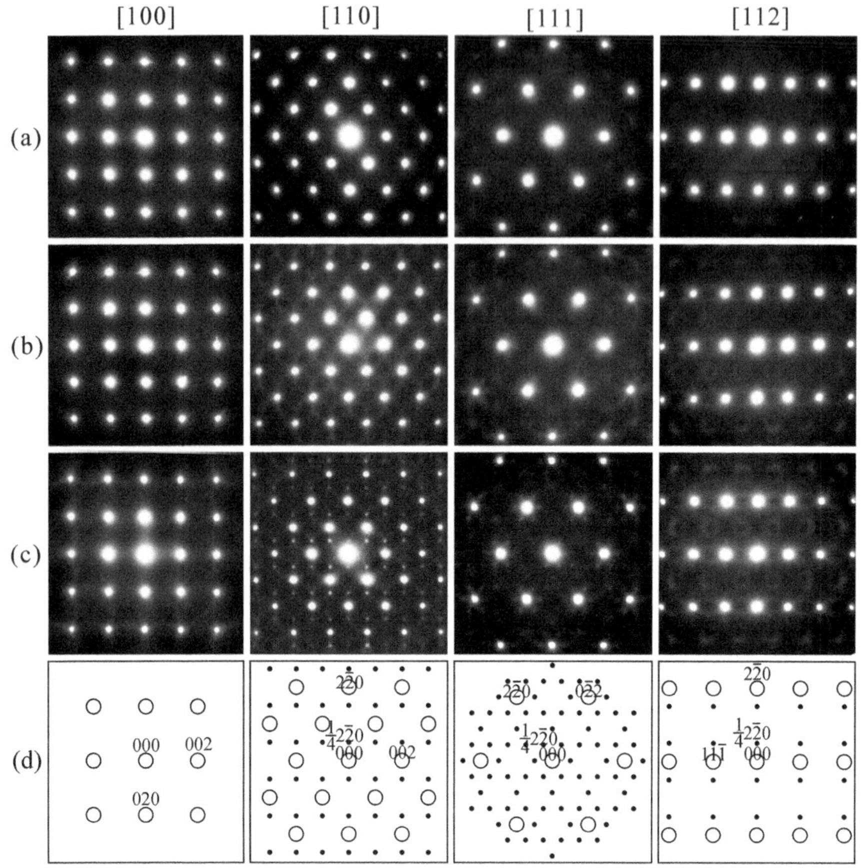

Fig. 11 SAED patterns of $Gd_xCe_{1-x}O_{2-x/2}$ samples with **a** $x = 0.1$, **b** $x = 0.2$, and **c** $x = 0.3$. **d** Key diagrams of these patterns [53]

The relative intensities of peaks B and C can be influenced by the crystal structure [55, 56], and the enhancement of peak B is due to the ordering of oxygen vacancies [57]. The spectra in Fig. 14 are normalized with respect to the intensity of peak C to compare the relative intensities of the samples. The integral intensities, I_B and I_C, were calculated using energy windows with a width of 2 eV, as illustrated by the inset in Fig. 14. The I_B/I_C ratios are 0.73, 0.78, and 0.82 for the samples with $x = 0.1$, 0.2 and 0.3, respectively, indicating the increase in the ordering of oxygen vacancies. Obviously, such increase is closely related to the domain development with increasing the doping concentration.

As mentioned above, the domain formation has been widely observed in various doped ceria. The TEM observation results of Y-doped ceria are given here to show the universal applicable of the results of Gd-doped ceria shown above. Figure 14 shows the HRTEM images of the Y-doped ceria. For the $Y_{0.1}Ce_{0.9}O_{1.95}$ sample

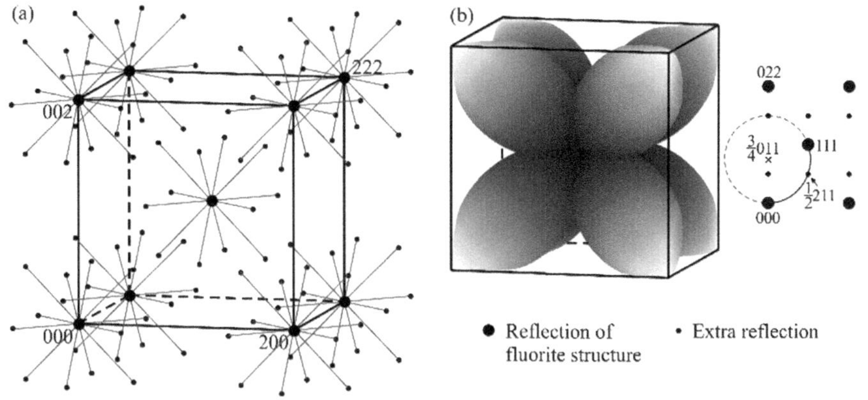

Fig. 12 Three-dimensional distribution of **a** the extra reflections and **b** the diffuse scatterings in a reciprocal unit cell. The arc for building the cylindroids is illustrated in image (**b**). (Reproduced from Refs. [22, 53], respectively)

Fig. 13 Elemental maps of Ce and Gd of $Gd_xCe_{1-x}O_{2-x/2}$ samples with **a** $x = 0.1$ and **b** $x = 0.3$ [23]

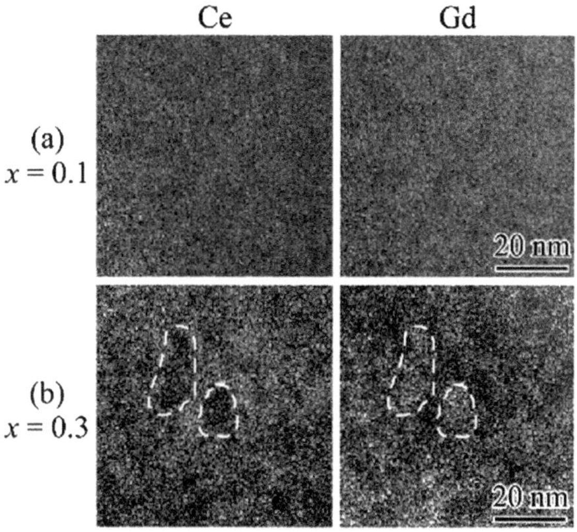

(Fig. 15a), dislocations and lattice distortions are observed in the ceria matrix with F-type structure. Larger distortions with dimensions up to 2 nm can also be seen. These lattice distortions are possibly induced by small clusters. In the image of $Y_{0.15}Ce_{0.85}O_{1.925}$ (Fig. 15b), nano-sized domains of about 5 nm in size can be seen. They appear as different contrast and lattice image from those of the matrix. When the doping concentration increases up to $x = 0.20$ (Fig. 15c) and $x = 0.25$ (Fig. 15d), the size and the distribution density of domains increase.

The corresponding [110] SAED patterns obtained from these samples sintered at 1400 °C are shown in Fig. 16. For $Y_{0.10}Ce_{0.90}O_{1.95}$ (Fig. 16a), the reflections of

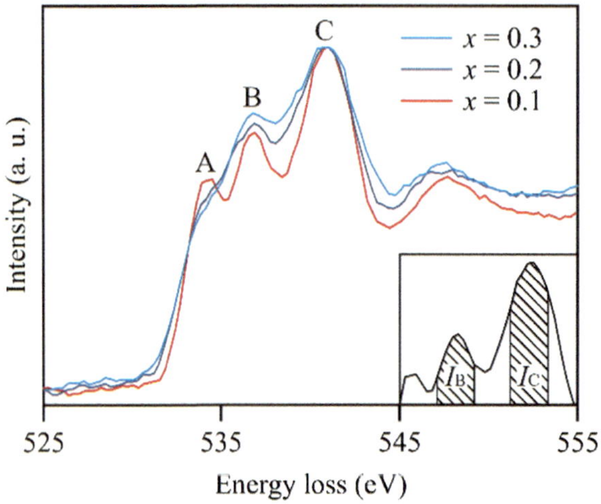

Fig. 14 Oxygen K-edge of $Ce_{1-x}Gd_xO_{2-x/2}$ samples. The inset illustrates the calculation of integral intensities of peaks B and C, I_B and I_C. (Reproduced from Ref. 23])

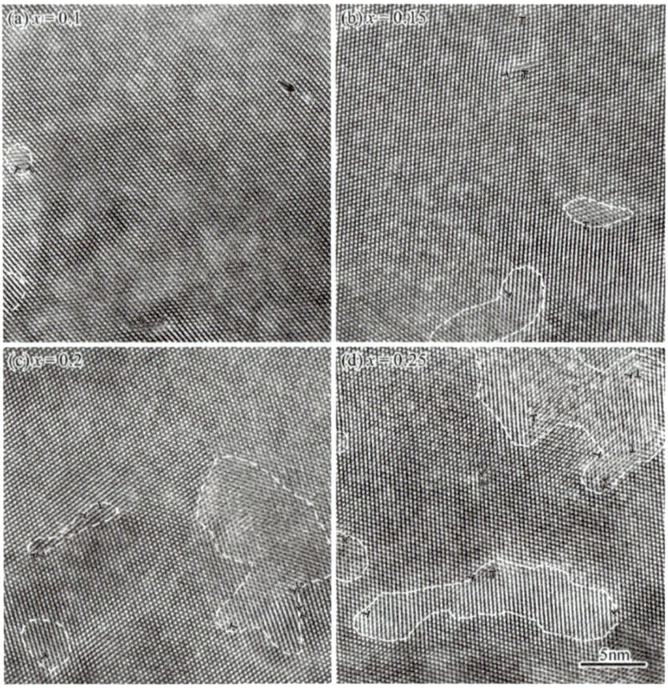

Fig. 15 [110] HRTEM images of $Y_xCe_{1-x}O_{2-x/2}$ electrolytes with **a** $x = 0.10$, **b** x = 0.15, **c** x = 0.20, **d** x = 0.25. (Reproduced from Ref. [4])

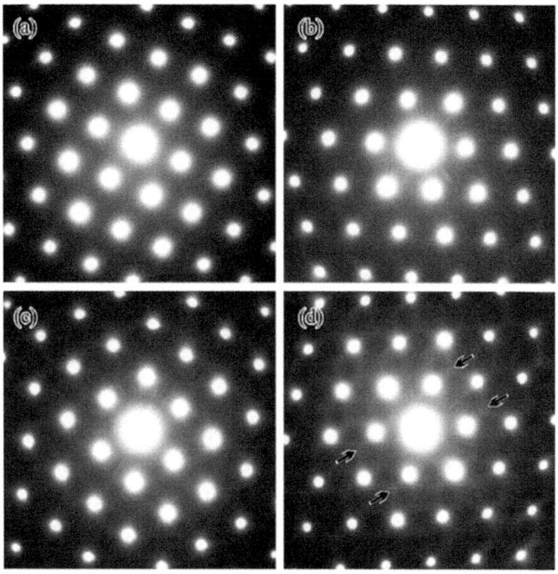

Fig. 16 [110] SAED patterns of $Y_xCe_{1-x}O_{2-x/2}$ electrolytes with **a** $x = 0.10$, **b** $x = 0.15$, **c** $x = 0.20$, **d** $x = 0.25$. The arrows in (d) indicate the position of extra diffraction spots. (Reproduced from Ref. [4])

F-type structure can be seen. Weak diffuse scattering appears for $Y_{0.15}Ce_{0.85}O_{1.925}$ (Fig. 16b) and $Y_{0.2}Ce_{0.8}O_{1.9}$ (Fig. 16c) samples. When the doping concentration increases to $x = 0.25$ (Fig. 16d), the contrast of the diffuse scattering increased, and weak extra spots are shown as marked by arrows. Again, these features in the SAED patterns are related to the development of nanosized domains.

3.4 Dopant Type Dependency of Cluster and Domain Formation

Since the domains are formed by the growth of defect clusters with increasing doping concentration, the dopant type dependency of the domain development can be explained by comparing the cluster growth tendency of different types of RE-doped ceria.

The study again starts from the simplistic cluster, $1V_{\ddot{O}}1M'_{Ce}$. The binding eneries of the clusters containing a first, a second, and a third nearest neighbor oxygen vacancy with respect to a substitutional cation, respectively, are compared in Fig. 17. It is obvious that smaller cations (Yb^{3+}, Y^{3+}, and Dy^{3+}) and larger cations (Gd^{3+} and Sm^{3+}) prefer first and second neighbor vacancies, respectively. This relationship between the binding energy and the dopant cation radius is consistent with the previous calculation [7, 8].

When the doping concentration increases, the probability of larger defect cluster increases. The structures of electrically neutral clusters containing one to four oxygen vacancies and two to eight M^{3+} cations are studied. In the clusters with relatively

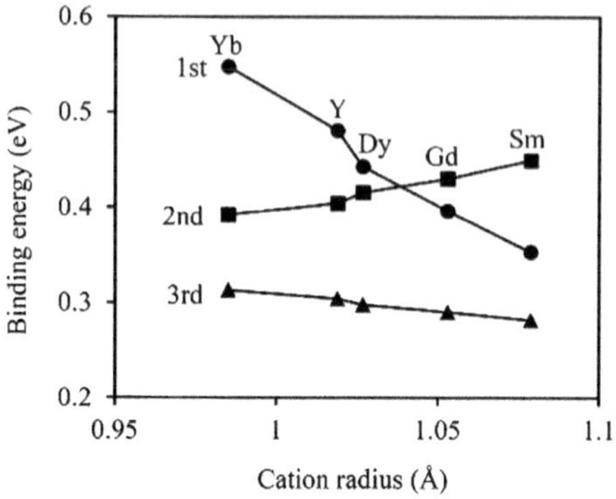

Fig. 17 Binding energy of cluster $1V_{\ddot{O}}1M'_{Ce}$ as a function of the cation radius. The effective ionic radii for VIII-coordinated cations are considered [58]. (Reproduced from Ref. [52])

high binding energies for all types of RE-doped ceria, including Y, Sm, Gd, Dy, and Yb-doped ceria, the vectors between adjacent oxygen vacancies are <1 10>/2, and their associated M^{3+} cations locate at the first or second nearest neighbor sites to the vacancies. The clusters in which the vacancies and the associated cations are separated by <113> /4 or <111 >/4, corresponding to larger or smaller cations, are named as type I or type II clusters, respectively. In all five doped ceria systems, the binding energy increases with increasing cluster size. As examples, the binding energies of clusters in Yb and Sm-doped ceria are shown in Fig. 18.

Since the formation of larger clusters can be considered as combination of smaller cluster units, the growth tendency of a cluster can be evaluated by comparing the formation energy or binding energy of the cluster with those of smaller clusters. This can be achieved by calculating.

$$\Delta E_{i-jk} = E_{\text{cluster}-j} + E_{\text{cluster}-k} - E_{\text{cluster}-i}, \quad (6)$$

or

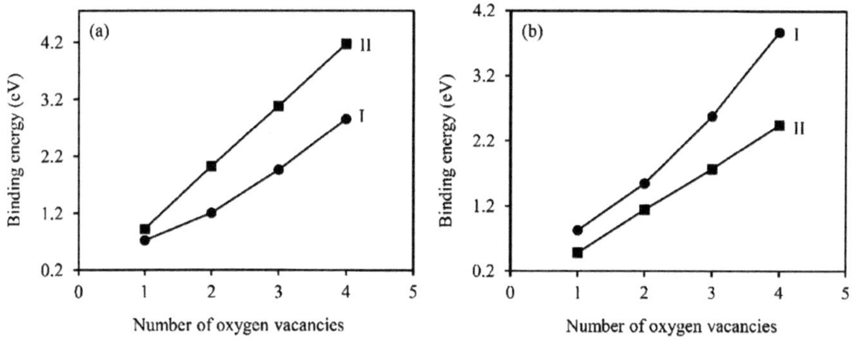

Fig. 18 Binding energy as a function of the number of oxygen vacancies in defect clusters of types I and II in **a** Yb-doped ceria and **b** Sm-doped ceria. (Reproduced from Ref. [52])

$$\Delta E_{i-jk} = E_{bi} - E_{bj} - E_{bk}, \tag{7}$$

where

$$i = j + k. \tag{8}$$

The subscripts *i*, *j*, and *k* indicate the number of oxygen vacancies in the clusters. A positive ΔE_{i-jk} value indicates a preference for the larger cluster containing *i* oxygen vacancies than two smaller clusters containing *j* and *k* oxygen vacancies. For the simplest case that $k = 1$, the values of ΔE_{2-11}, ΔE_{3-21}, and ΔE_{4-31} are calculated, respectively, and compared in Fig. 19.

For the clusters of type II, the ΔE_{i-jk} values are positive and decrease with increasing the cation radius. For the clusters of types I, ΔE_{2-11} values are negative, but most values of ΔE_{3-21} and ΔE_{4-31} are positive. Therefore, the growth of the clusters of types I from $1V_{\ddot{O}}2M'_{Ce}$ to $2V_{\ddot{O}}4M'_{Ce}$ is hard, while it becomes easier when they grow larger.

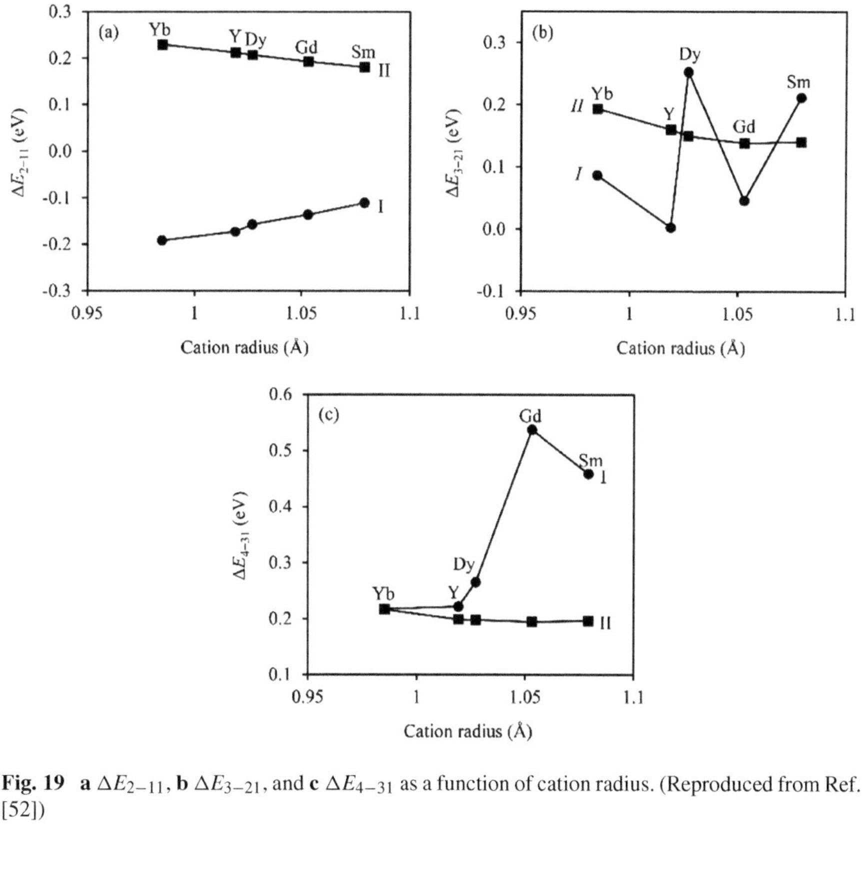

Fig. 19 **a** ΔE_{2-11}, **b** ΔE_{3-21}, and **c** ΔE_{4-31} as a function of cation radius. (Reproduced from Ref. [52])

The effect of dopant type on the growth of defect clusters can also be seen in Fig. 19. When the clusters grow large, the values of ΔE_{4-31} of the types II and I clusters are considered respectively for smaller cations (Yb and Y) and larger cations (Dy, Sm, and Gd). The ΔE_{4-31} values are in the order of Gd > Sm > Dy > Yb > Y, showing that the clusters in Gd and Sm doped ceria are easy to grow larger.

In our previous study [22], it has been shown that the ionic conductivity of doped ceria can be significantly affected by dopant type. The decrease in the conductivity of heavily doped ceria with increasing doping concentrations is calculated by.

$$\Delta(\log\sigma)_{25-15} = \log\sigma_{25} - \log\sigma_{15}, \tag{9}$$

where σ_{15} and σ_{25} is the conductivity of 15at.% and 25 at.% doped ceria. It is in the order of Gd > Sm > Dy > Y > Yb [22]. This result has been explained by the formation of the domains. A fast development of the domains corresponds to a larger value of $\Delta(\log\sigma)_{25-15}$. Because the domain formation results in the increase in the ordering level of oxygen vacancies, the domain development can be evaluated by

$$\Delta(I_B/I_C)_{25-15} = (I_B/I_C)_{25} - (I_B/I_C)_{15}, \tag{10}$$

where I_B and I_C are the integral intensities of peaks B and C in EELS spectra. The growth capability of larger cluster, i.e., ΔE_{4-31} values, are compared with previous results of $\Delta(I_B/I_C)_{25-15}$ and $\Delta(\log\sigma)_{25-15}$ in Fig. 20. As we expect, the $\Delta(I_B/I_C)_{25-15}$ and $\Delta(\log\sigma)_{25-15}$ values increase with increasing ΔE_{4-31} value.

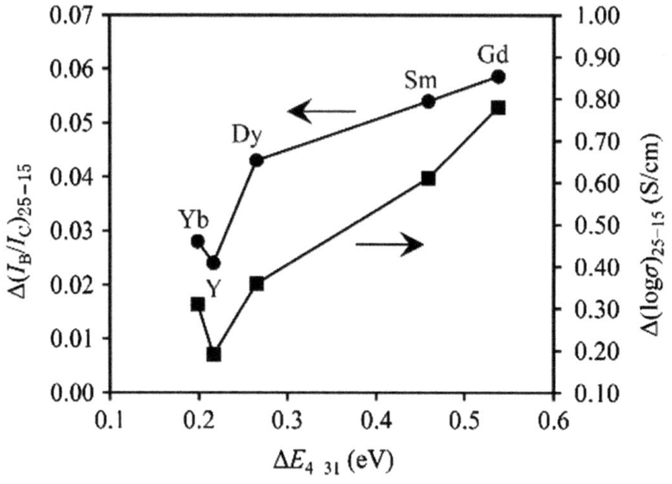

Fig. 20 Correlation of the values of ΔE_{4-31}, $\Delta(I_B/I_C)$ and $\Delta(\log\sigma)]$ (Reproduced from Ref. [52])

3.5 The Formation Mechanism of Nano-sized Domain

According to the observations, the phase transformation from F-type structure to C-type structure in doped ceria is not a sudden change, but gradually takes place from clusters to domains, and then to C-type precipitates with increasing the doping concentration. Because the cluster formation is the first step of the microstructural evolution, a comprehensive understanding of the domain formation and the phase transformation from the F-type structure to C-type structure can be achieved based on the cluster structures.

The substitution of Gd^{3+} for Ce^{4+} in the ceria creates charge-compensating vacancy. In the dilutely doped ceria, these point defects are randomly distributed with few interactions. As the doping level increases, the concentration of defects increases, and more defect-defect interactions occur. When two adjacent vacancies aggregate, they are located at the sites separated by <110>/2. Then, with more nearby vacancies aggregating, basic stable structures, isosceles triangle structure, and tetrahedral structure, will form. When the vacancy number increases up to 6, a distinct symmetric dumbbell structure will form by further combination of the basic structures. This dumbbell structure can act as the building block for larger cluster formation.

Since the vacancy ordering is generally associated with cation ordering, the enhanced vacancy ordering in the net of dumbbell structures will increase the ordering level of accompanied dopants. The combination of short-range ordering of dumbbell structures will evolve into long-range ordering (i.e., superstructure) and subsequently develop into the entire ordered structure, i.e., the C-type structure. The domain structure can be considered as an intermediate state between the F- and C-type structures. In addition, the domain growth, in terms of the formation of a series of dumbbell structures, will facilitate the phase transformation.

3.6 Improvement of Ionic Conductivity by Controlling Domain Formation

Since the domains, as the clusters, are formed by the aggregation of vacancies and cations, their formation can be effectively restrained by controlling the diffusion in the preparation process of the ceria electrolytes. Obviously, a simple method is lowering the sintering temperature. It should be noted that the grain size (d_g) of sintered samples is also reduced by decreasing the temperature.

Figure 21 shows Arrhenius plots for DC electrical conductivity of $Y_xCe_{1-x}O_{2-x/2}$ electrolytes with $x = 0.15$ and 0.25. These samples have different grain sizes, which are obtained by sintering at 950 to 1500 °C. It can be seen that the conductivity of $Y_{0.25}Ce_{0.75}O_{1.875}$ electrolytes shows a considerably change as the grain size increases from 90 nm to 0.9 μm, while this change for $Y_{0.15}Ce_{0.95}O_{1.925}$ samples is not as fast.

Figure 22 summarizes the conductivity of $Y_xCe_{1-x}O_{2-x/2}$ electrolytes at 500 °C and the calculated activation energy as a function of grain size. There exists a grain

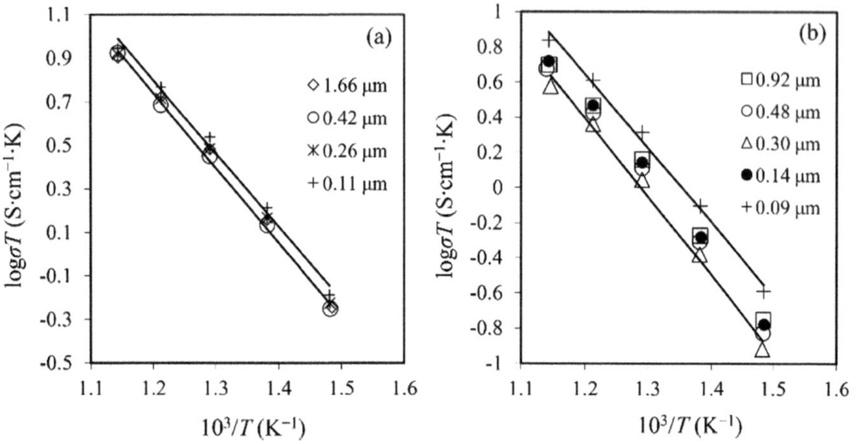

Fig. 21 Arrhenius plots for ionic conductivity of **a** $Y_{0.15}Ce_{0.85}O_{1.925}$ and **b** $Y_{0.25}Ce_{0.75}O_{1.875}$ electrolytes. The average grain size dg is as shown (Reproduced from Ref. [4])

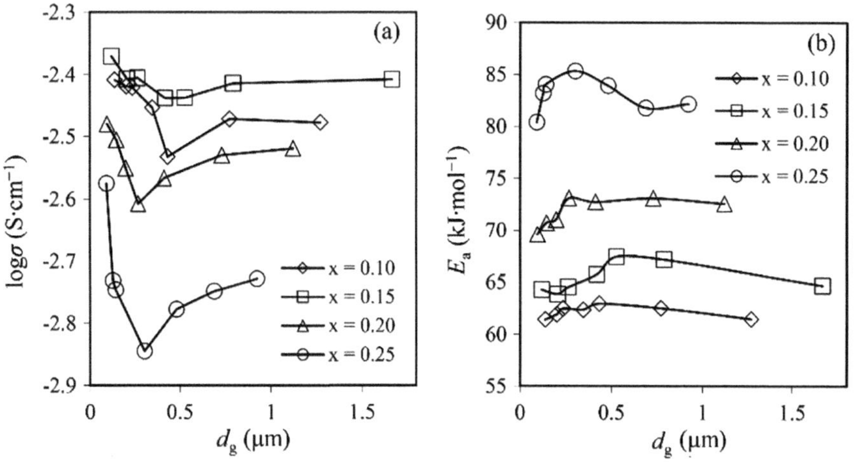

Fig. 22 **a** Conductivity at 500 °C and **b** activation energy E_a of $Y_xCe_{1-x}O_{2-x/2}$ electrolytes as a function of average grain size d_g. (Reproduced from Ref. 4])

size that corresponds to the minimum conductivity and the maximum activation energy for each doping concentration. In the region containing coarser grains, the conductivity decreases with decreasing grain size. This result is consistent with most previous studies [59, 60], and this phenomenon has been well explained by effect of grain boundaries based on the space charge model [61]. In the region containing finer grains, however, the conductivity increases with decreasing grain size.

According to these results, we believe that the grain sizes have two impacts on the electrolytic properties. First, the effect of space charge layers of the grain boundaries leads to a decrease in the conductivity with decreasing d_g. Second, the density of domains decreases with decreasing d_g and this reduction of the density of domains helps to improve the electrolytic properties. Therefore, the ultimate conductivity and activation energy should reflect the balance of this competition. It is obvious that, for the electrolytes with coarse grains, the effect of space charge layers dominates the conductivity, and for the electrolytes with fine grains, the disappearance of domains dominates.

The experimental evidence leads us to believe that the formation of domains is restrained by decreasing the grain size is mainly because of inhibition of diffusion. It is well known that the migration of atoms is exponentially proportional to temperature. Therefore, with decreasing sintering temperature, the mobility of dopant cations in the ceria decreases and, hence, the segregation of dopant cations decreases.

4 Summary and Prospects

The study of short-range ordered structures in RE-doped ceria as electrolyte in SOFC is introduced in this chapter. Through systematic computer simulation combined with electron microscopy analysis, the formation mechanism and the structural features of these subtle microstructures are revealed.

In RE-doped ceria, when the concentration of dopants is lower than the solubility, clusters and domains of aggregated dopant cations and oxygen vacancies developed continuously with increasing the dopant the concentration. A unique dumbbell structure formed with six oxygen vacancies in a ceria matrix has been identified as the building block for defect cluster growth and sequentially for the domain formation and phase transformation from a F-type structure to C-type precipitates. As such, the phase transformation as established on the atomic scale occurs through this defect cluster growth.

An example of Y-doped ceria sintered at different temperature shows that the domain formation can be restrained by low temperature sintering. The electrolytic properties may be adjusted through careful control of domain formation, doping level, and grain size.

Much more work needs to be carried out as logical extension of the works reviewed here. The present models of the short-range ordered structures are mainly derived from computer simulations. Direct observation and demonstration of these structures at atomic scale are still lacking. Although the negative impact of the clusters and domains on the electrolytic properties of doped ceria has been verified, the quantitative relationship between the formation of the short-range ordered structure and the material properties. To control the formation of domain, novel sintering methods, such as spark plasma sintering [17], have been suggested. However, this sintering technique requires special equipment. The sample size is also limited. Therefore, the simple and effective means to control the domain formation need to be further developed.

It is worth to note that the atomistic simulations of defect clusters reviewed in the present work were conducted based on energy minimization in static lattice model using an empirical pair potential to describe the short-range atomic interaction. The potential parameters were obtained by fitting to experimental results or to accurate quantum mechanical calculations. The same method has been successfully used to study the crystal properties, defect structures, and surface structure. However, the empirical potentials usually lack certain fundamental properties required for the accurate description of chemical bonding. Moreover, it often cannot guarantee the validity of the calculation, and hence it is necessary to verify the applied empirical potential and results. In contrast, first principles calculations based on the laws of quantum mechanics provide detailed insight into the mechanical, electronic, optical, and magnetic properties of materials. However, the calculations are time consuming. With the improvement in hardware and software tools, the first principles calculations will gain more importance for dealing with increasingly complex systems. Furthermore, the hybrid techniques that combine the empirical interatomic potential models and first principles calculations will develop for the research on fuel cell materials. They use, for example, the empirical interatomic potentials to provide the defect structure models, and the first principles calculations to yield higher precision in the result and to analyze other properties, such as catalytic properties of electrode materials. The computational efficiency and accuracy of the hybrid methods fall between the first principles methods and the empirical methods. It is expected that the hybrid methods will play an important role in the molecular-level modeling of SOFCs in the near future.

References

1. Steel BCH, Heinzel A (2001) Materials for fuel-cell technologies. Nature 414:345–352
2. Bevan DJM, Summerville E (1979) Mixed rare earth oxides. In: Gschneidner KA, Eyring L (eds) Handbook on the physics and chemistry of rare earths. Elsevier North-Holland Publishing Company, Oxford, pp 401–524
3. Inaba H, Tagawa H (1996) Ceria-based solid electrolytes. Solid State Ionics 83:1–16
4. Ou DR, Mori T, Ye F, Takahashi M, Zou J, Drennan J (2006) Microstructures and electrolytic properties of yttrium-doped ceria electrolytes: Dopant concentration and grain size dependences. Acta Mater 54:3737–3746
5. Catlow CRA (1981) Defect clustering in nonstoichiometric oxides. In: Sørensen OT (ed) Nonstoichiometric oxides. Academic Press, New York (NY), pp 61–98
6. Wang DY, Park DS, Griffith J, Nowick AS (1981) Oxygen-ion conductivity and defect interactions in yttria-doped ceria. Solid State Ionics 2:95–105
7. Minervini L, Zacate MO, Grimes RW (1999) Defect cluster formation in M_2O_3-doped CeO_2. Solid State Ionics 116:339–349
8. Butler V, Catlow CRA, Fender BEF, Harding JH (1983) Dopant ion radius and ionic conductivity in cerium dioxide. Solid State Ionics 8:109–113
9. Wu ZH, Cormack AN (2003) Explanation of discrete electric relaxation peak of CeO_2:Gd^{3+}. Radiat Eff Defects Solids 158:295–298
10. Ball JA, Grimes RW, Price DW (2005) A pair sum approximation for predicting defect cluster binding energies in fluorites. Model Simul Mater Sci Eng 13:1353–1363

11. Gerhardt-Anderson R, Nowick AS (1981) Ionic-conductivity of CeO_2 with trivalent dopants of different ionic-radii. Solid State Ionics 5:547–550
12. Kilner JA, Brook RJ (1982) A study of oxygen ion conductivity in doped nonstoichiometric oxides. Solid State Ionics 6:237–252
13. Kilner JA, Waters CD (1982) The effects of dopant cation oxygen vacancy complexes on the anion transport-properties of nonstoichiometric fluorite oxides. Solid State Ionics 6:253–259
14. Kilner JA (1983) Fast anion transport in solids. Solid State Ionics 8:201–207
15. Ye F, Ou DR, Mori T (2012) Microstructural evolution in a CeO_2-Gd_2O_3 system. Microsc Microanal 18:162–170
16. Mori T, Drennan J, Lee JH, Li JG, Ikegami T (2002) Oxide ionic conductivity and microstructures of Sm- or La-doped CeO_2-based systems. Solid State Ionics 154–155:461–466
17. Mori T, Kobayashi T, Wang Y, Drennan J, Nishimura T, Li JG, Kobayashi H (2005) Synthesis and characterization of nano-hetero-structured Dy doped CeO_2 solid electrolytes using a combination of spark plasma sintering and conventional sintering. J Am Ceram Soc 88:1981–1984
18. Ou DR, Mori T, Ye F, Zou J, Auchterlonie G, Drennan J (2007) Microstructural inhomogeneity in holmium-doped ceria and its influence on the ionic conduction. J Electrochem Soc 154:B616–B622
19. Ye F, Mori T, Ou DR, Takahashi M, Zou J, Drennan J (2007) Ionic conductivities and microstructures of ytterbium-doped ceria. J Electrochem Soc 154:B180–B185
20. Ou DR, Mori T, Ye F, Kobayashi T, Zou J, Auchterlonie G, Drennan J (2006) Oxygen vacancy ordering in heavily rare-earth-doped ceria. Appl Phys Lett 89:171911–13
21. Ou DR, Mori T, Ye F, Zou J, Auchterlonie G, Drennan J (2007) Evidence of intragranular segregation of dopant cations in heavily yttrium-doped ceria. Electrochem Solid State Lett 10:P1-3
22. Ou DR, Mori T, Ye F, Zou J, Auchterlonie G, Drennan J. Oxygen-vacancy ordering in lanthanide-doped ceria: dopant type dependence and structure model. Phys Rev B 2008;77:024108–1–8.
23. Ye F, Mori T, Ou DR, Zou J, Auchterlonie G, Drennan J (2008) Compositional and structural characteristics of nano-sized domains in gadolinium-doped ceria. Solid State Ionics 179:827–831
24. Ye F, Mori T, Ou DR, Zou J, Auchterlonie G, Drennan J (2007) Compositional and valent state inhomogeneities and ordering of oxygen vacancies in terbium-doped ceria. J Appl Phys 101:113528–15
25. Schmitt R, Nenning A, Kraynis O, Korobko R, Frenkel AI, Lubomirsky I, Haile SM, Rupp JLM (2020) A review of defect structure and chemistry in ceria and its solid solutions. Chem Soc Rev 49:554–592
26. Kraynis O, Timoshenko J, Huang JH, Singh H, Wachtel E, Frenkel AI, Lubomirsky I (2019) Modeling strain distribution at the atomic level in doped ceria films with extended X-ray absorption fine structure spectroscopy. Inorg Chem 58:7527–7536
27. Kossoy A, Frenkel AI, Wang Q, Wachtel E, Lubomirsky I (2010) Local structure and strain-induced distortion in $Ce_{0.8}Gd_{0.2}O_{1.9}$. Adv Mater 22:1659–1662
28. Wang Y, Kageyama H, Mori T, Yoshikawa H, Drennan J (2006) Local structures around Y and Ce cations in 10 mol% Y_2O_3 doped ceria ceramics by EXAFS spectroscopy. Solid State Ionics 177:1681–1685
29. Panlener RJ, Blumenthal RN, Garnier JE (1975) Thermodynamic study of nonstoichiometric cerium dioxide. J Phys Chem Solids 36:1213–1222
30. Rupp JLM, Fabbri E, Marrocchelli D, Han JW, Chen D, Traversa E, Tuller HL, Yildiz B (2014) Scalable oxygen-ion transport kinetics in metal-oxide films: impact of thermally induced lattice compaction in acceptor doped ceria films. Adv Funct Mater 24:1562–1574
31. Pikalova EY, Murashkina AA, Maragou VI, Demin AK, Strekalovsky VN, Tsiakaras PE (2011) CeO_2 based materials doped with lanthanides for applications in intermediate temperature electrochemical devices. Int J Hydrogen Energy 36:6175–6183

32. Ando M, Oikawa I, Ohki S, Tansho M, Shimizu T, Maekawa H, Kiyono H (2012) Oxygen-17 NMR study of defects in $LnO_{(1.5)}$-CeO_2 (Ln = La and Nd). Chem Lett 42:57–59
33. Heinmaa I, Joon T, Kooskora H, Pahapill J, Subbi J (2010) Local structure and oxygen ion dynamics in La doped ceria: O-17 NMR study. Solid State Ionics 181:1309–1315
34. Kim N, Stebbins JF (2007) Vacancy and cation distribution in yttria-doped ceria: an Y-89 and O-17 MAS NMR study. Chem Mater 19:5742–5747
35. Artini C, Pani M, Carnasciali MM, Plaisier JR, Costa GA (2016) Lu-, Sm-, and Gd-doped ceria: a comparative approach to their structural properties. Inorg Chem 55:10567–10579
36. Artini C, Pani M, Carnasciali MM, Buscaglia MT, Plaisier JR, Costa GA (2015) Structural features of Sm- and Gd-doped ceria studied by synchrotron X-ray diffraction and mu-Raman spectroscopy. Inorg Chem 54:4126–4137
37. McBride JR, Hass KC, Poindexter BD, Weber WH (1994) Raman and X-ray studies of $Ce_{1-x}RE_xO_{2-y}$, where RE=La, Pr, Nd, Eu, Gd, and Tb. J Appl Phys 76:2435–2441
38. Horlait D, Claparede L, Clavier N, Szenknect S, Dacheux N, Ravaux J, Podor R (2011) Stability and structural evolution of $Ce^{IV}{}_{1-x}Ln^{III}{}_xO_{2-x/2}$ solid solutions: a coupled mu-Raman/XRD approach. Inorg Chem 50:7150–7161
39. Li JG, Ikcgami T, Mori T, Wada T (2001) Reactive $Ce_{0.8}RE_{0.2}O_{1.9}$ (RE = La, Nd, Sm, Gd, Dy, Y, Ho, Er, and Yb) powders via carbonate coprecipitation 1 synthesis and characterization. Chem Mater 13:2913–2920
40. Mendelson MI (1969) Average grain size in polycrystalline ceramics. J Am Ceram Soc 52:443–446
41. Garvie LAJ, Buseck PR (1999) Determination of Ce^{4+}/Ce^{3+} in electron-beam-damaged CeO_2 by electron energy-loss spectroscopy. J Phys Chem Solids 60:1943–1947
42. Egerton RF (1996) Electron energy-loss spectroscopy in the electron microscope. Plenum Press, New York
43. Vyas S, Grimes RW, Gay DH, Rohl AL (1998) Structure, stability and morphology of stoichiometric ceria crystallites. J Chem Soc Faraday Trans 94:427–434
44. Grimes RW, Busker G, McCoy MA, Chroneos A, Kilner JA, Chen SP (1997) The effect of ion size on solution mechanism and defect cluster geometry. Bunsen Ges Phys Chem 101:1204–1210
45. Minervini L, Grimes RW, Sickafus KE (2000) Disorder in pyrochlore oxides. J Am Ceram Soc 83:1873–1878
46. Stanek CR, Minervini L, Grimes RW (2002) Nonstoichiometry in $A_2B_2O_7$ pyrochlores. J Am Ceram Soc 85:2792–2798
47. Binks DJ, Grimes RW (1993) Incorporation of monovalent Ions in ZnO and their influence on varistor degradation. J Am Cerm Soc 76:2370–2372
48. Dick BG, Overhauser AW (1958) Theory of the dielectric constants of alkali halide crystals. Phys Rev 112:90–103
49. Mott NF, Littleton MJ (1938) Conduction in polar crystals I: electrolytic conduction in solid salts. Trans Faraday Soc 34:485–491
50. Gale JD, Rohl AL (2003) The general utility lattice program (GULP). Mol Simul 29:291–341
51. Ye F, Mori T, Ou DR, Cormack AN, Lewis RJ, Drennan J (2008) Simulation of ordering in large defect clusters in gadolinium-doped ceria. Solid State Ionics 179:1962–1967
52. Ye F, Mori T, Ou DR, Cormack AN (2009) Dopant type dependency of domain development in rare-earth-doped ceria: an explanation by computer simulation of defect clusters. Solid State Ionics 180:1127–1132
53. Ye F, Mori T, Ou DR, Zou J, Drennan J (2009) A structure model of nano-sized domain in Gd-doped ceria. Solid State Ionics 180:1414–1420
54. Li ZP, Mori T, Ye F, Ou DR, Zou J, Drennan J (2011) Structural phase transformation through defect cluster growth in Gd-doped ceria. Phys Rev B 84:180201–180215
55. Soldatov AV, Ivanchenko TS, Della Longa S, Kotani A, Iwamoto Y, Bianconi A (1994) Crystal-structure effects in the Ce L_3-edge X-ray-absorption spectrum of CeO_2: multiple-scattering resonances and many-body final states. Phys Rev B 50:5074–5080

56. Douillard L, Gautier M, Thromat N, Henriot M, Guitter MJ, Duraud JP, Tourillon G (1994) Local electronic structure of Ce-doped Y_2O_3: an XPS and XAS study. Phys Rev B 49:16171–16180
57. Travlos A, Boukos N, Apostolopoulos G, Dimoulas A (2003) Oxygen vacancy ordering in epitaxial layers of yttrium oxide on Si (001). Appl Phys Lett 82:4053–4055
58. Shannon RD (1976) Revised effective ionic radii and systematic studies of interatomic distances in halides and chalcogenides. Acta Cryst A32:751–767
59. Tian C, Chan SW (2000) Ionic conductivities, sintering temperatures and microstructures of bulk ceramic CeO_2 doped with Y_2O_3. Solid State Ionics 134:89–102
60. Tschöpe A, Sommer E, Birringer R (2001) Grain size-dependent electrical conductivity of polycrystalline cerium oxide: i experiments. Solid State Ionics 139:255–265
61. Guo X (1995) Physical origin of the intrinsic grain-boundary resistivity of stabilized-zirconia: role of the space-charge layers. Solid State Ionics 81:235–242

Computational Approach for Designing of Electrode Interfaces in Fuel Cell

Michihisa Koyama

Abstract The efficiency of fuel cells is largely governed by the characteristics of the electrode interface. The electrode interface extends from the atomic scale to several tens of micrometers. To design such electrodes that span multiple scales in a sophisticated manner, it is important to utilize advanced simulation methodologies. In this chapter, an example of a multi-scale, multi-physics simulation of a solid oxide fuel cell anode is introduced, followed by an introduction of studies on the stability and activity of electrocatalysts in polymer electrolyte fuel cells based on massively parallel first-principles calculations. It is emphasized to consider the real electrode structure for the simulation in both cases.

Keywords Multi-scale simulation · Multi-physics simulation · First-principles · Massively parallel computing · Real-system structure

1 Multi-scale, Multi-physics Simulation of Solid Oxide Fuel Cell Anode

The SOFC anode is a porous cermet that consists of metal and ceramic oxide. Electrons conduct through the metal, and oxide ions diffuse through the oxide. It is commonly known that the fuel diffusing through the pores reacts with oxide ions to form H_2O and electrons at the TPB where the metal, oxide, and gas-phase meet. While the electrochemical reaction itself occurs near the three-phase interface on the nanometer scale, ion conduction, and gas diffusion proceed on the scale of a few to several tens of micrometers. Inevitably, elucidating the electrode reaction mechanism means the elucidation of multi-physics phenomena that proceed on multi-scales as illustrated in Fig. 1.

M. Koyama (✉)
Research Initiative for Supra-Materials, Shinshu University, 4-17-1 Wakasato, Nagano, Japan
e-mail: koyama_michihisa@shinshu-u.ac.jp

T. Mori et al. (eds.), *Design of Active Key Interfaces in Fuel Cells*,
https://doi.org/10.1007/978-981-97-5706-0_5

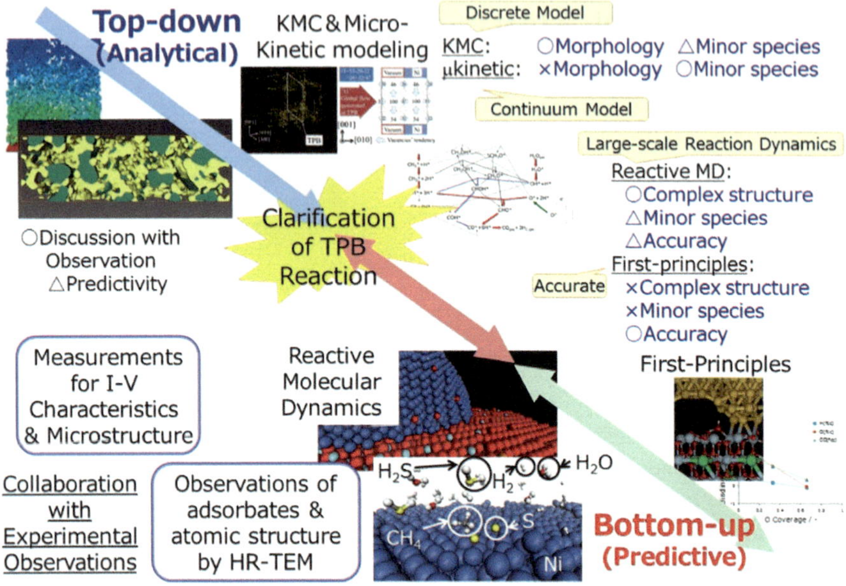

Fig. 1 Concept of multi-scale, multi-physics simulation scheme for SOFC anode

1.1 Difficulties in Elucidating the Anodic Reaction Mechanism

Multi-scale simulation as a concept can be found in many contexts. However, successful examples can be found in strength materials, etc., [1] where physical phenomena dominate the overall characteristics. On the contrary, multi-scale simulations of systems containing chemical reactions remain open. This is partly because one needs to bridge the gap between continuum and discrete simulations by its nature. Here the author would like to share an example of multi-scale, multi-physics in SOFC anode reaction. Atomistic simulation is typically conducted by first-principles calculations based on density functional theory [2–11].

The anode reaction in a hydrogen-fueled SOFC is represented by the following equation.

$$H_2 + O^{2-} \rightarrow H_2O + 2e^- \quad (1)$$

Although the reaction is simple, the three-dimensional expansion of the reaction field at the porous electrodes complicates the phenomenon, and experiments have been conducted using model electrode systems with two-dimensional Ni-patterned electrodes fabricated on single-crystal substrates [12–15]. Also, microkinetic simulations have been conducted to elucidate the reaction mechanism that explains the activity of the model electrode system [16–20]. The issues in these previous efforts can be categorized as follows.

(i) Large differences in anode activity are found in model electrode experiments.
(ii) Predictive or transferrable simulations have not been achieved.
(iii) It is difficult to obtain reliable parameter values representing each elementary process.

The first issue means that the results of Mizusaki et al. [12] and the other results [13–15] differ by about three orders of magnitude when we compare the activity per three-phase boundary (TPB) in a model electrode system. The fact that the same electrolyte material set was used to prepare Ni-patterned electrodes on a single crystalline YSZ (Yttria Stabilized Zirconia) substrate, and that research groups with sufficient experience reported very different activities, cannot be explained by experimental errors in cell setup or gas atmosphere. The second issue is related to microkinetic simulations. Quantitative simulation of a specific experimental result can be found [16–18]; however, such simulation fails to reproduce the experimental results reported by other researchers and sometimes the experimental results under different gas atmospheres. The first reason for the lack of predictability of the simulations is that the factors that explain the large gaps in the measurements are not sufficiently clarified or discussed. Besides, the rate-limiting process is often assumed a priori, and some of the parameters are used as fitting parameters, resulting in physically meaningless values. The third and last one can be understood from an exhaustive survey of the literature [2–11, 21, 22]. One can find a large difference in the reported values for the same elementary process. Further, data from first-principles calculations show differences of several tens of kJ/mol depending on the model structure used for the calculations, the level of theory and approximation, and the settings used. This results in large differences in parameters under SOFC operating temperatures. A possible approach in this situation is to simulate, for example, the average, minimum, and maximum values of the reported values, and then look at the sensitivity of the simulation to find a set of parameters that can reasonably explain the difference in the actual measurements. However, if this is conducted for n parameters, 3^n calculations are required, which is not easy to handle for a target system consisting of several dozen parameters.

1.2 Combined Top-Down and Bottom-Up Approach

The author and coworkers decided to take a combined top-down and bottom-up approach considering the current situation, where there are diverse differences in the reported model experiments as well as in the results of first-principles calculations that are used as inputs for microkinetic simulation as illustrated in Fig. 2 [19].

As a top-down approach, a lattice Boltzmann method (LBM)-based simulation is utilized, adopting the three-dimensional reconstruction of the porous anode structure observed by using FIB-SEM (Focused-ion beam scanning electron microscopy). The

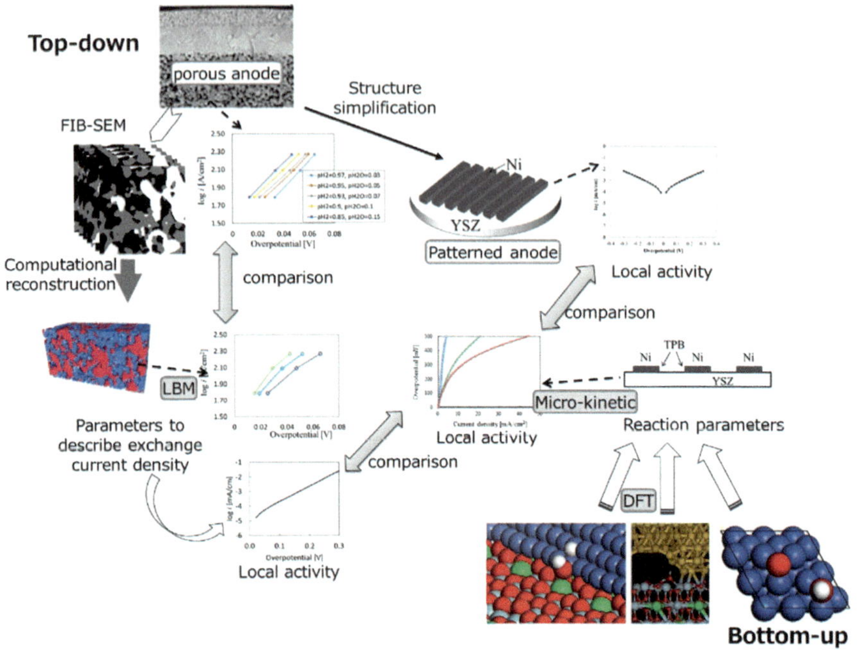

Fig. 2 The scheme of top-down and bottom-up approaches. Reprinted with permission from *ECS Trans.*, **78**, (2017). Copyright 2017 Electrochemical Society

local TPB activity, which explains the porous electrode properties, is determined in an inverse analytical manner to derive the activity per TPB length that can be directly compared with both microkinetic simulations and patterned electrode measurements. The bottom-up approach utilizes microkinetic simulations with elementary process parameters based on first-principles calculations, etc., which was conventionally used. The difference with literature is that physical meaning is much focused to realize simulations maintaining predictivity and transferability, thus avoiding the use of fitting parameters adopting first-principles results as possible.

The local TPB activity derived from the top-down approach was estimated as the activity between Mizusaki et al. [12] and the other researchers [13–15]. Therefore, the reason for the exceptionally low activity of Mizusaki et al. among the reports on the patterned anode is not due to any inadequacy in the experiments, but rather to the possibility that there are physical and chemical factors explaining the difference in the activity.

1.3 Exhaustive Microkinetic Simulation Based on First-Principles Calculations

For the microkinetic simulation, we used the simulator by Kohno et al. [18], which considers eight elementary processes of surface reactions and three elementary processes of charge-transfer reaction. A basic input set was prepared as input to the microkinetic simulation after an exhaustive survey of the literature and the process of scrutiny described in the following [20]. The purpose was to reduce the cost of calculations of the huge number of combinations. This directly means the need to minimize the huge simulation cost, because each of the eight surface and three charge-transfer elementary processes has a pre-exponential factor, an activation barrier, an enthalpy change, and an entropy change, meaning that there are 44 parameters in the reaction alone.

The parameters derived from first-principles calculations are reported to have different values depending on the surface orientation, for example, and different reactivities of the TPB are reported depending on the atomic-level local structure, such as charge transfer reactions on Y and Zr atoms in YSZ at the TPB. In actual porous anodes, various surface orientations and TPB structures will coexist, and they cannot be uniquely determined. Therefore, the values of the most stable surface were set as the standard values for the surface reaction. To explore the parameter values that can explain the experimentally observed TPB activity, the range of parameters reported in the literature was considered. The pre-exponential factor is generally known to be about k_bT/h, and therefore we confirmed that those did not deviate significantly from this standard value. The reported values of molecular dynamics and experimental measurements were referred for the surface diffusion coefficients.

Besides, it is necessary to maintain thermodynamic consistency. It should be pointed out that the energies of each elementary process are not independent of each other, and the energy difference between the initial and final states must be constant regardless of the path, due to Hess's law. Therefore, the same paper should be referred to for the surface reaction and the TPB reaction, respectively.

In the end, a set of 16 basic inputs was prepared as parameters for surface and charge-transfer reactions. Three different values were set for each of the seven surface diffusion coefficients, and 30 different overpotentials were set to derive current–voltage characteristics. In total, $16 \times 3^7 \times 30 = 2{,}099{,}520$ exhaustive simulations were performed.

Figure 3 shows the results of the exhaustive simulation, together with the results of Mizusaki et al. [12], Bieberle et al. [13], and the results calculated by the LBM simulation from the top-down approach.

It can be seen that even simulations using first-principles derived parameters can yield current–voltage characteristic results with a range of as much as four orders of magnitude, depending on the choice of parameters. In some cases, the results of microkinetic simulations are almost consistent with the results of LBM by the top-down approach. On the other hand, the results of Mizusaki et al. [12] are smaller than the lower bound of the simulation results obtained, and the results of Bieberle et al.

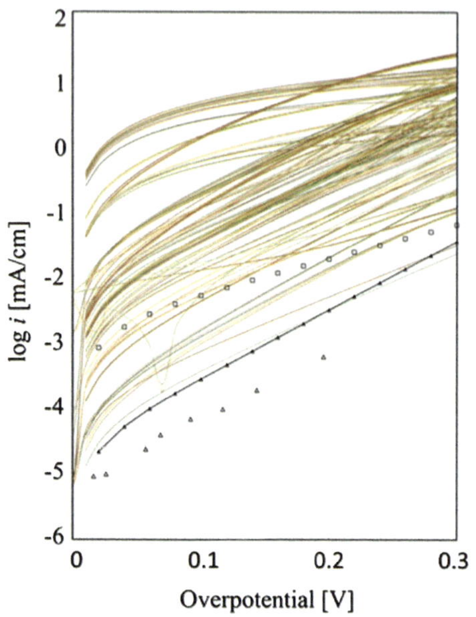

Fig. 3 The calculated overpotential characteristics by microkinetic simulation together with that derived from LBM simulation (the closed triangles with solid line) and experimental data of patterned anodes (the open triangles are from Mizusaki et al. [12] and open squares are from Bieberle et al. [13]). Reprinted with permission from *ECS Trans.*, **78**, (2017). Copyright 2017 Electrochemical Society

[13] are within the range of the simulation results. Then, a set of parameters that can reasonably explain those observations was explored, referring to the shape and slope of the current–voltage curve and the value of the exchange current density estimated from Tafel's equation. Here, "reasonably explain" refers to a search within the range of parameters reported in the literature, without using parameter values that deviate significantly from those by first-principles calculations reported. Figure 4 shows an example of the simulation results [20] compared with the observation by Mizusaki et al. [12]. It can be seen that the partial pressure dependence of the measured current–voltage characteristics was quantitatively expressed by using the parameters of the elementary processes in the range of those reported previously. Parameter sets that provide good descriptions of the properties by Yao et al. [15] and Bieberle et al. [13] were also identified as shown in Fig. 5. It was clarified that the diffusion of oxygen on the Ni surface, as well as oxygen migration reaction from YSZ to Ni surface, determine the high activity of the TPB observed in their studies.

1.4 Real-System Structure for Triple-Phase Boundary Reaction

It is suggested from the microkinetic simulation that the activity varies greatly depending on the rate-limiting process at the three-phase interface. So, what does the reaction at the TPB look like? Although many reaction analyses using simple structural models based on first-principles calculations have been reported, the dynamic

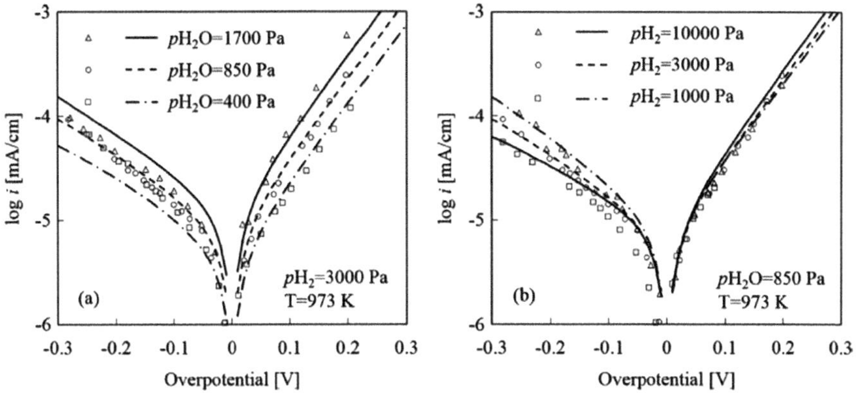

Fig. 4 Current density−overpotential curves under different partial pressures of **a** water and **b** hydrogen. The open symbols are the experimental data from Mizusaki et al. [12]

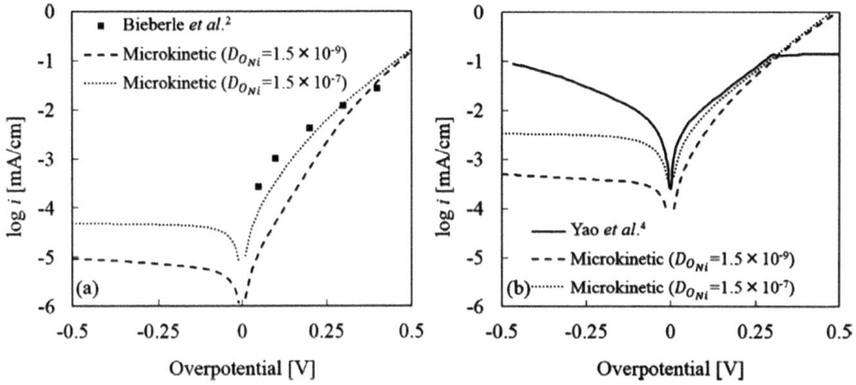

Fig. 5 Theoretical curves compared with experimental results by **a** Bieberle et al. [13] and **b** Yao et al. [15]. Details are found in [20]

picture and the effects of the local structure at finite temperature have not been clarified. Besides, the structure of the TPB is not clear, to begin with.

TPB has been one of the central issues in SOFC anode development, therefore, many efforts have been devoted to TPB, such as the experimental three-dimensional reconstruction on the porous anodes and kinetic modeling on the patterned anodes. However, the spatial scale of those works is limited to submicrometer, and there has been no report on the realistic atomic structure of TPB. The atomic-scale works can only be found in the theoretical simulations adopting non-realistic models mainly limited by high computational cost for computation. Furthermore, the simulation results from different TPB structure models are often discrepant as discussed in preceding sections of this chapter, which indicates the TPB structure determines the TPB activity.

Liu et al. clarified the structure of TPB by atomic resolution electron microscopy and the TPB reaction by a reactive force field (ReaxFF) molecular dynamics simulation [23]. First, the atomic resolution observation of the Ni/YSZ/pore TPB was challenged. They demonstrate the methodology of TEM observation and successfully observed TPB structures in an atomic resolution as shown in Fig. 6.

Then, large-scale reaction dynamics were challenged adopting the real-system structure in Fig. 6. The high computational cost of first-principles methods typically used to study chemical reactions hinders their application to TPB structures observed by atomic resolution microscopy. Therefore, ReaxFF was adopted, however, the previous parameterization appeared to be inaccurate and did not work effectively for the observed TPB structures. Many efforts have been made to develop a tool

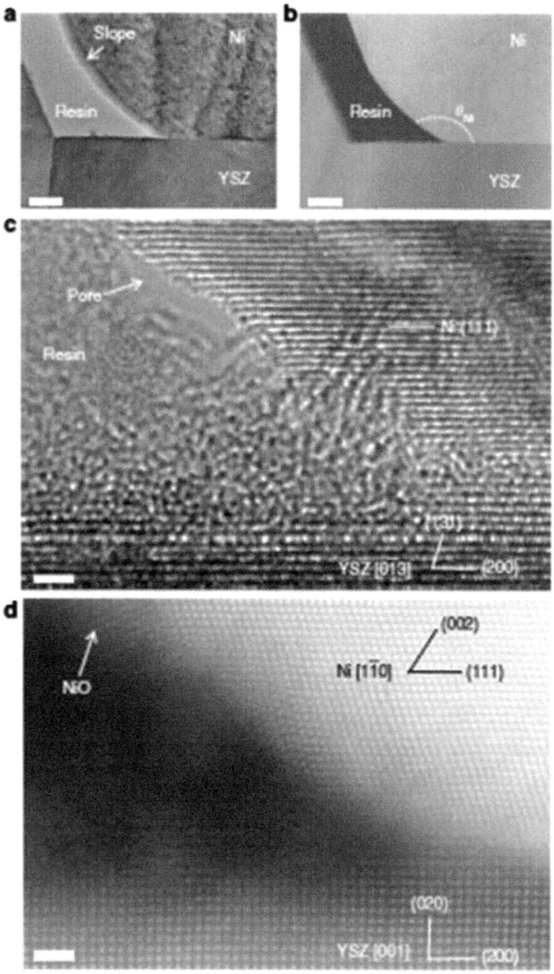

Fig. 6 One of an atomic resolution TPB structures observed in the porous anode. **a** Low-magnification transmission electron microscopy (TEM) image at YSZ[0–13] zone axis. The arrow points out the slope feature of the Ni surface in this observation direction. **b** Low-magnification high-angle annular dark-field–scanning TEM (HAADF-STEM) image at YSZ[001] zone axis. The dashed arc stands for the contact angle (θ_{Ni}). **c** High-resolution TEM image at YSZ[0–13] zone axis. The arrow denotes a pore after electron beam damage of the resin. **d** HAADF-STEM image at YSZ[001] zone axis. The arrow points to a thin layer of NiO on the Ni surface. Scale bars in **a**, **b** 50 nm; in **c**, **d** 1 nm

to efficiently support the parameterization of ReaxFF and to determine an accurate set of parameters for TPB. From the ReaxFF molecular dynamics simulations, three distinct reaction pathways are identified. A typical hydrogen oxidation reaction (HOR) pathway, i.e., diffusion of dissociated hydrogen on Ni surface to react with oxide ion at TPB, was observed (Fig. 7a). Further, two bulk pathways were identified; H diffusion through Ni bulk to Ni/YSZ interface (Fig. 7b) and O migration from YSZ to Ni surface through bulk transfer (Fig. 7c). Distinct reaction pathways and the activity of different TPB structures were observed in the ReaxFF simulations, which provides new insight into the understanding of the TPB reactions and verifies the importance of local structure.

The results from ReaxFF molecular dynamics simulations based on the real-system TPB structures suggest that reaction pathways that are completely unpredictable from static reaction analyses based on static first-principles calculations. From the first-principles-based microkinetic simulation discussed in the previous section, oxygen-related processes are identified to be important. Here it is noted that the bulk oxygen pathway is not considered in the microkinetic simulation. It is suggested from the ReaxFF simulations that the bulk oxygen pathway should be considered in the microkinetic simulation in future studies. Further, it has been reported that a small portion of impurity elements segregate in TPB. In the future, it is expected to conduct dynamics simulations that consider the effects of such impurities to clarify the dominant factors that lead to high TPB activity.

2 Real-System All-Electron Calculation of Polymer Electrolyte Fuel Cell Electrocatalyst

2.1 Real-System Structure for All-Electron Calculations

Supported metal nanoparticles are widely used as heterogeneous catalysts for various applications such as automotive exhaust treatment, steam reforming, and electrocatalysis for polymer electrolyte fuel cell (PEFC). A variety of alloy nanoparticles are used to enhance catalytic activity. There are various types of alloy nanoparticles, such as solid solution, skin structure, eutectic, and intermetallic compound, depending on the composition and the arrangement of each element. Therefore, alloys with various compositions and structures have been proposed to meet the desired functions. Along with the recent advancements in both theory, computers, and parallel-computing, expectations for theoretical elucidation and prediction of the origin of catalytic activity and long-term stability have increased in recent years. Nanoparticles of several nanometers are used in real-system catalysts. To avoid high computational costs, however, cluster models of a few tens of atoms or slab models are often used. It has been pointed out that in a cluster model of a few tens of atoms, the electronic states become discrete and do not become continuous as in a metal [24]. Further, it is necessary to consider the actual nanoparticle structure to discuss

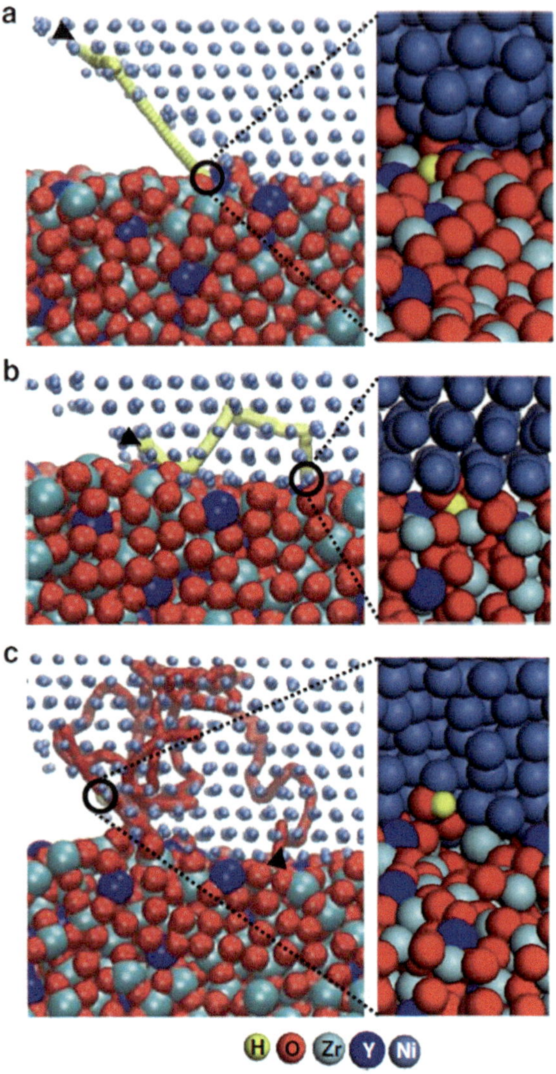

Fig. 7 Reaction dynamics simulations for triple-phase boundary-1. **a** Hydrogen oxidation reaction (HOR)-I pathway. **b** HOR-II pathway. The trajectory of hydrogen (H) starts from the arrow, and the circle denotes the final position where OH species is formed after a simulation period of 700 ps. The local environment near the formed OH species is presented in the magnified image, which is linked to the circle by dotted lines. **c** HOR-III pathway. A trajectory of oxygen (O) starts from the arrow, and the circle denotes the final position where OH species is formed after a simulation period of 700 ps. The local environment near the formed OH species is presented in the magnified image, which is linked to the circle by dotted lines. Trajectories are shown along the Ni[1–10] and YSZ[001] directions

the stability of nanoparticles. Therefore, the use of massively parallel computation of real-system structures is essential.

Figure 8 shows the major published results of massively parallel first-principles calculations of nanoparticles [24–53]. The largest calculation [25] reached 2406 atoms, or ca. 4.4 nm. Focusing on the binary alloy system, benchmark calculation is conducted for a nanoparticle of 1289 atoms [30], and the stability and electronic properties were discussed for 711 atomic systems [34–37]. It is possible to theoretically elucidate the properties of real-system nanoparticles, which will provide clues to understand and design their catalytic functions.

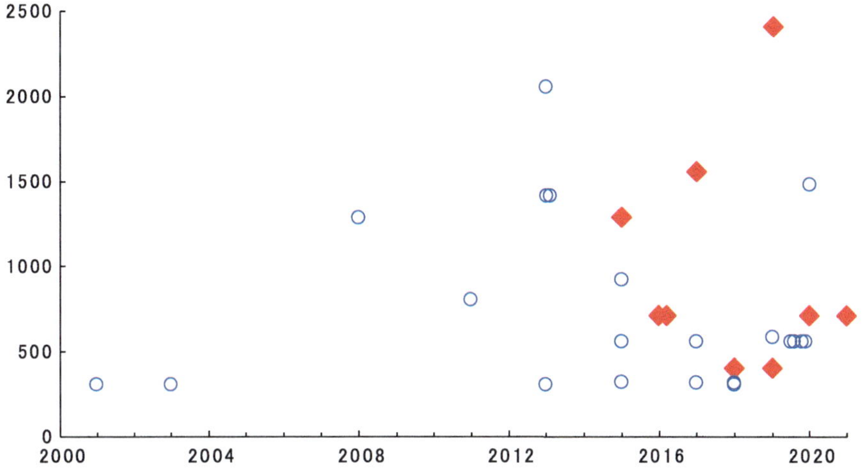

Fig. 8 Major published results of massively parallel first-principles calculations of nanoparticles. Filled diamonds are from the author's group and the open circles are from other groups

2.2 Electronic Structure of PdPt Alloy Nanoparticles

It is known that Pd and Pt form a solid solution alloy at high temperatures such as 800 °C, but they are immiscible at room temperature [54]. Therefore, the Pd–Pt combination is expected to be promising in terms of activity and stability among core–shell alloy catalysts aimed at reducing the amount of Pt usage in the PEFC cathode [55]. On the other hand, it has been reported that solid solution PdPt alloys are formed by hydrogen treatment of PdPt core–shell particles [56]. It is not easy to investigate the thermodynamics of nanoparticles experimentally. In this section, a theoretical study investigating the electronic state and stability of PdPt core–shell nanoparticles and solid-solution alloy nanoparticles [34] will be introduced. Models of core–shell and solid-solution structures of PdPt alloys were prepared to investigate the electronic structure and stability of the various structures as shown in Fig. 9. Two types of shell thicknesses, one and two layers, were prepared, and the Pt/Pd composition in the solid solution structure corresponded to the composition of the core–shell structure. Also, a phase-separated structure was prepared.

Figure 10 shows the charge distribution in the PdPt nanoparticles obtained from first-principles calculations. The charge distribution in the solid solution structure and monometallic nanoparticles is similar; the surface is negatively charged, and the subsurface layer is positively charged. On the other hand, a unique charge distribution is observed in the core–shell structure. This can be understood from the difference in the electronegativity of Pd and Pt. Pd has lower electronegativity and charge transfer from Pd to Pt occurs. Thus, there is also charge transfer at the interface between Pd and Pt in the core–shell structure, in addition to the charge transfer from the surface second layer to the surface. Therefore, it has a unique charge distribution and is therefore expected to lead to unique functions.

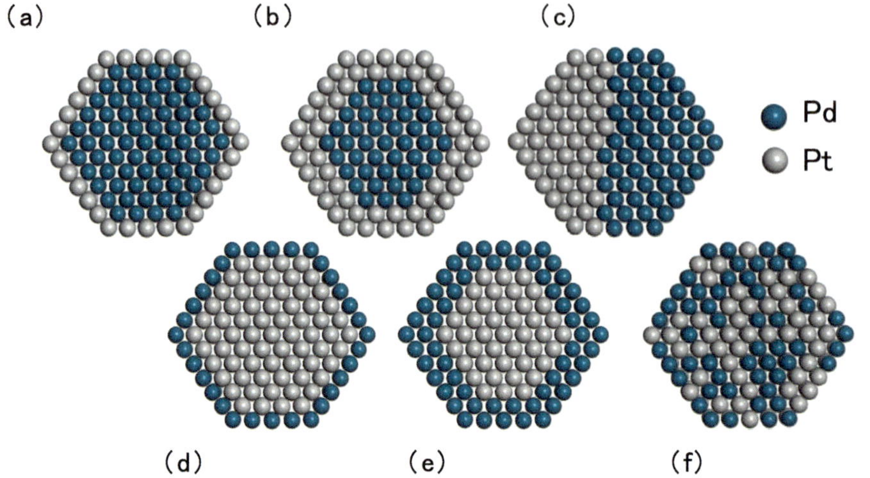

Fig. 9 Models of PdPt alloys investigated. **a** 1-layer Pd–Pt core–shell, **b** two-layer Pd–Pt core–shell, **c** phase-separated, **d** 1-layer Pt–Pd core–shell, **e** 2-layer Pt–Pd core–shell, and **f** solid solution configurations

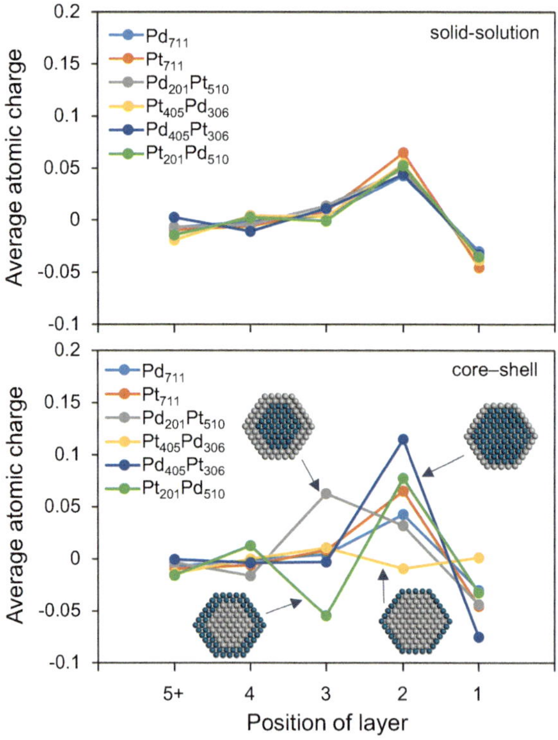

Fig. 10 Charge distributions in the PdPt nanoparticles. The layer position of 1 corresponds to the surface layer

2.3 Thermodynamic Stability of PdPt Alloy Nanoparticle

To discuss the thermodynamic stability of the alloy nanoparticles, the excess energy [57] was calculated based on the results of first-principles calculations. The excess energy is obtained from the following equation.

$$E_{\mathrm{exc}} = \frac{1}{711}\left(E\left(\mathrm{Pd}_{N_{\mathrm{Pd}}}\mathrm{Pt}_{N_{Pt}}\right) - \frac{N_{\mathrm{Pd}}}{711}E(\mathrm{Pd}_{711}) - \frac{N_{\mathrm{Pt}}}{711}E(\mathrm{Pt}_{711})\right) \tag{2}$$

The excess energy is defined as the difference of the energy of the alloy nanoparticles with respect to the energy of single-element nanoparticles of the same size; when the excess energy is negative, the alloy is stable. Figure 11a shows the excess energies of PdPt nanoparticles at 0 K. Among the nanoparticles with different configurations, the Pt shell was calculated to be the most unstable, followed by the solid solution, and the Pd shell was calculated to be the most stable. It was found that the phase-separated structure lies between Pt-shell and solid solution configurations. This order corresponded to the order of Pd occupancy on the surface and was considered to be due to the lower surface energy of Pd than Pt.

Then, to discuss the effect of temperature, entropy was considered. As for the effect of temperature, the enthalpy of vibration, the entropy of vibration, and the configurational entropy were considered. Among these, the configurational entropy has been identified to be a major determinant of the stability of various alloy configurations [58]. The vibrational contributions were estimated by referring to the previous report [59], and the configurational entropy is calculated according to Boltzmann's equation as follows:

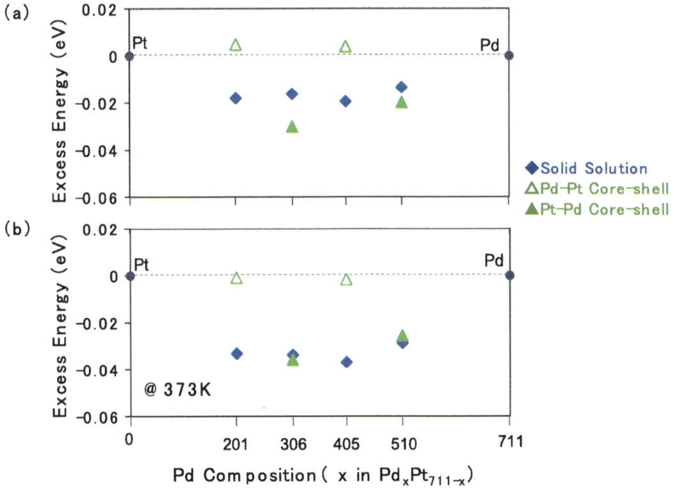

Fig. 11 The excess energies of PdPt nanoparticles at **a** 0 K and **b** 373 K

$$S = k \ln W \tag{3}$$

where k is Boltzmann's constant and W is the degree of configurational freedom. The configuration entropy is zero for both single-element and core–shell systems because the configuration degree of freedom is unity. Figure 11b shows the excess free energy of nanoparticles calculated at 373 K. The effect of the configurational entropy stabilizes the solid solution structure, and the excess energy becomes comparable to that of the Pd shell structure that is most stable at 0 K. A further increase in temperature will stabilize the solid solution structure more to become the most stable among all configurations. From the obtained first-principles calculations, it is expected that the Pd–Pt core–shell particles of about 3 nm will undergo a structural change to a more stable solid solution structure during long-term operation at the operating temperature of the PEFC.

2.4 Thermodynamic Stability of Pt-Skin Alloy Nanoparticle

The discussion of the thermodynamic stability of PdPt alloy nanoparticles in the previous section was extended to Pt skin catalysts, one of the most promising electrocatalysts for PEFCs [35, 36]. Models of Pt_3M (M=Co, Ni, Cu) with the number of atoms of 201, 405, and 711 were prepared. The excess energies were calculated for each size with different atomic configurations of solid solution, single-layer Pt-skin, and two-layer Pt skin as shown in Fig. 12. The Pt skin structures were found to be more stable than the solid solution structure where M atoms also exist on the surface at 0 K. The configurational entropy of the skin structure is smaller than that of the solid solution structure because the composition of the surface is only Pt. However, it was clarified that the skin structure is more stable even when the entropy term at

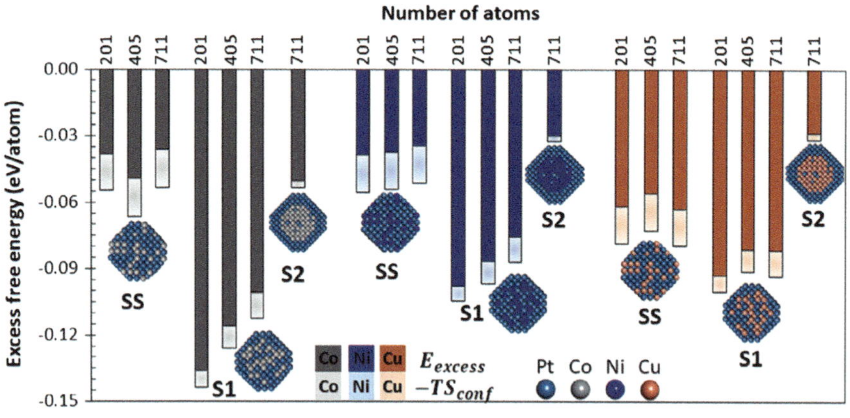

Fig. 12 The excess energies of Pt_3M nanoparticles at 353 K

353 K, which is the typical operating temperature of PEFC, is taken into account. First-principles calculations show that the Pt-skin catalyst can exist stably in the solid solution structure even during long-term operation at the operation temperature although the effects of surface reactions or other dynamic factors were not considered in the calculation.

2.5 Effect of Support on the Electronic Structure of Pt Nanoparticle

In actual PEFCs, the electrocatalyst is used in a supported form. To reveal the electronic structure of the supported electrocatalysts, it is necessary to incorporate a sufficient size of support for the metal nanoparticles into the calculation, leading to an extreme increase in the computational cost. Furthermore, the interface structure between the metal nanoparticles and support has a high degree of freedom, which poses a high barrier to conduct first-principles calculations. Rivera et al. conducted the massively parallel first-principles calculations of Pt supported on SnO_2 with different nanoparticle sizes as shown in Fig. 13 [25]. Figure 14 shows the electronic structures of supported Pt nanoparticles. The charge distribution of the subnanometer cluster, i.e., Pt_4, is rather simple; Top Pt atom is negatively charged while the remaining three Pt atoms at the interface become highly positive. When we focus on the Pt nanoparticle of 2.5 nm, i.e., Pt_{233}, on SnO_2, surface Pt atoms in the top, second, third, and fourth layers are negatively charged with a similar charge value. Atoms at the interface are positive, while the charge value is much smaller than those in Pt_4 or Pt_{13} clusters. From this calculation, one can understand the necessity of real-scale structure models to discuss the electronic structure of supported Pt nanoparticles too.

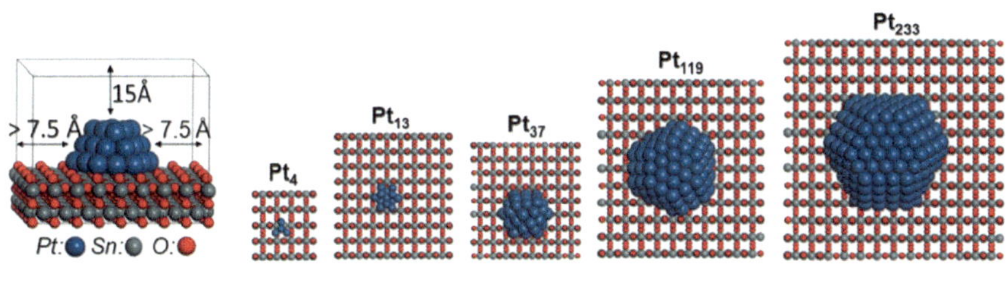

Fig. 13 Models of supported Pt nanoparticles employed in the calculations

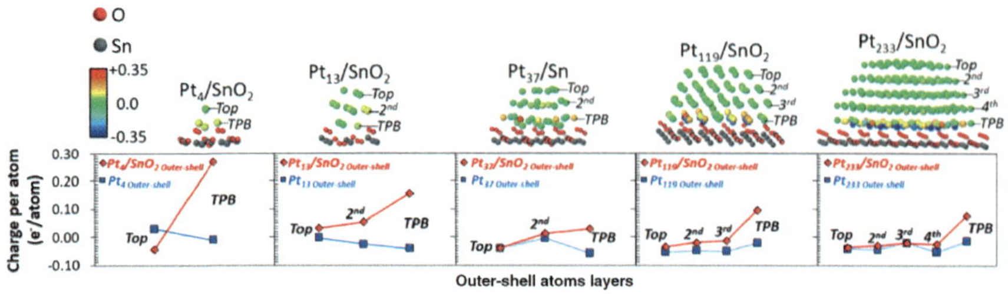

Fig. 14 Charge distributions in supported Pt nanoparticles. Charge of the outer-shell atoms of isolated and SnO_2 supported Pt nanoparticles are plotted as blue squares and red diamonds, respectively

3 Future Perspectives

In this chapter, an example of a multi-scale, multi-physics simulation of a SOFC fuel pole is discussed. Multi-scale, multi-physics simulations involving electrochemical reactions are no longer conceptual but have been demonstrated to be feasible in practice. It was also shown that it is important to understand the reaction mechanism taking into account the real-system TPB structure at the atomistic level. Each of these breakthroughs took several years to complete. For research and development for large-scale penetrations of SOFC into society, which is advancing at an accelerated rate, it is desirable to achieve a rapid material design that goes beyond the elucidation of reaction mechanisms. The realization of more advanced simulations incorporating high-throughput screening and machine learning will be a remaining challenge for future research.

Examples of first-principles calculations of the real-system structure of PEFC cathode catalysts were also discussed. Although we should acknowledge the usefulness of studies using simple cluster and slab models, it was shown that the inherent stability and function of nanoparticles can be clarified by using models of real-system structures. It was also presented an example of clarifying the electronic structures of nanoparticles on support. As future challenges, it is important not only to expand the approach to various combinations of alloys and support materials but also to consider the effects of surface adsorbed species and dynamics. For this purpose, as in the case of SOFC, it is essential to realize high-throughput simulations.

In both cases, it goes without saying that close collaborations with experimental observations are important. Figure 15 shows the future perspectives of the computational study of real-system electrode interface in fuel cells. In the future, the author dreams that the electronic state, dynamics, and reaction activity of various electrodes will be clarified one after another based on the actual electrode interface structure and will serve as a basis for the development of new materials.

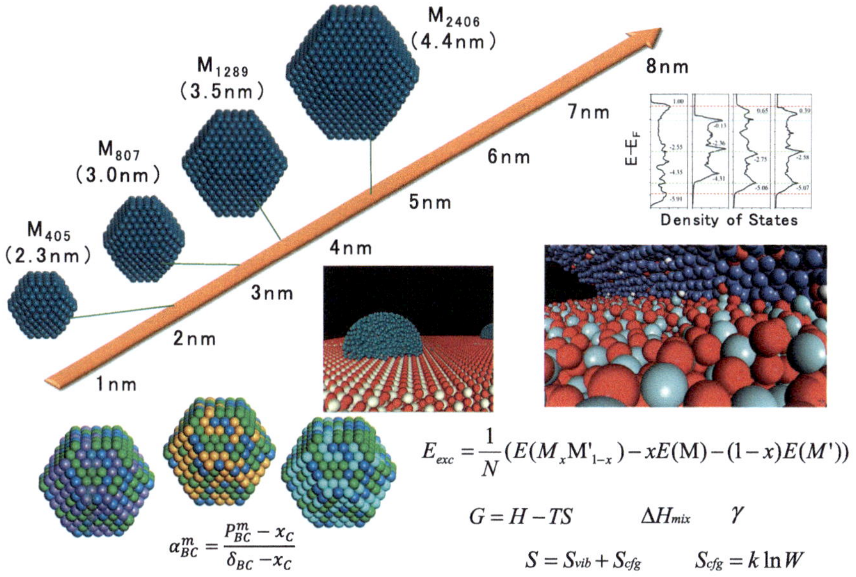

Fig. 15 Future perspectives of the computational study of real-system electrode interface

References

1. For example, Bacigalupo A, Morini L, Piccolroaz A (2014) Effective elastic properties of planar SOFCs: a non-local dynamic homogenization approach. Int J Hydrogen Energy 39:15017–15030
2. Gorski A, Yurkiv C, Starukhin D, Volpp H-R (2011) H_2O chemisorption and H_2 oxidation on yttria-stabilized zirconia: density functional theory and temperature-programmed desorption studies. J Power Sources 196:7188–7194
3. Chaopradith DT, Scanlon DO, Catlow CRA (2015) Adsorption of water on yttria-stabilized zirconia. J Phys Chem C 119:22526–22533
4. Ammal SC, Heyden A (2012) Combined DFT and microkinetic modeling study of hydrogen oxidation at the Ni/YSZ anode of solid oxide fuel cells. J Phys Chem Lett 3:2767–2772
5. Liu S, Ishimoto T, Monder DS, Koyama M (2015) First-principles study of oxygen transfer and hydrogen oxidation processes at the Ni−YSZ-gas triple phase boundaries in a solid oxide fuel cell anode. J Phys Chem C 119:27603–27608
6. Blaylock DW, Ogura T, Green WH, Beran GJO (2009) Computational investigation of thermochemistry and kinetics of steam methane reforming on Ni(111) under realistic conditions. J Phys Chem C 113:4898–4908
7. Wang GC, Tao SX, Bu XH (2006) A systematic theoretical study of water dissociation on clean and oxygen-preadsorbed transition metals. J Catal 244:10–16
8. Shishkin M, Ziegler T (2010) Hydrogen oxidation at the Ni/yttria stabilized zirconia interface: a study based on density functional theory. J Phys Chem C 114:11209–11214
9. Shishkin M, Ziegler T (2014) Direct modeling of the electrochemistry in the three-phase boundary of solid oxide fuel cell anodes by density functional theory: a critical overview. Phys Chem Chem Phys 16:1798–1808
10. Cucinotta CS, Bernasconi M, Parrinello M (2011) Hydrogen oxidation reaction at the Ni−YSZ anode of solid oxide fuel cells from first principles. Phys Rev Lett 107:206103

11. Fu Z, Wang M, Zuo PX, Yang Z, Wu R (2014) Importance of oxygen spillover for fuel oxidation on Ni/YSZ anodes in solid oxide fuel cells. Phys Chem Chem Phys 16:8536–8540
12. Mizusaki J, Tagawa H, Saito T, Yamamura T, Kamitani K, Hirano K, Ehara S, Takagi T, Hikita T, Ippommatsu M et al (1994) Kinetic studies of the reaction at the nickel pattern electrode on YSZ in H_2–H_2O atmospheres. Solid State Ionics 70–71:52–58
13. Bieberle A, Meier LP, Gauckler LJ (2001) The electrochemistry of Ni pattern anodes used as solid oxide fuel cell model electrodes. J Electrochem Soc 148:A646–A656
14. Utz A, Störmer H, Leonide A, Weber A, Ivers-Tiffée E (2010) Degradation and relaxation effects of Ni patterned anodes in H_2–H_2O atmosphere. J Electrochem Soc 157:B920–B930
15. Yao W, Croiset E (2015) Stability and electrochemical performance of Ni/YSZ pattern anodes in H_2/H_2O atmosphere. Can J Chem Eng 93:2157–2167
16. Goodwin DG, Zhu H, Colclasure AM, Kee RJ (2009) Modeling electrochemical oxidation of hydrogen on Ni–YSZ pattern anodes. J Electrochem Soc 156:B1004–B1021
17. Vogler M, Bieberle-Hutter A, Gauckler L, Warnatz J, Bessler WG (2009) Modelling study of surface reactions, diffusion, and spillover at a Ni/YSZ patterned anode. J Electrochem Soc 156:B663–B672
18. Kohno H, Liu S, Ogura T, Ishimoto T, Monder DS, Karan K, Koyama M (2013) Detailed transport-reaction models for SOFC Ni–YSZ patterned anodes: a critical inquiry. ECS Trans 57(1):2821–2830
19. Liu S, Liu S, Saha LC, Iskandarov AM, Jiao Z, Hara S, Ishimoto T, Tada T, Umeno Y, Shikazono N, Matsumura S, Koyama M (2017) Multi-scale, multi-physics approach for solid oxide fuel cell anode reaction. ECS Trans 78(1):2835–2844
20. Liu S, Muhammad A, Mihara K, Ishimoto T, Tada T, Koyama M (2017) Predictive micro–kinetic model for solid oxide fuel cell patterned anode: based on extensive literature survey and exhaustive simulations. J Phys Chem C 121:19069–19079
21. Hecht ES, Gupta GK, Zhu H, Dean AM, Kee RJ, Maier L, Deutschmann O (2005) Methane reforming kinetics within a Ni–YSZ SOFC anode support. Appl Catal A 295:40–51
22. Cao GX, Nabighian E, Zhu XD (1997) Diffusion of hydrogen on Ni(111) over a wide range of temperature: exploring quantum diffusion on metals. Phys Rev Lett 79:3696–3699
23. Liu S-S, Saha LC, Iskandarov A, Ishimoto T, Yamamoto T, Umeno Y, Matsumura S, Koyama M (2019) Atomic structure observations and reaction dynamics simulations on triple phase boundaries in solid-oxide fuel cells. Commun Chem 248
24. Li L, Larsen AH, Romero NA, Morozov VA, Glinsvad C, Abild-Pedersen F, Greeley J, Jacobsen KW, Nørskov JK (2013) Investigation of catalytic finite-size-effects of platinum metal clusters. J Phys Chem Lett 4:222–226
25. Rocabado DSR, Ishimoto T, Koyama M (2019) The effect of $SnO_2(110)$ supports on the geometrical and electronic properties of platinum nanoparticles. SN Appl Sci 1:1485
26. Ruiz-Serrano Á, Skylarisa C-K (2013) A variational method for density functional theory calculations on metallic systems with thousands of atoms. J Chem Phys 139:054107
27. Nanba Y, Ishimoto T, Koyama M (2017) Structural stability of ruthenium nanoparticles: a density functional theory study. J Phys Chem C 121:27445–27452
28. Farkaš B, de Leeuw NH (2020) Towards a morphology of cobalt nanoparticles: size and strain effects. Nanotechnol 31:195711
29. Laasonen K, Panizon E, Bochicchio D, Ferrando R (2013) Competition between icosahedral motifs in AgCu, AgNi, and AgCo Nanoalloys: a combined atomistic–DFT study. J Phys Chem C 117:26405–26413
30. Kent PRC (2008) Computational challenges of large-scale, long-time, first-principles molecular dynamics. J Phys Conf Ser 125:012058
31. Ishimoto T, Inadomi Y, Honda H, Koyama M (2015) Parallel performance analysis for electronic structure calculation of metal nanoparticles. J Comput Chem Jpn 14:52–53
32. Li L, Abild-Pedersen F, Greeley J, Nørskov JK (2015) Surface tension effects on the reactivity of metal nanoparticles. J Phys Chem Lett 6:3797–3801
33. Shao M, Peles A, Shoemaker K (2011) Electrocatalysis on platinum nanoparticles: particle size effect on oxygen reduction reaction activity. Nano Lett 11:3714–3719

34. Ishimoto T, Koyama M (2016) Electronic structure and phase stability of PdPt nanoparticles. J Phys Chem Lett 7:736–740
35. Nanba Y, Rivera DS, Ishimoto T, Koyama M (2016) First-principles study on alloy nanoparticles for polymer electrolyte fuel cell catalyst. ECS Trans 75(14):717–721
36. Rivera Rocabado DS, Nanba Y, Koyama M (2020) Electronic structure and phase stability of Pt_3M (M = Co, Ni, and Cu) bimetallic nanoparticles. Comput Mater Sci 184:109874
37. Nanba Y, Koyama M. Element-based generalized coordination number for predicting the oxygen binding energy on Pt_3M (M = Co, Ni, or Cu) alloy nanoparticles. ACS Omega accepted
38. Calle-Vallejo F, de Morais RF, Illas F, Loffreda F, Sautet P (2019) Affordable estimation of solvation contributions to the adsorption energies of oxygenates on metal nanoparticles. J Phys Chem C 123:5578–5582
39. Ellaby T, Briquet L, Sarwar M, Thompsett D, Skylaris C-K (2019) Modification of O and CO binding on Pt nanoparticles due to electronic and structural effects of titania supports. J Chem Phys 151:114702
40. Aarons J, Verga LG, Hine NDM, Skylaris C-K (2019) Atom-projected and angular momentum resolved density of states in the ONETEP code. Electron Struct 1:035002
41. Foster DM, Pavloudis Th, Kioseoglou J, Palmer RE (2019) Atomic-resolution imaging of surface and core melting in individual size-selected Au nanoclusters on carbon. Nat Commun 10:2583
42. Liu S, Huang S (2019) Structure engineering of Cu-based nanoparticles for electrochemical reduction of CO_2. J Catal 375:234–241
43. Aarons J, Jones L, Varambhia A, MacArthur KE, Ozkaya D, Sarwar M, Skylaris C-K, Nellist PD (2017) Predicting the oxygen-binding properties of platinum nanoparticle ensembles by combining high-precision electron microscopy and density functional theory. Nano Lett 17:4003–4012
44. Back S, Yeom MS, Jung Y (2015) Active sites of Au and Ag nanoparticle catalysts for CO_2 electroreduction to CO. ACS Catal 5:5089–5096
45. Ishimoto T, Koyama M (2018) Theoretical study of tetrahedral site occupation by hydrogen in Pd nanoparticles. J Chem Phys 148:034705
46. Nanba Y, Koyama M (2019) NO adsorption on 4d and 5d transition metal (Rh, Pd, Ag, Ir, and Pt) nanoparticles: density functional theory study and supervised learning. J Phys Chem C 123:28114–28122
47. Soini TM, Ma X, Aktürk OÜ, Suthirakun S, Genest A, Rösch N (2016) Extending the cluster scaling technique to ruthenium clusters with hcp structures. Surf Sci 643:156–163
48. Chepkasov IV, Visotin MA, Kovaleva EA, Manakhov AM, Baidyshev VS, Popov ZI (2018) Stability and electronic properties of PtPd nanoparticles via MD and DFT calculations. J Phys Chem C 122:18070–18076
49. Mokkath JH (2017) Evolution of the magnetic and optical properties in $Co_{core}Au_{shell}$ and $(CoRh)_{core}Au_{shell}$ core-shell nanoparticles. J Phys Chem C 121:24798–24803
50. Verga LG, Aarons J, Sarwar M, Thompsett D, Russell AE, Skylaris C-K (2018) DFT calculation of oxygen adsorption on platinum nanoparticles: coverage and size effects. Faraday Discuss 208:497–522
51. Anderson RM, Zhang L, Loussaert JA, Frenkel AI, Henkelman G, Crooks RM (2013) An experimental and theoretical investigation of the inversion of Pd@Pt Core@Shell dendrimer-encapsulated nanoparticles. ACS Nano 10:9345–9353
52. Krüger S, Vent S, Nörtemann F, Staufer M, Rösch N (2001) The average bond length in Pd clusters Pdn, n = 4–309: a density-functional case study on the scaling of cluster properties. J Chem Phys 115:2082
53. Nava P, Sierka M, Ahlrichs R (2003) Density functional study of palladium clusters. Phys Chem Chem Phys 5:3372–3381
54. Massalski TB, Okamoto H, Subramanian PR, Kacprzak L (eds) (1990) Binary alloy phase diagrams, 2nd edn. ASM International, Ohio, USA
55. For example, Jung N, Chung DY, Ryu J, Yoo SJ, Sung YE (2014) Pt-based nanoarchitecture and catalyst design for fuel cell applications. Nano Today 9:433–456

56. Kobayashi H, Yamauchi M, Kitagawa H, Kubota Y, Kato K, Takata M (2010) Atomic-level Pd−Pt alloying and largely enhanced hydrogen-storage capacity in bimetallic nanoparticles reconstructed from core/shell structure by a process of hydrogen absorption/desorption. J Am Chem Soc 132:5576–5577
57. Ferrando R, Fortunelli A, Rossi G (2005) Quantum effects on the structure of pure and binary metallic nanoclusters. Phys Rev B 72:085449
58. Ishimoto T, Koyama M (2020) First-principles study of the stability of bimetallic PdPt nanoparticles under a finite temperature. Preprint at arXiv:2007.06137
59. Liu HB, Pal U, Perez R, Ascencio JA (2006) Structural transformation of Au−Pd bimetallic nanoclusters on thermal heating and cooling: a dynamic analysis. J Phys Chem B 110:5191–5195

First Principles Simulation of Microstructures for Metals/Alloy Materials in Fuel Cell

Ke Tong, Fei Ye, and Zhipeng Li

Abstract The metal/alloy materials have been widely used in fuel cell applications. One of these materials is austenitic steel, which can be used as the interconnector or bipolar plate. The microstructures are elucidated through microscopy observation combined with first principle calculations. It was noted that, in the high nitrogen austenitic steel, there were short-range ordered structures formed through nano-scale clusters. The simulation of such clusters was conducted thereby, and the formation features were analyzed to understand the atomic structures and phase stability. The cluster structure is affected by both the local lattice distortion around the cluster and the Coulombic interaction in the cluster. It was suggested that the stabilization of the metastable phase is mainly a chemically driven mechanism by Cr and N atoms forming short-range ordered clusters.

Keywords Microstructure · Ordering · Cluster · Stainless steel · Simulation · Fuel cell

K. Tong (✉)
Center for High Pressure Science, State Key Laboratory of Metastable Materials Science and Technology, Yanshan University, No. 438 West Hebei Avenue, Qinhuangdao 066004, China
e-mail: tongke@ysu.edu.cn

F. Ye
Department of Materials Science and Engineering, Southern University of Science and Technology, 1088 Xueyuan road, Nanshan district, Shenzhen 518055, China
e-mail: yef3@sustech.edu.cn

Z. Li (✉)
Institute of Flexible Electronics, Northwestern Polytechnical University, 1 Dongxiang Road, Chang'an District, Xi'an 710072, China
e-mail: iamzpli@nwpu.edu.cn

T. Mori et al. (eds.), *Design of Active Key Interfaces in Fuel Cells*,
https://doi.org/10.1007/978-981-97-5706-0_6

1 Introduction

The components in fuel cell stacks involve a variety of materials. Great efforts have been made to develop new materials or improve the properties of existing materials for the fuel cell application. As a common saying, the properties of materials are closely related to their microstructure. The microstructures of materials can be studied from atomic scale to micron or even millimeter scale. With the development of material characterization and computer simulation technologies, material research and development has been able to be carried out on atomic and nano scales, so that the ability of material optimization and design is greatly improved.

In this chapter, the research work of microstructures of the metal/alloy materials in fuel cells, as well as the influence of microstructures on the relative properties, are reviewed. The austenitic nitrogen steel, which can be used as the interconnect or bipolar plate in proton exchange membrane fuel cell (PEMFC), has been introduced in this chapter. In order to increase surface hardness, reduce wear rate, and enhance fatigue life without any loss of the corrosion resistance of austenitic stainless steel, a supersaturated nitrided layer has been produced by plasma immersion ion implantation at low processing temperatures in a range of 230–450 °C [1].

It was well known that the mechanical and electrochemical properties of solid solutions are mainly determined by the behavior of solutes and dislocation properties. In high nitrogen steel, the distribution of solute atoms is far from that typical for dilute solid solutions where solute atoms do not interact with each other. The atoms of alloying elements reveal a tendency to have some preference in relation to their nearest atomic neighborhood. The addition of nitrogen into Fe–Cr–Ni austenitic steel has profound effects on the mechanical and chemical properties. These properties are closely related to the atomic ordering of the solute elements. The subtle microstructures at atomic and nano scales of these materials are elucidated thereafter in this chapter.

2 Experimental and Theoretical Methods

2.1 Experimental Methods

AISI 304L austenitic stainless steel was selected as the raw material. The main chemical compositions (wt. %) were Cr (18.11), Ni (8.14), Mn (1.04), Si (0.43), N (0.06), P (0.03), C (0.02), and the balance of Fe. A plasma source ion nitriding apparatus with an electron cyclotron resonance microwave plasma source was used to prepare the supersaturated nitrided layer at 400 °C for 4 h. After grinding, mechanical polishing, and following ion-thinning, the thin foils of the nitrided layer with a thickness of 30 μm were prepared for TEM observation. The TEM observations were performed on FEI Tecnai $G^2$20 and FEI Tecnai F30 microscopes.

2.2 Theoretical Methods

2.2.1 Simulation of Austenitic Nitrogen Steel

The first-principles calculations of the austenitic nitrogen stainless steel (Fe–Cr–Ni–N system) were conducted using the Vienna *ab initio* simulation package [2]. It is a plane wave pseudo-potential code and performs fully self-consistent density functional theory calculations to solve the Kohn–Sham equations. The projector-augmented wave method was used to describe the electron–ion interaction [3]. The generalized gradient approximation with the Perdew and Wang functional was used as the exchange–correlation functional [4]. To describe the magnetization of the system exactly, the Vosko-Wilk-Nusair interpolation was also applied [5].

Because Ni atoms are in the metallic state and their interactions with other atoms are similar to Fe atoms in the austenitic nitrogen steel, the interaction between Ni and N can be neglected in the research of the short-range ordered cluster structure, and the Fe–Cr–N alloy system was investigated. The structures of short-range ordered Fe–Cr–N clusters were calculated in $4 \times 4 \times 4$ face-centered cubic (f.c.c.) supercells of 256 (+1, +2, …) atoms. The atomic positions were free to relax while the supercell dimensions were fixed at their equilibrium values. An energy cutoff of 400 eV, energy convergence of 10^{-4} eV/cell, and $2 \times 2 \times 2$ k-point Monkhorst–Pack grid were chosen for the electronic self-consistency calculation. The structures were considered to be relaxed when the forces on all atoms had fallen below 0.03 eV/Å. Convergence test calculations showed that these parameters achieved accuracy in the electronic energy below 2 meV/atom, which was sufficiently small for the present study.

Localized moments with a set of stable, ordered collinear magnetic structures have achieved significant success in modeling the paramagnetism of Fe–Cr–Ni alloy [6]. This most stable ordered magnetic state was named antiferromagnetic double layer collinear magnetic reference state (denoted as afmD), in which the local magnetic moments of Fe atoms on every two (001) layers were in the same direction as shown in Fig. 1. The calculations performed in the most stable ordered magnetic state of f.c.c. Fe are introduced in this chapter. In fact, it has been verified that general conclusions for the structures of the Fe–Cr-N clusters can be deduced from the calculations performed in the afmD state [7].

The relative stability of the clusters was also examined by comparing their binding energies. The binding energy is defined as

$$E_b\left(\mathrm{Fe}_i\mathrm{Cr}_j\mathrm{N}_k\right) = jE_f(\mathrm{Cr}) + k\,E_f(\mathrm{N}) - E_f\left(\mathrm{Fe}_i\mathrm{Cr}_j\mathrm{N}_k\right), \quad (1)$$

where i, j, and k are the number of Fe, Cr, and N atoms in the cluster. The terms on the right side of the equation are the formation energy of isolated substitutional Cr, isolated interstitial N, and the cluster. They are defined as

$$E_f(\mathrm{Cr}) = E(\mathrm{Fe}_{m-1}\mathrm{Cr}) - (m-1)/mE(\mathrm{Fe}_m) - \left(1/m'\right)E(\mathrm{Cr}_{m'}), \quad (2)$$

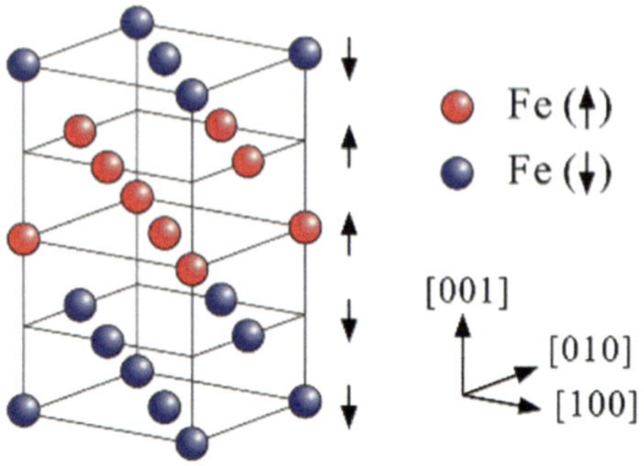

Fig. 1 The afmD structure model in which the Fe atoms in two f.c.c. unit cells have up, up, down, down ordering of magnetic moments in adjacent magnetic planes. (Reproduced from Ref. [8])

$$E_f(\mathrm{N}) = E(\mathrm{Fe}_m\mathrm{N}) - E(\mathrm{Fe}_m) - (1/2)E(\mathrm{N}_2), \quad (3)$$

$$E_f(\mathrm{Fe}_i\mathrm{Cr}_j\mathrm{N}_k) = E(\mathrm{Fe}_{m-j}\mathrm{Cr}_j\mathrm{N}_k) - (m-j)/mE(\mathrm{Fe}_m) - (j/m')E(\mathrm{Cr}_{m'}) - (k/2)E(\mathrm{N}_2), \quad (4)$$

where $E(\mathrm{Fe}_{m\text{-}j}\mathrm{Cr}_j\mathrm{N}_k)$ is the total energy of the system with a cluster. $E(\mathrm{Fe}_m)$, $E(\mathrm{Cr}_{m'})$ and $E(\mathrm{N}_2)$ are total energies of pure substances as the reference states, that is to say, the Fe in the corresponding magnetic state, Cr in its body-centered cubic (b.c.c.) antiferromagnetic ground state, and N_2 molecule in vacuum. m is the number of host Fe atoms (256 atoms) and m' is the number of Cr atoms in the crystal used as reference state.

3 Results and Discussion

3.1 Microstructural Observations of Clusters in High Nitrogen Austenitic Stainless Steel

The structural features of high nitrogen austenitic stainless steel were investigated by TEM. Strip-shaped regions with a width of 3–25 nm can be seen from the bright-field image (Fig. 2a) and dark-field image (Fig. 2b). In addition, abundant structural information was also exhibited in the SAED patterns. There are five typical features in Fig. 3a–c, marked by large circles, smaller circles, black spots, squares, and triangles, respectively. The large bright spots identified by the large circles in Fig. 3d–f are the reflections belonging to the f.c.c. matrix. Additionally, three sets of smaller patterns can also be observed in Fig. 3a–c. The spots identified by the squares in Fig. 3d are attributed to the twins occasionally formed in the f.c.c. matrix. The spots identified by the smaller circles in Fig. 3d–f are the forbidden spots of the f.c.c. structure. These forbidden spots have been reported in previous studies [9, 10], possibly formed due to N atoms at specific octahedral sites [9–11]. The spots identified by the black spots in Fig. 3d and e are attributed to the strip-shaped regions shown in Fig. 2. According

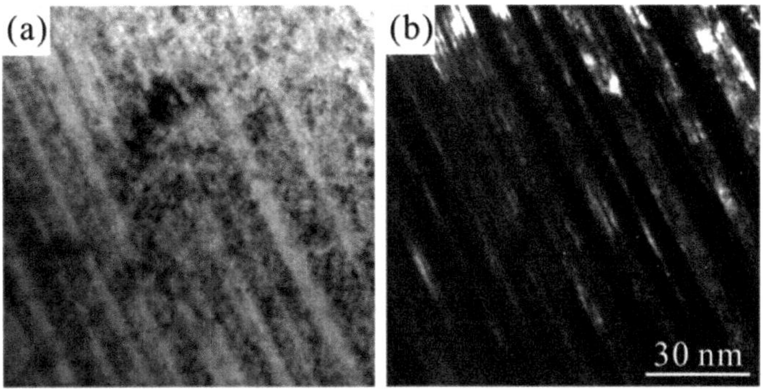

Fig. 2 **a** Bright-field image and **b** the corresponding dark-field image taken with the incident beam parallel to <110> $_{f.c.c.}$ zone axis and extra reflection $(001)_{h.c.p.}$ (Reproduced from Ref. [12])

to the periodicity of the reflections, the black spots overlapped with the large circles were drawn correspondingly in Fig. 3e and f. Finally, several very weak spots can also be noticed in Fig. 3c (identified by triangles in Fig. 3f). They could be the high-order Laue-zone diffraction spots of the strip-shaped regions.

Inspired by reflections from the strip-shaped regions in the SAED patterns, it is reasonable to believe that there is an ordered structure different from the f.c.c. matrix. Thus, the atomic structure of the strip-shaped regions was further explored using HRTEM. The results are given in Fig. 4. Obviously, the Fourier transform image (Fig. 4b) of the square area in Fig. 4a contains a set of spots, and these spots are consistent with the spots in Fig. 3a (as identified by the black spots in Fig. 3d). Moreover, most strip-shaped regions essentially consist of a series of highly dense SFs on every second $\{111\}_{f.c.c.}$ atomic plane, as shown in Fig. 4a. Then, the proportion of the strip-shaped region was counted from a large number of TEM images, and then, the SFs probability could be estimated. The estimated SFs probability is about 0.17, indicating the SFs form every sixth $\{111\}_{f.c.c.}$ planes. This result is basically consistent with the SF probability used for fitting the XRD patterns [8, 13].

Based on the HRTEM study, a microstructure model considering clustered SFs embedded in f.c.c. structure was built for the regions (Fig. 4c). In this model, the metal atoms, Fe, Ni, and Cr, randomly occupy the lattice sites, and N atoms randomly occupy the octahedral interstitial sites. It can be seen that the arrangement of SFs on every second {111}f.c.c. plane apparently produces a local hexagonal-close-packing (h.c.p.) stacking ABABAB. It is noted that this structure is the same as that of the ε-martensite Fe_2N_{1-x} with space group $P6_3/mmc$.

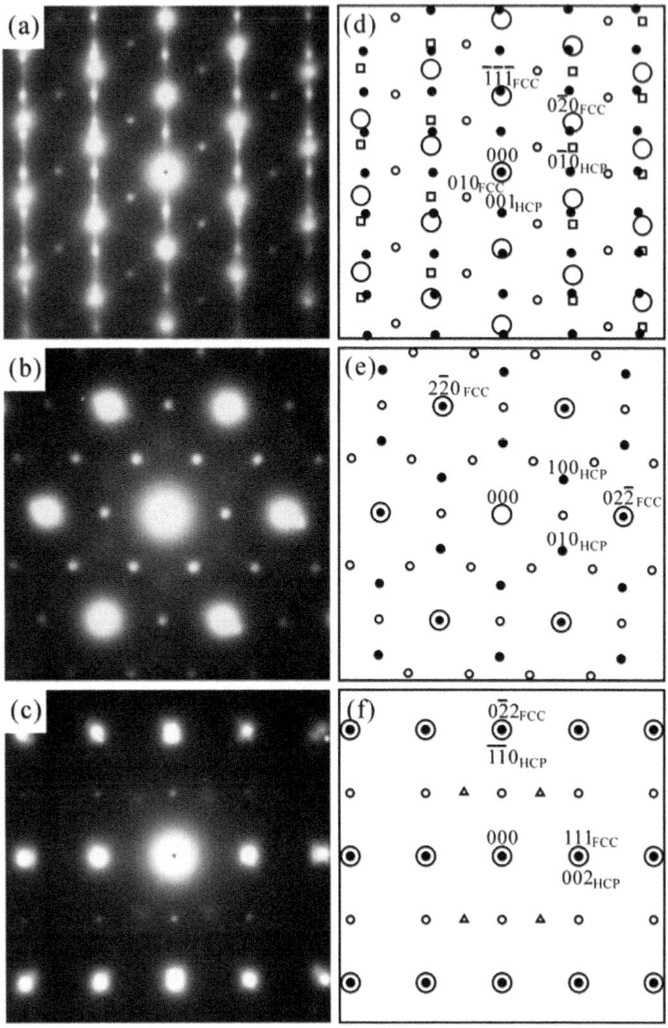

Fig. 3 SAED patterns from the area in Fig. 2 with zone axes of **a** $[10\bar{1}]$, **b** [111], and **c** $[2\bar{1}\bar{1}]$zone axis. **d**–**f** are the corresponding schematic diagrams. (○) Reflections of the f.c.c. structure; (□) Reflections of the twins of f.c.c. structure; (●) Reflections of the strip-shaped regions; (△) High-order Laue-zone diffraction spots of the strip-shaped regions

3.2 *Short-Range Ordered Structure in High Nitrogen Austenitic Stainless Steel*

Generally, the N atom would occupy the octahedral interstitial site in the crystal lattice of the austenitic nitrogen steel when its concentration is less than the solubility limit. The short-range ordering of N atoms in austenitic nitrogen steel has

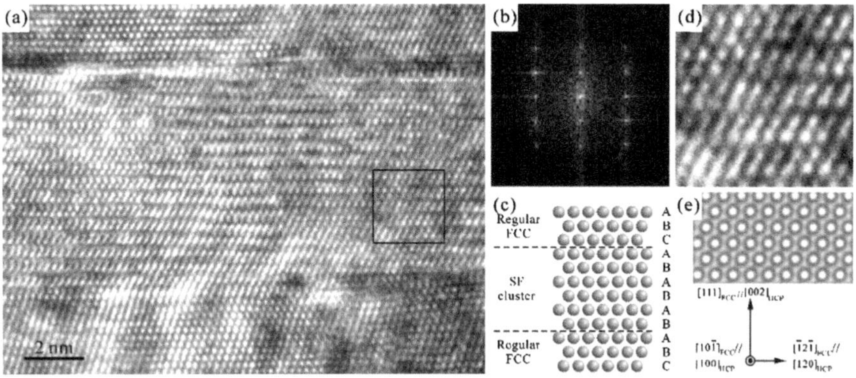

Fig. 4 **a** HRTEM image of a strip-shaped region with highly dense stacking faults observed along $[10\bar{1}]_{f.c.c.}$ ($[100]_{h.c.p.}$) direction. **b** Fourier transform image of the square area in (**a**). **c** A structure model of clustered SFs embedded in f.c.c. structure. **d** Enlarged HRTEM image of the square area in (**a**). **e** Simulated HRTEM image. The simulation was conducted by applying the parameters including the acceleration voltage of 300 kV, spherical aberration coefficient of 0.5 mm, chromatic aberration coefficient of 1.4 mm, specimen thickness of 59 nm, and focus value of −42 nm

been studied by computer simulations. Monte Carlo simulations concluded that the ordering was confined to the Cr and N atoms in the Fe–Cr–Ni–N alloy system [14, 15]. Furthermore, an octahedral structure model was constructed, in which an N atom was at the center of the octahedral and Fe, Cr, and Ni atoms randomly at the neighboring sites. Obviously, it is debatable that the random distribution of metal atoms in this structure model is due to the well-known strong interaction between Cr and N atoms.

As mentioned above, since Cr atoms appear to be in the Cr–N bond and Ni atoms are in the metallic state like Fe in the austenitic nitrogen steel, the interaction between Ni and N can be neglected in the study of the short-range ordered cluster structure. Therefore, the Fe–Cr–N alloy system was actually considered for simplification. The Fe–Cr–N cluster structures were systematically investigated by first-principles calculation. The detailed configurations of the octahedral clusters and their combinations were examined. The local lattice distortion around the clusters and charge transfer in the clusters were qualitatively evaluated to understand their effects on the cluster configuration.

3.2.1 Structures of $Fe_{6-n}Cr_nN$ Octahedral Clusters

Since the octahedral clusters can be formed in the austenitic nitrogen steel, it is suggested that the short-range ordered structure in the high nitrogen steel ought to take the octahedral cluster into consideration. Moreover, these octahedral clusters may combine into structures with larger scales with an increase in the N concentration. Firstly, the preferred occupation and formation energies of isolated Cr and N atoms in f.c.c. Fe were calculated. For isolated Cr atom, a substitutional site is preferred and the corresponding energy is 0.260 eV. For isolated N atom, it clearly favors octahedral interstitial site, and the solute energy is –0.269 eV. These results are consistent with the previous calculations using the same afmD model within an error of 0.013 eV [15].

Then, the Cr–N interaction was examined by evaluating the binding energies of the Cr–N clusters in which the Cr and N atoms are located at the first, second, and third nearest-neighbor (1nn, 2nn, and 3nn) sites to each other, and the results are listed in Table 1. The binding energy always decreases with their distance, no matter whether the Cr and N atoms are on the planes with the same or different magnetic moments. The 1nn clusters have relatively high binding energy, while the binding energies of the 2nn and 3nn clusters are close to zero. It means that the interaction between Cr and N atoms mainly localizes within 1.864 Å, i.e., in the 1nn shell of the N atom. Based on the octahedral structure model proposed by Monte Carlo calculations [14, 15], the 1nn Cr–N cluster structure can be described as a Fe_5Cr_1N octahedron. When the number of Cr atoms increases up to 6, they still prefer the 1nn sites to the N atom. Therefore, these octahedral clusters can be denoted as $Fe_{6\text{-}n}Cr_nN$ ($0 \leq n \leq 6$) clusters.

Furthermore, the Cr atoms are not randomly located at the six 1nn sites of the N atom but have specific configurations due to the strong interaction between Cr and N atoms. When the number of Cr atoms is between 2 and 4, these configurations can be classified into two types according to the sites of Cr atoms. One type possesses more conjugated Cr atoms in the configurations, and another has less conjugated Cr atoms, as shown in Fig. 5a. In this chapter, conjugated Cr atoms are at the opposite octahedron vertexes, i.e., along <100> direction. From the viewpoint of energy, binding energies of the $Fe_{6\text{-}n}Cr_nN$ clusters increase with n, indicating that the clusters with more Cr atoms are energetically favored. Furthermore, when n is between 2 and 4, the clusters with more conjugated Cr atoms have higher binding energy than those with less conjugated Cr atoms, as shown in Fig. 5b. Therefore, the clusters in which the Cr atoms distribute in pairs around the N atom are energetically favored.

Generally, the formation of $Fe_{6\text{-}n}Cr_nN$ clusters would lead to a local lattice distortion around the clusters, the variation of projected density of states (PDOSs), and charge transfer in the clusters. In this chapter, the lattice distortion was evaluated by calculating the increment of the volume (ΔV) of atomic coordination polyhedron

Table 1 The formation energies E_f and the binding energies E_b of the 1nn, 2nn, and 3nn Cr–N clusters. The Cr and N atoms are located on the planes with the same or different magnetic moments. [7] (Reproduced from Ref.)

	Direction	Distance (Å)	E_b (eV) Same magnetic moments	E_b (eV) Different magnetic moments
1nn	<001>/2	1.864	0.223	0.259
2nn	<111>/2	3.067	0.044	0.022
3nn	<012>/2	3.918	0.036	0.013

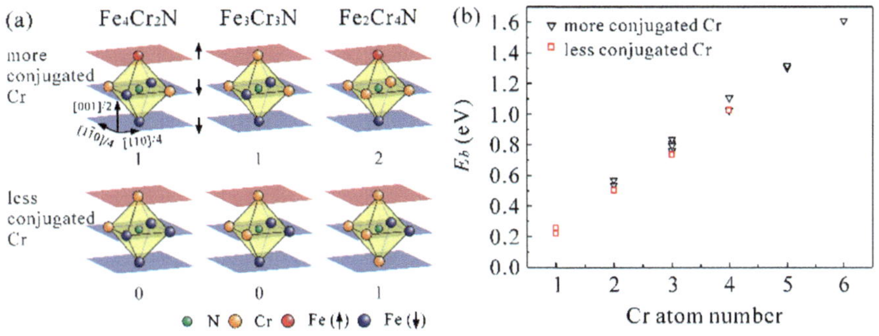

Fig. 5 **a** The $Fe_{6\text{-}n}Cr_nN$ clusters are classified into two types according to more or less conjugated Cr atoms. The numbers of conjugated Cr atoms are given below the clusters. **b** The binding energy E_b as a function of the number of Cr atoms in a $Fe_{6\text{-}n}Cr_nN$ cluster. (Reproduced from Ref. [7])

(ACP) as a function of the number of Cr atoms. The ΔV was defined as follows:

$$\Delta V = V_{relaxed} - V_{ref}, \tag{5}$$

where $V_{relaxed}$ and V_{ref} are the volumes of ACP in the relaxed structure and reference structure, respectively. The relaxed supercell structure containing only Fe in respective magnetic states was chosen as the reference structure. The ACPs formed by 1nn, 2nn, and 3nn atoms surrounding the N atom were considered. The average ΔV values for 1nn, 2nn, and 3nn ACPs were calculated by averaging the values of the $Fe_{6\text{-}n}Cr_nN$ clusters with all the possible configurations (Fig. 6). For 1nn and 2nn ACPs, these average ΔV values are all positive and increase monotonously. Compared with the ΔV of 2nn ACP and 3nn ACP, the ΔV of 1nn ACP is the smallest, implying no obvious lattice distortion in the 1nn shell around the N atom. In contrast, the ΔV and its variation gradient of 2nn ACP are much larger, indicating the lattice distortion is mainly located in the 2nn shell around the N atom. The ΔV values of 3nn ACP are between those of 1nn and 2nn ACPs, and the curve is relatively flat. The increase in ΔV with the number of Cr atoms corresponds to larger lattice distortion and, hence, lower energetical preference. Interestingly, it is inverse to the effect of the number of Cr atoms, as shown in Fig. 5b. These results indicate that the lattice distortion is not the main factor determining the number of Cr atoms in the $Fe_{6\text{-}n}Cr_nN$ cluster.

In the following, the energetical preference of cluster is tried to understand from the viewpoint of the electron. Figure 7 shows the PDOSs of the Fe_4Cr_2N clusters, in which the Cr atoms are at the opposite vertexes or at one edge of the octahedron. When the octahedral cluster contains the conjugated Cr atoms (Fig. 7a), the majority hybridization is the *spd* hybridization among the N $2p_x$ orbital, Cr 4 s orbital, and Cr 3 d_{x^2} orbital, indicating a Cr–N σ bonding molecular orbital is formed. When the Cr atoms occupy one edge of the octahedron, the *spd* hybridization among the N $2p_x$ orbital, Cr 4s orbital, and Cr 3 d_{x^2} orbital is not dominant, as shown in Fig. 7b, meaning that π bond characteristic is more dominant than σ bond characteristic in this case.

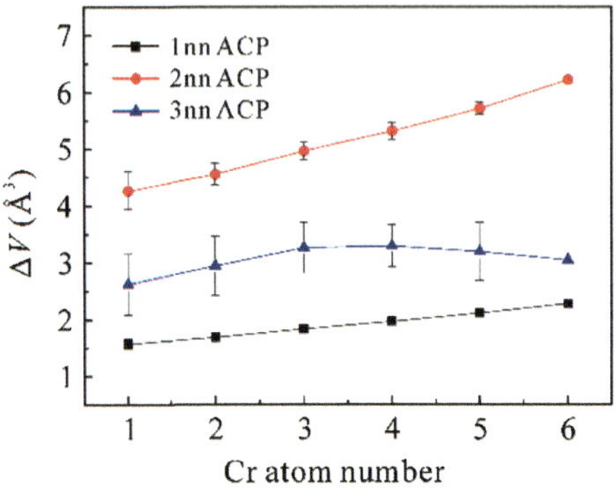

Fig. 6 The average ΔV for the 1nn, 2nn, and 3nn ACPs as a function of the number of Cr atoms in a $Fe_{6-n}Cr_nN$ cluster. The error bars represent the deviation from the average values [7]

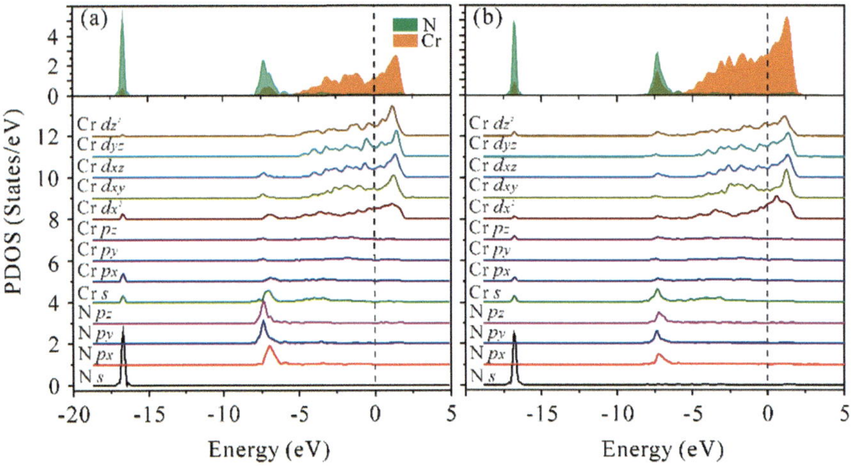

Fig. 7 The PDOSs of the Fe_4Cr_2N clusters in which the Cr atoms are **a** at the opposite vertexes and **b** at one edge of the octahedron, respectively. The local electron density of states of Cr *s*, *p*, *d*-orbitals and N *s*, *p*-orbitals were deconvoluted into sub-orbitals. The Fermi energy is chosen as 0 eV [7]

Furthermore, we quantitatively examined the charges of the atoms in the $Fe_{6-n}Cr_nN$ clusters using Bader analysis [16], and the results are listed in Table 2. Since there is more than one Fe or Cr atom in the clusters, the average charges of these atoms are given. The average charge of Fe atoms obviously decreases with Cr atoms from 0.211*e* to 0.163*e*, while that of Cr atoms decreases slightly from 0.830*e* to 0.808*e*. At the same time, the N atom in the $Fe_{6-n}Cr_nN$ cluster takes more electrons, and the charge varies from −1.559*e* to −1.802*e*. Consequently, the Coulombic attraction between Cr and N atoms becomes stronger when the number of Cr atoms in the cluster increases. Therefore, the $Fe_{6-n}Cr_nN$ cluster prefers to contain more Cr atoms.

Table 2 The average Bader charges difference (in *e*) of the Fe, Cr, and N atoms in the $Fe_{6\text{-}n}Cr_nN$ cluster. (Reproduced from Ref. [7])

n	Fe	Cr	N
0	0.211	–	−1.559
1	0.182	0.830	−1.593
2	0.170	0.831	−1.646
3	0.167	0.823	−1.709
4	0.149	0.817	−1.753
5	0.163	0.812	−1.793
6	–	0.808	−1.802

3.2.2 Combination of $Fe_{6\text{-}n}Cr_nN$ Octahedral Clusters

The octahedral clusters may further combine into clusters with a larger scale as the N concentration increases. Under the constraint of the f.c.c. lattice structure itself, only two combination modes are possible, namely, corner-shared mode and edge-shared mode. In addition, it also needs to explore which atom (Fe or Cr atom) prefers to be located at the shared sites. Finally, for the combination of octahedral clusters, N atoms in different octahedral clusters may be located on the planes with the same or different magnetic moments.

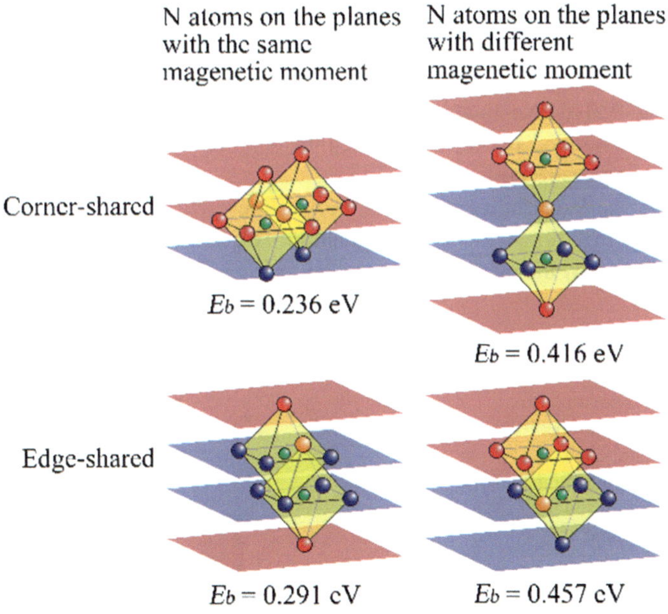

Fig. 8 The schematic illustrations of the combination modes of the $Fe_{6\text{-}n}Cr_nN$ octahedral clusters. The Cr atom is at the shared sites of the combined structures. (Reproduced from Ref. [7])

Firstly, the combinations of two $Fe_{6-n}Cr_nN$ clusters with only one Cr atom were studied to uncover the preferential combination mode and the locations of Cr and N atoms. As shown in Fig. 8, the binding energies for the edge-shared mode and the Cr atoms at the shared sites are higher. Moreover, the binding energies for the N atoms on the planes with different magnetic moments are much higher than those for the N atoms on the planes with the same magnetic moments. Therefore, the edge-shared mode is a more favorable mode for combining the octahedral clusters. In this mode, Cr atoms energetically prefer to locate at the shared edge, while N atoms prefer to locate on the planes with different magnetic moments.

When there are more $Fe_{6-n}Cr_nN$ octahedral clusters in the combined structures, the combination possibly contains both corner-shared and edge-shared modes. A parameter named edge-shared percentage (ESP) was calculated by ESP $= n_{edge}/n_{total}$ where n_{edge} is the number of shared edges, and n_{total} is the total number of shared corners and shared edges in a combined structure. To eliminate the effect of Cr atoms on the combination mode, only the combinations with one Cr atom at the shared sites were considered. The binding energies of the combinations of up to six $Fe_{6-n}Cr_nN$ clusters are shown in Fig. 9. The binding energies increase with the cluster number and reach the summit for two or three $Fe_{6-n}Cr_nN$ clusters, then decrease for more octahedral clusters. The decrease is because the number of Cr atoms is limited to only one. It is expected that the cluster with more Cr atoms has a greater energetic preference as shown in the following section. Figure 9 also indicates that for each cluster number in the combination, combinations with higher ESP value usually have higher binding energies, and this feature becomes more apparent when the combinations include more $Fe_{6-n}Cr_nN$ clusters. Therefore, the combined structures with larger scales have a higher tendency to be formed by the edge-shared mode. In addition, Fig. 9 also investigated the effect of the local magnetic moment. It can be seen that the combined structures in which the N atoms on the planes with different magnetic moments are preferred.

Similarly, the local lattice distortion around these clusters, the PDOSs, and the charge transfer in the clusters were then quantitatively evaluated. The lattice distortion was also assessed by the increment of the ACP volume, ΔV, and the ACP was defined

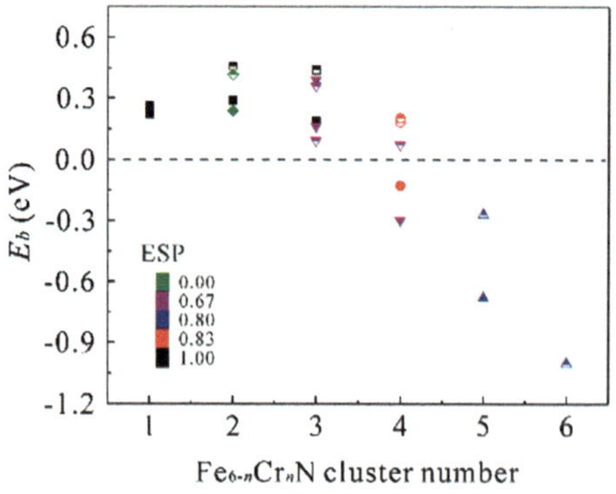

Fig. 9 The binding energy E_b of the combined structures as a function of the $Fe_{6-n}Cr_nN$ cluster number. Solid (open) symbols represent the N atoms in the combined structures on the planes with the same (different) magnetic moments. The color of the symbols represents the ESP of the combined structures (Reproduced from Ref. [7])

around the N atoms in a combined structure. Taking the combined structures with only one Cr atom as an example, the ΔV values of 1nn to 3nn ACPs were analyzed, and the average ΔV values are shown in Fig. 10. The average ΔV values are all positive and increase with the cluster number. The ΔV values of 2nn ACP are the largest, meaning that the lattice distortion is mainly located in the 2nn ACPs. These results agree with those of the isolated $Fe_{6\text{-}n}Cr_nN$ clusters shown in Fig. 6. For each cluster size, the ΔV values decrease with ESP, as shown in Fig. 10b–d, except for occasional deviation. Therefore, the edge-shared mode can improve the energetical preference of the combined structures by decreasing the lattice distortion.

Then, the PDOSs of both the corner-shared mode and edge-shared mode were calculated. A much stronger hybridization can be observed in edge-shared mode, which may be the reason that the edge-shared mode is energetically favored. As for the preference for Cr atom at the shared sites, it can be understood using the results of Bader analysis [16], which are listed in Table 3. More electrons were transferred from Cr atoms to N atoms, which is consistent with the charge transfer in isolated $Fe_{6\text{-}n}Cr_nN$ clusters, and results in the strong Coulombic attraction between the Cr

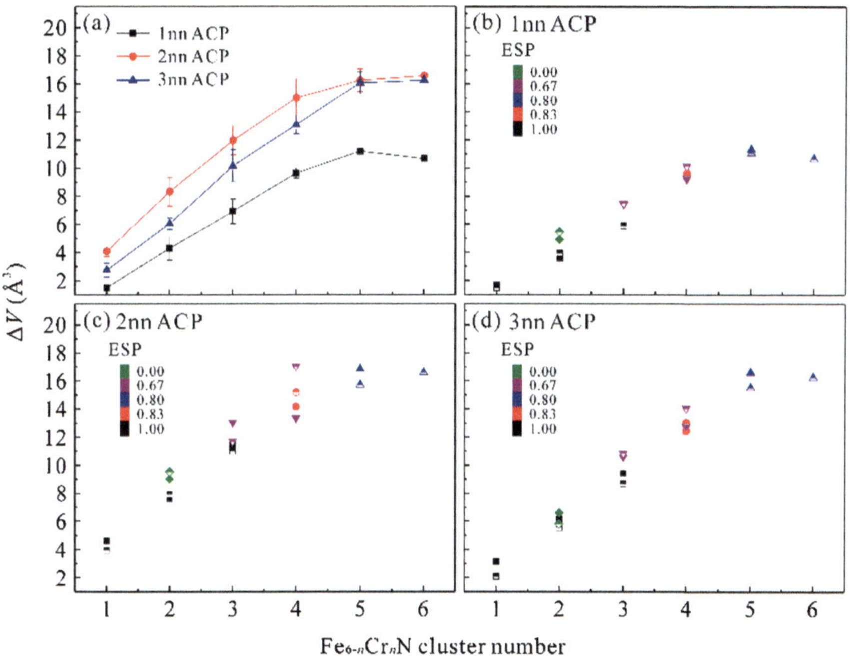

Fig. 10 **a** The average ΔV for the 1nn to 3nn ACPs as functions of the number of the Fe6-nCrnN cluster. The error bars represent the deviation from the average values. **b–d** The ΔV of 1nn, 2nn, and 3nn ACPs as functions of the Fe6-nCrnN cluster number. Solid (open) symbols represent the N atoms in the combined structures on the planes with the same (different) magnetic moment. The color of the symbols in **b–d** represents the ESP values of the combined structures. (Reproduced from Ref. [7])

Table 3 The average Bader charges difference (in *e*) of the Fe, Cr, and N atoms in the combinations of two $Fe_{6-n}Cr_nN$ clusters with only one Cr atom. The values shown in the bracket are the average Bader charges of shared Fe atoms (Reproduced from Ref. [7])

	Shared atom	Same magnetic moment			Different magnetic moment		
		Fe	Cr	N	Fe	Cr	N
Corner-shared	Cr	0.175	0.930	−1.541	0.147	0.968	−1.548
Corner-shared	Fe	0.198 (0.429)	0.864	−1.521	0.203 (0.433)	0.835	−1.575
Edge-shared	Cr	0.198	1.024	−1.505	0.222	1.031	−1.556
Edge-shared	Fe	0.211 (0.333)	0.832	−1.477	0.246 (0.422)	0.832	−1.505

and N atoms. Obviously, the shared site of the combined structure is beneficial for the charge transfer and Coulombic interaction. The average transferred electrons from metal atoms to N atoms for the edge-shared mode are more than those for the corner-shared mode when the Cr atom is located at the shared sites. Moreover, the values for the N atoms located on the planes with different magnetic moments are larger than those with the same magnetic moments. Consequently, the stronger Coulombic attraction between Cr and N atoms results in the greater preference of the combinations, in which the $Fe_{6-n}Cr_nN$ clusters are combined by the edge-shared mode, and the N atoms are located on the planes with different magnetic moments.

In the above calculations, the Cr and N concentrations in the supercell are up to 2.335 and 3.759 at. %, respectively, and the calculated Cr and N concentrations are much lower than those in high nitrogen steel. The higher concentrations lead to an avalanche-like increase in the number of Cr atoms, octahedral clusters in the combined structures, and the number of possible atomic configurations. Despite this, the atomic configurations of these combined structures can still be regarded as combinations of smaller clusters. These smaller clusters include not only the octahedral clusters but also the clusters formed by combinations of the octahedral clusters. For the smaller clusters formed by combinations of the octahedral clusters, the majority of Cr atoms are located at the shared sites so that the Cr and N interactions are confined within the combined structures. Therefore, when such smaller clusters further combine into a larger cluster, the interaction among them is relatively weak. It means that the estimated binding energies E_{eb}, which was calculated by simply summing the binding energies of the smaller clusters, would be a reasonable approximation of the binding energy of the larger cluster. In fact, both E_b and E_{eb} of the larger clusters containing up to 8 $Fe_{6-n}Cr_nN$ clusters were calculated, and the results are shown in Fig. 11. It can be seen that the data is basically linearly distributed in the figure, indicating the estimated energies are consistent with the calculated energies. This result confirms that the interaction among the smaller clusters is relatively weak.

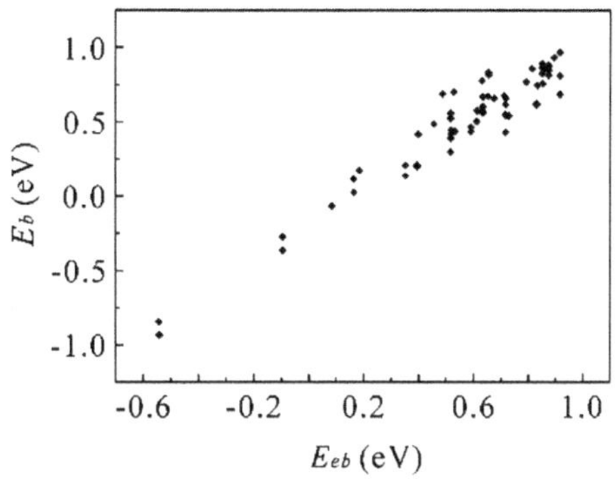

Fig. 11 The calculated binding energies versus the estimated binding energies (E_{eb}) of larger combined structures containing up to 8 $Fe_{6\text{-}n}Cr_nN$ clusters. (Reproduced from Ref. [7])

4 Summary and Prospects

The investigation of microstructures in high nitrogen austenitic steel as the interconnector or the bipolar plate in fuel cell application is introduced. Through systematical computer simulation based on first-principles calculation combined with electron microscopy analysis, the formation mechanism and the structural features of these subtle microstructures are revealed.

For high nitrogen austenitic steel, the high-density stacking faults are typical structural character at the nano- and micro-scales, and at the atomic scale, it has enormous short-range ordered structures, which can be considered as a combination of the $Fe_{6\text{-}n}Cr_nN$ octahedral clusters. The combination prefers the edge-shared mode. Cr atoms prefer the shared sites of octahedral clusters, and N atoms prefer to be located on planes with different magnetic moments. Obviously, various cluster structures can be formed by the combination of the octahedral clusters in terms of these regularities. These results also suggest that the phase stabilization is based on a chemically-driven mechanism by Cr and N atoms forming short-range ordered clusters in the steel.

Further work needs to be carried out as a logical extension of the works reviewed here. The present models of the microstructures are mainly derived from first-principles calculation. Direct observation and demonstration of these structures at the atomic scale need more comprehensive studies. For high nitrogen steels, although it is widely accepted that the existence of nitrogen can improve corrosion resistance, the relationship between the clusters and the properties of the high nitrogen steel has not been comprehensively studied. All these will become important research directions for future applications of these materials in fuel cell applications.

References

1. Dong H (2010) S-phase surface engineering of Fe-Cr, Co-Cr and Ni-Cr alloys. Int Mater Rev 55:65–98
2. Kresse G, Furthmuller J (1996) Efficient iterative schemes for ab initio total-energy calculations using a plane-wave basis set. Phys Rev B 54:11169–11186
3. Blöchl PE (1994) Projector augmented-wave method. Phys Rev B 50:17953–17979
4. Perdew JP, Chevary JA, Vosko SH, Jackson KA, Pederson MR, Singh DJ, Fiolhais C (1992) Atoms, molecules, solids, and surfaces-applications of the generalized gradient approximation for exchange and correlation. Phys Rev B 46:6671–6687
5. Vosko SH, Wilk L, Nusair M (1980) Accurate spin-dependent electron liquid correlation energies for local spin-density calculations—a critical analysis. J Can Phys 58:1200–1211
6. Klaver TPC, Hepburn DJ, Ackland GJ (2012) Defect and solute properties in dilute Fe-Cr-Ni austenitic alloys from first principles. Phys Rev B 85:174111–123
7. Tong K, Ye F, Wang YK (2019) Short-range ordered structure and phase stability of supersaturated nitride layer on austenitic stainless steel. Acta Mater 175:314–323
8. Lei MK, Liang J (2010) X-ray diffraction of high nitrogen face centred cubic phase formed on nitrogen modified austenitic stainless steel. Surf Eng 26:305–311
9. Meletis EI, Singh V. Jiang JC (2002) On the single phase formed during low-temperature plasma nitriding of austenitic stainless steels. J Mater Sci Lett 21:1171–1174
10. Stroz D, Psoda M (2010) TEM studies of plasma nitrided austenitic stainless steel. J Microsc 237:227–231
11. Jiang JC, Meletis EI (2000) Microstructure of the nitride layer of AISI 316 stainless steel produced by intensified plasma assisted processing. J Appl Phys 88:4026–4031
12. Tong K, Ye F, Che H, Lei M, Miao S, Zhang C (2016) High-density stacking faults in a supersaturated nitrided layer on austenitic stainless steel. J Appl Cryst:49
13. Borgioli F, Fossati A, Galvanetto E, Bacci T, Pradelli G (2006) Glow discharge nitriding of AISI 316L austenitic stainless steel: influence of treatment pressure. Surf Coat Tech 200:5505–5513
14. Grujicic M, Owen WS (1992) A chemical interstitial order - based model for nitrogen strengthening of Fe-Ni-Cr austenite. Calphad 16:291–299
15. Grujicic M, Owen WS (1995) Models of short-range order in a face-centered cubic Fe-Ni-Cr alloy with a high concentration of nitrogen. Acta Metall Mater 43:4201–4211
16. Henkelman G, Arnaldsson A, Jonsson H (2006) A fast and robust algorithm for Bader decomposition of charge density. Comput Mater Sci 36:254–360

Microanalysis

In Situ TEM Analysis

Takayoshi Tanji and Takafumi Ishida

Abstract *In situ* observations of nano- or sub-nano scale structures at the interface between the electrode and electrolyte of an oxide-ion conductor by transmission and scanning transmission electron microscopes are described. After the introduction of several characteristic techniques, the development of instruments for *in situ* observations is described. Electron holography can visualize the electrostatic potential inside a material. Shifts of such potentials resulting from the shifts of oxygen distribution are observed at high temperatures in a vacuum and an oxygen atmosphere. Electron energy loss spectroscopy identifies the cations responsible for the redox reactions in ion conductors. Comparing half-cell models (tungsten/oxide-ion conductor/platinum) and complete cell models (platinum/oxide-ion conductor/platinum), we recognize clearly the activeness of platinum electrodes by electron holography and electron energy loss spectroscopy.

Keywords Analytical electron microscopy · Electron holography · Electron energy loss spectroscopy · *In Situ* observation · Environmental electron microscope

1 Introduction

To improve the efficiency of solid oxide fuel cells (SOFCs), it is important to reveal many of the effects occurring on the interface between the electrode and electrolyte. The so-called overpotential effect is a typical problem with SOFCs. In this case, the drawable voltage decreases from the value expected in a stable state as the operating current is increased. One apparent origin of this effect is pile-up of oxygen ions near the anode or a lack of oxygen ions near the cathode.

T. Tanji (✉) · T. Ishida
Nagoya University, Furo-Cho, Chikusa, Nagoya 464-8601, Japan
e-mail: ttanji@me.com

T. Ishida
e-mail: ishida.takafumi.b5@f.mail.nagoya-u.ac.jp

T. Mori et al. (eds.), *Design of Active Key Interfaces in Fuel Cells*,
https://doi.org/10.1007/978-981-97-5706-0_7

If an external electric field is applied to solid electrolytes in a vacuum, oxygen ions are drawn toward the anode. The movement of oxygen ions shifts the electrostatic potential, and the ion flow might be blocked unless new oxygen ions are introduced from the cathode. This phenomenon is very similar to the overpotential effect in a working fuel cell.

Using electron microscopy, we observed electrostatic inner potentials at the interface of an oxide-ion conductor and a platinum electrode, and the electronic states of the constituent elements. To observe the atomic, elemental and electrostatic structures and the electronic states, we used high-resolution electron microscopy, electron holography, energy dispersive X-ray spectroscopy (EDX), and electron energy loss spectroscopy (EELS).

To simulate fuel cells, model SOFC cells have to be heated in two different gas atmospheres simultaneously and the electric current extracted. However, it is very hard to achieve this in the small space of a specimen chamber of an electron microscope. Therefore, in the results shown here, specimens were heated and an external potential was applied in a vacuum or in an oxygen atmosphere. This condition is close to the charging state of an SOFC or the operation conditions of solid oxide electrolysis cells (SOECs). New specimen holders have been developed for 200 kV and 1000 kV transmission electron microscopes that can apply external voltages of −1.5 V to + 1.5 V to specimens at high temperatures up to 800 °C in a vacuum and in an oxygen atmosphere.

The information obtained in this study should contribute to the development of highly efficient SOFCs.

2 Instrumentation

There are four typical types of electron microscope: a conventional transmission type electron microscope (TEM or CTEM), a scanning electron microscope (SEM), a scanning transmission electron microscope (STEM), and a reflection electron microscope (REM). In the observation of nanometer or sub-nanometer structures of specimens, TEM or STEM have advantages. Here, the "structures" mean not only atomic structures but also the electronic structures, distribution of elements, and electrostatic potential distribution inside the specimen.

Electron microscopes are also classified for their electron source types, namely, whether a microscope is equipped with a thermal emission electron gun, a Schottky type thermal emission gun (sometimes called a pointed filament or Hibi filament), a LaB_6 thermal emission electron gun, a Schottky type thermal field emission gun, or a cold field emission gun. Schottky type thermal field emission guns and cold field emission guns are suitable for ultra-high-resolution microscopy and electron holography because of their higher coherency. Sometimes electron microscopes are also classified for their accelerating voltage: low voltage microscopes (acceleration voltage Va $\lesssim$ a few kV), standard voltage microscopes, and high voltage microscopes (HVEM, Va $\gtrsim$ 500 kV).

Many of today's electron microscopes have many functions besides imaging, e.g., electron holography, electron energy loss spectroscopy (EELS), and X-ray spectroscopy. In observations by TEM/STEM, thinning of the specimen is necessary. There are many techniques for preparing thin specimens. Recently, preparation using a focused ion beam (FIB) has become very popular. This method provides fine treatments of specimens of less than several tens of micrometers after thinning in the same space. It is sometimes called micro-probing or micro-sampling.

HVEM has a big advantage in the observation of thicker specimens. A microscope with a 1000 kV acceleration voltage can observe a specimen more than three times thicker than a 200 kV microscope can [1]. This means that HVEM can observe specimens close to actual conditions because thinner specimens sometimes show different characteristics from bulk specimens. Moreover, HVEM has a much larger specimen chamber than 200 kV TEM/STEM, so it is easier to install some systems for specimen cooling, applying external electromagnetic fields, gas introduction, etc.

2.1 *Electron Holography*

2.1.1 Principles

It has been long known that a cylindrical electromagnetic field works as a convex lens and can enlarge small objects, which is the basis of the electron microscope. However, it is impossible to form a concave lens from a cylindrical field. Therefore, so-called spherical aberrations of electron lenses can be reduced by using strong fields but cannot be eliminated completely. Scherzer pointed out in 1942 that in electron microscopy the residual aberration determines the resolution limit of the microscope [2].

Denis Gabor invented holography in 1948 to solve this blurring effect [3]. In this scheme, the aberration blurring of the electron micrograph is eliminated by light optics using opposite aberration coefficient glass lenses, i.e., concave lenses applied a posteriori.

Unfortunately, in the time of Gabor, there were no highly coherent electron sources, such as today's Schottky type or cold type field emission guns (FEG). In 1979, Tonomura et al. successfully installed a cold type FEG to a TEM [4] and realized electron holography as an actual tool to eliminate aberrations in high-resolution electron microscopic images with light optics [5] and digital processing [6], and also to visualize electromagnetic fields [7].

In almost all applications of electron holography, so-called an image hologram recorded in the image plane is adopted by using a cold type FE-TEM or Schottky type FE-TEM with one or two electron biprisms. Figure 1 shows typical electron optics for recording electron holograms. The electron biprism developed by Möellenstedt et al. [8] consists of two grounded electrodes and a very fine filament electrode, which creates interference of two electron waves. By using an electron biprism, an object wave modulated in a specimen, due to the material's inner potential or an

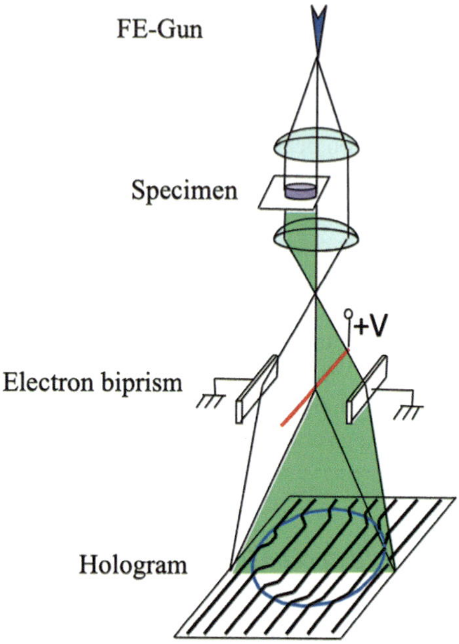

Fig. 1 Ray diagram for recording an off-axis electron hologram by TEM with an electron biprism

electromagnetic field, interferes with an unmodulated reference wave and creates an interference pattern, that is, a hologram. Many filament electrodes mainly used today are made of platinum, provided as a Wollaston wire [9], or of quartz glass coated with conductive materials [10]. The filaments used are typically 0.3–2 μm in diameter.

An image hologram *h(r)* is represented as an interference fringe pattern. The objective wave $\psi(\boldsymbol{r})$, whose amplitude *a(r)* and phase $\phi(\boldsymbol{r})$ are modulated by the electromagnetic potential in the specimen,

$$\psi(\boldsymbol{r}) = a(\boldsymbol{r})\exp\{i\phi(\boldsymbol{r})\}\exp(-i\boldsymbol{k}\cdot\boldsymbol{r}) \tag{1}$$

interferes with a reference wave having uniform wave planes, $\exp(i\boldsymbol{k}\cdot\boldsymbol{r})$, as

$$\begin{aligned} h(\boldsymbol{r}) &= \left|\psi(\boldsymbol{r})\exp(\pi i\boldsymbol{u}_0\cdot\boldsymbol{r}) + \psi^*\exp(-\pi i\boldsymbol{u}_0\cdot\boldsymbol{r})\right|^2 \\ &= 1 + |\psi(\boldsymbol{r})|^2 + 2a(\boldsymbol{r})cos\{\phi(\boldsymbol{r}) + 2\pi iu_0\boldsymbol{r}\} \end{aligned} \tag{2}$$

where $|\boldsymbol{u}_0| = \frac{\alpha_0}{\lambda}$, $\alpha_0 << 1$ is the deflection full angle by the electron biprism and ψ^* is the complex conjugate of ψ.

The reconstruction process is as follows:

$$\begin{aligned} I(\boldsymbol{u}) &= \mathrm{FT}\{h(\boldsymbol{r})\} \\ &= \delta(\boldsymbol{u}) + \mathrm{FT}\{\psi(\boldsymbol{r})\} * \mathrm{FT}\left\{\psi^*(\boldsymbol{r})\right\} + \mathrm{FT}\{\psi(\boldsymbol{r})\} * \delta(\boldsymbol{u}-\boldsymbol{u}_0) \\ &+ \mathrm{FT}\left\{\psi^*(\boldsymbol{r})\right\}\delta(\boldsymbol{u}+\boldsymbol{u}_0) \end{aligned} \tag{3}$$

$$\psi(\boldsymbol{r}) = \mathrm{FT}^{-1}[\mathrm{FT}\{\psi(\boldsymbol{r})\} * \delta(\boldsymbol{u} - \boldsymbol{u}_0)] \exp(-2\pi i \boldsymbol{u}_0 \cdot \boldsymbol{r}) \tag{4}$$

and

$$\begin{aligned} a(\boldsymbol{r}) &= |\psi(\boldsymbol{r})| \\ \phi(\boldsymbol{r}) &= \tan^{-1}\left[\frac{Im\{\psi(r)\}}{Re\{\psi(r)\}}\right] \end{aligned} \tag{5}$$

where FT, FT^{-1}, and $\tan^{-1}$ show the Fourier transform, inverse Fourier transform, and arc tangent, respectively.

2.1.2 Observation of Electromagnetic Fields

The purpose of Gabor's invention was to restore high-resolution electron-microscopic images blurred due to the spherical aberration of electron lenses using magnetic fields. Tonomura et al. have shown that electron holography is also a powerful tool for observing electromagnetic fields.

Figure 2 shows the principle of observing electric and magnetic fields. In an electric field, the equiphase lines in the image plane show the projection of electric equipotential planes, while in a magnetic field they show the lines of magnetic force. In a small magnetic particle, a single-domain state will become most stable below a certain critical size [11]. Figure 3 shows an interference image of a Ba-ferrite particle with a single magnetic domain. The contour lines correspond to the lines of magnetic force emitted from the particle into the vacuum. The figure shows clearly that the magnetization inside the particle is aligned in one direction.

The magnetic structures of many interesting materials have been observed, for instance, cross-sections of Co/Cu multilayered films, which show giant magneto-resistance (GMR) [12] and a chain of magnetic crystals in the bacterial strain Magnetospirillum magnetotacticum [13] is interesting, too. Visualizing the distribution of dopant elements on p–n junctions in semiconductors [14] is one of the most hopeful applications of electron holography. Figure 4 shows an example of an n-MOSFET showing the distribution of doped Bollon into a n-type semiconductor.

2.2 *Electron Energy Loss Spectroscopy*

Electrons passing through a specimen lose their energy by exciting core electrons and/ or plasmons, which are quanta of plasma oscillations of free electrons. Möllenstedt et al. have utilized the chromatic aberration of a magnetic lens in the energy analysis [15]. By applying this type of analyzer, Watanabe et al. [16] observed the dispersion of plasmons, which had been predicted only theoretically. Today, magnetic sector type analyzers (Fig. 5) are more popular, and systems combining quadrupole magnetic

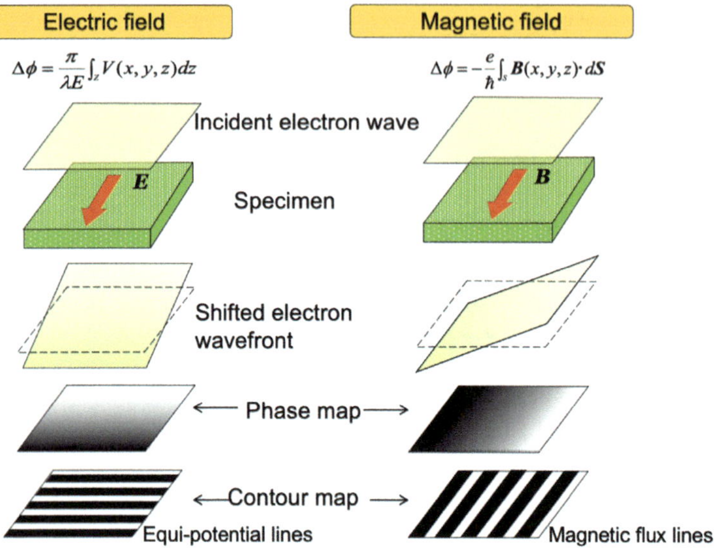

Fig. 2 Phase shift of electron waves and contour maps due to an electric field and a magnetic field. In the case of the electric field, the contour lines are parallel to the equipotential lines, while for the magnetic field they are parallel to the magnetic flux lines

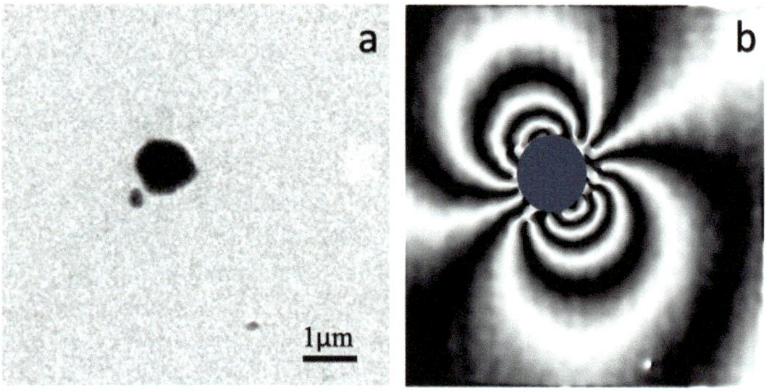

Fig. 3 Single magnetic domain particle: **a** normal TEM image, **b** electron holographic interferogram representing the lines of magnetic force

lenses and parallel detectors [17] have been installed in columns of TEM/STEMs in many laboratories. EELS spectra provide important information on the electronic states of elements. Therefore, it is expected to be observed the redox reaction using EELS in conjugation with electron holography by *in situ* electron microscopy.

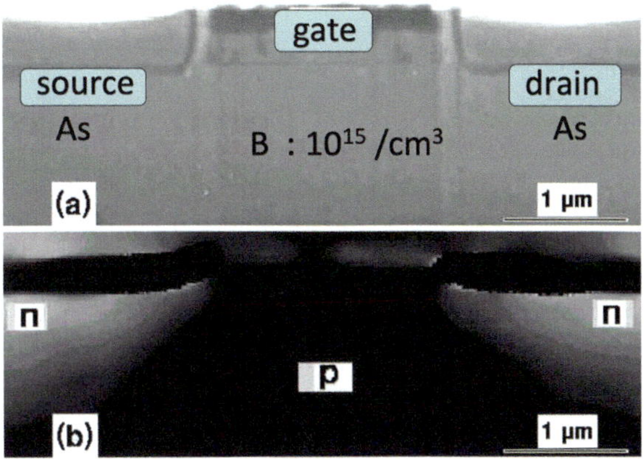

Fig. 4 Observation of a MOSFET: **a** normal TEM image, **b** phase distribution representing the inner potential, i.e., dopant distribution (courtesy of Drs. T. Hirayama and K. Yamamoto)

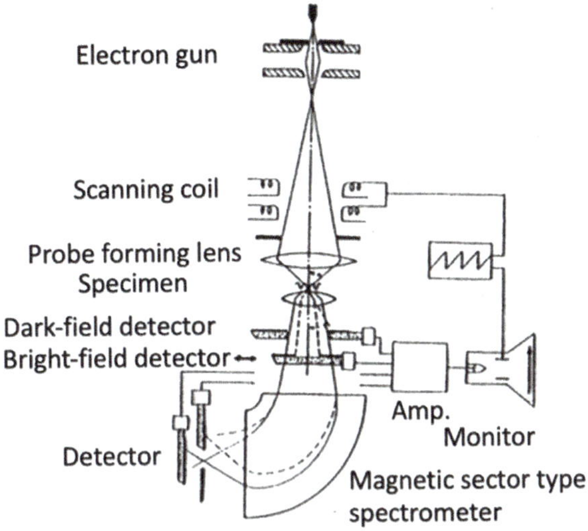

Fig. 5 Schematic of a scanning transmission electron microscope equipped with an electron energy spectrometer of magnetic sector type

2.3 Energy Dispersive X-ray Spectroscopy

Elemental micro-analysis techniques by X-ray spectroscopy have been established with two kinds of spectrometer: wavelength dispersive X-ray (WDX) analyzers and energy dispersive X-ray (EDX or EDS) analyzers [18]. In many cases, instruments dedicated to electron probe micro-analysis (EPMA) have been adopted for WDX, which has a higher energy resolution and has occasionally been installed in TEM systems [19]. However, EDX analyzers are more commonly used because of their easy installable compact size and easy treatment for TEM analysis.

Figure 6 shows a typical semiconductor detector of an EDX analyzer. Ealy apparatuses required thin carbon or beryllium windows and liquid nitrogen to protect the Li doped Si detector. Therefore, elements heavier than Be could not be analyzed.

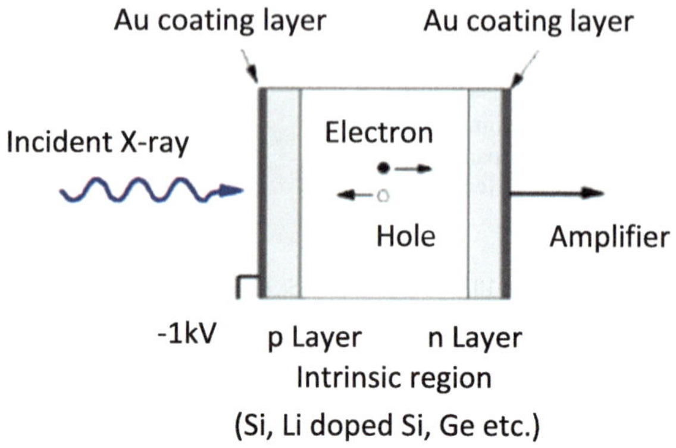

Fig. 6 Semiconductor detector for energy dispersive X-ray spectroscopy

Analyzers using highly purified Si elements in high vacuum do not need film windows or coolant, and can be used to analyze any element heavier than helium. Today's silicon drift type detectors (SDDs) can efficiently detect generated electrons and enable high speed signal response and work at about −15 °C by Peltier cooling [20].

Using EELS or EDX, we can obtain a distribution map of elements. Owing to its intense signal, EDX is better than EELS for obtaining clear maps. On the other hand, EELS can reveal the electronic states of component atoms.

3 Environmental Transmission Electron Microscopy

To observe the actual phenomena occurring in SOFCs, conditions similar to those in the cell have to be reproduced in the microscope. Therefore, environmental transmission electron microscopes (E-TEMs) are indispensable for in situ observations in a gas atmosphere. E-TEM observations of crystal growth were conducted by the inventor of the electron microscope, E. Ruska [21], and systems have been established since the 1960s [22–26]. Since the early twenty-first century, E-TEMs with differential evacuation systems have been available commercially. Moreover, many kinds of environmental cells installable in TEM have been developed. Aberration collected E-TEM has improved the study of crystal growth considerably [27] Closed type cells are more convenient to install than open-type cells which sometimes require a differential pumping system. For example, electron holography requires a very highly coherent electron source, i.e., a cold or thermal field emission (FE) gun. However, they have to be operated in a high vacuum of 10^{-8} Pa or better. This condition has restricted the application of FE-TEMs to *in situ* observations, as for E-TEMs. The holographic TEM HF-2000 used here was reconstructed as an E-TEM by installing a new gas introduction system in the specimen chamber, as shown in Fig. 7 [28].

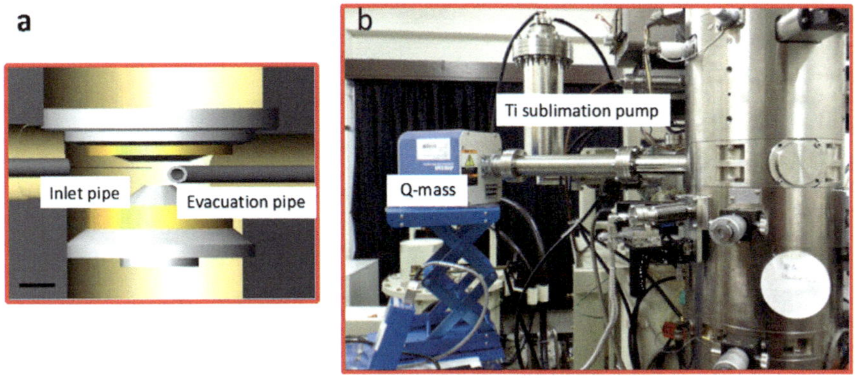

Fig. 7 Restructured holographic FE-TEM (Hitachi HF-2000). **a** Inlet and evacuation pipes installed into the objective gap. **b** Extra Ti sublimation pump and quadrupole mass analyzer

3.1 Environmental Transmission Electron Microscope

Many E-TEMs are operated at a medium accelerating voltage of 100–300 kV. JEM1000K RS at Nagoya University is a high voltage environmental TEM/STEM, in which a capsule-type retractable specimen chamber is installed in the objective gap, and a specimen holder is inserted into this chamber, as shown in Fig. 8 [29]. The microscope column is evacuated differentially by three sputter ion pumps, six turbo molecular pumps, and a diffusion pump. Many kinds of single gas or mixed gases can be introduced into the capsule up to 1.3×10^4 Pa, and an Au lattice image of 0.14 nm spacing is observable in this environment.

3.2 Development of Specimen Holders for in Situ Observations

A specimen holder for TEM is a key component for in situ observations of materials. Special specimen holders have been developed for dedicated applications, such as in gases, liquids, and for applying potentials [30]. We have also developed specimen holders to observe the electrode reactions of SOFCs, since 2006. In this section, we will describe our progress and introduce a specimen holder for *in situ* observations. Figure 9 shows a newly developed voltage applicable heating specimen holder [31] for observing electrode reactions at interfaces through the shift of the electric potential distribution, generated by the oxygen ions' distribution [32]. To observe the SOFC reaction at the nano-scale, the specimen holder has four electrodes in the holder tip. Two electrodes are for heating the specimen and the other two are for applying an external voltage. In the first produced holder, the specimen was mounted on a Ta thin belt heater directly and Joule-heated.

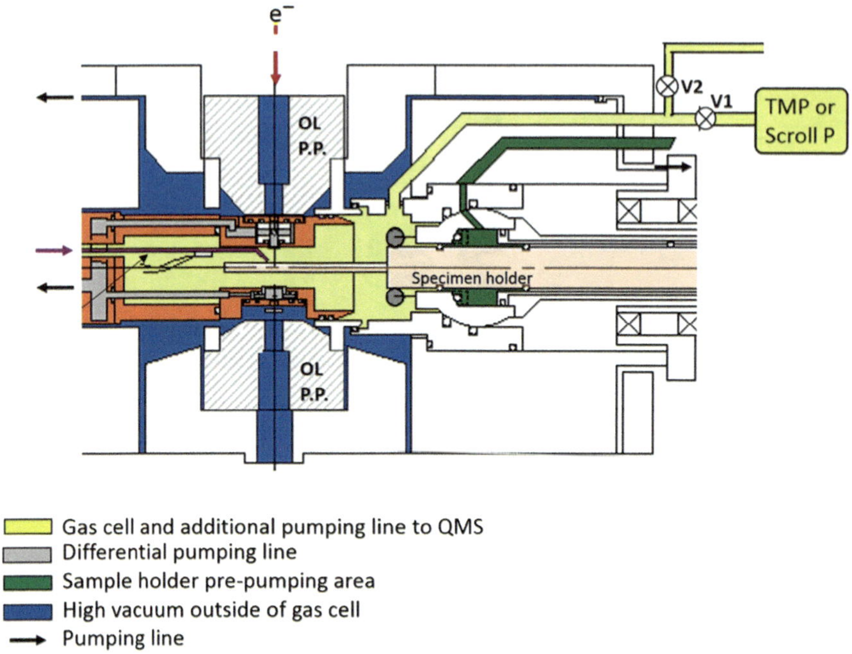

Fig. 8 Environmental HVEM (JEOL JEM-1000 K RS). The environmental cell is retractable and specimens can be observed in a high vacuum

Fig. 9 Schematic of the tip of a newly developed voltage applicable heating specimen holder (1st version)

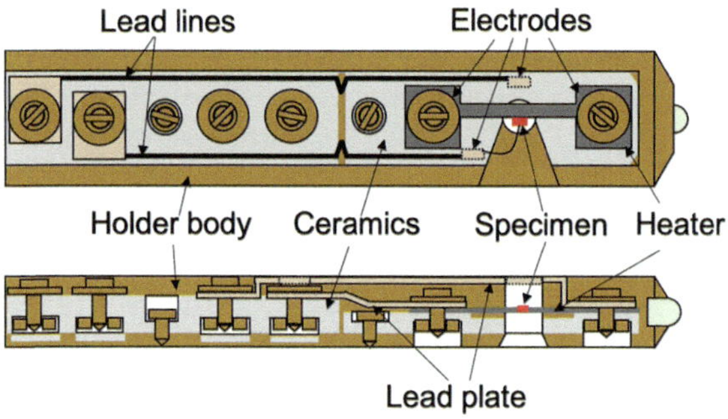

Voltage application [32] and high temperature [28] experiments using this holder were performed to determine the appropriate material for the Joule heater. Mechanical and electronical stability are required for the heating and voltage application to the specimen in gas environments. A new heater made of a Ni–Cr thin belt was found to satisfy this requirement. In addition, TEM analysis, which can provide the oxygen distribution at the nano-scale and the electronic states of the elements, requires different made TEM instruments from that for electron holography, which could not be used in the configuration of the first holder.

To overcome the above limitation, the voltage applicable heating specimen holder was improved [33]. The new holder developed has a removable heater unit on which a specimen is attached. This unit can be set on any specimen holder which satisfies the standard of manufacture. Thus, the crystal structure, electric field, and electron energy loss spectra in the identical area of the same specimen can be analyzed with various made electron microscopes.

Photographs of the tip of the voltage applicable heating specimen holder and the removable heater unit are shown in Fig. 10. The holder has three electrodes for heating and applying voltage to the specimen. The electrons scattered by the specimen pass through the grounded copper tube and the charging effect of a base insulator by scattered electrons is prevented. Figure 10c shows a photograph of the heater unit made of a NiCr alloy. The heater has high oxidation resistance and high stability in an oxygen atmosphere at high temperatures. The electrodes for the voltage application and heating are isolated electrically by insulator parts to maintain strength. The heater unit has a protuberance at the center of the heater part to guide the appropriate position of a specimen. More details are described in Ref. [33].

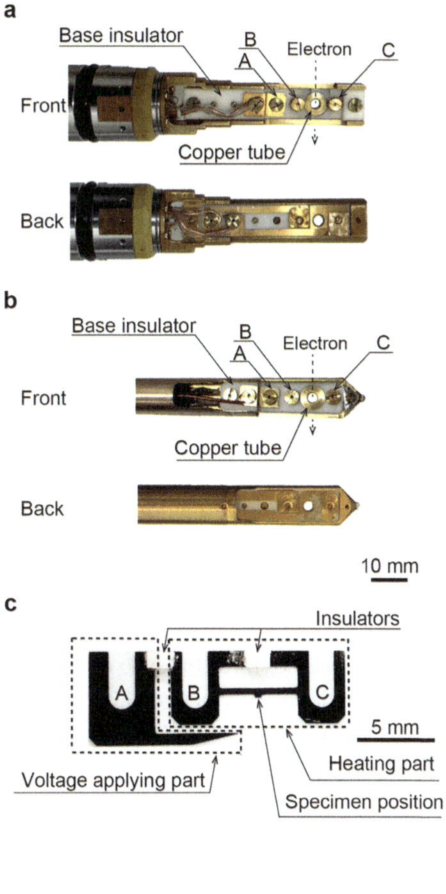

Fig. 10 Photograph of the tips of voltage applicable heating specimen holders. **a** is for JEM1000K RS and **b** for HF-2000. **c** is a removable heater unit. The parts indicated **A** is for applying a potential, and **B** and **C** are electrodes for heating. The heater unit allows observation of the same specimen by different transmission electron microscopes which have special functions

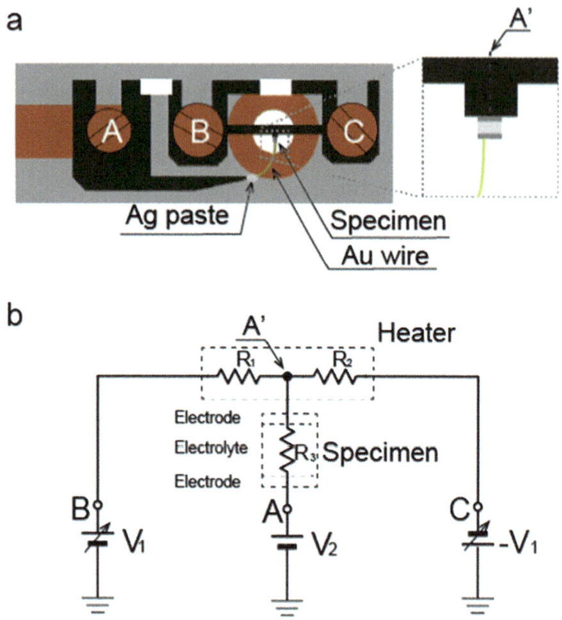

Fig. 11 Heating and applying a voltage with three electrodes. **a** Scheme of electrodes and the heater unit. **b** Circuit showing the heater operated with $2V_1$ while point A' is kept at the ground level and V_2 is applied to the specimen

The first voltage applicable heating holder had four electrodes, whereas the number of electrodes in the new specimen holder has been reduced to three, resulting in easier transfer of the heat unit. Two voltage suppliers can control the temperature and apply the potential to the specimen. In the configuration shown in Fig. 11, the center of the heater, which is the position of the SOFC specimen, is kept to ground level. Thus, only one electrode is enough to apply the potential. Therefore, only three electrodes are required to realize the application of an external voltage at a high temperature and observe oxygen ion migration and electrode reactions in SOFC specimens.

This new holder system with a removable heater unit (i.e., specimen cassette) has made possible the common use of FIB (Hitachi FB-2100), high voltage E-TEM/STEM (JEOL, JEM 1000K RS), and a holographic TEM with a cold FEG (Hitachi HF-2000) of different manufactures. Specimens can be heated up to 1000 °C and an electric voltage −5 V to +5 V can be applied in a gas atmosphere including oxygen.

Figure 12 shows temperatures of the heater unit as a function of heater current. The specimen holder with a heater unit was tested by flowing a current in an oxygen atmosphere of 0.5 Pa. We repeated the temperature measurement in the heating and cooling processes three times to confirm that the thermal characteristic is almost the same in each measurement and the heater can be reused several times without large deformation. The large difference in temperatures between the center and end of the heater is suppressed by the low thermal coefficient of Ni–Cr.

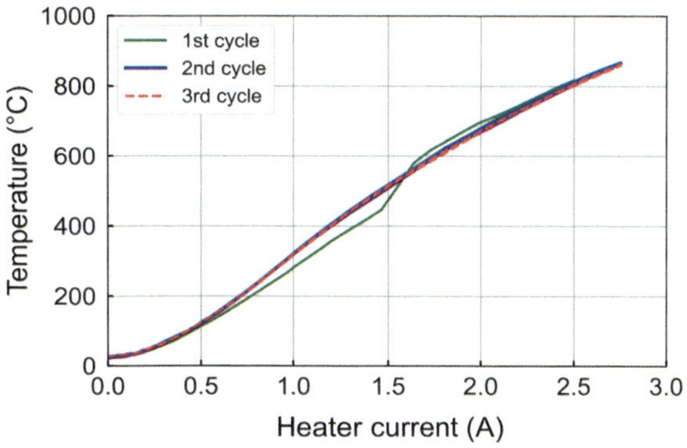

Fig. 12 Temperature of the heater versus current

Several companies [34] have commercialized specimen holders which can heat and/or apply a voltage to specimens, equipped with chips produced by micro-electromechanical systems (MEMS) technology. The MEMS chips can include electrical wires on small areas such as the tip sections of TEM specimen holders. Although they have made *in situ* observations easier, the potential of the commercialized specimen holders is still unknown for the observation of SOFCs.

4 Specimen Preparation

All SOFC specimens for TEM observations are prepared by a two-step process. Firstly, a bulk SOFC structure is prepared by pulsed laser deposition (PLD). Then, a cross-sectional TEM specimen is fabricated by FIB equipped with a micro-sampling system. The details of the specimen preparation for in situ observation of SOFCs are described below.

4.1 Pulsed Laser Deposition

Bulk SOFC samples are prepared by the PLD method. The target materials are sputtered using a high-power ultraviolet (UV) laser. Solid-state lasers and excimer lasers are often used for the high-power UV laser. Short wavelengths in the range 200–400 nm are usually used to ablate the target materials. The ions and atoms ablated from the materials are deposited onto a substrate. Further details of the sputtering mechanism are described in the textbook Ref. [35]. The PLD method. is often used to grow epitaxial thin films of solid oxide electrolytes for SOFCs.

In the following experiment, bulk samples of half and complete SOFCs were fabricated on substrates of ~3 × 3 mm^2 area by the PLD method using a Nd:YAG laser (LOTIS TII, LS-2147, $\lambda = 266$ nm: 4th harmonic wave) with a repetition rate

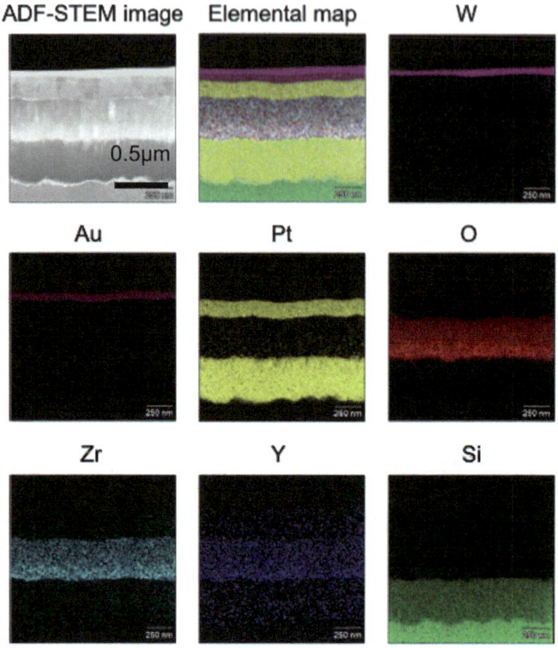

Fig. 13 ADF-STEM image and EDX elemental maps by the complete SOFC structure model of YSZ electrolyte and Pt electrodes

of 10 Hz. The base pressure of the PLD chamber was under 5×10^{-6} Pa at room temperature. The partial pressure of oxygen was kept to 4×10^{-2} Pa to minimize the number of vacancies in the oxide electrolyte. The electrodes and the electrolytes were made of pure Pt or Ni and of GDC or YZS, respectively.

Figure 13 shows an example annular dark field (ADF)-STEM image and STEM-EDX mapping of the cross-section of an SOFC fabricated by the PLD method. The specimen has a complete cell structure, consisting of a YSZ electrolyte sandwiched by two Pt electrodes. The W and Au are protection layers for FIB processing, and the Si layer is a substrate. The elemental maps indicate the cell structure is successfully fabricated by using the PLD method.

4.2 Focused Ion Beam (FIB) Fabrication

Figure 14a–d shows scanning ion microscope (SIM) images of an SOFC specimen obtained using a preparation process with an FIB instrument (Hitachi FB-2100). The heater unit is installed on a grounded dedicated FIB specimen holder.

A tungsten layer is deposited on the bulk specimen to protect the structure during cross-sectioning by FIB. In order to obtain a specimen with a small volume, a SOFC bulk sample is trenched as shown in Fig. 14a. A small piece of SOFC is picked up by a micro-sampling technique and connected to the electrode with a thin gold wire

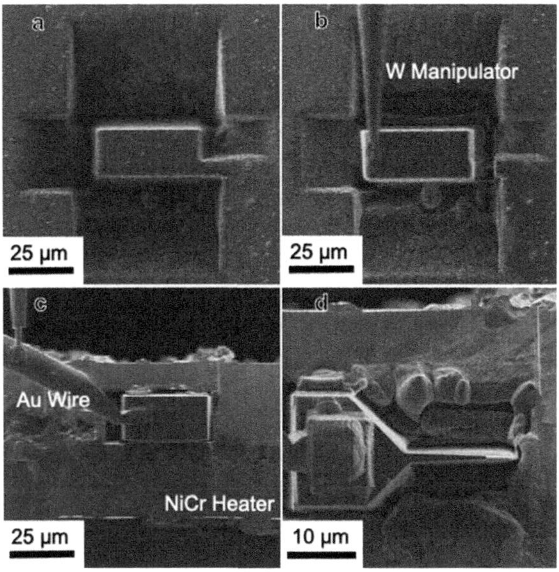

Fig. 14 Specimen preparation process by FIB. **a** Milling a deep trench, **b** picking up the specimen by the micro-sampling method, **c** mounting the specimen piece and connecting with a thin gold wire, **d** final thinning

(Fig. 14b, c). The specimen is further thinned and formed into a wedge shape by 40 keV Ga ion beams (Fig. 14d). All surfaces are cleaned by a low voltage Ga ion beam to protect from shorts in the electrical circuit.

Finally, the heater unit with the specimen is safely moved to the voltage applicable heating specimen holder using a dedicated stage system.

Conventional specimens prepared by FIB have a dumbbell shape, which is unsuitable for voltage applied experiments, because it is difficult to clean the surfaces of such a dumbbell-shaped specimen. The surfaces of a wedge-shaped specimen, shown in Fig. 14d, are easier to be flattened and cleaned using a low voltage ion beam. In addition, we can check the condition of the surface using an SIM image with a tilting specimen stage. These are key points for *in situ* experiments of SOFCs by E-TEM.

5 Observation of the Interface Between Electrodes and Electrolytes

GDC and YSZ are typical materials for SOFC electrolytes. GDC is a mixture conductor, that is electron- and oxygen ion-conductor. YSZ is a pure oxygen ion conductor. Some *in situ* observations of GDC and YSZ by electron holography and electron energy loss spectroscopy are shown in this section.

5.1 Electron Holography of GDC

A model cell of Pt (60 nm)/GDC (2 μm)/Pt (20 nm)/Si (substrate) is illustrated in Fig. 15. The cell was observed by electron holography with an applied external electric potential in a vacuum. The observation focused on the interface layer between the GDC electrolytes and Pt electrodes, as shown by the TEM image. The phase shifts corresponding to the inner potential distributions at the interface are shown in Fig. 16. Figure 16a shows the distribution of the electron phase modulated only by the mean inner potential of the Pt and GDC, and Fig. 16b shows the distributions with -1.0 V applied to the cathode. However, the effect of the external potential on the phase distribution is not clear. The difference between the phases in **a** and **b** makes clear the effect of the external voltage, as shown in **c**.

The phase distribution without an external voltage shown in Fig. 17a was subtracted from those with applied voltages, and Fig. 17b, c shows the resulting differences. External voltages from -1.0 V to $+ 1.0$ V were applied in the sequence shown in Fig. 17d. The shape of the phase shift can be understood as follows: At the interface, electric double layers are observed of about 2 nm thickness, considering the resolution of the electron holograms. In the electrolyte of the cathode side, O^{2-} ions are removed from near the interface and pile up in the area 15–25 nm away from the interface. They accumulate around 5–10 nm on the anode side when an external voltage is applied at room temperature, as shown in Fig. 18. This occurs because oxygen ions cannot be provided on the cathode side in a vacuum, and oxygen ions cannot be removed out by combining with positive ions, H^+, CO^{2+}, etc., on the anode side [32].

At higher temperatures we observe redox reactions. In the case of a half cell, W/ GDC/Pt, the reduction process.

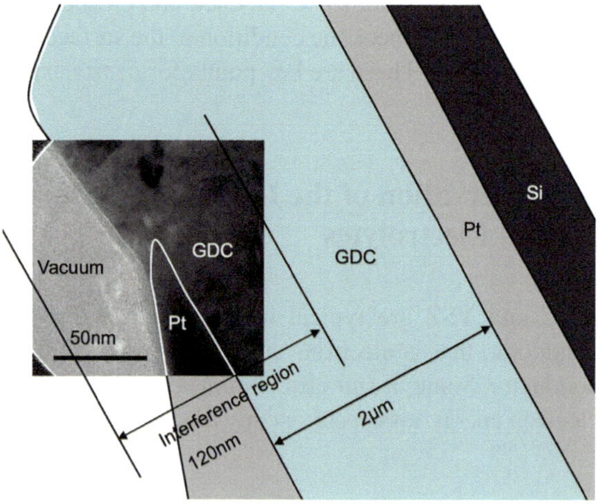

Fig. 15 Schematic of a Pt/ GDC/Pt model cell prepared by the PLD method. The inset is a TEM image of the area observed by electron holography

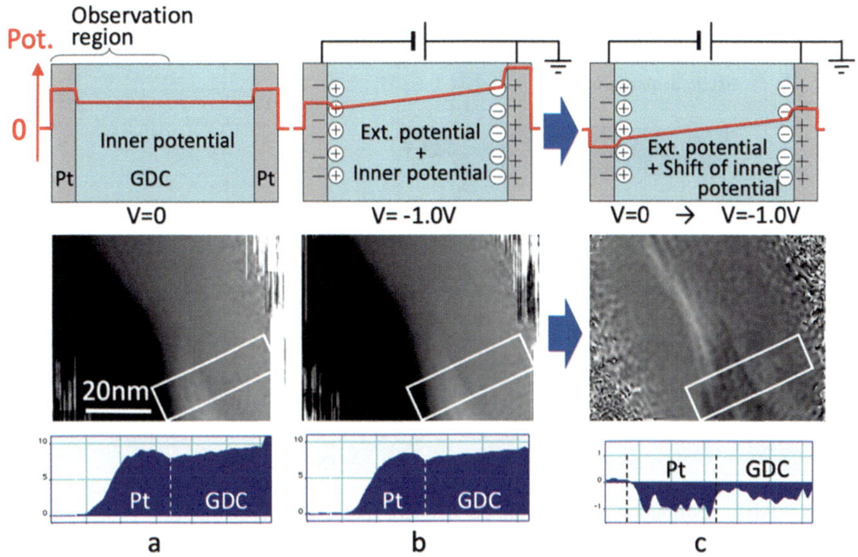

Fig. 16 Schemes of expected potential distributions (red lines) in the cell with applying external voltages (top layer). Reconstructed phase images (middle layer) and their line-profiles along the long sides of the rectangular frames (bottom layer). They are averaged along the short sides of the frames. **a** without an external field, **b** with −1.0 V applied to the Pt electrode, **c** subtraction of (**a**) from (**b**)

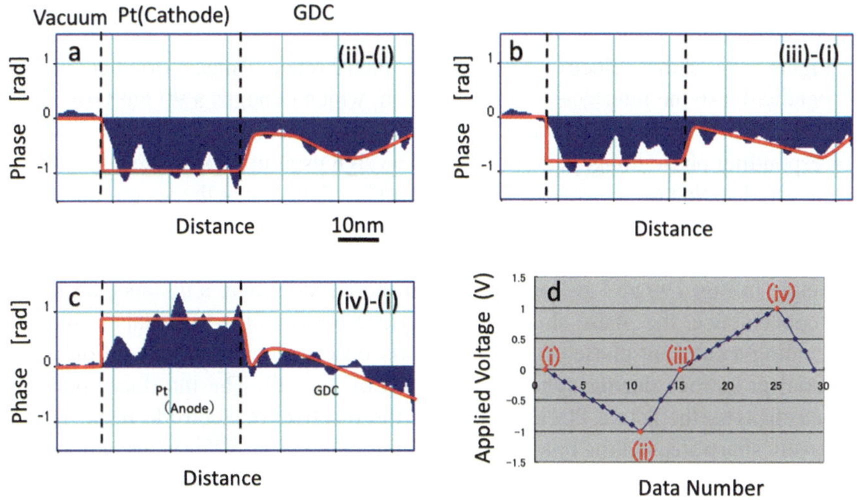

Fig. 17 The line-profiles along the long sides of the rectangular frames in Fig. 16 are shown in (**a**), (**b**), and (**c**). They were subtracted the value as in Fig. 16 (**c**). **d** Application sequence of the voltage. **b** and **c** correspond to points (ii) and (iii), respectively

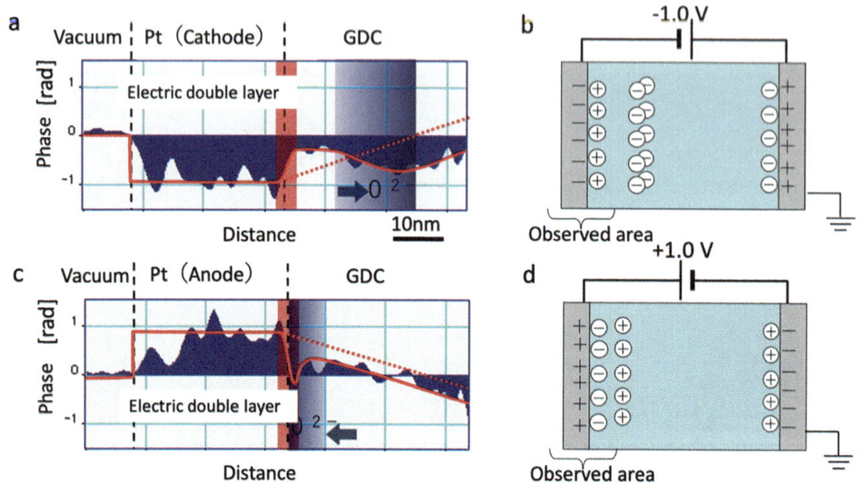

Fig. 18 Phase distribution on the cathode side (**a**), and expected distribution of oxygen ions (negative charges) and oxygen vacancies (positive charges) (**b**) and those of anode side (**c**, **d**), of the specimen shown in Fig. 16. Electric double layers on both sides are observed

$$2O^{2-} \rightarrow 4e^{-} + O_2 \tag{6}$$

is observed with increasing specimen temperature on the Pt catalysis side, resulting in a decrease of the electrostatic potential. This reaction is also called an oxygen enhanced reaction, OER.

Figure 19a shows electron-holographic interference fringes oriented almost perpendicular to the interfaces in the specimen, which contains a 90 nm GDC electrolyte layer, an active Pt electrode, and an inactive W electrode. Figure 19b shows a corresponding phase image reconstructed. The sign used in Figs. 19, 20, 21, and 23 are selected so that the phase in a vacuum is higher than that in the specimen, which corresponds to the potential for electrons.

Figure 20a shows a phase profile extracted from the region marked by a white rectangle in Fig. 19b and averaged along its shorter side (over a distance of 5 nm). For convenience, the phase shift at the interface between the materials is set to 0 (by adding a constant phase offset) to simplify observations of the evolution of the material properties during dynamic operation of the cell. The interfaces between the electrolyte, the active Pt electrode, and the inactive W electrode result in two relatively sharp steps in the phase profile, owing to the large differences in the mean inner potentials.

Minor variations in the phase profile are likely to be associated with slight specimen thickness variations, dynamical contributions to the phase shift in the polycrystalline material and statistical noise. The slight phase gradient in the electrolyte layer in Fig. 20a may be associated with the fact that the ceria-based oxide can be reduced by electron beam illumination. This reaction can start at room temperature in the

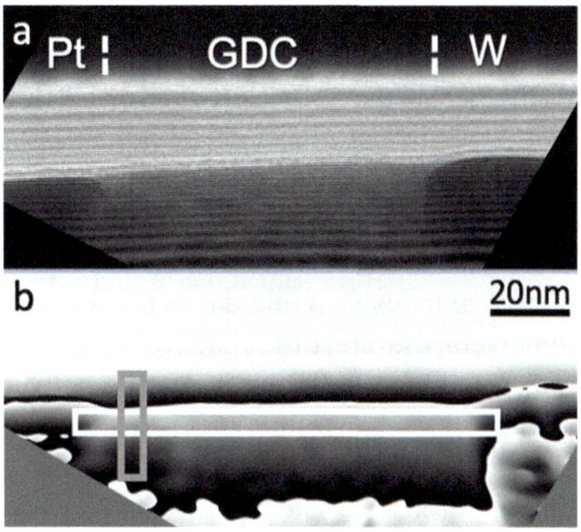

Fig. 19 Electron holography of a half SOFC cell Pt/GDC/W. **a** electron hologram with interference fringes perpendicular to the interfaces. **b** reconstructed phase image

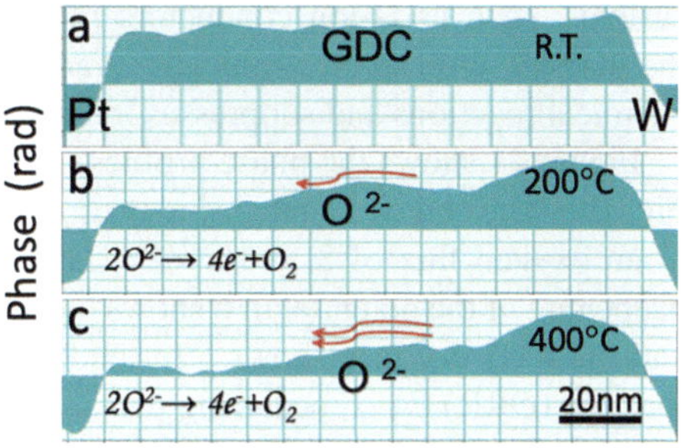

Fig. 20 Reconstructed phase image from the area indicated by the white rectangle in Fig. 19. **a** Reference phase profile at R.T. in a vacuum. **b, c** Phase profiles at 200 °C and 400 °C show the O^{2-} ion conduction in the electrolyte from the W side to the Pt side as shown with red arrows

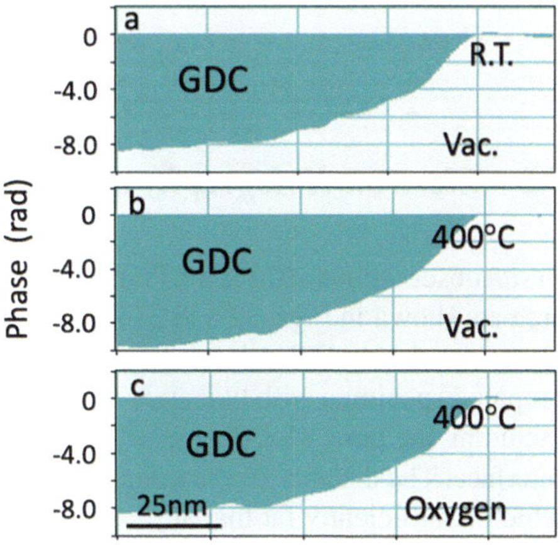

Fig. 21 Phase profiles of the area indicated the gray rectangle in Fig. 19 in different environments. **a** Reference profile acquired at R.T., **b** at 400 °C in a vacuum, and **c** at 400 °C in an oxygen flux

presence of an intense electron beam. Therefore, care is required in the irradiation by an electron beam. In order to minimize the influence of the electron beam on the results, all of the cells examined need to be under similar illumination conditions before the start of *in situ* observations.

Figure 20b, c shows phase profiles measured at 200 °C and 400 °C. The profiles were extracted from the phase image in the same white rectangular areas shown in Fig. 19b. When the temperature is raised, the potential distribution in the cell changes inhomogeneously. While the step in the potential at the active interface decreases at higher temperature, little change is observed at the inactive W electrode side, resulting in a steep slope in the phase profile between the two electrodes. This is thought to result from a change in oxygen vacancy concentration between the two electrodes, and can be explained by the catalytic selectivity of the Pt interface, resulting in O^{2-} ion migration in a system, as indicated with red arrows, that is charge-neutral overall. A similar behavior is expected in ionically conducting materials in other dynamic systems, such as H-ion conductors [36] and Li-ion conductors [37]. In contrast, the inactive W is an electronically conductive electrode, but acts as an obstacle for chemical reactions and suppresses ionic activity.

Figure 21 shows the phase profiles in the area indicated with a gray rectangle in Fig. 19b. The cell was heated slowly up to 400 °C in a vacuum and was then exposed to an oxygen flux of 3×10^{-2} Pa·ℓ/s at the same temperature in the TEM. The deepening of the potential in **b** shows the same reduction reaction Eq. 6 as in Fig. 20, and the recovery in **c** shows a re-oxidation process, that is, the oxygen reduction reaction (ORR) Eq. 7, occurring just at the Pt interface [38]:

$$O_2 + 4e^- \rightarrow 2O^{2-} \quad (7)$$

5.2 Electron Holography of YSZ

In situ observations at the YSZ/Pt interface and inside an oxide with no catalytic metal area are shown in Figs. 22 and 23 [39]. A specimen with two different thinned areas was prepared by FIB micro-sampling and tilting techniques. Figure 22 shows the prepared specimen with two areas of interest for *in situ* electron holography studies, including the pure YSZ (henceforth referred to as the interior) and the area near the interface. The distance of the thinned edge area from the Pt interface is almost 4 μm, which is sufficiently far that any catalytic effect can be neglected. The simultaneous preparation of the interior and interface areas makes it possible to assess the effect of the Pt interface on the solid oxide in various environments.

Electron holograms at the interface and interior areas are shown in Fig. 22. The specimen was gradually heated from room temperature to 400 °C in a high vacuum of 8×10^{-6} Pa and finally exposed to a flux of pure oxygen at a few tens of mPa·ℓ/sec at the same temperature, in a holographic TEM(HF-2000) with a gas inlet system. As

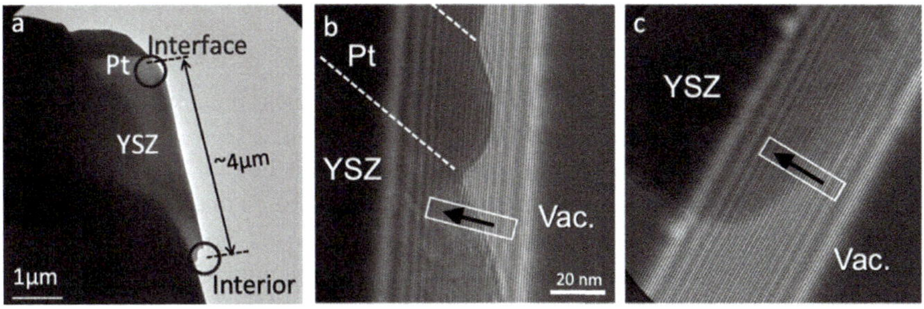

Fig. 22 **a** Low magnification TEM image of a specimen for *in situ* electron holography observations with two thinned areas, at the Pt interface and far from the interface (interior). The black circles approximately show the field of view for electron holography. **b, c** Holograms recorded at **b** the Pt interface area and **c** at the pure YSZ edge area (interior)

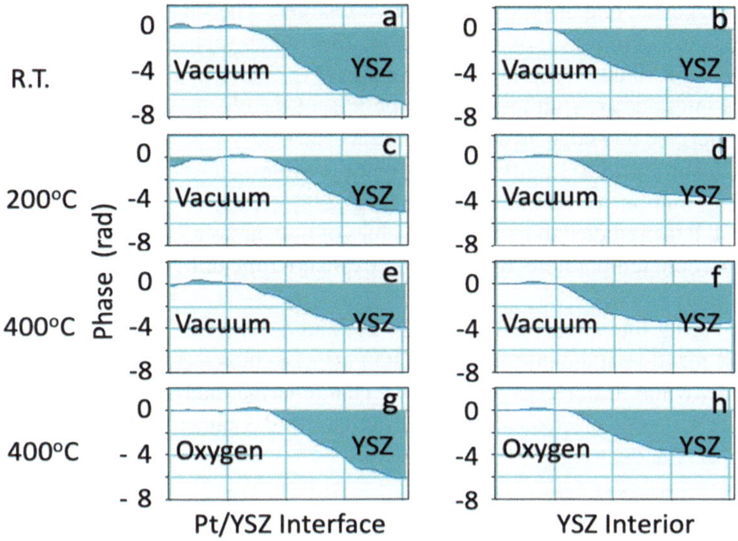

Fig. 23 Phase shift profiles at the interface area (left column) and pure oxide interior area (right column) in different environments: **a**, **b** at R.T., **c**, **d** at 200 °C, **e**, **f** at 400 °C in a vacuum and **g**, **h** at 400 °C in an oxygen atmosphere

the interface and interior areas have different geometries, two different observation orientations were required at every intended temperature and oxygen pressure.

From the reconstructed phase images of the interface and interior (pure YSZ) regions in various environments, profiles were extracted and plotted along the directions of the arrows in the rectangles in Fig. 22b and c, after averaging along the shorter sides of the rectangles. The phase profiles are summarized in Fig. 23. The profiles in the left column are those of the interface area and the right column profiles are of the interior edge area. The phase-shift value of the YSZ close to the interface at room temperature (Fig. 23a) is almost twice as large as that in the interior area (Fig. 23b). This difference is due to the difference in thickness.

Upon heating the specimen to 200 °C and 400 °C, a clear reduction in the phase shift values of YSZ is recognized at the Pt interface area (Fig. 23 c, e), which is considerably different from the minor changes in the pure oxide interior region (Fig. 23d, f). This clearly shows the catalytic effect of the Pt interface on the oxide electrolyte. This reduction in the phase shift values of YSZ is associated with increasing negative charges inside the oxide, which represents oxygen ion migration from the oxide toward the interface and out of the specimen near the Pt interface in the vacuum, that is, an oxygen enhancement reaction (OER), Eq. 6. In addition, this increase of negative charge in the material suggests that the oxide was reduced during the heating.

A reverse behavior in the phase shift of the oxide close to the Pt interface was observed just after the specimen was exposed to an oxygen flux at 400 °C (Fig. 23g), which suggests compensation of negative charges. The phase shift shows similar values for the initial solid oxide state (Fig. 23a). As expected, the minor changes in the phase shift value of YSZ at the interior area (Fig. 23d, f, h) are negligible compared with the interface area, although the directions of the changes are consistent in both the Pt-interface and edge areas.

The resulting phase shift profiles at the Pt-interface areas in a high vacuum (Fig. 23e) and an oxygen atmosphere (Fig. 23g) clearly suggest that ORR (Eq.7) occurred. It is predicted that the phase values of both areas recover to the exact initial values at room temperature after a longer exposure time or by increasing the oxygen flux pressure.

Comparing the results of GDC (Fig. 21) and YSZ (Fig. 23), we recognize that the OER reaction at the interface of GDC/Pt (Eq. 6) increases the positive charge in the oxide, while it increases the negative charge at the interface of YSZ/Pt. This is because GDC is an electron and oxygen-ion mixture conductor, while YSZ is a pure oxygen ion conductor.

5.3 TEM–EELS of GDC

EELS is one of the most powerful techniques for investigating electronic states of elemental atoms. *In situ* EELS is extremely useful for identifying cations that undergo a change in state of oxidation during a reaction and for obtaining information about the mechanism of the reaction. Figure 24 shows TEM images of a cross-sectional specimen of a Pt/GDC/Pt complete cell and Fig. 25 shows EDX elemental mapping a and profiles b. It is clear that a layer of GDC is sandwiched between two layers of Pt. EELS spectra acquired close to the interfaces of a specimen heated from room temperature to 400 °C in a vacuum are shown in Fig. 26. Magnified regions of the spectra, showing the energy-loss near-edge structure for Ce and Gd, are shown in Fig. 26b, c, respectively. The spectrum from the electrolyte has the structure expected for $Gd_2O_3 + CeO_2$ at room temperature.

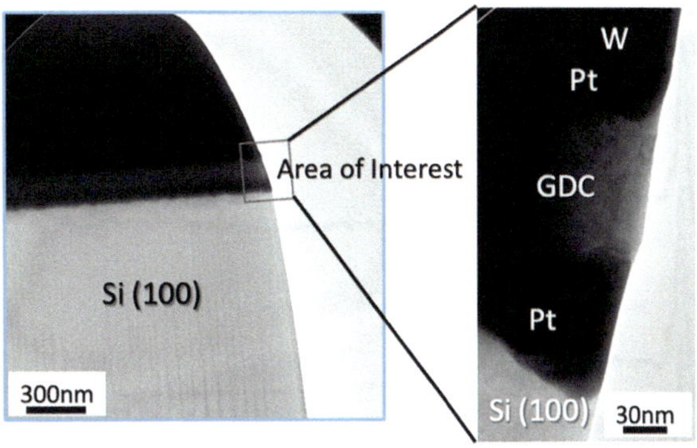

Fig. 24 TEM image of a full model cell of Pt/GDC/Pt on Si(100). W is a protective layer for FIB thinning

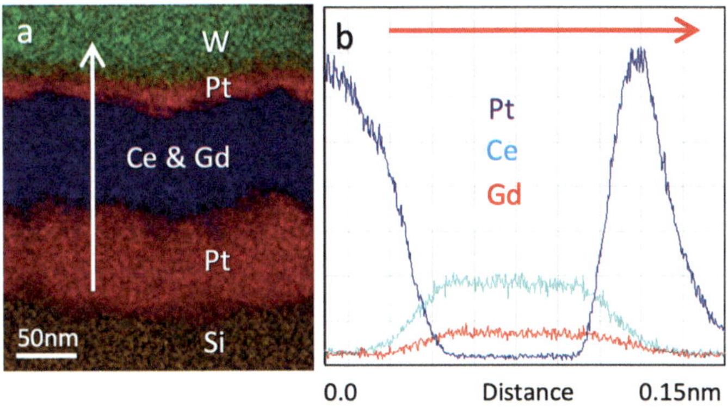

Fig. 25 Elemental mapping and profiles of a complete cell of Pt/GDC/Pt on a Si substrate by EDX. The profiles are measured along the direction of the arrows

There is no significant change when the specimen temperature is increased to 200 °C, despite the fact that changes in the mean inner potential of the GDC were observed by electron holography at this temperature. Instead, a change in the electronic state of the electrolyte is observed at 400 °C, with the Gd-M edge undergoing a chemical shift toward higher energy losses by up to about 3 eV. A similar chemical shift was reported for the Ce-M edge in the reverse direction during the reduction of Ce from tetravalent Ce^{4+} to trivalent Ce^{3+} [40]. Such a reduction would result in a reversal of the M4 and M3 white line intensities and a decrease in the satellite peaks, which are evident just after the white lines in the Ce^{4+} state [41, 42]. Neither change is observed for Ce in **b**. Furthermore, *in situ* electron holography results suggest that the solid oxide is oxidized rather than reduced at elevated temperature. Therefore, the observed chemical shift of Gd may suggest the presence of Gd in a new oxidation state of $Gd^{(4-x)+}$ ($0<x<1$). This oxidation state, which is supported by the electron holographic phase profiles described in the above section, is rare in Gd

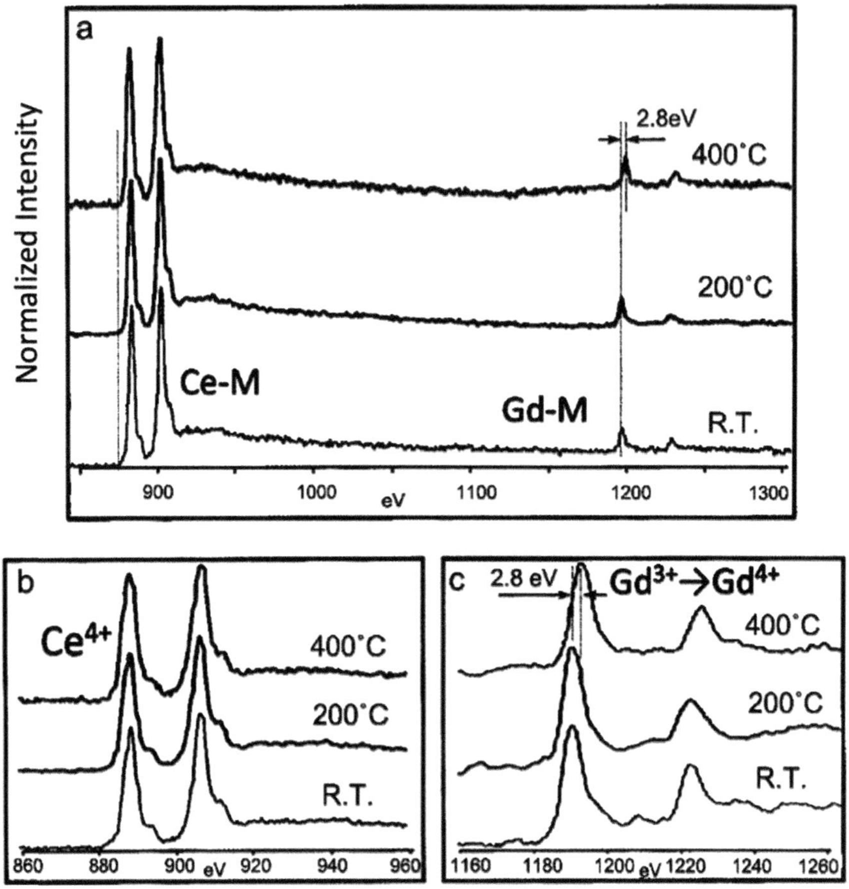

Fig. 26 EELS spectra by TEM at an interference region at room temperature, 200 °C and 400 °C. **a** Wide range spectra showing peaks of Ce-M and Gd-M. **b** The peak height ratio of the Ce-M edges and their shapes does not change with temperature, while **c** the Gd-M edges exhibit a change of peak height ratio and show an energy shift at 400 °C

oxide [43, 44]. Its identification suggests a significant role for dopant cations in the present electrolyte material, in which the host cations are expected to be responsible for ionic conduction and redox reactions.

These electron holography and EELS results suggest the importance of material selection in the optimization of SOFC design and properties.

5.4 TEM–EELS of YSZ at High Temperatures in a Vacuum

Investigations of the O-K edges in ZrO_2 and YSZ samples with varying compositions report that O-K near edge structures are very sensitive to local crystalline structures and dopant contents. The local microstructures of Pt/YSZ and the proton conductive oxide $SrZr_{0.9}Y_{0.1}O_{3-a}$ (SZO) have been studied by high resolution TEM [36]. There are also several reports of theoretical investigations.

of Pt/ZrO_2. There are only a few reports, however, on analytical TEM of Pt/YSZ or Pt/ZrO_2 interfaces. Srot et al. [45] performed an analytical study on Pt/YSZ at R.T., characterizing the basic features of the interfaces.

Figure 27a shows the cross-sectional TEM image of a specimen with the areas of interest indicated by circles, where the spectra were acquired in different environments. Figures 27b and 28 are EELS spectra acquired from the YSZ near the interface region with a Pt layer (circle I) and the region about 1.5 μm from the interface (circle II). The measurements were performed by 1000 kV E-TEM/STEM (JEM 1000 K RS).

The outer shells of Zr and Y (i.e., the L edges) are considered to monitor any probable partial changes in these cations' oxidation states. An imaging filter (Gatan GIF) with 2048 channels sets the energy dispersion to 0.3 eV per channel so that the spectra cover both elements' L edges. The energy resolution in this experiment is almost 1.1 eV, according to the zero-loss spectra full width at half maximum method (FWHM). Figure 27b shows a typical spectrum of the YSZ structure including Y-$L_{2,3}$ and Zr-$L_{2,3}$ white lines. The pre-edge background of this spectrum at the Y-L edges was subtracted, and extracted signals were calibrated according to the onset of the Y-$L_{2,3}$ white line at 2080 eV. Measurements were made on the two areas of interest while heating the specimen from R.T. to 100, 300, and 500 °C.

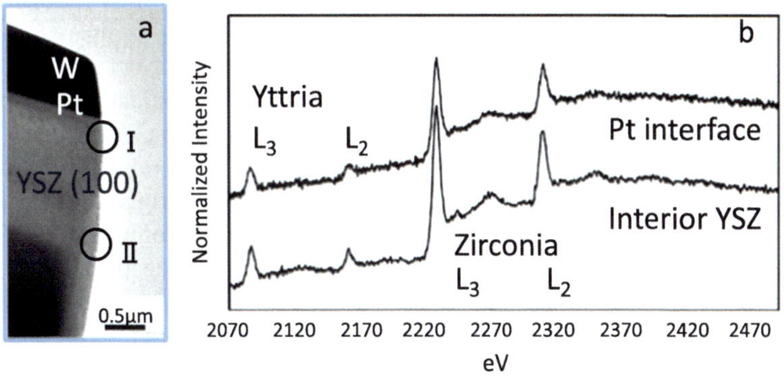

Fig. 27 **a** Cross section of a half cell, YSZ/Pt. **b** Electron energy loss spectra of yttria-stabilized zirconia from the Pt interface and the interior pure oxide areas. Both spectra show typical structures of yttria and zirconia, with a negligible effect from the Pt background in the spectra acquired from the area close to the Pt interface

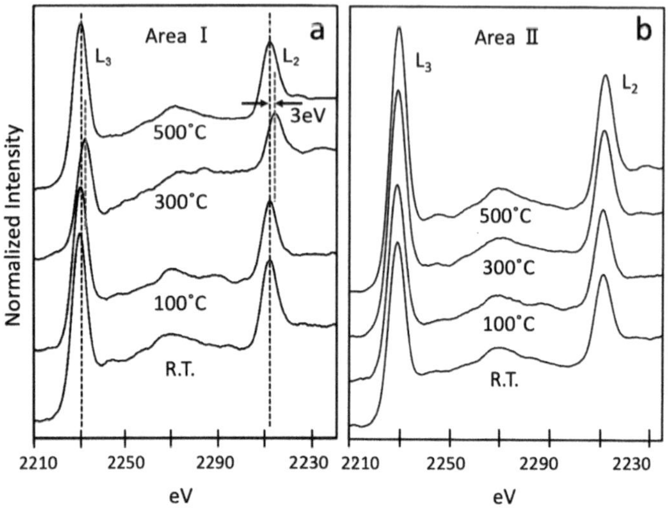

Fig. 28 **a** EELS spectra of Zr-L in YSZ from the interior pure oxide area I at different temperatures. No discernable change was seen in the electronic structure of the Zr from room temperature to 500°CC. **b** Spectra from area II. A reversal was observed in the L_3 and L_2 white line intensity ratios at 300 °C, accompanied by a considerable chemical shift in the spectra

No strong change is detected in the Y-L edges at any position and in any environment. Hence, all spectra were calibrated according to the onset of the Y-L_3 line, and the pre-edge Zr-L_3 background was subtracted from every spectrum. Figure 28 summarizes all the processed spectra (background subtracted) at different temperatures. There is no strong change from R.T. up to 500 °C in Fig. 28b, while there is a significant difference at 300 °C in the spectra acquired from the Pt interface area, as shown in Fig. 28a, where the L_3 and L_2 white lines intensity ratio changes, accompanied by a considerable chemical shift of almost 3 eV compared with the spectra acquired at R.T. and 100 °C. The phenomenon then recovers at 500 °C. This change at 300 °C is associated with a change in the zirconium oxidation state, where the valence state of Zr^{4+} was reduced to a metastable oxidation state of Zr^{3+}. This effect is consistent with oxygen deficiency in the bulk zirconia oxidation process [46] and is associated with the role of oxygen vacancies in YSZ in triggering a reduction reaction in commercial SOFC anode material [47]. This result completely confirmed the in situ electron holography data. It also explains that the negative charge found at high temperatures in a high vacuum was caused by a reduction in the zirconium oxidation state.

Although YSZ is a very good ionic conductor at high temperature, it is also well known to be an electronic insulator [48, 49]. Hence, OER occurred during the heating process and oxygen ions were released from the Pt interface as neutral molecules, and the compensation for the remaining charges reduced the ionic valence of the Zr. This is consistent with increasing the negative charge in electron holography profiles as mentioned in Sect. 7.5.2. It is important to note that 300 °C is not a sufficiently high temperature to activate oxygen ion movement inside the bulk zirconia structure through the oxygen vacancies introduced by the doped yttria. Moreover, it is reasonable that at 500 °C, which may be considered to be the onset of the thermal range of ionic conductivity in zirconia, oxygen ions would partially migrate toward the interface area due to the chemical potential developed at the Pt interface, and the reduced oxidation state of Zr (Zr^{3+}) would return back to the steady state of Zr (Zr^{4+}). This can be explained by the anodic reaction in real SOFCs with YSZ electrolytes, in which oxygen ions transfer the carrier charge to the anode and finally to the current collector electrode at temperatures higher than 500 °C.

It appears that *in situ* electron holography has a higher sensitivity than other analytical electron microscopy techniques for detecting small changes inside objects, as it can reveal the confirmed trend of the reaction event at relatively low temperatures.

6 EELS of YSZ with a Voltage Applicable Heating Specimen Holder

The oxygen behavior was observed at the interface between an electrode and electrolyte in nano-scale reacting areas. This section describes the observation and measurement of the oxygen distribution at the interface of an SOFC specimen which consists of Pt electrodes and a YSZ electrolyte using the voltage applicable heating specimen holder by EELS in both the TEM- and STEM-modes of E-HVEM.

6.1 Optimization of Experimental Conditions

Determining the experimental conditions is important for observing the oxygen distribution at an interface during electrode reactions. Oxygen ions play an important role in the ion conduction and electrode reactions of SOFCs. In other words, the above phenomena can be observed with an oxygen concentration around the interface of active sites. In order to clearly measure and observe the oxygen distribution, it is necessary that the structure of the SOFC specimens is stable without degradation of the materials through the experiment.

Figure 29 shows TEM images of a cross-sectional SOFC specimen observed before and after heating. In many cases, Pt atoms of the electrode diffuse into the adjoining layer in an oxygen atmosphere. The structural change and atomic diffusion obstruct the electrode reactions. The diffusion of the Pt atoms depends on the oxygen partial pressure and the specimen temperature, as shown in Fig. 30. In the following experiments, the optimized environmental conditions to ensure no structural changes were set to be an oxygen atmosphere of 0.5 Pa and temperature of 400 °C. STEM-EELS measurements can be performed stably in this condition.

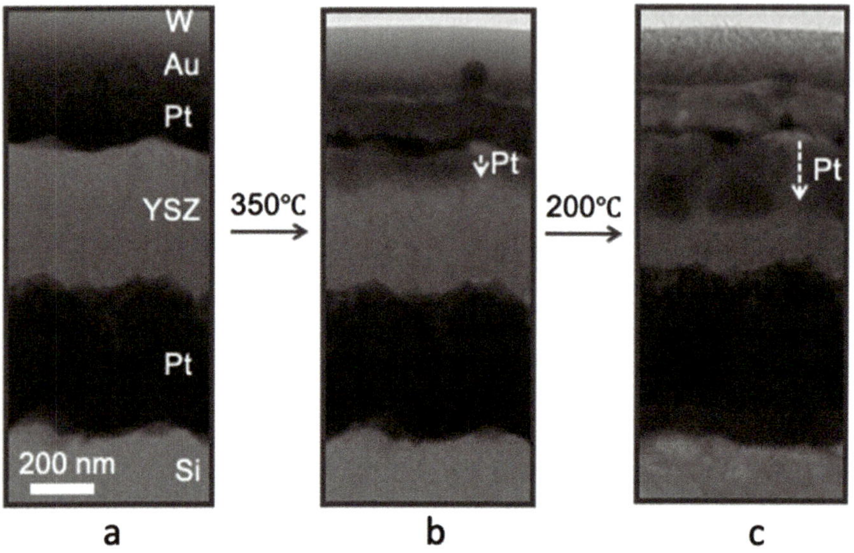

Fig. 29 TEM images of a specimen acquired before and after heating in an oxygen atmosphere of 0.1 Pa. The specimen temperature is 50 °C (**a**) 400 °C (**b**), 600 °C (**c**)

Fig. 30 Conditions of structural stability in SOFC structures. Diffusion of Pt atoms is promoted by a high oxygen partial pressure and high temperature

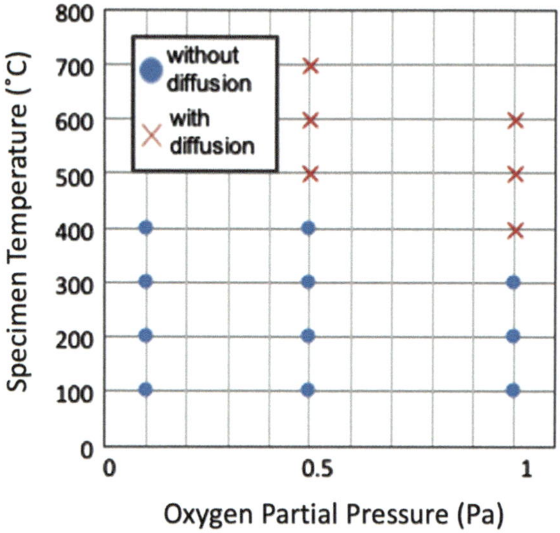

6.2 *Observation by TEM–EELS*

Figure 31a shows a TEM image of a specimen consisting of Pt electrodes and a YSZ electrolyte with a protective layer, with a voltage applied to the specimen. Results of

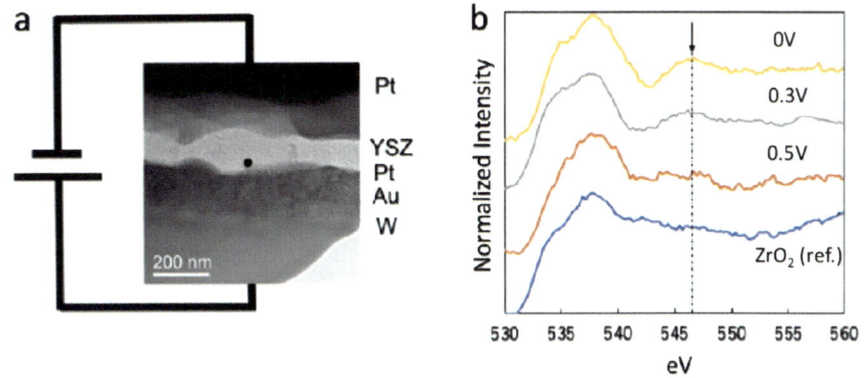

Fig. 31 **a** TEM image of an SOFC specimen, **b** O-K edge spectra in YSZ and $Zr0_2$. The O-K edges of YSZ were acquired at the black dot in (**a**). The arrow in **b** indicates the peak of the O-K edge

an in situ TEM–EELS measurement are shown in Fig. 31b, at 500 °C in an oxygen atmosphere of 0.5 Pa. The oxygen K edges in YSZ are compared with the acquired O-K edge of ZrO_2 as a reference.

The changes of the O-K edge spectra due to voltage application imply an increase of oxygen ions near the interface. When a voltage of +0.5 V is applied, the O-K edge changes at the peak indicated by the arrow in Fig. 31b. Considering that the shape of the spectrum is close to that of ZrO_2, we can see that oxygen ions of the YSZ increased at the interface by the applied positive voltage. A detailed discussion of the change of the O-K edge spectrum is described in [33]. This result shows that fluctuations of the oxygen ions during the electrode reaction at the interface can be detected with EELS spectra.

TEM–EELS measurement in this experiment did not provide precise information on the position because a TEM image and an EELS spectrum cannot be acquired simultaneously. Detailed information regarding oxygen ion concentration at the interface in an activated SOFC specimen can be obtained by STEM-EELS measurements, which provide the simultaneous measurement of a STEM image and EELS spectra. On the other hand, TEM–EELS measurements can be carried out easily with the structural stability condition described in the previous section. The selection of TEM and STEM modes depends on the measurement targets.

6.3 *Observation by STEM-EELS*

Figure 32a shows an annular dark field (ADF)-STEM image of the specimen. The E-HVEM was operated in STEM mode at an acceleration voltage of 1000 kV. Oxygen distribution images were reconstructed from the two-dimensional data including an O-K edge EELS spectrum acquired by STEM-EELS. The observation conditions were as follows: The specimen temperature, voltages applied to the specimen, and

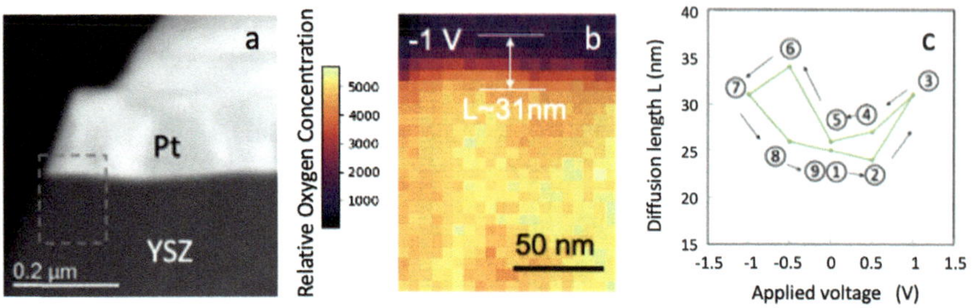

Fig. 32 **a** ADF-STEM image of an SOFC specimen, **b** example oxygen distribution around the Pt/YSZ interface and **c** applied voltage dependence of the diffusion length. The dashed box in **a** is the acquisition area of STEM-EELS. The Pt electrodes have bias potentials: −1 V in the case of (**b**). The numbers in **c** indicate the acquisition order of EELS spectra. The diffusion lengths are measured from the oxygen concentration at the interface obtained from the EELS spectra

the O_2 partial pressure were 400 °C, −1 to 1 V, and 0.5 Pa, respectively. For these conditions, the structure is confirmed to be stable. With an applied external voltage to the Pt electrode at the upper side in Fig. 32a, the oxygen concentration map obtained around the Pt/YSZ interface is shown in Fig. 32b. The interior of the YSZ has a high oxygen concentration, whereas the oxygen around the Pt/YSZ interface decreases from that in YSZ to Pt. To estimate the influence of the electrode reactions on the oxygen ion distribution, the length L shown in Fig. 32b is defined as a diffusion length. Figure 32c shows the applied voltage dependence of L. L extends for both positive and negative applied voltages. The change of L is reversible by changing the applied voltage. It is found that L extends ~6 nm by a positive external voltage to the Pt electrode and ~9 nm by a negative voltage. This result implies that diffusion lengths are changed by electrode reactions and oxygen ion migrations. When a positive voltage is applied to a Pt electrode, Pt oxides are formed during an anodic reaction at the interface, and then L lengthens. When a negative voltage is applied to the electrode, oxygen ions move to the interior of the YSZ from the Pt/YSZ interface. Thus, an extension of L is observed in the same way. The result for the anodic reaction supports the explanation by impedance spectroscopy [50].

6.4 *Observation by STEM-EELS with Electrochemical Measurements*

Figure 33 shows *in situ* STEM-EELS with the electrochemical measurement of an SOFC specimen around the interface. To measure the current generated by electrolysis, a pico-ammeter was connected between the Pt electrode and the power supply. The specimen was heated to 400 °C while the pressure in the specimen chamber was kept at 0.5 Pa oxygen. An external voltage was also applied to the Pt electrode up to +2 V.

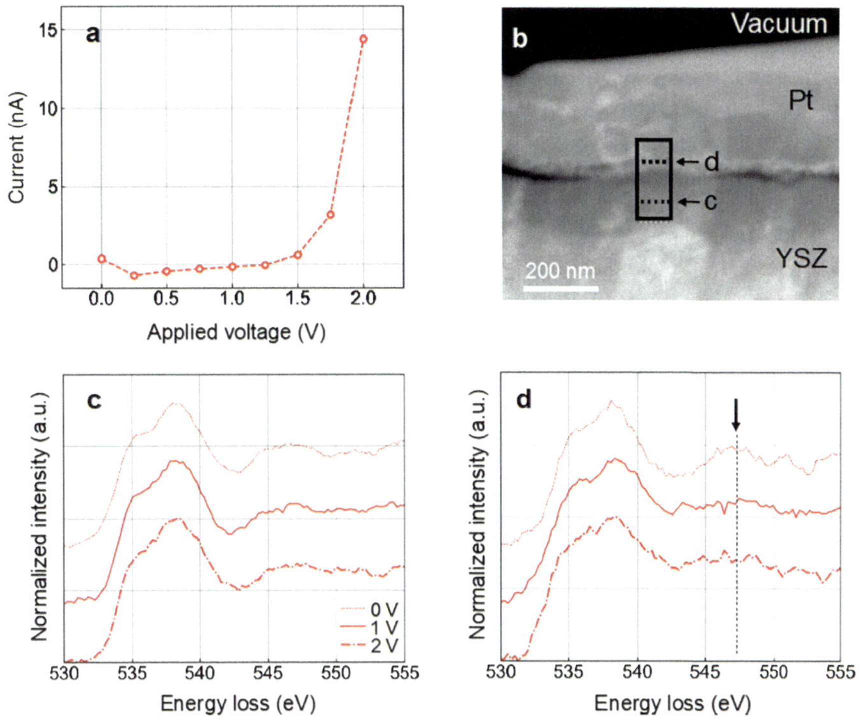

Fig. 33 **a** Electrolysis current as a function of the applied voltage to the bottom electrode, **b** ADF-STEM image of the YSZ/Pt interface and **c**, **d** EELS spectra of O-K edges at the interior of the YSZ and the YSZ/Pt interface, respectively. The EELS spectra of (c) and (d) were taken at the positions of the dashed lines indicated by arrows in (**b**)

Figure 33a shows the electrolysis current as a function of the voltage applied to the platinum electrode, which has the typical features of an electrolysis current [51]. This means that the Pt/YSZ/Pt system functions as an electrolysis cell, appropriately. The decomposition voltage, that is the minimum voltage for the onset of electrolysis, is found to be approximately 1.6 V in this figure. Based on electro-chemistry, an electrical double layer is formed by applying a voltage under 1.6 V at the interface. This is consistent with the electron holographic results shown in Sect. 7.5.1. Figure 33**c** and **d** shows the O-K edge spectra extracted from the area marked with dashed lines in Fig. 33b. The O-K edge spectra of the interior (Fig. 33c) do not change by applying a voltage, while the features of the interface spectra (Fig. 33d) clearly show a change at the peak indicated by the arrow in Fig. 33d. This result agrees with the TEM–EELS measurements in the previous Sect. 5.2. The electrolysis current and these EELS spectra indicate that part of the O-K edge spectra shape is altered by the oxygen ion concentration in the YSZ electrolyte.

Furthermore, the correlation between the electrolysis current and the O-K edge spectra shape suggests the following phenomena: In the case of no extra voltage, oxygen ions in the YSZ electrolyte are stationary and the peak shape does not change. Increasing the voltage up to the decomposition voltage, oxygen ions move to the interface and form an electrical double layer. This phenomenon is seen as a decrease in the peak in the O-K edge. When the voltage exceeds the decomposition voltage, the peak in the O-K edge is close to that of the no extra voltage case. This implies that the electrical double layer is resolved, and electrolysis reactions are promoted by the applied voltage. The above phenomena, which generally occur in an electrolysis cell of liquid, were observed in a solid electrolysis cell as well. Therefore, the oxygen ion concentration around an interface can be revealed effectively.

7 Summary and Perspectives

In this chapter, we have described analytical microscopy by TEM and STEM and its applications to interfaces between electrodes and electrolytes. Electron holography, EELS, and EDX are characteristic techniques, besides high-resolution imaging, for the observation of nano-scale structures, including atomic structures, electronic structures, distributions of elements, and electrostatic potential distributions inside materials. Observing these structures will help improve the performance of SOFCs.

Note that we have not introduced any observations of SOFCs here, but have only modeled SOECs. This is because for SOFCs we need to introduce two kinds of gases independently under a high specimen temperature, which requires a sufficiently large specimen chamber. HVEM and aberration corrected TEM/STEMs having large specimen chambers will improve *in situ* experiments and make even operando observations easier, which are performed under actual operation conditions. Their outcomes should contribute to the development of new SOFC materials.

Funding Support This interactive project was supported in part by the Global Research Center for Environment and Energy based on Nanomaterials Science (GREEN, from FY2010 to FY2018).

References

1. Fujita H, Tabata T (1973) Voltage dependence of the maximum observable thickness by electron microscopy up to 3 MV. Jpn J Appl Phys 12:471–472
2. Scherzer O (1949) The theoretical resolution limit of the electron microscope. J Appl Phys 20:20–29
3. Gabor D (1949) A new microscopic principle. Nature 161:7778
4. Tonomura A, Matsuda T, Endoh J, Todokoro H, Komoda T (1979) Development of a field emission microscope. J Electron Microsc 28:1–11
5. Tonomura A, Matsuda T, Endo J (1979) Spherical-aberration correction of an electron lens by holography. Jpn J Appl Phys 18:1373–1378
6. Tanji T, Urata K, Ishizuka K, Ru Q, Tonomura A (1993) Observation of atomic surface potentials by electron holography. Ultramicroscopy 49:259–264
7. Tonomura A, Matsuda T, Endo J, Arii T, Mihama K (1980) Direct observation of fine structure of magnetic domain walls by electron holography. Phys Rev Lett 44:1430–1433

8. Möllenstedt G, Düker H (1956) Beobahtungen und Messungen an Biprisma-Interferenzen mit Elektronenwellen. Z Phys 145:377–397
9. Wollaston WH (1813) A method of drawing extremely fine wires. Philos Trans R Soc Lond:114–8
10. Yada K, Shibata K, Hibi T (1973) A high resolution electron interference microscope and its application to the measurement of mean inner potential. J Electron Microsc 22:223–230
11. Hirayama T, Ru Q, Tanji T, Tonomura T (1993) Observation of magnetic-domain states of barium ferrite particles by electron holography. Appl Phys Lett 63:418–420
12. Tanji T, Hasebe S, Nakagami K, Yamamoto K, Ichihashi M (2004) Observation of magnetic multilayers by electron holography. Microsc Microanal 10:146–152
13. Dunin-Borkowski RE, McCartney MR, Frankel RB, Bazylinski DA, Pósfai M, Buseck PR (1998) Magnetic microstructure of magnetotactic bacteria by electron holography. Science 282:1868–1870
14. Rau WD, Schwander P, Baumann FH, Höppner W, Ourmazd A (1999) Two-dimensional mapping of the electrostatic potential in transistors by electron holography. Phys Rev Lett 82:2614–2617
15. Möllenstedt G (1949) The electrostatic lens as a high resolving power velocity analyst. Optik 5:499–518
16. Watanabe H (1956) Experimental evidence for the collective nature of the characteristic energy loss of electrons in solids. J Phys Soc Jpn 11:112–119
17. Krivanek OL, Ahn CC, Keeney RB (1987) Parallel detection electron spectrometer using quadrupole lenses. Ultramicrosc 22:103–115
18. Botton G (2007) Analytical electron microcopy. In: Hawkes PW, Spence JCH (eds) Science of microscopy, vol I. Springer, New York, pp 273–405
19. Terauchi M, Koike M, Fukushima K, Kimura A (2010) Development of wavelength-dispersive soft X-ray emission spectrometers for transmission electron microscopes—an introduction of valence electron spectroscopy for transmission electron microscopy. J Electron Microsc 59:251–261
20. Strüder S, Soltau H (1995) High resolution silicon detectors for photons and particles. Radiat Prot dosim 61:39–46
21. Ruska E (1942) Über die Linsen hochauflösender Elektronenmikroskope. Archiv f. Elektrotechnik 36:431–454
22. Ito T, Hiziya K (1958) A specimen reaction device for the electron microscope and its applications. J Electron Microsc 6:4–8
23. Hashimoto H, Tanaka K, Yoda E (1958) A specimen treating device at high temperature for the electron microscope. J Electronmicroc 6:8–11
24. Hashimoto H (1960) Crystal growth of oxide and sulphide. J Electronmicroc 9:130–137
25. Hashimoto H, Naiki T, Eto T, Fujiwara K (1968) High temperature gas reaction specimen chamber for an electron microscope. Jpn J Appl Phys 5:946–962
26. Allinson DL (1970) Environmental cell for use in a high voltage electron microscope. In: Favard P (ed) Electron microscopy (Proceedings of the 7th international congress electron microscopy, Aug 30–Sep 5 1970; Grenoble (France)). French Society of Microscopies, Paris, pp 169–170
27. Takeda S, Yoshida H (2013) Atomic-resolution environmental TEM for quantitative in-situ microscopy in materials science. Microscopy 62:193–203
28. Tavabi AH, Yasenjiang Z, Tanji T (2011) In situ off-axis electron holography of metal-oxide hetero-interfaces in oxygen atmosphere. J Electron Microsc 60:307–314
29. Tanaka N, Usukura J, Kusunoki M, Saito Y, Sasaki K, Tanji T, Muto S, Arai S (2013) Development of an environmental high-voltage electron microscope for reaction science. Microsc 63:205–225
30. Jungjohann K et al (2016) *In situ* operando. In: Carter CB, Williams DB (eds) Transmission electron microscopy: diffraction, imaging, and spectrometry. Springer, pp 17–80
31. Moritomo H, Ohura M, Tanji T, Enomoto S (2008) New specimen holder with 4 electrodes. In: Tanaka N, Smith D, Baba, N, Midgley P, Lichte H, Mayer J, Colliex C, Etheridge J (eds) Proceedings of 16th international microscopy congress vol II; Sapporo, Japan, 3–8 Sept 2006. Elsevier, Amsterdam, p 1154

32. Tanji T, Moritomo H, Shimura T, Kato T, Hirayama T (2007) Electron holography of a hetero-interface in a solid oxide fuel cell. Microsc Microanal 13:1224–1225
33. Ishida T, Hiroshima H, Higuchi K, Tomita M, Saitoh K, Tanji T (2020) Development of a voltage applicable heating specimen holder for in situ observations of solid oxide fuel cells with an environmental transmission electron microscope. Surf Interface Anal 52:584–590
34. E.g. DENSsolusions (https://denssolutions.com) and Hummingbird Scientific (http://hummingbirdscientific.com)
35. Chrisey DB, Hubler GK (eds) (1994) Pulsed laser deposition of thin films. Wiley, Now York
36. Matsumoto H, Furuya Y, Okada S, Tanji T, Ishihara T (2007) Effect of dispersion of nanosize platinum particles on electrical conduction properties of proton-conducting oxide $SrZr_{0.9}Y_{0.1}O_{3-a}$. Electrochem Solid State Lett 10:11–3
37. Yamamoto K, Iriyama Y, Asaka T, Fujita H, Fisher CAJ, Nonaka K, Sugita Y, Ogumi Z (2010) Dynamic visualization of the electric potential in an all-solid-state rechargeable lithium buttery. Angew Chem Int Ed 49:4414–4417
38. Tavabi AH, Arai S, Muto S, Tanji T, Dunin-Brokowski RE (2014) In situ transmission electron microscopy of ionic conductivity and reaction mechanisms in ultrathin slid oxide fuel cells. Micros Microanal 20:1817–1825
39. Tavabi AH, Arai S, Tanji T (2012) In situ analytical electron microscopy studies of redox reactions at a YSZ/Pt interface. Microsc Microanal 18:538–544
40. Crozier PA, Wang R, Sharma R (2008) In situ environmental TEM studies of dynamic changes in cerium-based oxide nanoparticles during redox processes. Ultramicroscopy 108:1432–1440
41. Garvie LAJ, Buseck PR (1999) Determination of Ce^{4+}/Ce^{3+} in electron-beam-damaged CeO_2 by electron energy-loss spectroscopy. J Phys Chem Solids 66:1943–1947
42. Arai S, Muto S, Murai J, Sasaki T, Ukyo Y, Kuroda K (2004) Valence change of cations in ceria-zirconia solid solution associated with redox reactions studied with electron energy-loss spectroscopy. Mater Trans 45:2951–2955
43. Landheer D, Gupta JA, Sproule GI, McCaffrey JP, Graham ML, Yang KC, Lu ZH, Lennard WN (2001) Characterization of Gd_2O_3 films deposited on Si(100) by electron-beam evaporation. J Electrochem Soc 148:G29-35
44. Adhikari R, Karmakar D, Ghatak J (2010) Gd-doped SnO_2 nanoparticles: structure and magnetism. J Magn Magn Mater 322:3631–3637
45. Srot V, Watanabe M, Scheu C, Van Aken PA, Salzberger U, Luessen B, Janek J, Rühle N (2010) Characterization of chemical composition and electronic structure of Pt/YSZ interfaces by analytical transmission electron microscopy. Soli State Ionics 181:1616–1622
46. Douglass DL, Wagner C (1966) The oxidation of oxygen-deficient zirconia and its relationship to the oxidation of zirconium. J Electrochem Soc 113:671–676
47. Jeangros Q, Faes A, Wagner JB, Hansen TW, Aschauer U, Van Herle J, Hessler-Wyser A, Dunin-Borkowski RE (2010) In situ redox cycle of a nickel-YSZ fuel cell anode in an environmental transmission electron microscope. Acta Mater 58:4578–4589
48. Mench MM (2008) Fuel cell engines. John Wiley and Sons Inc., Hoboken NJ
49. O'Hayre R, Cha SW, Colella W, Prinz FB (2009) Fuel cell fundamentals, 2nd edn. John Wiley and Sons Inc., Hoboken NJ
50. Pöpke PH, Mutoro E, Raiß C, Luerßen B, Amati M, Abyaneh MK, Gregoratti L, Janek J (2011) The role of platinum oxide in the electrode system $Pt(O_2)$/yttria-stabilized zirconia. Electrochim Acta 56:10668–10675
51. Hamann C, Hamnett A, Vielstich W (2007) Electrochemistry, 2nd edn. Wiley-VCH, Weinheim

TEM Characterisation of Electrode–Electrolyte Interfaces in Solid-State Fuel Cells and Batteries

Lingbing Ran, Emily R. Cooper, Shiwei Tao, and Ruth Knibbe

Abstract Solid heterointerfaces are a common source of weakness in solid oxide cells (SOCs) and solid-state batteries (SSBs). Although pristine interfaces may be produced during synthesis, the quality of these interfaces may degrade during operation due to chemical and electrochemical instabilities or high operating temperatures (for SOCs). As these interfacial changes are often limited in scale (nm to μm), bulk characterisation techniques often cannot track and trace these changes. In light of this, transmission electron microscopy (TEM), plays a unique role in interface characterisation. This chapter explores solid hetero-interfacial characterisation in both SOCs and SSBs using advanced TEM techniques. This includes high resolution scanning TEM (HR-STEM), electron energy loss spectroscopy (EELS) and in-operando TEM. Furthermore, new and emerging techniques including cryo-TEM, 4D-scanning TEM (4D-STEM) and developing in-operando TEM methods are also discussed.

Keywords Solid interfaces · Solid oxide cells · Solid-state batteries · TEM · In-operando · HRSTEM · EELS · Cryo-TEM

1 Introduction

Solid heterointerfaces are a common source of weakness in solid oxide cells (SOCs) and solid-state batteries (SSBs). SOCs and SSBs are both electrochemical systems which are constructed of an anode, solid electrolyte and a cathode. Both systems can be operated reversibly—the SSB being charged and discharged and the SOC can be operated as a fuel cell (SOFC) or an electrolysis cell (SOFC). However, chemical and electrochemical instabilities and high temperatures (for SOFCs) in these systems put the heterointerfaces under duress. Understanding the driving forces and mechanisms associated with these interfaces is key in designing new and improved solid-state energy conversion systems.

L. Ran · E. R. Cooper · S. Tao · R. Knibbe (✉)
School of Mechanical and Mining Engineering University of Queensland, St Lucia, QLD 4072, Australia
e-mail: ruth.knibbe@uq.edu.au

T. Mori et al. (eds.), *Design of Active Key Interfaces in Fuel Cells*,
https://doi.org/10.1007/978-981-97-5706-0_8

Electron microscopy is a key characterisation tool to characterise solid electrode–electrolyte interfaces. Standard scanning electron microscopy (SEM) can be used to reveal complex microstructures, and transmission electron microscopy (TEM) can provide a higher-resolution insight into microstructures, coupled with information on crystal structure, disorder and valence states. More recently, the development of in situ and in-operando TEM characterisation has opened the possibility to observe materials under dynamic conditions. This chapter will focus on advanced TEM characterisation of solid electrode–electrolyte interfaces including high resolution scanning TEM, electron energy loss spectroscopy and in-operando TEM. Furthermore, new and emerging techniques including cryo-TEM, 4D-scanning TEM (4D-STEM) and in-operando TEM methods will be discussed.

One major benefit of TEM is the relatively high spatial resolution that can be obtained. This allows research to focus on nanoscale features that cannot be collected through other techniques. For example, the observation of interfacial defects will have a direct impact on the strength and long-term stability of solid electrode–electrolyte interfaces. However, the high resolution also carries a limitation as the TEM operator needs to be judicious in selecting images or videos that are representative of the entire sample. Using TEM in conjunction with techniques such as X-ray photoelectron spectroscopy (XPS) and X-ray diffraction (XRD) are key in providing a detailed insight into these interfaces.

1.1 SOC Introduction

A solid oxide cell (SOC) is an electrochemical cell with a porous fuel electrode, dense ceramic electrolyte and porous oxygen electrode. SOCs can be operated either in fuel cell or electrolysis mode. During fuel cell mode (SOFC), the cell is supplied with a fuel (e.g. hydrogen, methane or a short-chain hydrocarbon) to the fuel electrode and oxygen, or air, to the air electrode. During SOFC operation, energy is produced. During electrolysis (SOEC) mode, the SOC fuel electrode is supplied with carbon dioxide or steam which is converted into carbon monoxide and hydrogen respectively—as such energy is supplied and converted to a fuel. The produced hydrogen can be subsequently directly used as a fuel and the carbon monoxide can be used to convert to synthetic fuels.

The typical operating temperature of a SOC is around 650–850 °C. This high operating temperature also allows for the use of non-precious metal catalysts and direct oxidation of hydrocarbon fuels. However, the high operating temperatures can present issues due to thermal mismatch at interfaces, long-term degradation of oxide materials and agglomeration of metal catalysts.

The first generation of SOCs was designed around a thick ceramic electrolyte—typically an yttria stabilised zirconia (YSZ) ceramic. The second generation of SOCs uses an anode supported configuration. In this configuration, the porous anode support is several hundred microns thick and the electrolyte is a dense, ca. 12 microns thick YSZ layer. The oxygen electrode for the second generation cells is around 30 microns typically made of a composite of YSZ and LSM ((La,Sr)MnO_3). Despite the thickness of these electrodes, the electrochemically active region is confined to several microns adjacent to the dense electrolyte.

Over the past decades there has been a push to decrease the SOC operating temperature, but this is limited by the sluggish kinetics of the oxygen electrode—either oxygen evolution reaction (OER) during electrolysis or oxygen reduction reaction (ORR) during fuel cell [1]. However, this reduction in operating temperature could make the 50,000 h design life for commercial SOCs more achievable.

1.2 SSB Introduction

SSBs have gained attention in the past decades due to their high safety and energy density. The SSB operation principle is the same as conventional liquid electrolyte batteries, but the flammable liquid electrolyte is replaced with a non-flammable solid electrolyte (SE). The use of the SE also allows for high specific capacity lithium or sodium metal to be used as the anode, which enhances the SSBs energy density. Although the high rigidity and mechanical strength of the SEs can prevent dendrite growth and penetration, the problem with dendritic growth isn't completely eliminated in SSBs [2].

Many kinds of SEs have been developed in the past few decades, for example, oxide [3–7], sulphide [8–10] and polymer-based [11–14]. After years of development, the ionic conductivity of some SEs can compete with the liquid electrolytes [10]. Despite this, the electrochemical performance, especially the rate performance, still needs further improvement because of the large interfacial resistance [15]. The poor wettability of SEs leads to the poor interface contact, which blocks ion transport. Besides, the unstable chemical or electrochemical interface can also degrade the electrochemical performance especially cycling performance.

1.3 Driving Forces for Degradation

Interfacial issues impede the electrochemical performance especially under high current densities for SOCs and under high-rates for SSBs. The driving forces for interfacial degradation can be grouped into physical contact, chemical stability and electrochemical stability.

1.3.1 Physical Contact

The physical contact includes two different regions—the electrolyte-active material interfaces in the composite electrodes, and the electrode–electrolyte interfaces. There are two levels of physical contacts—the initial contact before operation and the long-term dynamic contact during operation. Both the mechanical properties of the SEs and the associated interface have an important effect on the evolution of the SOC or the SSB performance.

The poor initial contact in SOCs and SSBs is mainly ascribed to the poor wettability of SEs due to the rigid properties. For example, oxide electrolytes such as $Li_{6.19}Al_{0.28}La_3Zr_2O_{12}$ and $LiTi_2(PO_4)_3$ show a high Young's Modulus (E) of 150 GPa [16] and 143 GPa respectively [17]. In comparison, the sulphide electrolytes, used in SSBs, show a low value of Young's Modulus of 20 GPa [18]. The reduced rigidity of the sulphide electrolytes makes it easier to form an intimate adhesion with the electrodes, but the poor fracture properties of these electrolytes means easily fracture when mechanically stressed [19]. The high Young's Modulus of the oxide ceramics reduces the intimacy of the active material-electrolyte interface resulting in slow ion transportation (Fig. 1a). To ensure good interface contact for high Modulus materials often high sintering temperatures are required, which can lead to interfacial reactions (See chemical stability).

The second physical contact to consider is the dynamic long-term interface stability which can impact cell durability. For SOC systems, the active material is only an electrocatalyst and as such does not physically expand and contract in the same way as SSB active materials. In SOC interfaces, interfacial stresses are caused due to temperature cycling during operation. The difference in thermal expansion coefficients of materials joined at the interface can lead to high interfacial stresses, which cause microcracks and in the worst-case scenario delamination.

For SSBs, the active material physically expands and contracts during charging/discharging. For example, the common cathode materials $LiCoO_2$ and $LiFePO_4$ expand ca. 2% [23] and 5% [24] respectively. However, some anode materials can

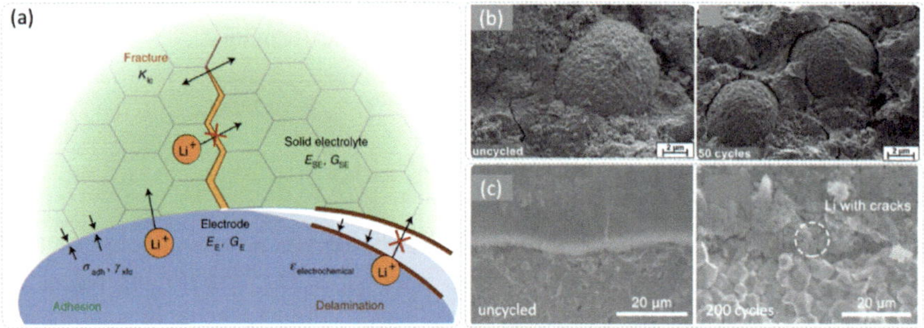

Fig. 1 **a** The mechanical properties (Young's Modulus E, shear moduli G, fracture toughness K_{Ic} of the solid electrolyte (SE) and electrode dictate the response of the system to the strain ($\varepsilon_{electrochemical}$) during cycling. The adhesion of different solid phases in the composite can be quantified through the adhesion strength σ_{adh} and interfacial energy γ_{xfc}. Reproduced with permission from Ref. [20]. **b** Scanning electron micrographs of the cathode composite of NCM811 and β-Li_3PS_4 before and after 50 cycles. Reproduced with permission from Ref. [21]. Copyright 2017 American Chemical Society **c** Morphology evolution of Li metal before and after 200 cycles in the Li–S battery. Reproduced with permission from Ref. [22]

expand dramatically, e.g. the Sn_4P_3 anode can expand up to 500% [25]. These cyclical expansion/contractions cause localised interfacial stresses, which similar to SOCs can lead to interfacial microcracks and possible delamination (Fig. 1a). Kovrver and co-workers observed that the physical contact between the active material particles and SEs was lost during cycling despite an initial intimate contact (Fig. 1b) [21]. This contact loss limits current flow through these regions which degrades the battery cycling life.

For SSBs there is a great interest in using Li and Na foils, as this will increase the battery energy density. However, metallic Li foil tends to pulverise during cycling, which leads to contact loss between the Li and the SE (Fig. 1c) [22]. Furthermore, SE flaws can act as nucleating sites for metallic Li, which can alter the electric field and in turn lead to Li filament growth. Eventually, this continued growth could short-circuit the SSB [26, 27].

The initial contact before operation and the long-term dynamic contact during operation in SOCs and SSBs is critical for performance. An initially intimate contact and a dynamically stable contact are necessary to prevent system degradation. Accordingly, it is of great importance to use effective characterisation techniques to observe both the initial and the dynamic (in situ) contact.

1.3.2 Chemical Stability

The chemical stability of the SE and associated interfaces is key to SOC and SSB performance. Interfacial reactions can impact both the electrochemical performance and the mechanical quality of the interface region. There are several causes leading to this chemical instability but can be grouped into chemical stability during production and from atmospheric conditions.

Chemical Stability—During Production

To ensure good physical contact high production temperatures are often required. However, these production temperatures can lead to interfacial reactions and changes in the bulk and surface stoichiometry. In SOCs, these interfacial reactions have been mitigated through the use of buffer layers. For example, YSZ electrode reacts readily with $(La,Sr)CoO_3$ (LSC) cathode materials to form $SrZrO_3$ or $LaZr_2O_7$. However, a CeO_2 buffer layer between chemical reaction will largely block this interfacial reaction—although Sr diffusion is still observed in the CeO_2 grain boundaries (Fig. 2) [28].

For SSBs reactions can occur between $Li_7La_3Zr_2O_{12}$ (LLZO) SEs and the $LiCoO_2$ active material. This reaction is encouraged at high temperatures, which reduces battery performance [29]. A similar result was also found when the mixture of the $LiNi_{0.5}Mn_{1.5}O_4$ active material and the $Li_{6.6}La_3Zr_{1.6}Ta_{0.4}O_{12}$ SE was sintered above 600 °C where Li_2CO_3, $La_2Zr_2O_7$ and $LaCoO_3$ interfacial products were found to form [30]. The Li-free reaction products tend to block the Li ion diffusion pathway—increasing the cell polarisation [31]. As such, although an intimate contact can be

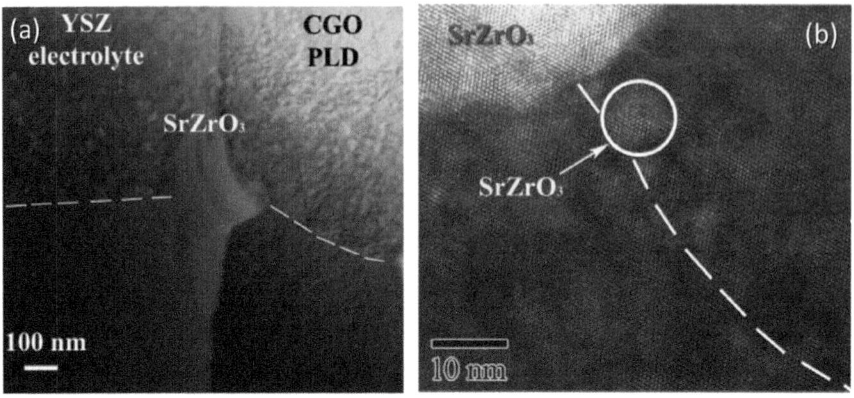

Fig. 2 TEM micrograph of SrZrO3 formation under a CeO_2 buffer layer with both large and small $SrZrO_3$ grains observed after sintering [28]

formed when co-sintering the electrode and electrolyte, the unfavourable products generated at high temperature is likely to impair the battery performance.

Chemical Stability—Atmospheric and Associated Impurities

Poor chemical stability of SE and associated interfaces can drive degradation. Although many SOC materials can tolerate atmospheric conditions, at the elevated operating temperatures minor gas impurities in the fuel or oxidant stream can lead to deleterious reactions. For SOCs, this is a large field of work and has been handled in other reviews [32].

Many SSB materials have poor stability in atmospheric conditions which may induce inferior electrochemical performance and safety hazards. For example, lithium sulphide electrolytes ($xLi_2S{\cdot}P_2S_5$) are well known for releasing H_2S when exposed to air [33] Although oxide electrolytes have been assumed to be stable in air, recent studies show that the surface impurities can be formed on the electrolyte when exposed to air. For example, Li_2CO_3 was found on the surface of a garnet-type LLZO electrolyte when it was exposed to air for two months [34]. These impurities can cause a large interfacial resistance, degrading the rate performance [35].

1.3.3 Electrochemical Stability

Electrochemical stability of electrodes, electrolytes and associated interfaces is critical for achieving a high-performance and durable SSBs and SOCs. Both the electrochemical window and the electrochemical reaction are involved to build an electrochemically stable interface. Although the fundamental electrochemical drivers are similar between SSBs and SOCs the specifics are different. As such, the electrochemical stability sections for SOCs and SSBs will be considered separately.

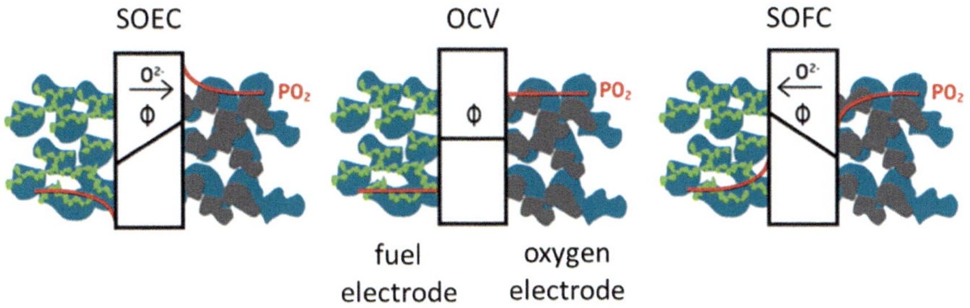

Fig. 3 SOC schematics showing the electrical and oxygen partial pressure gradients in SOEC, OCV and SOFC mode [36]

The electrochemical operation of SOCs impacts the oxygen partial pressures in the electrodes and more specifically at the electrode–electrolyte interface. At open circuit voltage, the oxygen partial pressure (pO_2) in the electrodes is defined by the gas composition, pressure and temperature. Assuming no gas leaking across, in or out of the SOC and no electronic short-circuiting through the electrolyte, the cell voltage is defined by the Nernst equation. However, as a current is drawn or applied across the SOC, the electrodes experience an overvoltage (η), which changes the average oxygen partial pressure inside, through and along the porous electrodes [36]. The new oxygen activity at the electrode–electrolyte interface can be estimated using the Nernst equation (Eq. 1, Fig. 3)

$$p_{O_2,\,\text{interface}} = \frac{p_{O_2,\,\text{bulk}}}{\exp\frac{4\eta F}{RT}} \tag{1}$$

where η is the electrode overpotential (V), $pO_{2,\text{bulk}}$ is the oxygen partial pressure supplied to the electrode, $pO_{2,\text{interface}}$ is the pO_2 at the electrode/electrolyte interface and F, R, T have the usual meaning.

This means during SOEC operation, the pO_2 at the fuel electrode–electrolyte interface becomes more negative, whereas the pO_2 at the oxygen electrode–electrolyte interface becomes more positive (in extreme cases is estimated to be around 100 bar). It is the oxygen activity at these interfaces due to the electrochemical processes which drives degradation by impacting stable valence state, oxygen vacancy concentration, cation interdiffusion, and destabilising compounds or facilitating the formation of a new phase.

Electrochemical Stability—SSBs

The electrolyte is unstable when operated beyond its electrochemical stability window. If this occurs this can lead to rapid degradation of the materials and in some cases lead to an SEI formation. Often as a result of the SEI formation, the reversible capacity is decreased due to Li consumption. It has been experimentally found that sulphide electrolytes are electrochemically stable up to 5 V vs. Li [37–39]. However, later calculation results revealed that these electrochemical windows were overestimated. Zhu et al. reported that sulphide electrolytes display a narrower intrinsic electrochemical stability window compared to the experimental results [40]. The enlarged electrochemical window previously reported was ascribed to the interphase overpotential formed during the decomposition reaction (Fig. 4b). Although this interphase formation protects the SE, the electrically insulating interphase caused inferior kinetics, which degraded the battery's rate performance.

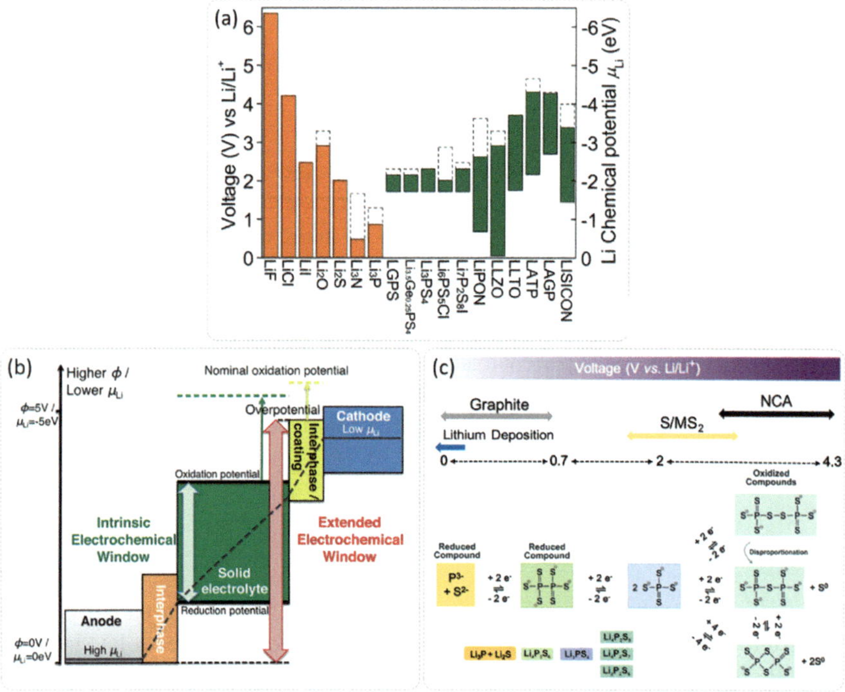

Fig. 4 **a** Electrochemical stability window of solid electrolytes (*green bar*) and other materials (*orange bar*). The oxidation potential to fully delithiate the material is marked by the dashed line. **b** Schematic diagram about the electrochemical window (*colour bars*) and the Li chemical potential profile (*black line*) in the all-solid-state Li-ion battery. The profile of chemical potential is schematic in this plot and may not be linear. The high µLi in the anode (*silver*) and low µLi in the cathode (*blue*) are beyond the stability window of the solid electrolyte (*green*). The observed nominal electrochemical window is extended by the overpotential (*dashed line*) and by the interphases (*orange* and *yellow*), which account for the gap of µLi between solid electrolyte and electrodes across the interfaces. Reproduced with permission from Ref. [40]. **c** Possible interfacial reactions of sulphide SEs at a wide range of operating voltages. Top: Voltage ranges when using graphite, Li metal deposition, layered oxide cathode (NCA), Li–S or metal sulphide (*arrows*). Bottom: The possible redox reactions at the corresponding voltage windows to illustrate the possible oxidation or reduction species. Reproduced with permission from Ref. [41]. Copyright 2017 American Chemical Society

Take $Li_{10}GeP_2S_{12}$ (LGPS), for example (Fig. 4a), the theoretical electrochemical stability window is 1.7–2.1 V. The lithiation products $Li_{15}Ge_4$, Li_3P and high-impedance Li_2S are produced when the reduction potential drops below 1.7 V vs. Li. On the cathode side, the LGPS can be oxidised to Li_3PS_4 and GeS_2 at potentials larger than 2.1 V vs. Li. These side products can cause the large charge transfer resistances. Other sulphide electrolytes such as Li_3PS_4, Li_4GeS_4, $Li_7P_3S_{11}$ have a similar reduction potential of 1.6–1.7 V vs. Li and an oxidation potential of 2–2.5 V vs. Li.

The oxide electrolytes possess a higher oxidation voltage than the sulphide electrolytes. For example, the LLZO, LISICON and $La_{1-x}Li_xTiO_3$ (LLTO) have an oxidation potential of 2.9, 3.4 and 3.7 V, respectively. For the oxide electrolytes, they tend to release O_2 above the oxidation potential [40]. However, slow kinetics are often involved in these oxidation processes, which limits the ion transport efficiency at high rates.

As mentioned above, the SEs begin to decompose when the operating voltage surpasses its intrinsic electrochemical stability window. However, the decomposition products are also related to the electrode materials. For example, the oxidative decomposition intends to be more severe at high voltage when coupled with $LiCoO_2$ or $LiNi_xMn_yCo_{1-x-y}O_2$ (NMC). As a result, an unfavourable thicker cathode electrolyte interphase (CEI) and a low Coulombic efficiency (CE) were presented. Moreover, the CEI was continually grew along with cycling and SE oxidation, resulting in the battery degradation [42]. In comparison, the solid-state batteries (SSBs) exhibit good cycling performance when adopting TiS_2 and S as cathode [43]. The mismatch of electrodes and electrolytes can be one driving force for the degradation.

The formed interphase on the anode side can be classified into three types (Fig. 5): thermodynamic stable interphase, mixed electronic and ionic conducting interphase, and the ionic conductive but electronic insulating interphase. The ideal scenario is the thermodynamically stable interphase as there is no reaction between electrode and SEs. However, this is not applicable for most of the electrode-electrolytes interface. A mixed conducting interphase allows chemical reactions between two materials caused by the thermodynamic driving force. The formed interphase might continue to grow until penetrating the bulk electrolyte, which likely leads to the self-discharge of the battery [46]. Wenzel et al. proved that the interphase layer formed after $Li_{0.35}La_{0.55}TiO_3$ (LLTO) contact with Li, consists of Ti_2O_3, TiO, Ti and Li_2O by adopting in situ XPS [44]. This resulting mixed conducting interphase layer is blamed for the degradation of battery performance. The electronic insulating but ionic conducting layer is also a favourable interface which possesses similar function of SEI in the liquid electrolyte. As such, the ionic conductivity of this interphase largely determines the battery performance [44].

1.3.4 Concluding Remarks

The driving force for interface degradation is a coupled effect of mechanical, chemical and electrochemical factors. Although computational models are powerful in

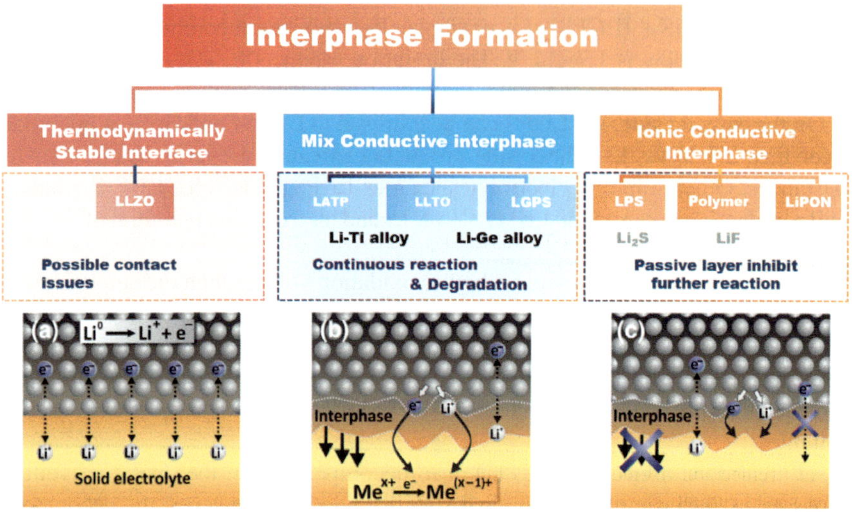

Fig. 5 Illustration of three types of interphases between SSE and metallic Li [44, 45]

understanding of these interfaces, this work needs to be supported with detailed experimental evidence. Advanced TEM techniques provide powerful tools to explore this hidden world.

2 TEM Techniques

New electron guns and detectors in TEMs are allowing researchers to observe materials at a higher resolution with a lower electron dose. Aberration corrected TEMs are becoming more common place—which further pushes the imaging resolution through correcting for spherical and/or chromatic aberrations in the electron beam. Such powerful tools can be used with standard TEM techniques such as electron diffraction, conventional imaging, EDS and EELS.

With aberration corrected TEMs, a larger emphasis has been placed on the improved quality and signal when operating in scanning TEM (STEM) mode. In high resolution STEM (HRSTEM) researchers can gain an atomic level insight into newly-designed materials. In comparison to conventional high-resolution TEM images, the interpretation of HRSTEM images can be relatively easy. For high angle annular dark field (HAADF) HRSTEM the intensity of atom columns is reflective of the average atomic number—the brighter columns are associated with heavier atoms. This still makes imaging of low atomic number atom columns, such as O, Li and Na, difficult or impossible.

Another powerful tool in the TEM for characterisation of complex metal oxides is EELS. In the simplest format, sample composition can be obtained using EELS.

However, EELS can also be used to assess valence states and coordination of various atoms/atom columns. In battery and fuel cell materials, this is used to look at transition metals, which may exist in a combination of valence states dependent on the state of charge or operating conditions.

These techniques have been used for ex situ characterisation to understand the structure of the pristine interface and also changes occurring at the interface during operation. There is also a growing body of work using in situ techniques. For studying interfaces this has been limited to environmental TEM (ETEM) work with either a H_2 or O_2 atmosphere. For fuel cell systems these atmospheres have been combined with temperature to track processes such as NiO reduction and oxidation or used in combination with EELS to track valence state changes of transition metals.

Electron holography is an advanced TEM which allows the electron beam phase to be captured (instead of just the beam intensity). With this additional information, imaging resolution can be pushed and information about magnetic and electric domains can be extracted. This has been used in LIBs to understand the charge distribution during lithiation [47] or to investigate the charge distribution at the solid electrode–electrolyte interface [48–53].

3 SOC Solid Electrolyte Interfaces

SOC materials degradation is dependent on the material origin, processing and the associated cell operating conditions [54–56]. The parameters that will drive degradation during SOC operation include the operating temperature, current density, cell polarisation, gas supplied (gas type & purity) and fuel utilisation. This section will start off with first discussing the oxygen electrode–electrolyte interface, followed by an exploration of the fuel electrode–electrolyte interface.

3.1 Fuel Electrode–Electrolyte Interface

The most common fuel electrode is a porous Ni-YSZ. When running as a SOFC, steam is produced in the fuel electrode and as such there are two key driving forces for interfacial degradation—the high oxygen partial pressure and the steam. If the SOFC is run under a high current density (typically leading to high fuel utilisation), the oxygen partial pressure and the steam concentration increase which encourages NiO formation and Ni volatilisation.

Another important degradation phenomenon in the Ni-YSZ fuel electrode is the Ni coarsening, which is dependent on the temperature, atmosphere, time and the catalysis support. Ni coarsening can lead to a reduced surface area (increasing the overpotential), but can also impact the Ni percolation (reducing the electrode electronic conductivity). The reduction of the Ni percolation is attributed to Ni particle grain growth, formation of volatile Ni species ($Ni(OH)_2$) and relocation of impurities to the Ni–Ni connection points.

There are several key TEM focused studies which have helped in understanding the mechanisms behind the fuel electrode–electrolyte interface. These studies have probed impurity segregation [57], the detailed nature of the Ni-YSZ interface [58, 59] and the dynamic processes associated with Ni oxidation [60]. More recently HR-TEM has been used to characterise new fuel electrode microstructures which can improve the electrode–electrolyte interface durability [61, 62].

Impurities are considered to be a major factor leading to large differences in experimental results of nominally similar materials in different labs. The dynamic nature of the electrode–electrolyte interfaces has been attributed to the segregation and contamination of the interface by fuel impurities, but also by the SOC balance of plant components such as Cr from the interconnects and SiO_2 from glass seals. In 2006, Liu et al. investigated impurity segregation at the Ni-YSZ electrode–electrolyte interface using a combination of FIB and TEM [57]. Figure 6 shows the Ni and YSZ grains from the fuel electrode and the YSZ electrolyte substrate. There are large amorphous regions throughout the electrode which were identified as SiO_2. The covalent bonding from these SiO_2 compounds has a chemical affinity for the ionic oxide surfaces and is often observed at the interface [54]. This impurity accumulation can impact the Ni percolation, but can also dramatically increase the cell polarisation.

The Ni-YSZ interfacial nature is key in controlling long-term Ni agglomeration and stability. Sasaki et al. used HRTEM in combination with EELS to characterise the atomic structure of the Ni-YSZ interface (Fig. 7) [59]. The group identified that Ni was epitaxially oriented to the YSZ surface with the following orientation relationship—(111)Ni (111)YSZ, [1-11]Ni [1-10]YSZ.

The interface region was atomically flat with no observed amorphous phase and with no observed dislocations—suggesting an incoherent interface. The lack of coherency is in agreement with the large lattice misfit associated with the observed Ni–YSZ plane alignment. Unfortunately, these incoherent interfaces are considered

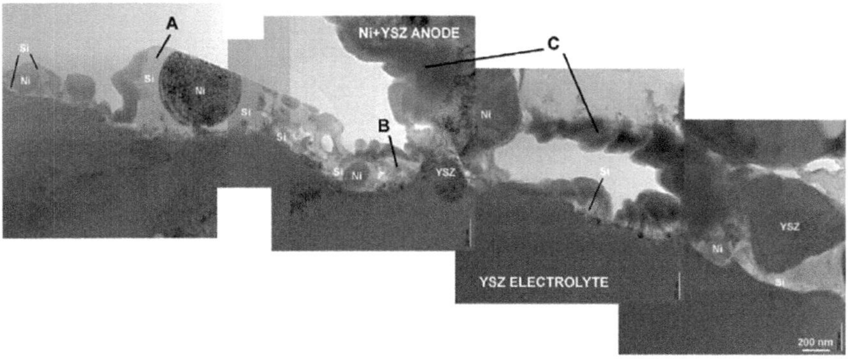

Fig. 6 TEM micrographs showing the entire length of the Ni-YSZ anode/YSZ electrolyte interface [57]

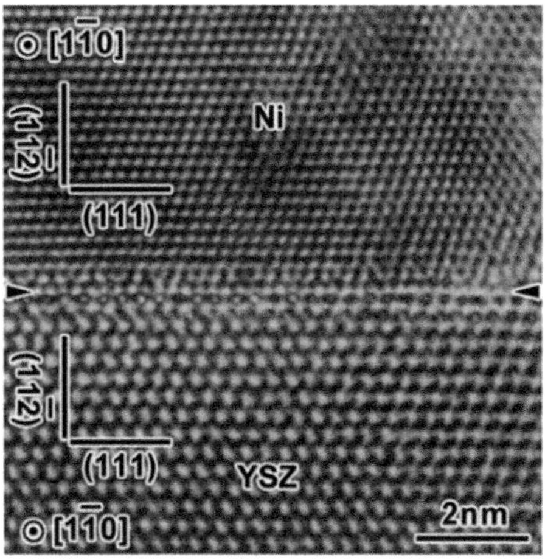

Fig. 7 HRTEM image of the Ni/YSZ interface with the Ni (111) plane epitaxially oriented with the YSZ (111) plane

weak. The HR-TEM images were confirmed using multislice simulations. However, it was not possible to determine if the YSZ termination layer (interface layer with the Ni) would be Zr or O.

To clarify the interface termination layer, the Ni-YSZ interface was investigated using EELS. From the O-K edge spectra, a number of differences in peak shape and energies were observed in the interface region. These differences were compared with the NiO EELS spectrum and attributed to bonding between Ni and O—indicating that the YSZ termination layer is O, rather than Zr. This interface termination layer is considered to favourable for creating a strong interface bonding at the Ni-YSZ interface, which offsets the weakness associated with incoherent interface.

More recently Liu et al. took a closer look at the Ni-YSZ interface of a real Ni-YSZ electrode using HR-TEM [58]. The motivation of this work was to understand the three phase boundary (TPB) on an atomic scale and to simulate the hydrogen evolution reaction (HOR) processes using realistic models. Using HR-STEM and HR-TEM two distinct structures were found with different contact angles of metal/oxide interfaces, metal surfaces and pore opening sizes which the authors identified as TPB-1 and TPB-2. A HAADF-STEM image of the TPB-1 interface is shown in Fig. 8. The YSZ is imaged down the [001] zone axis with a (200) terminating plane—which is exposed to the gas phase. The YSZ (200) plane interfaces with the Ni (111) plane. The opening angle of the pore (θ_{pore}) between the Ni and YSZ surfaces is approximately 35°. The θ_{pore} for TPB-2 is significantly larger ca. 103°. From MD modelling, the higher HOR activity is associated with TPB-1. Three HOR pathways are identified: two types of hydrogen diffusion processes, and one type of oxygen migration process which is a new pathway. In light of this insight, the microstructures developed should encourage this type of wetting to produce high-performance electrodes.

The previous TEM investigations have focussed on ex situ TEM methods to probe the Ni-YSZ interface. However, due to the electrochemical driving forces

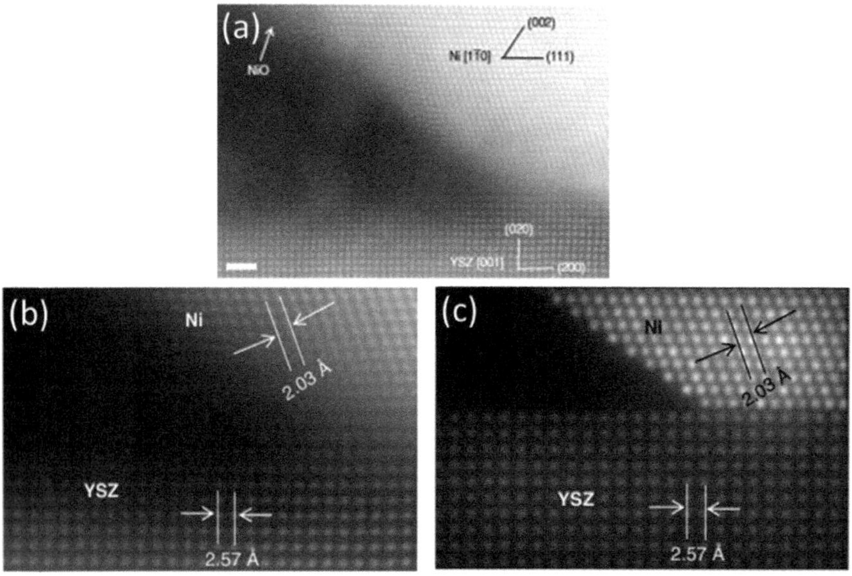

Fig. 8 **a** HAADF-HRSTEM image down the YSZ[001] zone axis showing the (200) termination YSZ planes interfacing with the Ni (111) planes. **b** filtered HAADF-HRSTEM image. **c** multislice simulation of the Ni-YSZ interface [58]

and high temperature, this interface is highly dynamic. As such, the possibility of exploring the interface in situ provides researchers with a more realistic insight into the dynamic processes. However, there are only a few dedicated instruments where this environmental TEM (ETEM) type of work is possible.

The first ETEM study on SOC materials looked at the reduction–oxidation (redox) behaviour of Ni [60]. If the pO_2 in the fuel electrode becomes too high the Ni in the Ni-YSZ fuel electrode can be converted to NiO. Understanding the Ni redox characteristics in Ni-YSZ electrodes is important in understanding the long-term SOC durability. In this work, Jeangros et al. used ETEM in combination with DFT calculations to study the redox stability of a Ni/YSZ fuel electrode. The work was conducted in a dedicated ETEM on a FIB lamella. The NiO reduction process was observed using an atmosphere of 1.4 mbar of H_2 in a temperature range of 300–500 °C. The Ni reoxidation process was conducted in 3.2 mbar in O_2 (Fig. 9). Bright field TEM images were acquired during the 2 h process.

The in situ observations show that the transfer of O_2 from NiO to YSZ triggers the reduction (NiO to Ni) reaction. This process creates oxygen vacancies in the NiO which activates the reduction reaction at the NiO/YSZ interface by the creation of hydrogen adsorption sites (Ni atoms surrounding oxygen vacancies).

During Ni reoxidation, a porous structure is formed which is attributed to the imbalanced mass transport between the outward Ni diffusion and the inward O diffusion in the oxidising Ni grains. Both the expansion of NiO during a redox cycle and the presence of stress in the YSZ are directly observed during the ETEM experiment.

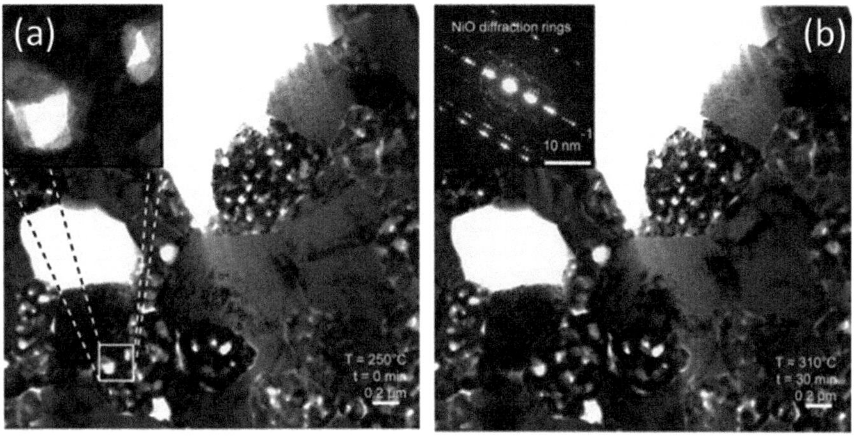

Fig. 9 Bright field ETEM images of Ni-YSZ during reoxidation **a** at 0 min at 250 °C and **b** at 30 min at 310 °C with embedded selected area diffraction showing NiO diffraction rings [60]

It is this process that accounts for the redox instability of the Ni–YSZ anode. Based on the present in situ ETEM observations, the group proposes a fine Ni microstructure with well-dispersed pores. This microstructure shows an improved redox tolerance and thus enhanced durability.

Exsolution:

To avoid the issue with the Ni agglomeration, alternative SOC electrodes have been considered. Titanate perovskites are attractive fuel electrode materials due to the good conductivity and tolerance to coking and redox. However, the electrocatalytic performance of these materials is low in comparison to Ni-based fuel electrodes. To improve the electrocatalytic performance these materials are often impregnated with nano-particles. However, these nano-particles show poor long-term stability. To solve this issue Neagu et al. (2015) demonstrated the use of direct electrocatalyst exsolution from a parent oxide matrix [62]. In this work Ni was exsolved from a $La_{0.4}Sr_{0.4}Ni_{0.03}Ti_{0.9}7O_{3-\gamma}$ parent oxide substrate.

Neagu et al. used HR-TEM in combination with diffraction to show that the Ni particles exsolved on the (110) termination plane with the particle embedded around 30% into the parent substrate (Fig. 10). From diffraction work the researchers concluded that the Ni (111) planes align with the perovskite (111) planes. Due to the mismatch between the parent perovskite phase and the Ni particle, dislocations are observed at the interface. It is proposed by the researchers that the strained and embedded nanoparticle mitigates long-term agglomeration and also hydrocarbon coking—a problem typically associated with nanoparticles.

He et al. extended the work by Neagu and presented a method of co-exsolution to produce both CeO_2 and Ni particles from a titanate perovskite substrate—$La_{0.8}Ce_{0.1}Ni_{0.4}Ti_{0.6}O_{3-\delta}$ (LCeNT) [61]. The strained faceted CeO_2 cubes are produced after an air-annealing process. It is shown that the CeO_2 morphology can

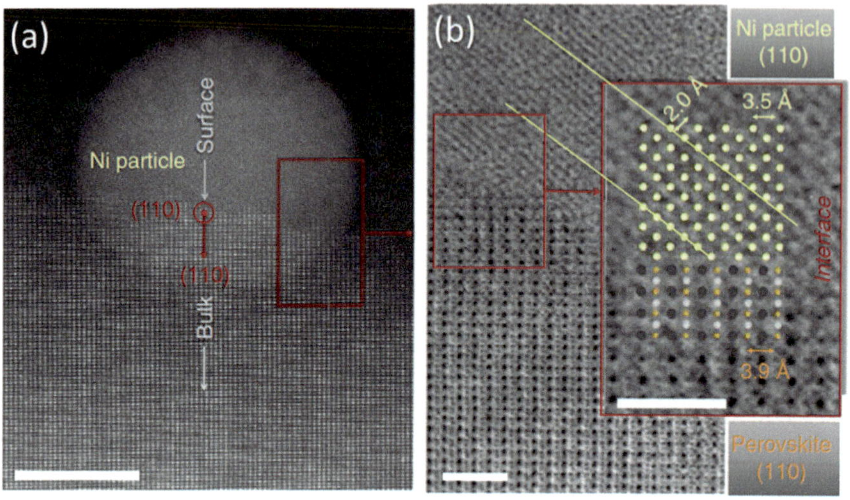

Fig. 10 Ni particle exsolved from the $La_{0.4}Sr_{0.4}Ni_{0.03}Ti_{0.97}O_{3-\gamma}$ parent oxide substrate showing crystallographic alignment [62]

be tuned by varying the annealing temperature, whilst exsolution of Ni nanoparticles forms subsequently following reduction.

The CeO_2 forms a socketed particle—similar to the Ni observed by Neagu with a strong crystallographic orientation with the oxide LCeNT substrate. The HRSTEM work shows a clean, amorphous free CeO_2-LCeNT interface showing a cube-on-cube epitaxy (Fig. 11). The HRSTEM also shows that the CeO_2 surface is terminated with the (111) facet. The authors report a strained CeO_2 crystal, but with limited evidence. This co-exsolved microstructure shows a good electrocatalytic performance with stable behaviour when operating in a H_2/CH_4 environment.

Several groups have worked on advanced TEM characterisation of the fuel electrode–electrolyte interface. The most recent works use advanced aberration corrected HR-STEM to image the interfacial structures. However, to date this has been limited to materials in the pristine state. To understand the long-term behaviour, these heterointerfaces need more dedicated TEM characterisation after operation or, where possible, in situ. Possibly due to the limited number of dedicated ETEMs, and the complexity of this work, there are only a few studies using this in situ ETEM to investigate the fuel electrode–electrolyte interface. In the future, it would be interesting to extend this work to understanding microstructure changes during long-term operation.

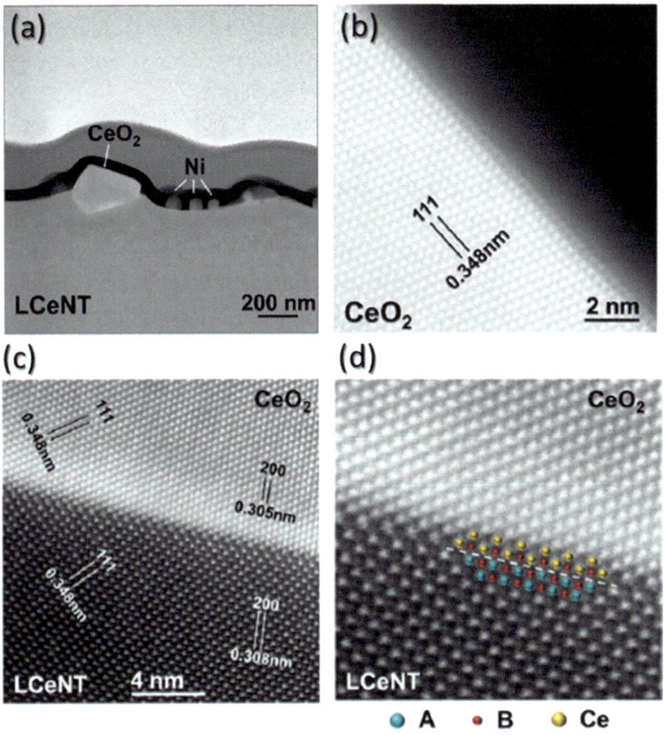

Fig. 11 Ni and CeO_2 particles co-exsolved from a—$La_{0.8}Ce_{0.1}Ni_{0.4}Ti_{0.6}O_{3-\delta}$ (LCeNT) parent oxide substrate with HR-STEM showing the crystallographic orientation between the exsolved CeO_2 and the parent LCeNT oxide [61]

3.2 *Oxygen Electrode–Electrolyte*

Oxygen electrodes are typically composed of electronic conductors, an electronic conductor/ionic conductor composite or mixed ionic and electronic conductors (MIEC) [63]. For the electronic or electronic/ionic composite oxygen electrodes, the ORR/OER is confined to the active three phase boundary. For MIEC oxygen electrodes the ORR/OER is expected to spread out over some of, or the entire gas exposed MIEC surface [63–65].

The high operating temperature and oxidising atmosphere makes metal oxides the favoured material for oxygen electrodes in SOCs. Most oxygen electrodes consist of mixed conducting perovskite type oxides of the general composition $ABO_{3-\delta}$. The A-site usually is occupied by a large tri-valent lanthanide ion (e.g. La^{3+}, Pr^{3+}, Sm^{3+}, Gd^{3+}) and often partially substituted by di-valent alkali earth ions (Ca^{2+}, Sr^{2+}, Ba^{2+}) to both increase electronic and ionic conductivity. The B-site is often occupied by one or more small tri- or tetra-valent 3-d transition metal ions ($Mn^{3+/4+}$, $Fe^{3+/4+}$, $Co^{3+/4+}$, $Ni^{3+/4+}$). The exact $ABO_{3-\delta}$ oxygen deficiency (δ) exists often at only one pO_2. The perovskite oxide is oxygen deficient at lower pO_2 and oxygen stoichiometric or cation deficient at higher pO_2. Materials which display a greater oxygen deficiency tend to be more active electrocatalysts, but are also less stable.

Operation of an SOC in SOFC mode causes a pO_2 decrease at the electrode–electrolyte interface. Conversely, operation of an SOC in SOEC mode causes a pO_2 increase at the electrode–electrolyte interface. Therefore, the oxygen activity, at the oxygen electrode–electrolyte interface, increases in SOEC mode and decreases in SOFC mode. These local pO_2 changes impact the oxygen deficiency (δ), electrode conductivity (ionic and electronic), electrocatalytic activity and mechanical stability, which affect both the performance and durability of the oxygen electrode and the associated interfaces.

Most advanced TEM studies use EELS to assist in understanding the oxidation states of the transition metal ions and the associated oxygen deficiency/stoichiometry. More recently, these investigations have been conducted under in situ conditions—giving researchers a more realistic insight into oxidation state changes during operation.

(La,Sr)MnO_3 (LSM) is a common oxygen electrode material. It exhibits good electronic, but poor ionic conductivity. Due to the transition metal in the perovskite B-site, LSM also has good electrocatalytic properties. However, the valence of the Mn transition metal is sensitive to changes in oxygen partial pressure. To improve the ionic conductivity LSM is often composited with YSZ.

Early EELS work was key in identifying the changes in the Mn valence state during operation. Backhaus-Ricoult completed a detailed study of the LSM-YSZ interface in SOFC using HR-TEM in combination with EELS (Fig. 12). All HRTEM images showed the LSM/3YSZ interfaces to be free of interfacial glass and amorphous phase with a transition zone between 1 and 4 nm.

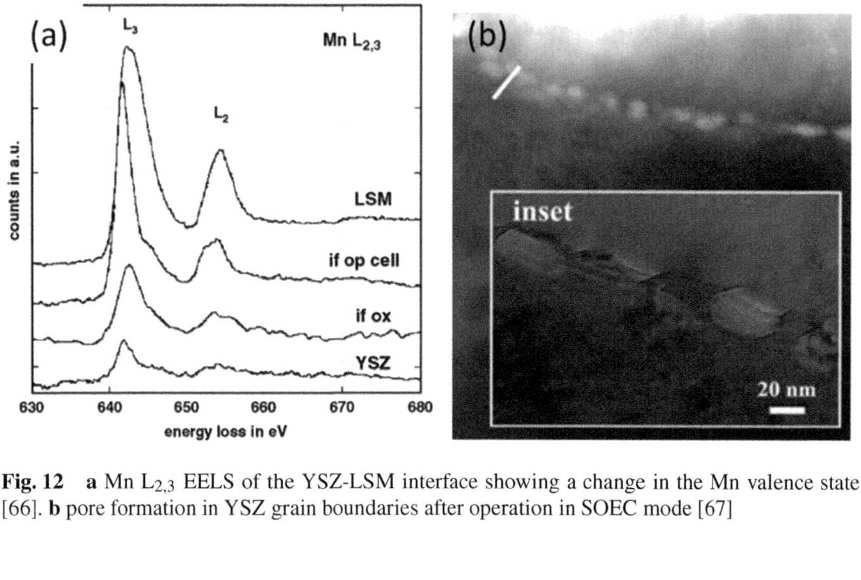

Fig. 12 **a** Mn $L_{2,3}$ EELS of the YSZ-LSM interface showing a change in the Mn valence state [66]. **b** pore formation in YSZ grain boundaries after operation in SOEC mode [67]

In the LSM structure, $Mn^{3+/4+}$ was found in the LSM bulk phase. In contrast, when LSM is in contact with YSZ, La and Mn were found to form a solution with the YSZ phase. Looking at the near edge fine structure of the EELS spectrum Backhaus-Ricoult shows a change of the Mn valence state across the interface. This is observed through changes in the peak energy and L_3/L_2 intensity ratios. In the LSM perovskite, the Mn $L_{2,3}$ peaks are located at 641.2 and 652.2 eV and the L_3/L_2 peak intensity ratio is around 2—reflecting an Mn oxidation state of $3+/4+$. In contrast, the Mn peak energy positions and L_3/L_2 peak intensity ratios (ca. > 4) in the zirconia phase, indicate Mn^{2+} in the YSZ solid solution. After sintering 'if ox' shows predominately a Mn^{3+} valence state, whereas after SOFC operation 'if op cell' the Mn valence decreases to predominately a Mn^{2+} valence state.

These Mn valence state changes are accompanied by changes in the oxygen near edge fine structure features. Quantification of the absorption edge height revealed that the absolute oxygen concentration at the interface was lowered compared to the bulk YSZ phase. This was interpreted as sign of an extremely high oxygen vacancy concentration and confirmed that the interfacial phase had a defective, over-doped fluorite structure. This Mn^{2+} enrichment increases the local oxygen vacancy concentration in YSZ, which in turn facilitates the ORR reaction (Eq. 2).

$$4Mn^{2+} + 2V_O + 1/2\ O_{2\,gas} \rightarrow 4\ Mn^{3+} + 2\ O^{2-} \tag{2}$$

As such, this HR-TEM–EELS study shows that at low oxygen partial pressures, Mn^{2+} interfacial defects dominate at the interface, whereas at high oxygen partial pressure, Mn^{4+} interfacial clusters are dominant interface defects. This facilitates the ORR kinetics in SOFC mode, but impedes the OER kinetics in SOEC. This understanding is fundamental in the problems encountered when operating in SOEC mode under high current densities or polarisations [67]. Under high current densities, oxygen has been found to precipitate in the SE grain boundaries, which is catastrophic for SOEC performance [67]. This has been attributed to the poor OER kinetics and associated cation migration when operating in SOEC mode. Graves et al. found that this cation migration behaviour is reversible and can be recovered by switching from SOEC mode to SOFC mode [68].

The work by Backhaus–Ricoult has recently been extended by He et al., who used a combination of FIB-STEM techniques to study the GDC/LSM and YSZ/LSM interfaces. For the LSM/YSZ interface, extensive Mn and limited La diffusion was observed in the YSZ grain boundaries and interfaces [69]. However, this is not observed with the LSM/GDC couple. Despite this lack of Mn in the GDC, the GDC was found to also show an activation process during SOFC operation.

One attractive industry-applicable cathode material for intermediate temperature SOCs is $(La_{0.6}Sr_{0.4})(Co_{0.2}Fe_{0.8})O_3$ (LSCF). However, if LSCF is used in conjunction with an YSZ, $SrZrO_3$ (SZ) forms. To reduce this issue, a gadolinia-doped ceria (GDC) diffusion barrier layer is incorporated.

Develos-Bagarinao investigated SZY (Y doped SrZrO3)-YSZ interfaces using HAADF in combination with EELS [70]. Both the SZY and YSZ exhibit a prominent O K-edge P_1, P_2 doublet in the EELS spectra. These peaks are attributed to the crystal

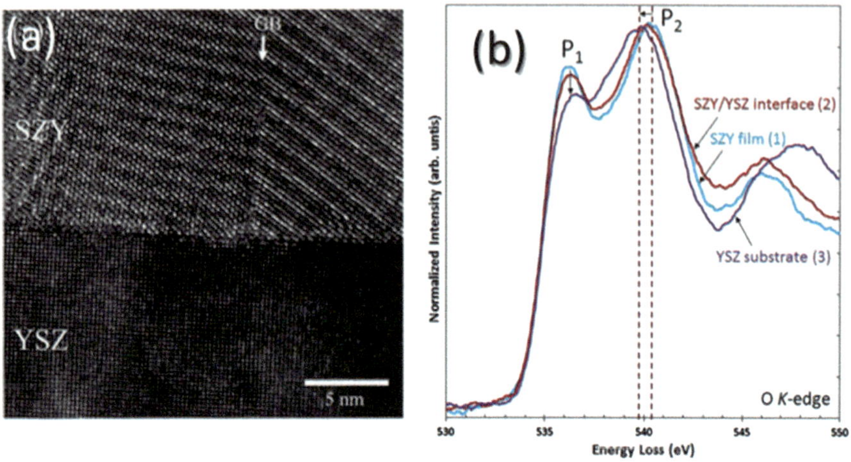

Fig. 13 **a** HRTEM of SZY-YSZ interface. **b** EELS SZY, SZY-YSZ interface and YSZ [70]

field splitting of the Zr 4d into e_g (P_1) and t_{2g} (P_2) orbitals [70]. However, the O doublet peaks from the YSZ are shifted towards lower energies in comparison to the SZY or interface O doublet peaks. Notably the YSZ substrate P_1 peak is remarkably depressed in comparison to that of the SZY film – suggesting a mixing of the e_g and t_{2g} states indicating oxygen disorder in the YSZ. The authors conclude that this kind of strained heterointerface is detrimental as it impedes the oxide ion migration (Fig. 13).

The EELS work discussed until this point has been conducted ex situ. To understand the changes in oxidation states under more realistic conditions ETEM characterisation is of great value. Oxidation states of cations in a perovskite structure play an important role for conductivity and electrocatalytic performance in SOCs. Although the bulk oxidation states can be calculated and are generally well understood for specific conditions, this is not the case for the heterointerfaces. Gaulandris et al. conducted an in situ ETEM investigation of a $La_{0.6}Sr_{0.4}CoO_{3-\delta}$ (LSC) and YSZ heterointerface using a symmetric SOC using STEM in conjunction with EELS. The experiments were performed at temperatures up to 600°C in high vacuum (ca. 10^{-7} mbar) and in 2 mbar O_2 [71]. Figure 14 shows the EELS of the Co and O in both oxygen and vacuum conditions as the temperature is ramped from room temperature up to 600 °C (Fig. 14). The authors analysed the data in two ways—by looking at the absolute EELS peak intensities (Fig. 14) and by monitoring the energy difference between the Co-L_3 and Co-L_2 white lines. Both analysis methods showed similar trends, i.e. as the temperature increased the Co oxidation state increased.

The in situ ETEM EELS results show that as the temperature is increased, the Co-$L_{3,2}$ edges broaden and EELS analysis shows that the Co is oxidised from 2+ to 3+—surprisingly in both the vacuum and oxygen atmospheres. The authors attribute this behaviour due to (1) the oxygen supply from the YSZ; (2) a strain relaxation at elevated temperatures which decreases the oxygen vacancy concentration and

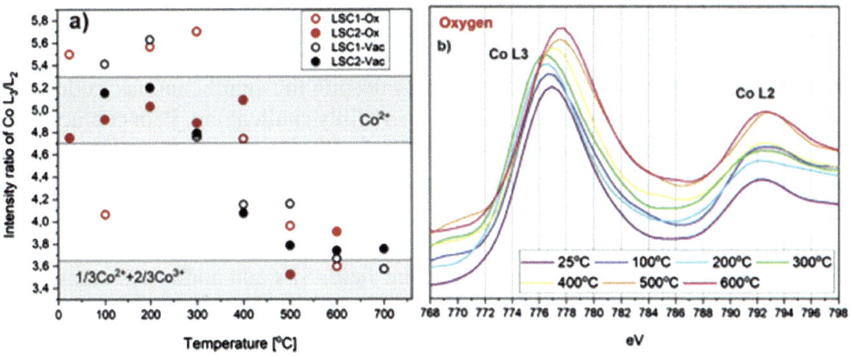

Fig. 14 **a** Change in the Co valence state as temperature is increased from low to high temperature in vacuum (black) and in oxygen (red). **b** Co EELS spectra taken during ETEM experiment in oxygen [71]

increases the Co oxidation state or (3) an instability of the LSC. This type of in situ work is critical in understanding complex behaviour at interfaces, which can have a dramatic impact on electrochemical performance.

3.3 Concluding Remarks

Advanced TEM characterisation has been invaluable when probing solid heterointerfaces in SOCs. HRTEM and associated characterisation has been conducted for several decades on these systems. More recently, the improved detectors now allow for HRSTEM. Combining these techniques with in situ ETEM characterisation has provided useful mechanistic insights for the SOC community.

4 Li Solid Electrolyte Interfaces

To further highlight the importance and applicability of TEM and related techniques to SSB applications; a summary of mechanistic and proof-of-concept works focusing on each of the four driving forces of battery degradation is given here starting with solid-state Li metal batteries (SSLMBs).

4.1 Physical Contact

Ensuring adequate interface contact in SSBs despite the significant electrode volumetric changes that occur during cycling is incredibly challenging. Poor contact leads to poor battery performance overall and can lead to local overpotentials that can seriously degrade electrolyte materials. In conventional liquid electrolyte cells, surface contact is always maintained throughout cycling. SEs, however, typically lack the flexibility or adhesiveness to keep good electrode contact long term. This has led to SE interface engineering remaining a popular field. SEs add additional complexity as most practical SEs are polycrystalline, and as such feature grain boundaries (GBs) which form homogenous interfaces in addition to the heterogeneous SE-electrode interfaces.

High resolution TEM images are useful in probing solid interfacial interactions—including GBs. Volumetric changes during cycling occur from active material expansions and contractions from ion insertion/extraction and have a significant impact on solid-state interfaces, both within composite electrode materials and at the electrode–electrolyte interface.

As an example, Notohara et al. (2018) investigated the effect of nanoporous carbon in SnO_2 anodes in SSLIBs using LiI-Li_2S-P_2S_5 as the electrolyte. Nanoporous carbon was hypothesised to provide sufficient room for volume changes and provide a more stable framework for interface contact [72]. After cycling, the electrolyte-anode interface was analysed using SEM and EDS-TEM. Surprisingly to the group, significant cracking and material separation was apparent in SEM images of the interface despite the introduction of the nanoporous carbon (Fig. 15a, b). Although, the interface was still of higher quality than in untreated samples [72, 73].

Further interface engineering is required in many electrode materials to suppress or accommodate for volumetric changes during cycling. Higher level atomic imaging techniques may shed further light on the nuances of volumetric cycling in SSLIBs.

4.2 Electrochemical Stability

Electrochemical instability and incompatibility in solid electrolyte materials has a major impact on the safety, longevity and electrical performance of the whole battery system. Electrochemical phenomena can also interfere with the visualisation of electrolyte materials and their interfaces when using EM technologies. Successfully collecting TEM data from in situ electrochemical systems can still be challenging, however modern electron holography (EH), differential phase contrast (DPC) and EELS methods are becoming increasingly useful and effective.

The formation of a space charge layer (SCL), whereby ions migrate from the electrolyte into the cathode material thus creating an ion deficient band in the electrolyte, has been observed for over 100 years [74]. It is known that SCLs are generally resistive and limit cycling capability of the cell however, the in situ visualisation of

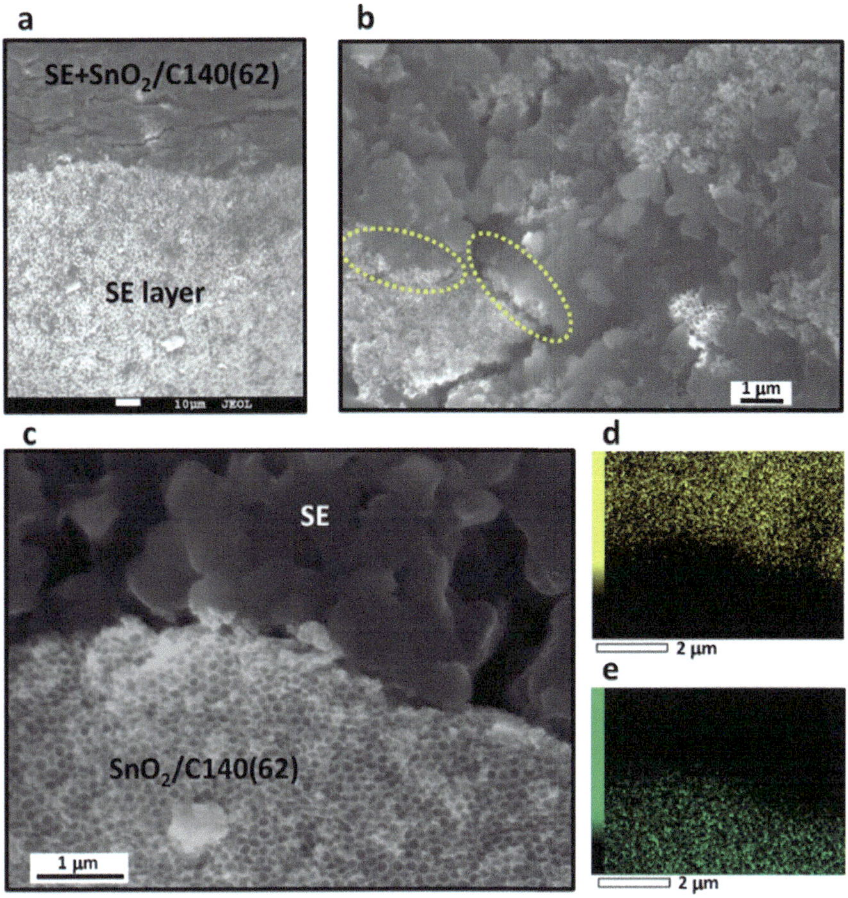

Fig. 15 Interface between SnO_2 with nanoporous carbon (SnO_2/C140(62)) and LiI-Li2S-P2S5 (SE). **a–c** SEM images of interface at varying magnifications, note the cracks in **b** (*yellow circles*). Nanopores are clearly visible in the anode in (**c**). **d**, **e** EDS mapping of S and Sn, respectively, from (**c**). Image from [72]

the formation of the SCL in a solid electrolyte has not been possible until recently [75]. With the advent of highly sensitive TEM techniques that can map atom and charge movement in situ such as EELS, DPC and EH methods, great progress has been made into fully understanding the formation, nature and impact of SCLs on SSLMBs.

In 2010, the Yamamoto group used EH-TEM to visualise the potential distribution in a SSLMB [48]. The goal was to map the potential across a $Li_{1+x+y}Al_yTi_{2-y}Si_xP_{3-x}O_{12}$ and $LiCoO_2$ interface using in situ TEM. By thinning a portion of the cathode-electrolyte interface in the test cell using a FIB, it was possible to conduct cyclic voltammetry, BF and EH-TEM using a holder with two fixed electrodes to allow voltages to be applied across the sample.

From the initial EH experiment, the Yamamoto group observed the change in potential gradient across the cathode and electrolyte depended on the magnitude of the applied voltage during charge/discharge processes. EELS investigations were then completed to confirm that the changing potential distribution was caused by the movement of Li ions between both materials. To do this, EEL spectra of Co were collected before and after charging in the thinned region. This allowed Co valence state changes (suggesting changes in intercalation) to be measured. The collected EEL spectra showed energy-loss peaks of Co increased by 0.4 eV after charging which indicated a corresponding increase in the valence state. An increase in valence state thereby suggested that Co^{3+} was partially oxidised to Co^{4+} as a consequence of the removal of Li^+ from Li_2CoO_2. The removal of these ions causes the local Fermi level to decrease [76]. The inner potential (V_0) measured in EH relates to volume average of the atomic electrostatic potential, and so its position in the valence band did not significantly vary between charge/discharge cycles as Li was much lighter than the elements present (see Fig. 16b). The stable V_0 position, but decreasing Fermi level with Li^+ extraction is what was highlighted with EELS. With Fermi levels aligned, the gradual change in relative V_0 values was captured (see Fig. 17b–i) [48].

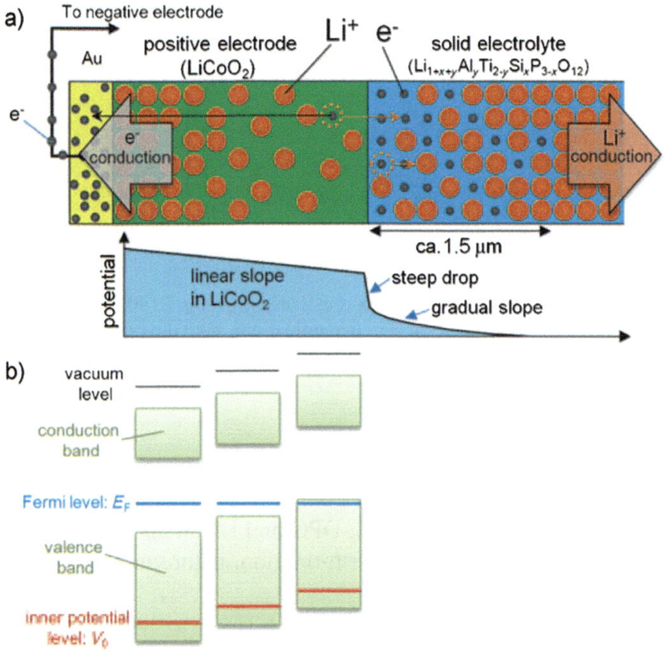

Fig. 16 Change in charge distributions lead to a change in Fermi levels. **a** Illustration of shift in Li^+ and electrons at interface. **b** Demonstration of shifting Fermi levels being represented in change of (stable) V_0 values. Figure from [48]

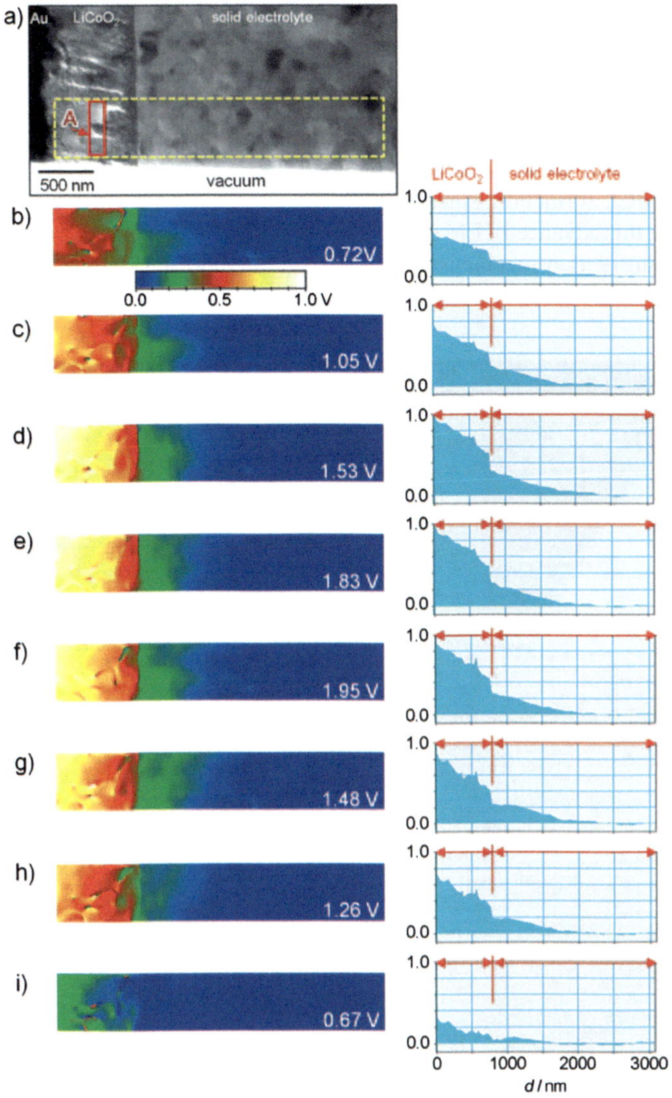

Fig. 17 EELS spectra from $LiCoO_2/Li_{1+x+y}Al_yTi_{2-y}Si_xP_{3-x}O_{12}$ interface. Ba BF-TEM of interface. **b–i** EELS data as potential images (*left*) and as line profiles (*right*), collected from the target region in (**a**) (*dashed line*). Figure from [48]

The Wang group (2020) took the visualisation of in situ SCL formation one step further with the use of DPC-STEM [77]. Their application was based on earlier work of the Haas group (2019), who used a segmented-detector DPC-STEM to produce electric vector field and charge density maps of a silicon p-channel metal–oxide–semiconductor (pMOS) device [78]. The method used by Haas et al. (2019) did not impose the same sample geometry restrictions as EH and produced data with higher spatial resolution [77]. Wang et al. (2020) used in situ DPC-STEM to visualise the charge-density distribution around a Li_2CoO_2 and Li_6PS_5Cl argyrodite interface with and without $BaTiO_3$ nanoparticle (BTO) additives (see Figs. 18 and 19). The overall performance (i.e. ionic conductivity, cycling performance and rate capability) in BTO samples was greater than those without. This was attributed to the suppression of extreme SCL formation, since the BTO was deposited only at the interface [77]. Once again, this advanced TEM characterisation work was imperative to facilitate the understanding of a key mechanism that impacts both short and long-term performance of SSLIBs.

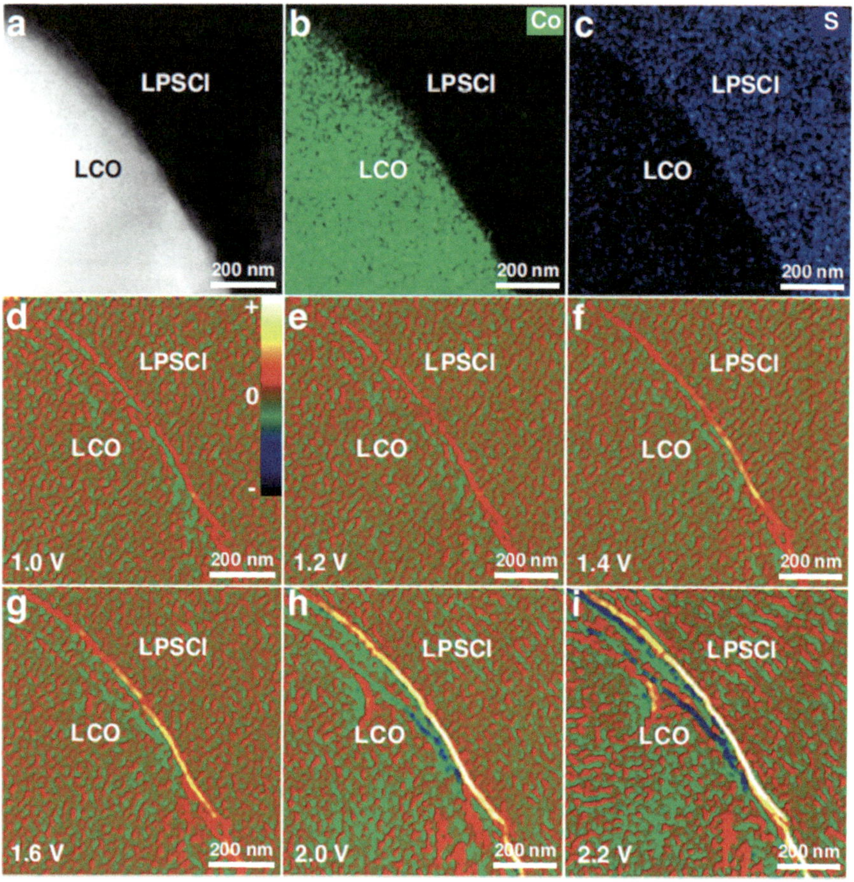

Fig. 18 Li_2CoO_2 and Li_6PS_5Cl interface. **a–c** HAADF-STEM images of interface with and without mappings. **d–i** In situ DPC-STEM images at varying bias voltages. Colour scale in **d** is the relative magnitude of -positive/negative charge density. Note the increasing polarisation at higher applied voltages. Figure from [77]

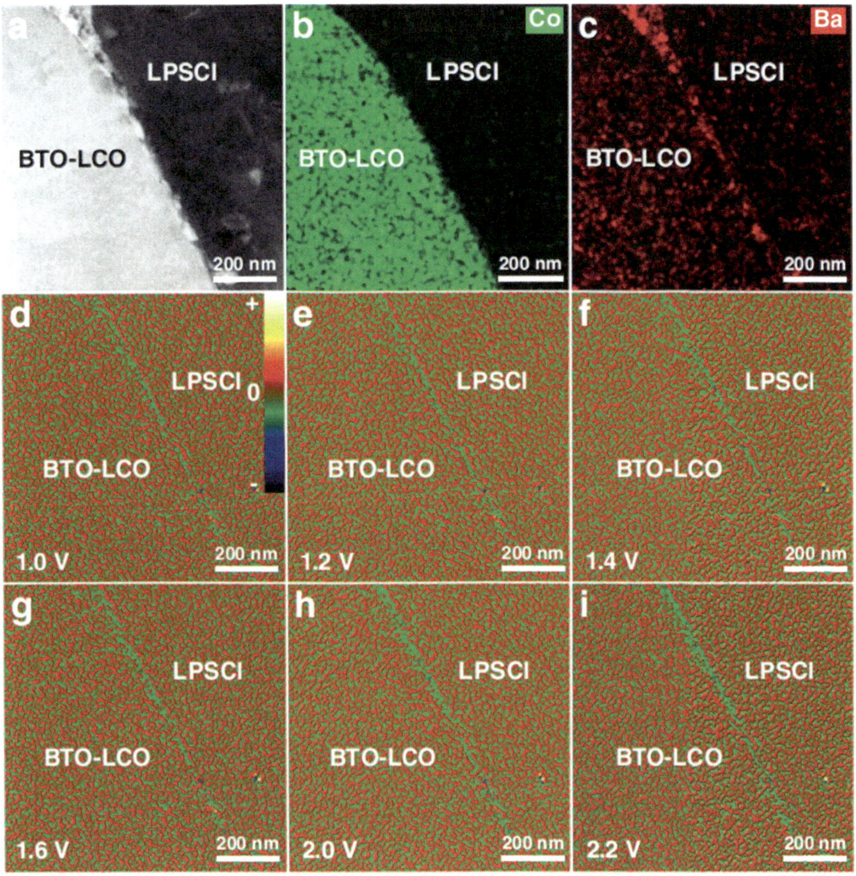

Fig. 19 BTO-Li_2CoO_2 and Li_6PS_5Cl interface. **a–c** HAADF-STEM images of interface with and without mappings. **d–i** In situ DPC-STEM images at varying bias voltages. Colour scale in **d** is the relative magnitude of positive/negative charge density. The applied voltage appears to have minimal impact on the charge distribution when BTO is added. Figure from [77]

The electrochemical stability window of SEs plays a critical role in determining the battery systems they are compatible with. Indeed, the electrochemical reactions that can occur outside of the stable window can cause significant degradation of the battery cell through the formation of insulating products. In a reverse process, some chemical reactions produce ionic and/or electronically insulative compounds at the heterogenic interfaces, which can impede accurate stability window measurements [79].

A thio-LISICON sample ($Li_{10}SnP_2S_{12}$) in a Li_3NbO_4-$LiCoO_2$/ $Li_{10}SnP_2S_{12}$/ Li-In cell was analysed using TEM by Vinado et al. (2018). The aim was to investigate the electrochemical stability and reactivity of the electrolyte at the $LiCoO_2$ cathodic interface that had been treated with a protective Li_3NbO_4 coating [79]. After 10 C/

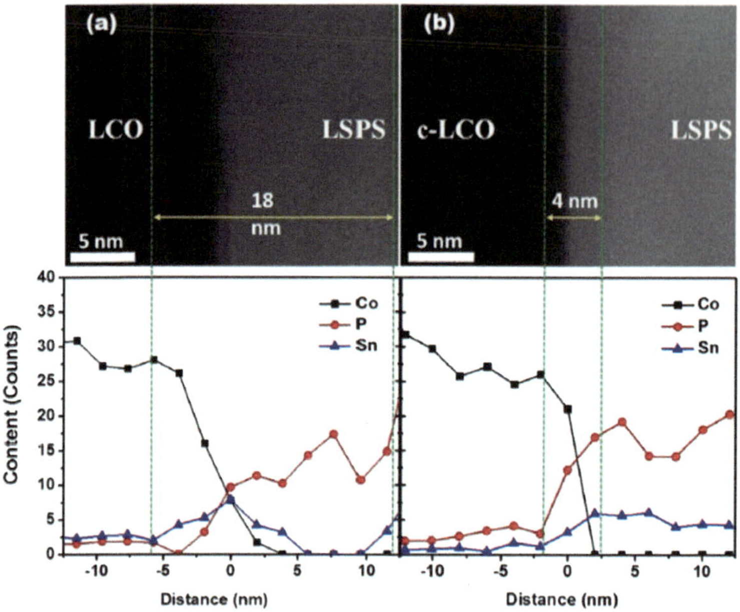

Fig. 20 Interfacial layer formed between $LiCoO_2$ and $Li_{10}SnP_2S_{12}$ with and without Li_3NbO_4 coating. **a** Uncoated sample showing interfacial layer of 18 nm (*dotted lines*), EDS scan shown below. **b** Li_3NbO_4 coated interface showing thinner interfacial layer at 4 nm, also with EDS data (*below*). Figure from [79]

10 cycles at 60 °C, the test cells were disassembled, and the cathode-electrolyte interface was analysed. TEM and EDS analysis showed a significant reduction in the interfacial layer when the cathode was coated (Fig. 20). This passivating layer was believed to be Li^+ and electronically resistive Co of Sn sulphides caused by electrolyte degradation at the interface. The important point from this research was the impact this layer had on the discharge voltage and specific capacity of the system. The more stable coated systems had higher cycling life, and higher discharge voltages at increased specific capacities.

4.3 *Chemical Stability*

The analysis of sulphide-based electrolytes presents a challenge due to the group's chemical reactivity and sensitivity to electron beams [77]. The moisture and Li/Na metal sensitivity of sulphide-based electrolytes is a major inhibitor to commercialisation of these highly conductive and malleable electrolyte materials.

Sulphide-based electrolytes release toxic H_2S gas in contact with water, posing a serious risk to the user and greatly reducing the ionic conductivity of the electrolyte. Tsukasaki et al. (2021) [80], aimed to better clarify this degradation process

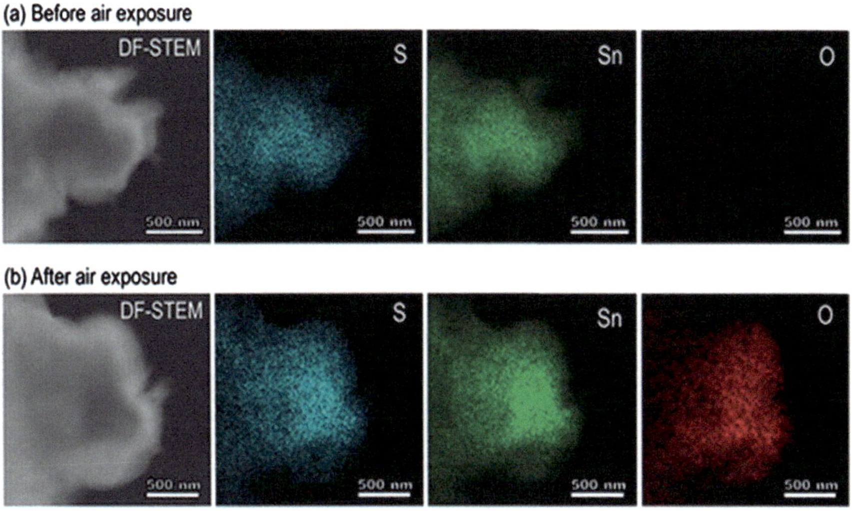

Fig. 21 DF-STEM and EDS mappings for Li_3SnS_4. From left to right, DF-STEM image, S, Sn and O elemental mappings for sample before air exposure (**a**) and after (**b**). Figure from [80]

by exposing a sulphide-based electrolyte (Li_3SnS_4) to air in a controlled in situ STEM experiment. By collecting dark field STEM (DF-STEM) and EDS spectra, the uniform reaction with air was confirmed as O was evenly dispersed throughout the sample (Fig. 21b) [80]. Crystalline regions of Li_3SnS_4 identified using DF-STEM appeared to fade into amorphous regions with increasing time or air partial pressure. This uniform reaction with air was still quite fast (less than 20 min) even using Li_3SnS_4, a supposedly more stable derivative of the standard Li_3PS_4 electrolyte. Such investigations highlight the ongoing challenge of degradation in sulphide-based electrolytes [80].

Many other SEs are susceptible to continuous reaction with Li anodes, forming increasingly thick separating layers and (sometimes rapidly) degrading battery performance. Polymer electrolytes fall into this category which greatly restricts the application of these relatively conductive and highly flexible materials. Polyethylene-oxide (PEO) is a very common polymer that is typically blended with electrolytic salts such as lithium bis(trifluoromethylsulphonyl) imide (LiTFSI) to create an SE. This polymer system is very straightforward and easy to synthesise, and so has been an attractive candidate for further research into additives that can slow or stop parasitic Li metal reactions.

In 2020, the Sheng group investigated one such additive, Li_2S, in a proof-of-concept work. The group used cryo-TEM and HR-TEM to analyse the Li deposition behaviour in the polymer electrolyte with and without the Li_2S additive. The gathered images show the reduction in Li crystalline deposition within the polymer when the additive was present (Fig. 22b, d), and the group attributed this to the increased production of LiF at the interface. LiF supposedly inhibited further reaction between

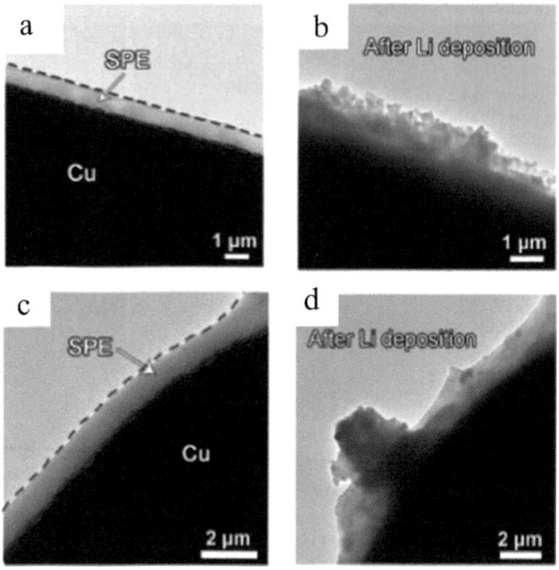

Fig. 22 Cryo-TEM images of PEO-LiTFSI electrolyte with and without Li_2S additive. Sample without additive before (**a**) and after (**b**) Li deposition cycle, notice Li penetration into the electrolyte causing significant changes to the surface. Sample with Li_2S additive before (**c**) and after (**d**) Li deposition. The surface of the electrolyte in (**d**) is smoother than in (**b**), but some uneven deposition is visible. Figure from [81]

the SE and Li metal [81]. Cryo-TEM is additionally useful for analysing polymer systems as it inhibits beam-induced damage.

Chemical incompatibilities between electrolytes, electrodes and the atmosphere are clearly still prevalent in modern high-performing SSEs. The use of TEM techniques to characterise and quantify these reactions has shed vast light on the details of these reactions and is useful in verifying proposed solutions. Increasingly high-resolution images, and novel techniques may accelerate our understanding and solution-finding processes to deliver incredibly stable, safe and high-performing SSEs sooner.

4.4 *Thermal Issues*

The impact of temperature on the performance of battery systems is well understood as a general concept, however the nuanced impacts in interfaces and electrolyte performance in SSBs is still under investigation. Thermal challenges arise as a secondary effect from other issues in the battery system (e.g. poor contact, resistance, electrochemical reactions, dendritic growth, incorrect charging load). Whilst the root for these problems should be addressed, solid-state battery systems must still be stable and safe in a wide range of operating temperatures. Solid-state batteries are generally more thermally stable than those using liquid electrolytes, however thermal challenges can still be present.

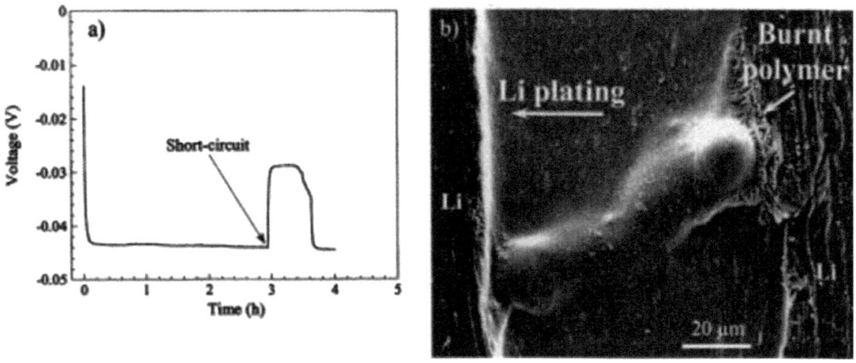

Fig. 23 Voltage profile and corresponding SEM image of Li/PEO-LiTFSI/Li cell. **a** Voltage profile of short circuit. **b** SEM image of cross section of symmetric cell after being short circuited. Li penetrated the polymer and caused a short circuit. Burnt polymer can be seen at the site of the short circuit (*right arrow*). Figure from [82]

In polymer electrolytes, where dendritic growth can still be an issue, short circuits can cause serious local thermal runaway due to rapid battery discharge. As a historical, but still relevant example of thermal electrolyte decomposition, Dollé et al. exposed a Li/PEO-LiTFSI/Li cell to an excessive current density of 0.5 mA cm^{-2}. This resulted not only in a short circuiting of the cell from dendritic growth, but also the combustion of the polymer electrolyte (Fig. 23b) [82]. Short circuits lead to local thermal runaway that can easily exceed the melting point of Li metal [82]. SEs such as the PEO-LiTFSI investigated by Dollé et al. are stable only against solid Li metal and are susceptible to react with molten Li metal at elevated temperatures [82]. The result is combustion or explosion as is common in conventional liquid electrolyte batteries. As such, the root causes for thermal runaway (namely dendritic growth) must be rectified.

Even inorganic battery materials are susceptible to thermal decomposition under the right conditions. Sharifi-Asl et al. (2017) observed the in situ degradation of $LiCoO_2$ cathode materials which resulted in the release of O_2 (creating an obvious safety concern around sensitive or combustible materials). By using HAADF-STEM methods, clear regional transitions from the layered to spinel structure at high temperatures were observed as well as void formation (Fig. 24). EELS data also revealed the high temperature release of O_2 which suggested a second transition from the spinel to rock salt formation [83].

While these temperatures for decomposition seem high, they are achievable under physical stresses, over-charging, environmental factors and uncontrolled reactions [79, 82]. As shown above these thermally driven reactions can cascade in thermal runaway leading to performance loss at a minimum and dangerous reactions in extreme cases. For this reason, not only should solid-state battery materials be stable in large temperature ranges, but the underlying causes for increasing cell temperature should be minimised in all cases.

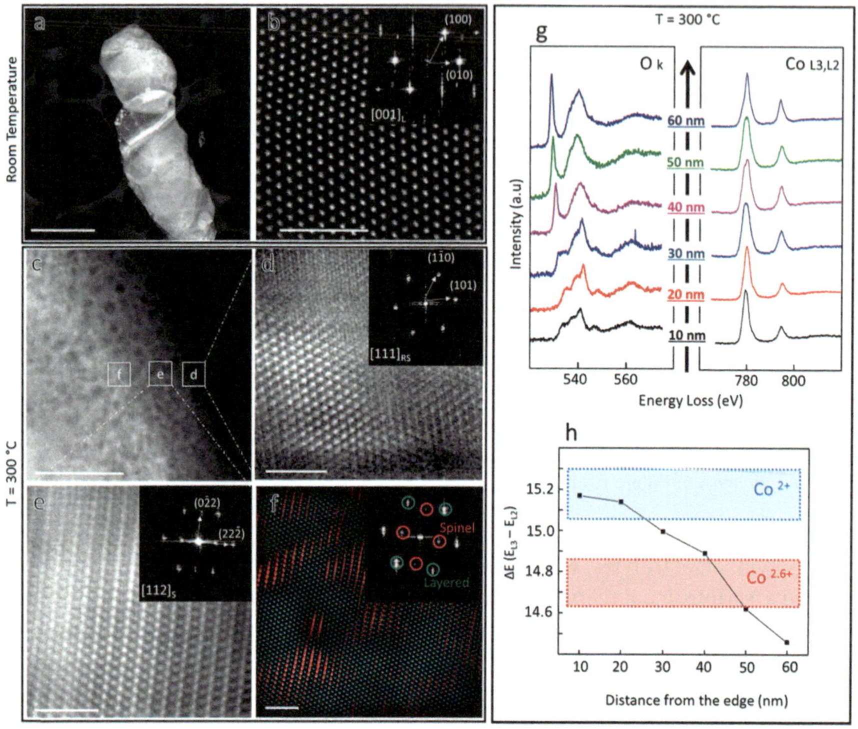

Fig. 24 HAADF-STEM images and EELS data from high temperature $LiCoO_2$ cathode sample. **a–f** HAADF-STEM images of sample at RT (**a**, **b**) and 300 °C (**c–f**). **e, f** Magnifications of the regions shown in **c**. **f** is coloured to show regions of spinel (*red*) and layered (*green*) compound. **g** EEL spectra from 300 °C sample at various distances from a grain edge. **h** Energy difference of Co L3 and L2 edges (ΔE) as a function of distance from grain edge, coloured regions show Co oxidation states. Figure from [83]

4.5 Concluding Remarks

The applications covered here only represent a small portion of all the high-quality SSLMB TEM characterisation work that is still ongoing. However, increasing attention is shifting away from Li-based technologies, to those based in Na.

5 Na Solid Electrolyte Interfaces

Although advanced TEM methods have been widely used to characterise electrode–electrolyte interfaces in solid-state Li-ion batteries (SSLIBs), to date there is limited work on characterising these interfaces in solid-state Na-ion batteries (SSSIBs). In this part, recent TEM work characterising the physical contact and electrochemical stability at the electrode–electrolyte interface are introduced.

5.1 Physical Contact

NASICON (**Na** **S**uper **I**onic **CON**ductor) is a common SE used for SSSIBs, however, the high interfacial resistance limits a good rate performance. The electrode–electrolyte interface plays a vital role in improving the power density of SSSIBs. The contact issues in NASICON are more prominent due to its rigidity. Lan et al. developed an in situ synthesis method to create a tight contact between the $Na_3V_2(PO_4)_3$ (NVP) active material and the NASICON-type $Na_3Si_2Zr_2PO_{12}$ (NZSP) SE by introducing a NVP precursor into the porous NZSP structure [84] The intimate contact between NZSP and NVP was confirmed by HAADF-STEM ([84]). The intimate contact between NZSP and NVP was confirmed by HAADF-STEM (Fig. 25a, b). Furthermore, no obvious voids were formed after 100 cycles (Fig. 25c), indicating that the intimate interface was well maintained.

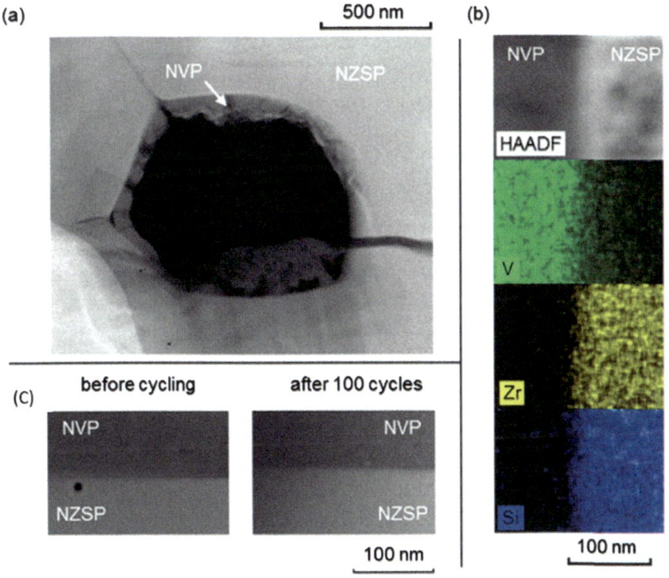

Fig. 25 Microscopic morphology and compositional characterisation near the NVP-NSZP interface. **a** Low magnification HAADF-STEM image of as-prepared NVP-NZSP composite collected from a typical micrometer-scale pore. **b** Enlarged HAADF-STEM image of the interface structure, along with 'fingerprinting' elemental maps of V for NVP and Zr, Si for NZSP. **c** Mesoscale comparison of the interface structure before (*left*) and after 100 cycles of charging and discharging (*right*) test at a current of 0.6 C. A black spot in the image was caused by electron-beam irradiation [84]. Copyright 2019 Elsevier

Apart from the in situ formed interphase reaction, a thin interlayer/coating layer can be introduced to form an intimate electrode–electrolyte interface. For example, Banerjee et al. confirmed that a thin Na_3SbS_4 layer (200 nm) (Fig. 26a) uniformly coated the $NaCrO_2$ using TEM characterisation of a FIB lamella [85]. The charge transfer resistance of the coated electrode was greatly reduced from 30 to 0.6 kΩ (Fig. 26c), corresponding to that of the mixed electrode with electrolyte particles. The high-quality coating layer makes this SSB competitive to a conventional sodium ion battery.

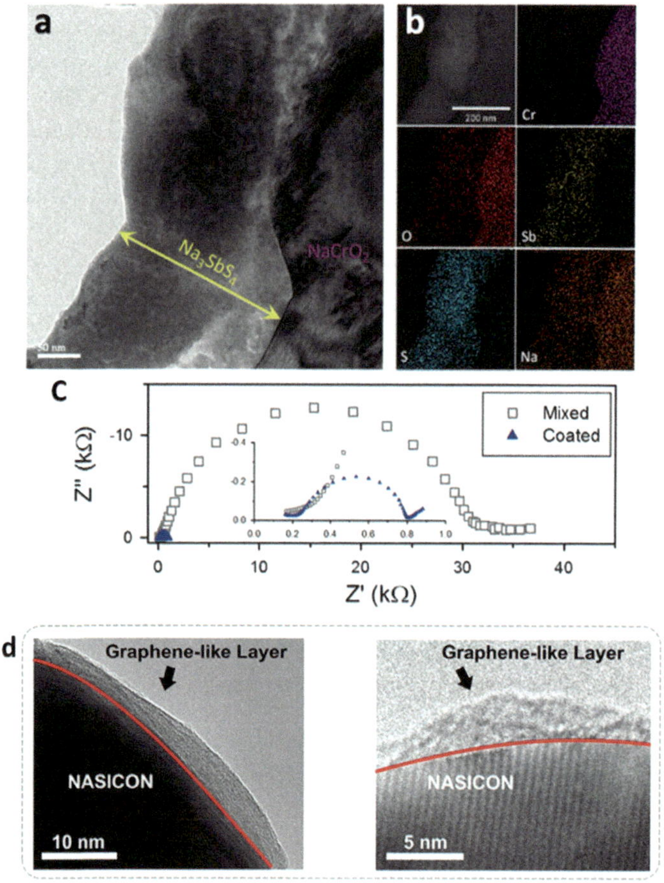

Fig. 26 Results of Na_3SbS_4–coated $NaCrO_2$ (NCO). **a** TEM image of FIB-sectioned Na_3SbS_4–coated NCO. **b** Annular dark-field (ADF) TEM image of FIB-sectioned Na_3SbS_4-coated NCO and its corresponding EDS elemental maps. **c** Nyquist plots of NCO/Na-Sn all-solid-state cells. Results of the mixed electrode and the Na_3SbS_4-coated NCO electrode are compared. **d** the TEM images of G-NASICON, revealing the graphene-like layer coating on NASICON electrolyte [85]. Copyright 2016 Wiley [86]. Copyright 2019 American Chemical Society

5.2 Electrochemical Stability

Recently, Matios et al. observed that an ultrathin graphene-like layer (around 3–4 nm) could be successfully introduced onto the NASICON surface (Fig. 26d). The surface defects of the graphene-like interlayer facilitated uniform Na deposition and prevented Na dendrite growth along the grain boundaries [86]. Moreover, the area specific resistance was reduced from 524 to 46 Ω cm^2, which is beneficial for the power performance by facilitating the Na^+ diffusion across the graphene coated NASICON-Na interface [86]. Moreover, the area specific resistance was reduced from 524 to 46 Ω cm^2, which is beneficial for the power performance by facilitating the Na^+ diffusion across the graphene coated NASICON-Na interface (Fig. 27).

Although SEs are better at suppressing Na dendrite nucleation and growth, this process is not completely eliminated. To inhibit dendrite growth, the dendrite nucleation and growth behaviour need to be further understood. This has recently been investigated by Li et al. with the aid of in situ TEM [87]. A nanobattery was

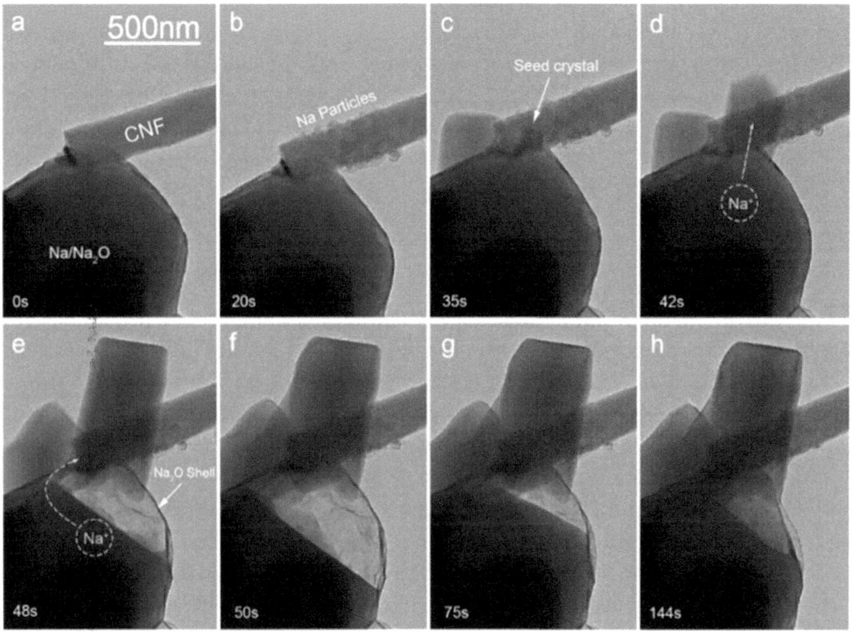

Fig. 27 In situ TEM observation of the Na deposition at the CNF/electrolyte interface and the corresponding morphological change of the Na electrode. **a–d** Particle growth at the interface without obvious dimensional change of the Na electrode. The nucleation of a Na seed (*white arrow*) in **c** and the Na^+ diffusion pathway in Na electrode during the seed growth (*yellow arrow*) in (**d**). **e, f** Drastic volume change of the Na electrode due to the Na loss upon the growth of a large cuboid-like crystal, **g, h** and its morphology recovery due to the continuous Na supply from nearby. The new Na^+ pathway after the void formation shown in **e** (*yellow arrow*) [87]. Copyright 2017 Elsevier

constructed with a Na counter electrode, a Na_2O solid electrolyte and a CNF working electrode with the aid of in situ TEM [87]. A nanobattery was constructed with a Na counter electrode, a Na_2O solid electrolyte and a CNF working electrode. During charging, the Na preferentially nucleated on the CNF surface mesoporous sites. However, the CNF interior space also accommodated Na nucleation, which plays an important role in dendrite-free growth. Based on these findings, CNF combined with Na metal could be adopted to eliminate dendrite growth in the SSSIBs.

Advanced TEM characterisation work will be needed to elucidate key mechanisms and issues at electrode–electrolyte interfaces in solid-state Na-ion batteries (SSSIBs). The ongoing development of cryo-TEM will be of great importance in characterising some of these air-sensitive materials.

6 Future Developments of TEM Characterisation of Solid Electrolyte Heterointerfaces

High-end TEM techniques including HRSTEM combined with EELS and in situ TEM have been successfully used to characterise solid electrode–electrolyte interfaces. However, new and emerging techniques including cryo-TEM, 4D STEM and in-operando TEM methods will potentially emerge as techniques of use in the future.

6.1 Cryo-TEM

Cryo-TEM is opening opportunities for imaging air-sensitive and beam-sensitive materials. For solid-state batteries, this will be of primary interest in characterising Na or Li dendritic growth at the electrode–electrolyte interface. The cryo-TEM instrument coupled with the development of cryo-TEM sample preparation workflows, allows for intermediate and metastable phases to be captured.

At room temperature, Li metal is very reactive and corrodes when exposed to air [89]. Futhermore Li has a low melting point and weak atomic bonding—making Li unstable when exposed to an electron beam [90]. Li et al. used cryo-TEM to observe Li dendrites formed on a Cu TEM grid [88]. The observed dendrites showed clear faceted growth and due to the cryo-TEM operation, it was possible to clearly observe the Li columns using high resolution imaging (Fig. 28). This improves on previous work completed under either non-cryo transfer conditions or imaged under standard TEM conditions. To date the use of cryo-TEM in alkali-ion batteries is limited to only a few studies including Li dendrite morphology characterisation [88, 91]; imaging of Li atomic columns [88] and characterisation of SEI phases [91–94].

More recently, the development of cryo-FIB techniques has opened the possibility of milling and reconstructing materials that previously were not possible to mill with a conventional FIB [91]. This will open many opportunities for both Na and Li SSBs.

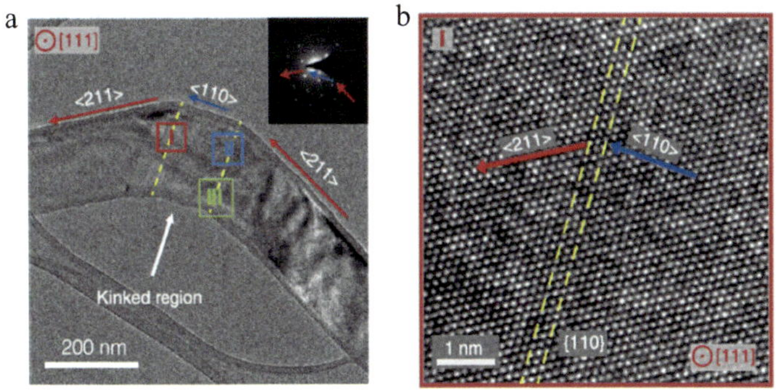

Fig. 28 Cryo-TEM atomic-resolution images of kinked Li metal dendrite and SEI interface. **a** TEM image of kinked Li metal dendrite that changes from a <211> to <110> and back to <211> growth direction. Inset shows corresponding SAD pattern. Note change in growth direction (*dashed yellow lines*). **b** Magnified images of red region I [88]

6.2 4D STEM

4D STEM is a developing field in electron microscopy and is an extension of the traditional STEM and diffraction techniques. The electron beam is rastered over the sample. At each position, a diffraction pattern is collected using a pixelated detector (often referred to as a direct electron detector (DED)). The DED have an improved quantum efficiency and signal-to-noise in comparison to their traditional predecessors. Furthermore, they can push imaging frame rates to 400 fps (frames per second). From these diffraction patterns detailed sample information including electric and magnetic field, strain, phase and orientation [95, 96] can be obtained. Furthermore, the use of electron ptychography has pushed the imaging resolution [97] with the possibility of observing light elements in LIB cathode materials [98]. Due to the nature of these detectors, the beam intensity can be reduced, which reduces beam-induced sample damage. 4D STEM will play a role in the characterisation of space charge layers in solid electrolyte interfaces and also in HR imaging.

7 Conclusions

Solid heterointerfaces are a point of weakness in both SOCs and SSBs. This weakness is due to a lack of physical contact, chemical or electrochemical reactions which can impede the charge transfer and electrochemical processes in these systems. Advanced TEM characterisation has played a key role in unveiling the detailed nature of these heterointerfaces. HRTEM and HRSTEM provide a detailed visual insight, and can be supported with complementary analytical TEM techniques including EDS, diffraction and EELS. EELS plays a particularly important role in electrochemical systems with complex oxides as information on oxidation states helps unravel the defect chemistry and associated electrocatalytic processes.

More recently EELS has been combined with in situ ETEM characterisation to provide a more realistic understanding of oxidation changes during operation. Similarly, the in situ EH work probing SSB heterointerfaces provides unparalleled information on the role of space-charge layers formed in SSBs. In the future, it will be interesting to see the in situ TEM work expand into in-operando TEM work—where both the correct operating environment and bias are used. As with all TEM characterisation methods, beam-induced effects need to be systematically studied and considered. However, with continued developments in low-dose detectors and cryo-TEM these issues can be alleviated.

References

1. Zhang Y, Knibbe R, Sunarso J, Zhong Y, Zhou W, Shao Z et al (2017) Recent progress on advanced materials for solid-oxide fuel cells operating below 500 °C. Adv Mater 29 (article number 1700132)
2. Hu Y-S (2016) Batteries: getting solid. Nat Energy 1:16042
3. Du YA, Holzwarth NAW (2007) Li Ion Diffusion mechanisms in the crystalline electrolyte γ-Li_3PO_4. J Electrochem Soc 154:A999
4. Deng Y, Eames C, Fleutot B, David R, Chotard J-N, Suard E, Masquelier C, Islam MS (2017) Enhancing the lithium ion conductivity in lithium superionic conductor (LISICON) solid electrolytes through a mixed polyanion effect. ACS Appl Mater Interfaces 9:7050–7058
5. Kumar B, Thomas D, Kumar J (2009) Space-charge-mediated superionic transport in lithium ion conducting glass–ceramics. J Electrochem Soc 156:A506–A513
6. Xiong S, Liu Y, Jankowski P, Liu Q, Nitze F, Xie K et al (2020) Design of a multifunctional interlayer for NASCION-based solid-state Li metal batteries. Adv Funct Mater 30 (article number 2001444)
7. López-Aranguren P, Reynaud M, Głuchowski P, Bustinza A, Galceran M, López del Amo JM et al (2021) Crystalline LiPON as a bulk-type solid electrolyte. ACS Energy Lett 6:445–450
8. Seino Y, Ota T, Takada K, Hayashi K, Tatsumisago M (2014) A sulphide lithium superion conductor is superior to liquid ion conductors for use in rechargeable batteries. Energy Environ Sci 7:627–631
9. Liang X, Wang L, Jiang Y, Wang J, Luo H, Liu C et al (2015) In-channel and In-plane Li ion diffusions in the superionic conductor $Li_{10}GeP_2S_{12}$ probed by solid-state NMR. Chem Mater 27:5503–5510
10. Zhou L, Assoud A, Zhang Q, Wu X, Nazar LF (2019) New family of argyrodite thioantimonate lithium superionic conductors. J Am Chem Soc 141:19002–19013
11. Lopez J, Mackanic DG, Cui Y, Bao Z (2019) Designing polymers for advanced battery chemistries. Nat Rev Mater 4:312–330
12. Lv F, Wang Z, Shi L, Zhu J, Edstrom K, Mindemark J, Yuan S (2019) Challenges and development of composite solid-state electrolytes for high-performance lithium ion batteries. J Power Sources 441 (article number 227175)
13. Chen F, Zha W, Yang D, Cao S, Shen Q, Zhang L, Sadoway DR (2018) All-solid-state lithium battery fitted with polymer electrolyte fnhanced by solid plasticizer and conductive ceramic filler. J Electrochem Soc 165:A3558–A3565
14. Ma J, Chen B, Wang L, Cui G (2018) Progress and prospect on failure mechanisms of solid-state lithium batteries. J Power Sources 392:94–115

15. Ran L, Baktash A, Li M, Yin Y, Demir B, Lin T et al (2021) Sc, Ge co-doping NASICON boosts solid-state sodium ion batteries' performance. Energy Storage Mater 40:282–291
16. Yu S, Schmidt RD, Garcia-Mendez R, Herbert E, Dudney NJ, Wolfenstine JB et al (2016) Elastic properties of the solid electrolyte $Li_7La_3Zr_2O_{12}$ (LLZO). Chem Mater 28:197–206
17. Deng Z, Wang Z, Chu I-H, Luo J, Ong SP (2015) Elastic properties of alkali superionic conductor electrolytes from first principles calculations. J Electrochem Soc 163:A67-74
18. Sakuda, A., Hayashi A, Tatsumisago M (2013) Sulfide solid electrolyte with favorable mechanical property for all-solid-state lithium battery. Sci Rep 3 (article number 2261)
19. McGrogan FP, Swamy T, Bishop SR, Eggleton E, Porz R, Chen X et al (2017) Compliant yet brittle mechanical behavior of $Li_2S–P_2S_5$ lithium-ion-conducting solid electrolyte. Adv Energy Mater **7** (article number 1602011)
20. Famprikis T, Canepa P, Dawson JA, Islam MS, Masquelier C (2019) Fundamentals of inorganic solid-state electrolytes for batteries. Nat Mater 18:1278–1291
21. Koerver R, Aygün I, Leichtweiß T, Dietrich C, Zhang W, Binder JO et al (2017) Capacity Fade in solid-state batteries: I interphase formation and chemomechanical processes in nickel-rich layered oxide cathodes and lithium thiophosphate solid electrolytes. Chem Mater 29:5574–5582
22. Lu Y, Huang X, Song Z, Rui K, Wang Q, Gu S et al (2018) Highly stable garnet solid electrolyte based Li-S battery with modified anodic and cathodic interfaces. Energy Storage Mater 15:282–290
23. Reimers JN, Dahn JR (1992) Electrochemical and in situ X-ray diffraction studies of lithium intercalation in Li_xCoO_2. J Electrochem Soc 139:2091–2097
24. Maxisch T, Ceder G (2006) Elastic properties of olivine Li_xFePO_4 from first principles. Phys Rev B 73 (article number 174112)
25. Liu J, Kopold P, Wu C, Van Aken PA, Maier J, Yu Y (2015) Uniform yolk–shell Sn_4P_3@C nanospheres as high-capacity and cycle-stable anode materials for sodium-ion batteries. Energy Environ Sci 8:3531–3538
26. Bucci G, Swamy T, Chiang Y-M, Carter WC (2017) Modeling of internal mechanical failure of all-solid-state batteries during electrochemical cycling, and implications for battery design. J Mater Chem A 5:19422–19430
27. Porz L, Swamy T, Sheldon BW, Rettenwander D, Frömling T, Thaman HL et al (2017) Mechanism of lithium metal penetration through inorganic solid electrolytes. Adv Energy Mater **7** (article number 1701003)
28. Knibbe R, Hjelm J, Menon M, Pryds N, Søgaard M, Wang HJ et al (2010) Cathode-electrolyte interfaces with CGO barrier layers in SOFC. J Am Ceram Soc 93:2877–2883
29. Kim KH, Iriyama Y, Yamamoto K, Kumazaki S, Asaka T, Tanabe K et al (2011) Characterization of the interface between $LiCoO_2$ and $Li_7La_3Zr_2O_{12}$ in an all-solid-state rechargeable lithium battery. J Power Sources 196:764–767
30. Miara L, Windmüller A, Tsai C-L, Richards WD, Ma Q, Uhlenbruck S et al (2016) About the compatibility between high voltage spinel cathode materials and solid oxide electrolytes as a function of temperature. ACS Appl Mater Interfaces 8:26842–26850
31. Vardar G, Bowman WJ, Lu Q, Wang J, Chater RJ, Aguadero A et al (2018) Structure, chemistry, and charge transfer resistance of the interface between $Li_7La_3Zr_2O_{12}$ electrolyte and $LiCoO_2$ cathode. Chem Mater 30:6259–6276
32. Ebbesen SD, Graves C, Hauch A, Jensen SH, Mogensen M (2010) Poisoning of solid oxide electrolysis cells by impurities. J Electrochem Soc 157:B1419–B1429
33. Muramatsu H, Hayashi A, Ohtomo T, Hama S, Tatsumisago M (2011) Structural change of $Li_2S–P_2S_5$ sulfide solid electrolytes in the atmosphere. Solid State Ion 182:116–119
34. Cheng L, Crumlin EJ, Chen W, QiaoR, Hou H, Lux SF et al (2014) The origin of high electrolyte–electrode interfacial resistances in lithium cells containing garnet type solid electrolytes. Phys Chem Chem Phys 16:18294–300
35. Sharafi A, Kazyak A, Davis AL, Yu S, Thompson T, Siegel DJ et al (2017) Surface chemistry mechanism of ultra-low interfacial resistance in the solid-state electrolyte $Li_7La_3Zr_2O_{12}$. Chem Mater 29:7961–7968

36. Knibbe R, Hauch A, Hjelm J, Ebbesen SD, Mogensen M (2011) Durability of solid oxide cells. Green 1:141–169
37. Kanno R, Murayama M (2001) Lithium ionic conductor Thio-LISICON: the Li_2S-GeS_2-P_2S_5 system. J Electrochem Soc 148 (article number A742)
38. Liu Z, Tang Y, Wang Y, Huang F (2014) High performance Li_2S–P_2S_5 solid electrolyte induced by selenide. J Power Sources 260:264–267
39. Kamaya N, Homma K, Yamakawa Y, Hirayama M, Kanno R, Yonemura M et al (2011) A lithium superionic conductor. Nat Mater 10:682–686
40. Zhu Y, He S, Mo Y (2015) Origin of outstanding stability in the lithium solid electrolyte materials: insights from thermodynamic analyses based on first-principles calculations. ACS Appl Mater Interfaces 7:23685–23693
41. Banerjee A, Wang X, Fang C, Wu CA, Meng YS (2020) Interfaces and interphases in all-solid-state batteries with inorganic solid electrolytes. Chem Rev 120:6878–6933
42. Koerver R, Walther F, Aygün I, Sann J, Dietrich C, Zeier WG et al (2017) Redox-active cathode interphases in solid-state batteries. J Mate Chem A 5:22750–22760
43. Trevey JE, Stoldt CR, Lee SJ (2011) High power nanocomposite TiS_2 cathodes for all-solid-state lithium batteries. J Electrochem Soc 158 (article number A1282)
44. Wenzel S, Leichtweiss T, Krüger D, Sann J, Janek J (2015) Interphase formation on lithium solid electrolytes—an in situ approach to study interfacial reactions by photoelectron spectroscopy. Solid State Ion 278:98–105
45. Chen R, Li O, Yu X, Chen L, Li H (2020) Approaching practically accessible solid-state batteries: stability issues related to solid electrolytes and interfaces. Chem Rev 120:6820–6877
46. Hartmann P, Leichtweiss T, Busche MR, Schneider M, Reich M, Sann J et al (2013) Degradation of NASICON-type materials in contact with lithium metal: formation of mixed conducting interphases (MCI) on solid electrolytes. J Phys Chem C 117:21064–21074
47. Gan Z, Gu M, Tang J, Wang C-Y, He Y, Wang KL et al (2016) Direct mapping of charge distribution during lithiation of Ge nanowires using off-axis electron holography. Nano Lett 16:3748–3753
48. Yamamoto K, Iriyama Y, Asaka T, Hirayama T, Fujita H, Fisher CAJ et al (2010) Dynamic visualization of the electric potential in an all-solid-state rechargeable lithium battery. Angew Chem Int Ed 49:4414–4417
49. Yamamoto K, Iriyama Y, Asaka T, Hirayama T, Fujita H, Nonaka K et al (2012) Direct observation of lithium-ion movement around an in-situ-formed- negative-electrode/solid-state-electrolyte interface during initial charge-discharge reaction. Electrochem Commun 20:113–116
50. Aizawa Y, Yamamoto K, Sato T, Murata H, Yoshida R, Fisher CAJ et al (2017) In situ electron holography of electric potentials inside a solid-state electrolyte: effect of electric-field leakage. Ultramicroscopy 178:20–26
51. Hirayama T, Aizawa Y, Yamamoto K, Sato T, Murata H, Yoshida R, Fisher CAJ et al (2017) Advanced electron holography techniques for in situ observation of solid-state lithium ion conductors. Ultramicroscopy 173:64–70
52. Wen Y, Chen X, Lu X, Gu L (2018) Interface charges boosted ultrafast lithiation in $Li_4Ti_5O_{12}$ revealed by in-situ electron holography. J Energy Chem 27:1397–1401
53. Nomura Y, Yamamoto K, Hirayama T, Ouchi S, Igaki E, Saitoh K (2019) Direct observation of a li-ionic space-charge layer formed at an electrode/solid-electrolyte interface. Angew Chem Int Ed 58:5292–5296
54. Irvine JTS, Neagu D, Verbraeken MC, Chatzichristodoulou C, Graves C, Mogensen MB (2016) Evolution of the electrochemical interface in high-temperature fuel cells and electrolysers. Nat Energy 1 (article number: 15014)
55. Traulsen ML, Chatzichristodoulou C, Hansen KV, Kuhn LT, Holtappels P, Mogensen MB (2015) Need for in operando characterization of electrochemical interface features. ECS Trans 66:3–20
56. Vels Jensen K, Primdahl S, Chorkendorff I, Mogensen MB (2001) Microstructural and chemical changes at the Ni/YSZ interface. Solid State Ion 144:197–209

57. Liu YL, Jiao C (2005) Microstructure degradation of an anode/electrolyte interface in SOFC studied by transmission electron microscopy. Solid State Ion 176:435–442
58. Liu SS, Saha LC, Iskandarov A, Ishimoto T, Yamamoto T, Umeno Y et al (2019) Atomic structure observations and reaction dynamics simulations on triple phase boundaries in solid-oxide fuel cells. Commun Chem 2 (article number: 48)
59. Sasaki T, Matsunaga K, Ohta H, Hosono H, Yamamoto T, Ikuhara Y (2004) Atomic and electronic structures of Ni/YSZ(111) interface. Mater Trans 45:2137–2143
60. Jeangros Q, Faes A, Wagner JB, Hansen TW, Aschauer U, Van herle J, et al (2010) In situ redox cycle of a nickel-YSZ fuel cell anode in an environmental transmission electron microscope. Acta Mater 58:4578–4589
61. He S, Li M, Hui J, Yue X (2021) In-situ construction of ceria-metal/titanate heterostructure with controllable architectures for efficient fuel electrochemical conversion. Appl Catal B:298 (article number 120588)
62. Neagu D, Oh TS, Miller DN, Ménard H, Bukhari SM, Gamble SR et al (2015) Nano-socketed nickel particles with enhanced coking resistance grown in situ by redox exsolution. Nat Commun 6 (article number: 8120)
63. Adler SB (2004) Factors governing oxygen reduction in solid oxide fuel cell cathodes. Chem Rev 104:4791–4843
64. Fleig J (2002) On the width of the electrochemically active region in mixed conducting solid oxide fuel cell cathodes. J Power Sources 105:228–238
65. Primdahl S, Mogensen M (2002) Mixed conductor anodes: Ni as electrocatalyst for hydrogen conversion. Solid State Ion 152–153:597–608
66. Backhaus-Ricoult M (2006) Interface chemistry in LSM-YSZ composite SOFC cathodes. Solid State Ion 177:2195–2200
67. Knibbe R, Traulsen ML, Hauch A, Ebbesen SD, Mogensen M (2010) Solid oxide electrolysis cells: degradation at high current densities. J Electrochem Soc 157:B1209–B1217
68. Graves C, Ebbesen SD, Jensen SH, Simonsen SB, Mogensen MB (2015) Eliminating degradation in solid oxide electrochemical cells by reversible operation. Nature Mater 14:239–244
69. He S, Chen K, Saunders M, Li J, Cui CQ, Jiang SP (2017) A FIB-STEM study of $La_{0.8}Sr_{0.2}MnO_3$ cathode and Y_2O_3-ZrO_2/Gd_2O_3-CeO_2 electrolyte interfaces of solid oxide fuel cells. J Electrochem Soc 164:F1437–F1447
70. Develos-Bagarinao K, Yokokawa H, Kishimoto H, Ishiyama T, Yamaji K, Horita T (2017) Elucidating the origin of oxide ion blocking effects at GDC/$SrZr(Y)O_3$/YSZ interfaces. J Mater Chem A 5:8733–8743
71. Gaulandris F, Kuhn LT, Wagner JB, Sanna S, Simonsen SB (2021) Co oxidation state at LSC-YSZ interface in model solid oxide electrochemical cell. Solid State Ion 359 (article number: 115531)
72. Notohara H, Urita K, Yamamura H, Moriguchi I (2018) High capacity and stable all-solid-state Li ion battery using SnO_2-embedded nanoporous carbon. Sci Rep 8 (article number: 8747)
73. Chen X, Kierzek K, Wilgosz K, Machnikowski J, Gong J, Feng J et al (2012) New easy way preparation of core/shell structured SnO_2@carbon spheres and application for lithium-ion batteries. J Power Sources 216:475–481
74. Gouy M (1910) Sur la constitution de la charge électrique à la surface d'un électrolyte. J Phys Theor Appl 9:457–468
75. Liu J, Yuan H, Liu H, Zhao CZ, Lu Y, Cheng XB et al (2021) Unlocking the failure mechanism of solid state lithium metal batteries. Adv Energy Mater (early view article number: 202100748)
76. Zhang C, Feng Y, Han Z, Gao S, Wang M, Wang P (2020) Electrochemical and structural analysis in all-solid-state lithium batteries by analytical electron microscopy: progress and perspectives. Adv Mater 32 (early view article number: 201903747)
77. Wang L, Xie R, Chen B, Yu X, Ma J, Li C et al (2020) In-situ visualization of the space-charge-layer effect on interfacial lithium-ion transport in all-solid-state batteries. Nat. Commun 11 (article number: 5889)

78. Haas B, Rouvière J-L, Boureau V, Berthier R, Cooper D (2019) Direct comparison of off-axis holography and differential phase contrast for the mapping of electric fields in semiconductors by transmission electron microscopy. Ultramicroscopy 198:58–72
79. Vinado C, Wang S, He Y, Xiao X, Li Y, Wang C et al (2018) Electrochemical and interfacial behavior of all solid state batteries using $Li_{10}SnP_2S_{12}$ solid electrolyte. J Power Sources 396:824–830
80. Tsukasaki H, Igarashi K, Wakui A, Yaguchi T, Nakajima H, Kimura T et al (2021) In situ observation of the deterioration process of sulfide-based solid electrolytes using airtight and air-flow TEM systems. J Microsc (on-line published corrected proof number: dfab022)
81. Sheng O, Zheng J, Ju Z, Jin C, Wang Y, Chen M et al (2020) In situ construction of a LiF-enriched interface for stable all-solid-state batteries and its origin revealed by cryo-TEM. Adv Mater 32 (on-line publication article number: 202000223)
82. Dollé M, Sannier L, Beaudoin B, Trentin M, Tarascon JM (2002) Live scanning electron microscope observations of dendritic growth in lithium/polymer cells. Electrochem Solid-State Lett 5:A286–A289
83. Sharifi-Asl S, Soto FA, Nie A, Yuan Y, Asayesh-Ardakani H, Foroozan T et al (2017) Facet-dependent thermal instability in $LiCoO_2$. Nano Lett 17:2165–2171
84. Lan T, Tsai CL, Tietz F, Wei XK, Dunin-Borkowski RE et al (2019) Room-temperature all-solid-state sodium batteries with robust ceramic interface between rigid electrolyte and electrode materials. Nano Energy 65 (article number: 104040)
85. Banerjee A, Park KH, Heo JW, Nam YJ, Moon CK, Oh SM et al (2016) Na_3SbS_4: a solution processable sodium superionic conductor for all-solid-state sodium-ion batteries. Angew Chem Int Ed 55:9634–9638
86. Matios E, Wang H, Wang C, Hu X, Lu X, Luo J et al (2019) Graphene regulated ceramic electrolyte for solid-state sodium metal battery with superior electrochemical stability. ACS Appl Mater Interfaces 11:5064–5072
87. Li X, Zhao L, Li P, Zhang Q, Wang M-S et al (2017) In-situ electron microscopy observation of electrochemical sodium plating and stripping dynamics on carbon nanofiber current collectors. Nano Energy 42:122–128
88. Li Y, Li Y, Pei A, Yan K, Sun Y, Wu C-L et al (2017) Atomic structure of sensitive battery materials and interfaces revealed by cryo–electron microscopy. Science 358:506–510
89. Markowitz MM, Boryta DA (1962) Lithium metal-gas reactions: interaction of lithium metal with air and its component gases. J Chem Eng Data 7:586–591
90. Egerton RF, Li P, Malac M (2004) Radiation damage in the TEM and SEM. Micron 35:399–409
91. Zachman MJ, Tu Z, Choudhury S, Archer LA, Kourkoutis LF (2018) Cryo-STEM mapping of solid–liquid interfaces and dendrites in lithium-metal batteries. Nature 560:345–349
92. Huang W, Attia PM, Wang H, Renfrew SE, Jin N, Das S et al (2019) Evolution of the solid-electrolyte interphase on carbonaceous anodes visualized by atomic-resolution cryogenic electron microscopy. Nano Lett 19:5140–5148
93. Wang X, Zhang M, Alvarado J, Wang S, Sina M, Lu B et al (2017) New insights on the structure of electrochemically deposited lithium metal and its solid electrolyte interphases via cryogenic TEM. Nano Lett 17:7606–7612
94. Xu Y, Wu H, He Y, Chen Q, Zhang JG, Xu W et al (2020) Atomic to nanoscale origin of vinylene carbonate enhanced cycling stability of lithium metal anode revealed by cryo-transmission electron microscopy. Nano Lett 20:418–425
95. Brunetti G, Robert D, Bayle-Guillemaud P, Rouvière JL, Rauch EF, Martin JF et al (2011) Confirmation of the domino-cascade model by $LiFePO_4/FePO_4$ precession electron diffraction. Chem Mater 23:4515–4524
96. Ophus C (2019) Four-dimensional scanning transmission electron microscopy (4D-STEM): from scanning nanodiffraction to ptychography and beyond. Microsc Microanal 25:563–583
97. Gao W, Addiego C, Wang H, Yan X, Hou Y, Ji D et al (2019) Real-space charge-density imaging with sub-ångström resolution by four-dimensional electron microscopy. Nature 575:480–484
98. Lozano JG, Martinez GT, Jin L, Nellist PD, Bruce PG (2018) Low-dose aberration-free imaging of Li-rich cathode materials at various states of charge using electron ptychography. Nano Lett 18:6850–6855

GPSR Compliance

The European Union's (EU) General Product Safety Regulation (GPSR) is a set of rules that requires consumer products to be safe and our obligations to ensure this.

If you have any concerns about our products, you can contact us on ProductSafety@springernature.com

In case Publisher is established outside the EU, the EU authorized representative is:

Springer Nature Customer Service Center GmbH
Europaplatz 3
69115 Heidelberg, Germany

Batch number: 10703869

Printed by Printforce, the Netherlands